DESIGNERS' GUIDES TO THE EUROCODES

DESIGNERS' GUIDE TO EN 1992-2
EUROCODE 2: DESIGN OF CONCRETE STRUCTURES

PART 2: CONCRETE BRIDGES

Eurocode Designers' Guide Series

Designers' Guide to EN 1990. Eurocode: Basis of Structural Design. H. Gulvanessian, J.-A. Calgaro and M. Holický. 0 7277 3011 8. Published 2002.

Designers' Guide to EN 1994-1-1. Eurocode 4: Design of Composite Steel and Concrete Structures. Part 1.1: General Rules and Rules for Buildings. R. P. Johnson and D. Anderson. 0 7277 3151 3. Published 2004.

Designers' Guide to EN 1997-1. Eurocode 7: Geotechnical Design – General Rules. R. Frank, C. Bauduin, R. Driscoll, M. Kavvadas, N. Krebs Ovesen, T. Orr and B. Schuppener. 0 7277 3154 8. Published 2004.

Designers' Guide to EN 1993-1-1. Eurocode 3: Design of Steel Structures. General Rules and Rules for Buildings. L. Gardner and D. Nethercot. 0 7277 3163 7. Published 2004.

Designers' Guide to EN 1992-1-1 and EN 1992-1-2. Eurocode 2: Design of Concrete Structures. General Rules and Rules for Buildings and Structural Fire Design. A.W. Beeby and R. S. Narayanan. 0 7277 3105 X. Published 2005.

Designers' Guide to EN 1998-1 and EN 1998-5. Eurocode 8: Design of Structures for Earthquake Resistance. General Rules, Seismic Actions, Design Rules for Buildings, Foundations and Retaining Structures. M. Fardis, E. Carvalho, A. Elnashai, E. Faccioli, P. Pinto and A. Plumier. 0 7277 3348 6. Published 2005.

Designers' Guide to EN 1995-1-1. Eurocode 5: Design of Timber Structures. Common Rules and for Rules and Buildings. C. Mettem. 0 7277 3162 9. Forthcoming: 2007 (provisional).

Designers' Guide to EN 1991-4. Eurocode 1: Actions on Structures. Wind Actions. N. Cook. 0 7277 3152 1. Forthcoming: 2007 (provisional).

Designers' Guide to EN 1996. Eurocode 6: Part 1.1: Design of Masonry Structures. J. Morton. 0 7277 3155 6. Forthcoming: 2007 (provisional).

Designers' Guide to EN 1991-1-2, 1992-1-2, 1993-1-2 and EN 1994-1-2. Eurocode 1: Actions on Structures. Eurocode 3: Design of Steel Structures. Eurocode 4: Design of Composite Steel and Concrete Structures. Fire Engineering (Actions on Steel and Composite Structures). Y. Wang, C. Bailey, T. Lennon and D. Moore. 0 7277 3157 2. Forthcoming: 2007 (provisional).

Designers' Guide to EN 1993-2. Eurocode 3: Design of Steel Structures. Bridges. C. R. Hendy and C. J. Murphy. 0 7277 3160 2. Forthcoming: 2007 (provisional).

Designers' Guide to EN 1991-2, 1991-1-1, 1991-1-3 and 1991-1-5 to 1-7. Eurocode 1: Actions on Structures. Traffic Loads and Other Actions on Bridges. J.-A. Calgaro, M. Tschumi, H. Gulvanessian and N. Shetty. 0 7277 3156 4. Forthcoming: 2007 (provisional).

Designers' Guide to EN 1991-1-1, EN 1991-1-3 and 1991-1-5 to 1-7. Eurocode 1: Actions on Structures. General Rules and Actions on Buildings (not Wind). H. Gulvanessian, J.-A. Calgaro, P. Formichi and G. Harding. 0 7277 3158 0. Forthcoming: 2007 (provisional).

Designers' Guide to EN 1994-2. Eurocode 4: Design of Composite Steel and Concrete Structures. Part 2: General Rules and Rules for Bridges. C. R. Hendy and R. P. Johnson. 0 7277 3161 0. Published 2006.

www.eurocodes.co.uk

DESIGNERS' GUIDES TO THE EUROCODES

DESIGNERS' GUIDE TO EN 1992-2 EUROCODE 2: DESIGN OF CONCRETE STRUCTURES

PART 2: CONCRETE BRIDGES

C. R. HENDY and D. A. SMITH

Published by ICE Publishing, One Great George Street, Westminster, London SW1P 3AA.

Full details of ICE Publishing sales representatives and distributors can be found at: www.icevirtuallibrary.com/info/printbooksales

First published 2007
Reprinted 2010
Reprinted with amendments 2013

Eurocodes Expert

Structural Eurocodes offer the opportunity of harmonized design standards for the European construction market and the rest of the world. To achieve this, the construction industry needs to become acquainted with the Eurocodes so that the maximum advantage can be taken of these opportunities

Eurocodes Expert is a new ICE and Thomas Telford initiative set up to assist in creating a greater awareness of the impact and implementation of the Eurocodes within the UK construction industry

Eurocodes Expert provides a range of products and services to aid and support the transition to Eurocodes. For comprehensive and useful information on the adoption of the Eurocodes and their implementation process please visit our website or email eurocodes@thomastelford.com

---

A catalogue record for this book is available from the British Library

ISBN: 978-0-7277-3159-3

© The authors and Thomas Telford Limited 2007

All rights, including translation, reserved. Except as permitted by the Copyright, Designs and Patents Act 1988, no part of this publication may be reproduced, stored in a retrieval system or transmitted in any form or by any means, electronic, mechanical, photocopying or otherwise, without the prior written permission of the Publisher, ICE Publishing, One Great George Street, Westminster, London SW1P 3AA.

This book is published on the understanding that the authors are solely responsible for the statements made and opinions expressed in it and that its publication does not necessarily imply that such statements and/or opinions are or reflect the views or opinions of the publishers. While every effort has been made to ensure that the statements made and the opinions expressed in this publication provide a safe and accurate guide, no liability or responsibility can be accepted in this respect by the authors or publishers.

Typeset by Academic + Technical, Bristol
Printed and bound by CPI Group (UK) Ltd, Croydon CR0 4YY

Preface

Aims and objectives of this guide
The principal aim of this book is to provide the user with guidance on the interpretation and use of EN 1992-2 and to present worked examples. It covers topics that will be encountered in typical concrete bridge designs and explains the relationship between EN 1992-2 and the other Eurocodes.

EN 1992-2 is not a 'stand alone' document and refers extensively to other Eurocodes. Its format is based on EN 1992-1-1 and generally follows the same clause numbering. It identifies which parts of EN 1992-1-1 are relevant for bridge design and adds further clauses that are specific to bridges. It is therefore not useful to produce guidance on EN 1992-2 in isolation and so this guide covers material in EN 1992-1-1 which will need to be used in bridge design.

This book also provides background information and references to enable users of Eurocode 2 to understand the origin and objectives of its provisions.

Layout of this guide
EN 1992-2 has a foreword, 13 sections and 17 annexes. This guide has an introduction which corresponds to the foreword of EN 1992-2, Chapters 1 to 10, which correspond to Sections 1 to 10 of the Eurocode and Annexes A to Q which again correspond to Annexes A to Q of the Eurocode.

The guide generally follows the section numbers and first sub-headings in EN 1992-2 so that guidance can be sought on the code on a section by section basis. The guide also follows the format of EN 1992-2 to lower levels of sub-heading in cases where this can conveniently be done and where there is sufficient material to merit this. The need to use several Eurocode parts can initially make it a daunting task to locate information in the order required for a real design. In some places, therefore, additional sub-sections are included in this guide to pull together relevant design rules for individual elements, such as pile caps. Additional sub-sections are identified as such in the sub-section heading.

The following parts of the Eurocode are intended to be used in conjunction with Eurocode 2:

EN 1990:	Basis of structural design
EN 1991:	Actions on structures
EN 1997:	Geotechnical design
EN 1998:	Design of structures for earthquake resistance
hENs:	Construction products relevant for concrete structures
EN 13670:	Execution (construction) of concrete structures

These documents will generally be required for a typical concrete bridge design, but discussion on them is generally beyond the scope of this guide.

In this guide, references to Eurocode 2 are made by using the abbreviation 'EC2' for EN 1992, so EN 1992-1-1 is referred to as EC2-1-1. Where clause numbers are referred to in the text, they are prefixed by the number of the relevant part of EC2. Hence:

- 2-2/clause 6.3.2(6) means clause 6.3.2, paragraph (6), of EC2-2
- 2-1-1/clause 6.2.5(1) means clause 6.2.5, paragraph (1), of EC2-1-1
- 2-2/Expression (7.22) means equation (7.22) in EC2-2
- 2-1-1/Expression (7.8) means equation (7.8) in EC2-1-1.

Note that, unlike in other guides in this series, even clauses in EN 1992-2 itself are prefixed with '2-2'. There are so many references to other parts of Eurocode 2 required that to do otherwise would be confusing.

Where additional equations are provided in the guide, they are numbered sequentially within each sub-section of a main section so that, for example, the third additional expression within sub-section 6.1 would be referenced equation (D6.1-3). Additional figures and tables follow the same system. For example, the second additional figure in section 6.4 would be referenced Figure 6.4-2.

Acknowledgements

Chris Hendy would like to thank his wife, Wendy, and two boys, Peter Edwin Hendy and Matthew Philip Hendy, for their patience and tolerance of his pleas to finish 'just one more section'.

David Smith would like to thank his wife, Emma, for her limitless patience during preparation of this guide. He also acknowledges his son, William Thomas Smith, and the continued support of Brian and Rosalind Ruffell-Ward from the very beginning.

Both authors would also like to thank their employer, Atkins, for providing both facilities and time for the production of this guide. They also wish to thank Dr Paul Jackson and Dr Steve Denton for their helpful comments on the guide.

Chris Hendy
David A. Smith

Contents

Preface		v
	Aims and objectives of this guide	v
	Layout of this guide	v
	Acknowledgements	vi
Introduction		1
	Additional information specific to EN 1992-2	2
Chapter 1.	**General**	3
	1.1. Scope	3
	1.1.1. Scope of Eurocode 2	3
	1.1.2. Scope of Part 2 of Eurocode 2	4
	1.2. Normative references	4
	1.3. Assumptions	4
	1.4. Distinction between principles and application rules	5
	1.5. Definitions	5
	1.6. Symbols	5
Chapter 2.	**Basis of design**	7
	2.1. Requirements	7
	2.2. Principles of limit state design	7
	2.3. Basic variables	7
	2.4. Verification by the partial factor method	9
	2.4.1. General	9
	2.4.2. Design values	9
	2.4.3. Combinations of actions	9
	2.5. Design assisted by testing	10
	2.6. Supplementary requirements for foundations	10
Chapter 3.	**Materials**	11
	3.1. Concrete	11
	3.1.1. General	11
	3.1.2. Strength	11
	3.1.3. Elastic deformation	14
	3.1.4. Creep and shrinkage	14
	3.1.5. Concrete stress–strain relation for non-linear structural analysis	19

		3.1.6.	Design compressive and tensile strengths	20
		3.1.7.	Stress–strain relations for the design of sections	21
		3.1.8.	Flexural tensile strength	22
		3.1.9.	Confined concrete	23
	3.2.	Reinforcing steel		23
		3.2.1.	General	23
		3.2.2.	Properties	23
		3.2.3.	Strength	23
		3.2.4.	Ductility	24
		3.2.5.	Welding	25
		3.2.6.	Fatigue	25
		3.2.7.	Design assumptions	25
	3.3.	Prestressing steel		25
		3.3.1.	General	25
		3.3.2.	Properties	26
		3.3.3.	Strength	27
		3.3.4.	Ductility characteristics	27
		3.3.5.	Fatigue	28
		3.3.6.	Design assumptions	28
	3.4.	Prestressing devices		29
		3.4.1.	Anchorages and couplers	29
		3.4.2.	External non-bonded tendons	29

Chapter 4.	**Durability and cover to reinforcement**			**31**
	4.1.	General		31
	4.2.	Environmental conditions		32
	4.3.	Requirements for durability		35
	4.4.	Methods of verification		36
		4.4.1.	Concrete cover	36

Chapter 5.	**Structural analysis**			**39**
	5.1.	General		39
	5.2.	Geometric imperfections		40
		5.2.1.	General (additional sub-section)	40
		5.2.2.	Arches (additional sub-section)	43
	5.3	Idealization of the structure		44
		5.3.1	Structural models for overall analysis	44
		5.3.2.	Geometric data	44
	5.4.	Linear elastic analysis		48
	5.5.	Linear elastic analysis with limited redistribution		49
	5.6.	Plastic analysis		52
		5.6.1.	General	52
		5.6.2.	Plastic analysis for beams, frames and slabs	52
		5.6.3.	Rotation capacity	53
		5.6.4.	Strut-and-tie models	56
	5.7.	Non-linear analysis		58
		5.7.1.	Method for ultimate limit states	58
		5.7.2.	Scalar combinations	60
		5.7.3.	Vector combinations	61
		5.7.4.	Method for serviceability limit states	62
	5.8.	Analysis of second-order effects with axial load		62
		5.8.1.	Definitions and introduction to second-order effects	62
		5.8.2.	General	63
		5.8.3.	Simplified criteria for second-order effects	64

		5.8.4.	Creep	69
		5.8.5.	Methods of analysis	70
		5.8.6.	General method – second-order non-linear analysis	70
		5.8.7.	Second-order analysis based on nominal stiffness	71
		5.8.8.	Method based on nominal curvature	76
		5.8.9.	Biaxial bending	80
	5.9.	Lateral instability of slender beams		80
	5.10.	Prestressed members and structures		81
		5.10.1.	General	81
		5.10.2.	Prestressing force during tensioning	82
		5.10.3.	Prestress force	83
		5.10.4.	Immediate losses of prestress for pre-tensioning	84
		5.10.5.	Immediate losses of prestress for post-tensioning	85
		5.10.6.	Time-dependent losses	90
		5.10.7.	Consideration of prestress in the analysis	95
		5.10.8.	Effects of prestressing at the ultimate limit state	96
		5.10.9.	Effects of prestressing at the serviceability and fatigue limit states	98
	5.11.	Analysis for some particular structural members		104

Chapter 6. **Ultimate limit states** — 105

	6.1.	ULS bending with or without axial force		105
		6.1.1.	General (additional sub-section)	105
		6.1.2.	Reinforced concrete beams (additional sub-section)	105
		6.1.3.	Prestressed concrete beams (additional sub-section)	118
		6.1.4.	Reinforced concrete columns (additional sub-section)	121
		6.1.5.	Brittle failure of members with prestress (additional sub-section)	126
	6.2.	Shear		131
		6.2.1.	General verification procedure rules	132
		6.2.2.	Members not requiring design shear reinforcement	133
		6.2.3.	Members requiring design shear reinforcement	140
		6.2.4.	Shear between web and flanges of T-sections	154
		6.2.5.	Shear at the interface between concrete cast at different times	158
		6.2.6.	Shear and transverse bending	160
		6.2.7.	Shear in precast concrete and composite construction (additional sub-section)	160
	6.3.	Torsion		166
		6.3.1.	General	166
		6.3.2.	Design procedure	167
		6.3.3.	Warping torsion	171
		6.3.4.	Torsion in slabs (additional sub-section)	172
	6.4.	Punching		175
		6.4.1.	General	175
		6.4.2.	Load distribution and basic control perimeter	176
		6.4.3.	Punching shear calculation	177
		6.4.4.	Punching shear resistance of slabs and bases without shear reinforcement	179
		6.4.5.	Punching shear resistance of slabs and bases with shear reinforcement	183
		6.4.6.	Pile caps (additional sub-section)	185
	6.5.	Design with strut-and-ties models		193
		6.5.1.	General	193

		6.5.2.	Struts	193
		6.5.3.	Ties	195
		6.5.4.	Nodes	196
	6.6.	Anchorage and laps		201
	6.7.	Partially loaded areas		201
	6.8.	Fatigue		208
		6.8.1.	Verification conditions	208
		6.8.2.	Internal forces and stresses for fatigue verification	208
		6.8.3.	Combination of actions	209
		6.8.4.	Verification procedure for reinforcing and prestressing steel	209
		6.8.5.	Verification using damage equivalent stress range	210
		6.8.6.	Other verification methods	212
		6.8.7.	Verification of concrete under compression or shear	213
	6.9.	Membrane elements		215

Chapter 7. Serviceability limit states — 225

- 7.1. General — 225
- 7.2. Stress limitation — 226
- 7.3. Crack control — 230
  - 7.3.1. General considerations — 230
  - 7.3.2. Minimum areas of reinforcement — 232
  - 7.3.3. Control of cracking without direct calculation — 234
  - 7.3.4. Control of crack widths by direct calculation — 237
- 7.4. Deflection control — 243
- 7.5. Early thermal cracking (additional sub-section) — 243

Chapter 8. Detailing of reinforcement and prestressing steel — 245

- 8.1. General — 245
- 8.2. Spacing of bars — 246
- 8.3. Permissible mandrel diameters for bent bars — 246
- 8.4. Anchorage of longitudinal reinforcement — 247
  - 8.4.1. General — 247
  - 8.4.2. Ultimate bond stress — 248
  - 8.4.3. Basic anchorage length — 248
  - 8.4.4. Design anchorage length — 249
- 8.5. Anchorage of links and shear reinforcement — 251
- 8.6. Anchorage by welded bars — 251
- 8.7. Laps and mechanical couplers — 252
  - 8.7.1. General — 252
  - 8.7.2. Laps — 252
  - 8.7.3. Lap length — 253
  - 8.7.4. Transverse reinforcement in the lap zone — 254
  - 8.7.5. Laps of welded mesh fabrics made of ribbed wires — 257
  - 8.7.6. Welding (additional sub-section) — 257
- 8.8. Additional rules for large diameter bars — 257
- 8.9. Bundled bars — 258
- 8.10. Prestressing tendons — 258
  - 8.10.1. Tendon layouts — 258
  - 8.10.2. Anchorage of pre-tensioned tendons — 259
  - 8.10.3. Anchorage zones of post-tensioned members — 262
  - 8.10.4. Anchorages and couplers for prestressing tendons — 271
  - 8.10.5. Deviators — 272

Chapter 9.	**Detailing of members and particular rules**		**275**
	9.1. General		275
	9.2. Beams		275
		9.2.1. Longitudinal reinforcement	275
		9.2.2. Shear reinforcement	278
		9.2.3. Torsion reinforcement	279
		9.2.4. Surface reinforcement	279
		9.2.5. Indirect supports	279
	9.3. Solid slabs		281
		9.3.1. Flexural reinforcement	281
		9.3.2. Shear reinforcement	282
	9.4. Flat slabs		282
	9.5. Columns		282
		9.5.1. General	282
		9.5.2. Longitudinal reinforcement	283
		9.5.3. Transverse reinforcement	283
	9.6. Walls		284
	9.7. Deep beams		284
	9.8. Foundations		285
	9.9. Regions with discontinuity in geometry or action		288
Chapter 10.	**Additional rules for precast concrete elements and structures**		**289**
	10.1. General		289
	10.2. Basis of design, fundamental requirements		289
	10.3. Materials		290
		10.3.1. Concrete	290
		10.3.2. Prestressing steel	290
	10.4. Not used in EN 1992-2		290
	10.5. Structural analysis		290
		10.5.1. General	290
		10.5.2. Losses of prestress	291
	10.6. Not used in EN 1992-2		291
	10.7. Not used in EN 1992-2		291
	10.8. Not used in EN 1992-2		291
	10.9. Particular rules for design and detailing		291
		10.9.1. Restraining moments in slabs	291
		10.9.2. Wall to floor connections	291
		10.9.3. Floor systems	291
		10.9.4. Connections and supports for precast elements	291
		10.9.5. Bearings	292
		10.9.6. Pocket foundations	293
Chapter 11.	**Lightweight aggregate concrete structures**		**295**
	11.1. General		295
	11.2. Basis of design		296
	11.3. Materials		296
		11.3.1. Concrete	296
		11.3.2. Elastic deformation	296
		11.3.3. Creep and shrinkage	297
		11.3.4. Stress strain relations for non-linear structural analysis	298
		11.3.5 Design compressive and tensile strengths	298
		11.3.6. Stress strain relations for the design of sections	298
		11.3.7. Confined concrete	298
	11.4. Durability and cover to reinforcement		298

	11.5. Structural analysis	298
	11.6. Ultimate limit states	298
	11.7. Serviceability limit states	302
	11.8. Detailing of reinforcement – general	302
	11.9. Detailing of members and particular rules	302
Chapter 12.	**Plain and lightly reinforced concrete structures**	**303**
Chapter 13.	**Design for the execution stages**	**307**
	13.1. General	307
	13.2. Actions during execution	308
	13.3. Verification criteria	309
	13.3.1. Ultimate limit state	309
	13.3.2. Serviceability limit states	309
Annex A.	Modification of partial factors for materials (informative)	311
Annex B.	Creep and shrinkage strain (informative)	313
Annex C.	Reinforcement properties (normative)	316
Annex D.	Detailed calculation method for prestressing steel relaxation losses (informative)	317
Annex E.	Indicative strength classes for durability (informative)	322
Annex F.	Tension reinforcement expressions for in-plane stress conditions (informative)	324
Annex G.	Soil-structure interaction	325
Annex H.	*Not used in EN 1992-2*	
Annex I.	Analysis of flat slabs (informative)	326
Annex J.	Detailing rules for particular situations (informative)	327
Annex K.	Structural effects of time-dependent behaviour (informative)	331
Annex L.	Concrete shell elements (informative)	344
Annex M.	Shear and transverse bending (informative)	346
Annex N.	Damage equivalent stresses for fatigue verification (informative)	356
Annex O.	Typical bridge discontinuity regions (informative)	362
Annex P.	Safety format for non-linear analysis (informative)	363
Annex Q.	Control of shear cracks within webs (informative)	364
References		**369**
Index		**371**

Introduction

The provisions of EN 1992-2 are preceded by a foreword, most of which is common to all Eurocodes. This *Foreword* contains clauses on:

- the background to the Eurocode programme
- the status and field of application of the Eurocodes
- national standards implementing Eurocodes
- links between Eurocodes and harmonized technical specifications for products
- additional information specific to EN 1992-2
- National Annex for EN 1992-2.

Guidance on the common text is provided in the introduction to the *Designers' Guide to EN 1990 – Eurocode: Basis of Structural Design*,[1] and only background information relevant to users of EN 1992-2 is given here.

It is the responsibility of each national standards body to implement each Eurocode part as a national standard. This will comprise, without any alterations, the full text of the Eurocode and its annexes as published by the European Committee for Standardization (CEN, from its title in French). This will usually be preceded by a National Title Page and a National Foreword, and may be followed by a National Annex.

Each Eurocode recognizes the right of national regulatory authorities to determine values related to safety matters. Values, classes or methods to be chosen or determined at national level are referred to as nationally determined parameters (NDPs). Clauses of EN 1992-2 in which these occur are listed in the *Foreword*.

NDPs are also indicated by notes immediately after relevant clauses. These Notes give recommended values. It is expected that most of the Member States of CEN will specify the recommended values, as their use was assumed in the many calibration studies made during drafting. Recommended values are used in this guide, as the National Annex for the UK was not available at the time of writing. Comments are made regarding the likely values to be adopted where different.

Each National Annex will give or cross-refer to the NDPs to be used in the relevant country. Otherwise the National Annex may contain only the following:[2]

- decisions on the use of informative annexes, and
- references to non-contradictory complementary information to assist the user to apply the Eurocode.

The set of Eurocodes will supersede the British bridge code, BS 5400, which is required (as a condition of BSI's membership of CEN) to be withdrawn by early 2010, as it is a 'conflicting national standard'.

Additional information specific to EN 1992-2

The information specific to EN 1992-2 emphasizes that this standard is to be used with other Eurocodes. The standard includes many cross-references to EN 1992-1-1 and does not itself reproduce material which appears in other parts of EN 1992. Where a clause or paragraph in EN 1992-2 modifies one in EN 1992-1-1, the clause or paragraph number used is renumbered by adding 100 to it. For example, if paragraph (3) of a clause in EN 1992-1-1 is modified in EN 1992-2, it becomes paragraph (103). This guide is intended to be self-contained for the design of concrete bridges and therefore provides commentary on other parts of EN 1992 as necessary.

The *Foreword* lists the clauses of EN 1992-2 in which National choice is permitted. Elsewhere, there are cross-references to clauses with NDPs in other codes. Otherwise, the Normative rules in the code must be followed, if the design is to be 'in accordance with the Eurocodes'.

In EN 1992-2, Sections 1 to 13 (actually 113 because clause 13 does not exist in EN 1992-1-1) are Normative. Of its 17 annexes, only its Annex C is 'Normative', as alternative approaches may be used in other cases. (Arguably Annex C, which defines the properties of reinforcement suitable for use with Eurocodes, should not be in Eurocode 2 as it relates to material which is contained in product standards.) A National Annex may make Informative provisions Normative in the country concerned, and is itself Normative in that country but not elsewhere. The 'non-contradictory complimentary information' referred to above could include, for example, reference to a document based on provisions of BS 5400 on matters not treated in the Eurocodes. Each country can do this, so some aspects of the design of a bridge will continue to depend on where it is to be built.

CHAPTER 1

General

This chapter is concerned with the general aspects of EN 1992-2, *Eurocode 2: Design of Concrete Structures. Part 2: Concrete Bridges*. The material described in this chapter is covered in section 1 of EN 1992-2 in the following clauses:

- Scope   *Clause 1.1*
- Normative references   *Clause 1.2*
- Assumptions   *Clause 1.3*
- Distinction between principles and application rules   *Clause 1.4*
- Definitions   *Clause 1.5*
- Symbols   *Clause 1.6*

1.1. Scope
1.1.1. Scope of Eurocode 2
The scope of EN 1992 is outlined in 2-2/clause 1.1.1 by reference to 2-1-1/clause 1.1.1. It is to be used with EN 1990, *Eurocode: Basis of Structural Design*, which is the head document of the Eurocode suite and has an Annex A2, 'Application for bridges'. ***2-1-1/clause 1.1.1(2)*** emphasizes that the Eurocodes are concerned with structural behaviour and that other requirements, e.g. thermal and acoustic insulation, are not considered.

2-1-1/clause 1.1.1(2)

The basis for verification of safety and serviceability is the partial factor method. EN 1990 recommends values for load factors and gives various possibilities for combinations of actions. The values and choice of combinations are to be set by the National Annex for the country in which the structure is to be constructed.

2-1-1/clause 1.1.1(3)P states that the following parts of the Eurocode are intended to be used in conjunction with Eurocode 2:

2-1-1/clause 1.1.1(3)P

EN 1990:	Basis of structural design
EN 1991:	Actions on structures
EN 1997:	Geotechnical design
EN 1998:	Design of structures for earthquake resistance
hENs:	Construction products relevant for concrete structures
EN 13670:	Execution (construction) of concrete structures

These documents will often be required for a typical concrete bridge design, but discussion on them is generally beyond the scope of this guide. They supplement the normative reference standards given in 2-2/clause 1.2. The Eurocodes are concerned with design and not execution, but minimum standards of workmanship and material specification are required to ensure that the design assumptions are valid. For this reason, 2-1-1/clause 1.1.1(3)P includes the European standards for concrete products and for the execution of concrete structures. ***2-1-1/clause 1.1.1(4)P*** lists the other parts of EC2.

2-1-1/clause 1.1.1(4)P

One standard curiously not referenced by EN 1992-2 is EN 15050: *Precast Concrete Bridge Elements*. At the time of writing, this document was available only in draft for comment, but its scope and content made it relevant to precast concrete bridge design. At the time of the review of prEN 15050: 2004, its contents were a mixture of the following:

- definitions relevant to precast concrete bridges
- informative design guidance on items not covered in EN 1992 (e.g. for shear keys)
- cross-reference to design requirements in EN 1992 (e.g. for longitudinal shear)
- informative guidance duplicating or contradicting normative guidance in EN 1992-2 (e.g. effective widths for shear lag)
- cross-reference to EN 13369: *Common rules for precast concrete products*
- requirements for inspection and testing of the finished product.

Comment was made that EN 15050 should not contradict or duplicate design requirements in EN 1992. If this is achieved in the final version, there will be little Normative in it for the designer to follow, but there may remain some guidance on topics not covered by EN 1992.

1.1.2. Scope of Part 2 of Eurocode 2

EC2-2 covers structural design of concrete bridges. Its format is based on EN 1992-1-1 and generally follows the same clause numbering as discussed in the Introduction to this guide. It identifies which parts of EN 1992-1-1 are relevant for bridge design and which parts need modification. It also adds provisions which are specific to bridges. Importantly, *2-1-1/clause 1.1.2(4)P* states that plain round reinforcement is not covered.

2-1-1/clause 1.1.2(4)P

1.2. Normative references

References are given only to other European standards, all of which are intended to be used as a package. Formally, the Standards of the International Organization for Standardization (ISO) apply only if given an EN ISO designation. National standards for design and for products do not apply if they conflict with a relevant EN standard. As Eurocodes may not cross-refer to national standards, replacement of national standards for products by EN or ISO standards is in progress, with a time-scale similar to that for the Eurocodes.

During the period of changeover to Eurocodes and EN standards it is possible that an EN referred to, or its national annex, may not be complete. Designers who then seek guidance from national standards should take account of differences between the design philosophies and safety factors in the two sets of documents.

Of the material and product standards referred to in 2-1-1/clause 1.2, Eurocode 2 relies most heavily on EN 206-1 (for the specification, performance, production and compliance criteria for concrete), EN 10080 (technical delivery conditions and specification of weldable, ribbed reinforcing steel for the reinforcement of concrete) and EN 10138 (for the specification and general requirements for prestressing steels). Further reference to and guidance on the use of these standards can be found in section 3, which discusses materials.

1.3. Assumptions

It is assumed in using EC2-2 that the provisions of EN 1990 will be followed. In addition, EC2-2 identifies the following assumptions, some of which reiterate those in EN 1990:

- Structures are designed by appropriately qualified and experienced personnel and are constructed by personnel with appropriate skill and experience.
- The construction materials and products are used as specified in Eurocode 2 or in the relevant material or product specifications.
- Adequate supervision and quality control is provided in factories, in plants and on site.
- The structure will be adequately maintained and used in accordance with the design brief.

- The requirements for construction and workmanship given in EN 13670 are complied with.

EC2-2 should not be used for the design of bridges that will be executed to specifications other than EN 13670 without a careful comparison of the respective tolerance and workmanship requirements. Slender elements in particular are sensitive to construction tolerances in their design.

1.4. Distinction between principles and application rules

Reference has to be made to EN 1990 for the distinction between 'Principles' and 'Application Rules'. Essentially, Principles comprise general statements and requirements that must be followed and Application Rules are rules that comply with these Principles. There may, however, be other ways to comply with the Principles and these methods may be substituted if it is shown that they are at least equivalent to the Application Rules with respect to safety, serviceability and durability. This, however, presents the problem that such a design could not then be deemed to comply wholly with the Eurocodes.

Principles are required by EN 1990 to be marked with a 'P' adjacent to the paragraph number. In addition, Principles can also generally be identified by the use of 'shall' within a clause, while 'should' and 'may' are generally used for Application Rules, but this is not completely consistent.

1.5. Definitions

Reference is made to the definitions given in clauses 1.5 of EN 1990 and further bridge-specific definitions are provided.

There are some significant differences in the use of language compared to British codes. These arose from the use of English as the base language for the drafting process, and the resulting need to improve precision of meaning and to facilitate translation into other European languages. In particular:

- 'action' means a load and/or an imposed deformation;
- 'action effect' and 'effect of action' have the same meaning: any deformation or internal force or moment that results from an action.

Actions are further subdivided into permanent actions, G (such as dead loads, shrinkage and creep), variable actions, Q (such as traffic loads, wind loads and temperature loads), and accidental actions, A. Prestressing, P, is treated as a permanent action in most situations.

The Eurocodes denote characteristic values of any parameter with a suffix 'k'. Design values are denoted with a suffix 'd' and include appropriate partial factors. It should be noted that this practice is different from current UK practice in concrete design, where material partial factors are usually included in formulae to ensure they are not forgotten. It is therefore extremely important to use the correct parameters, duly noting the suffix, to ensure that the material partial factors are included when appropriate.

1.6. Symbols

The symbols in the Eurocodes are all based on ISO standard 3898: 1987.[3] Each code has its own list, applicable within that code. Some symbols have more than one meaning, the particular meaning being stated in the clause. There are a few important changes from previous practice in the UK. For example, an x–x axis is along a member and subscripts are used extensively to distinguish characteristic values from design values. The use of upper-case subscripts for γ factors for materials implies that the values given allow for two types of uncertainty, i.e. in the properties of the material and in the resistance model used.

CHAPTER 2

Basis of design

This chapter discusses the basis of design as covered in section 2 of EN 1992-2 in the following clauses:

- Requirements   *Clause 2.1*
- Principles of limit state design   *Clause 2.2*
- Basic variables   *Clause 2.3*
- Verification by the partial factor method   *Clause 2.4*
- Design assisted by testing   *Clause 2.5*
- Supplementary requirements for foundations   *Clause 2.6*

2.1. Requirements

2-1-1/clause 2.1.1 makes reference to EN 1990 for the basic principles and requirements for the design process for concrete bridges. This includes the limit states and combination of actions to consider, together with the required performance of the bridge at each limit state. These basic performance requirements are deemed to be met if the bridge is designed using actions in accordance with EN 1991, combination of actions and load factors at the various limit states in accordance with EN 1990, and the resistances, durability and serviceability provisions of EN 1992.

2-1-1/clause 2.1.1

2-1-1/clause 2.1.3 refers to EN 1990 for rules on design working life, durability and quality management for bridges. Design working life predominantly affects calculations on fatigue and durability requirements, such as concrete cover. The latter is discussed in section 4 of this guide. Permanent bridges have an indicative design life of 100 years in EN 1990. For political reasons, it is likely that the UK will adopt a design life of 120 years in the National Annex to EN 1990 for permanent bridges for consistency with previous national design standards.

2-1-1/clause 2.1.3

2.2. Principles of limit state design

The principles of limit state design are set out in section 3 of EN 1990. They are not specific to the design of concrete bridges and are discussed in reference 1.

2.3. Basic variables

Actions to consider
2-1-1/clause 2.3.1.1(1) refers to EN 1991 for actions to consider in design and also refers to EN 1997 for actions arising from soil and water pressures. Actions not covered by either of these sources may be included in a Project Specification.

2-1-1/clause 2.3.1.1(1)

2-1-1/clause 2.3.1.2 and 2-1-1/clause 2.3.1.3 cover thermal effects and differential settlements respectively, which are 'indirect' actions. These are essentially imposed deformations rather than imposed forces. The effects of imposed deformations must also always be checked at the serviceability limit state so as to limit deflections and cracking – *2-1-1/clause 2.3.1.2(1)* and *2-1-1/clause 2.3.1.3(2)* refer. Indirect actions can usually be ignored for ultimate limit states (excluding fatigue), since yielding of overstressed areas will shed the locked-in forces generated by imposed deformation. However, a certain amount of ductility and plastic rotation capacity is required to shed these actions and this is noted in *2-1-1/clause 2.3.1.2(2)* and *2-1-1/clause 2.3.1.3(3)*. A check of ductility and plastic rotation capacity can be made as described in section 5.6.3.2 of this guide. The same clauses also note that indirect actions should still be considered where they are 'significant'. The examples given are where elements are prone to significant second-order effects (particularly slender piers) or when fatigue is being checked. For most bridges, these will be the only situations where indirect actions need to be considered for ultimate limit states, providing there is adequate ductility and rotation capacity to ignore them in other cases.

2-1-1/clause 2.3.1.2(1)
2-1-1/clause 2.3.1.3(2)

2-1-1/clause 2.3.1.2(2)
2-1-1/clause 2.3.1.3(3)

Imposed deformations covered by the above discussions include those from:

- Thermal effects – variable action
- Differential settlement – permanent action
- Shrinkage – permanent action, covered by 2-1-1/clause 2.3.2.2
- Creep – permanent action, covered by 2-1-1/clause 2.3.2.2.

Secondary effects of prestress are not dealt with in the same way as the above imposed deformations because tests have shown that they remain locked in throughout significant rotation up to failure. Consequently, *2-1-1/clause 2.3.1.4* does not contain similar provisions to those above and secondary effects of prestress are always considered at the ultimate limit state.

2-1-1/clause 2.3.1.4

Material and product properties

2-1-1/clause 2.3.2.2(1) and *(2)* relate to the treatment of shrinkage and creep at serviceability and ultimate limit states respectively and make similar requirements to those for thermal effects and settlements discussed above. *2-1-1/clause 2.3.2.2(3)* requires creep deformation and its effects to be based on the quasi-permanent combination of actions, regardless of the design combination being considered.

2-1-1/clause 2.3.2.2(1) and (2)
2-1-1/clause 2.3.2.2(3)

Geometric data

Generally, the dimensions of the structure used for modelling and section analysis may be assumed to be equal to those that are put on the drawings. The exceptions to this rule are:

(1) Member imperfections due to construction tolerances – these need to be accounted for where departure from the drawing dimensions leads to additional effects, such as additional bending moments in slender columns under axial load (imperfections are discussed in section 5.2 of this guide).
(2) Eccentricities of axial load – a minimum moment from eccentricity of axial load has to be considered in the design of beam-columns according to 2-1-1/clause 6.1(4), but this is not additive to the moments from imperfections.
(3) Cast in place piles without permanent casing – the size of such piles cannot be accurately controlled so *2-1-1/clause 2.3.4.2(2)* gives the following diameters, d, to be used in calculations based on the intended diameter, d_{nom}, in the absence of specific measures to control diameter:

2-1-1/clause 2.3.4.2(2)

$d_{nom} < 400\,\text{mm}$ $d = d_{nom} - 20\,\text{mm}$

$400 \leq d_{nom} \leq 1000\,\text{mm}$ $d = 0.95 d_{nom}$

$d_{nom} > 1000\,\text{mm}$ $d = d_{nom} - 50\,\text{mm}$

2.4. Verification by the partial factor method
2.4.1. General
2-1-1/clause 2.4.1(1) refers to section 6 of EN 1990 for the rules for the partial factor method. They are not specific to the design of concrete bridges and are discussed in reference 1.

2-1-1/clause 2.4.1(1)

2.4.2. Design values
Partial factors for actions
Partial factors for actions are given in EN 1990 and its Annex A2 for bridges, together with rules for load combinations. EC2-1-1 defines further specific load factors to be used in concrete bridge design for shrinkage, prestress and fatigue loadings in its clauses 2.4.2.1 to 2.4.2.3. The values given may be modified in the National Annex. The recommended values are summarized in Table 2.4-1 and include recommended values for prestressing forces at SLS from 2-1-1/clause 5.10.9. They apply unless specific values are given elsewhere in EC2-2 or the National Annexes.

Table 2.4-1. Recommended values of load factors – may be modified in National Annex

Action	ULS unfavourable (adverse)	ULS favourable (relieving)	SLS unfavourable (adverse)	SLS favourable (relieving)	Fatigue
Shrinkage	$\gamma_{SH} = 1.0$	0	1.0	0	1.0 if unfavourable 0 if favourable
Prestress – global effects	$\gamma_{P,unfav} = 1.3$ (See Note 1)	$\gamma_{P,fav} = 1.0$ (See Note 4)	(See Note 2)	(See Note 2)	1.0
Prestress – local effects	$\gamma_{P,unfav} = 1.2$ (See Note 3)	$\gamma_{P,fav} = 1.0$	(See Note 2)	(See Note 2)	1.0
Fatigue loading	–	–	–	–	$\gamma_{F,fat} = 1.0$

Notes
(1) In general, *2-1-1/clause 2.4.2.2(1)* requires $\gamma_{P,fav}$ to be used for prestressing actions at the ultimate limit state. The use of $\gamma_{P,unfav}$ in *2-1-1/clause 2.4.2.2(2)* relates specifically to stability checks of externally prestressed members. In previous UK practice, the equivalent of $\gamma_{P,unfav}$ was also used in checking other situations where prestress has an adverse effect (e.g. where draped tendons have an adverse effect on shear resistance) so this represents a relaxation.
(2) *2-1-1/clause 5.10.9* gives factors that differ for pre-tensioning and post-tensioning and also for favourable and unfavourable effects.
(3) This value of $\gamma_{P,unfav}$ applies to the design of anchorage zones. For externally post-tensioned bridges, it is recommended here that the characteristic breaking load of the tendon be used as the ultimate design load, as discussed in section 8.10.3 of this guide.
(4) This value applies to the prestressing force used in ultimate bending resistance calculation. For internal post-tensioning, the prestrain used in the bending calculation should correspond to this design prestressing force, as discussed in section 6.1 of this guide.

2-1-1/clause 2.4.2.2(1)
2-1-1/clause 2.4.2.2(2)

Material factors
2-1-1/clause 2.4.2.4 defines specific values of material factor for concrete, reinforcement and prestressing steel to be used in concrete bridge design, but they may be modified in the National Annex. These are summarized in Table 2.4-2. They do not cover fire design. The material factor values assume that workmanship will be in accordance with specified limits in EN 13670-1 and reinforcement, concrete and prestressing steel conform to the relevant Euronorms. If measures are taken to increase the level of certainty of material strengths and/or setting out dimensions, then reduced material factors may be used in accordance with Annex A.

2-1-1/clause 2.4.2.4

2.4.3. Combinations of actions
Combinations of actions are generally covered in Annex A2 of EN 1990, as stated in Note 1 of *2-1-1/clause 2.4.3(1)*, but fatigue combinations are covered in 2-2/clause 6.8.3. For each

2-1-1/clause 2.4.3(1)

Table 2.4-2. Recommended values of material factors

Design situation	γ_C for concrete	γ_S for reinforcing steel	γ_S for prestressing steel
ULS persistent and transient	$1.5^{(2)}$	1.15	1.15
ULS accidental	$1.2^{(2)}$	1.0	1.0
Fatigue	1.5	1.15	1.15
SLS	$1.0^{(1)}$	$1.0^{(1)}$	$1.0^{(1)}$

Notes
(1) Unless stated otherwise in specific clauses (*2-1-1/clause 2.4.2.4(2)*).
(2) Increase by a recommended factor of 1.1 for cast in place piles without permanent casing (*2-1-1/clause 2.4.2.5(2)*).

2-1-1/clause 2.4.2.4(2)
2-1-1/clause 2.4.2.5(2)

permanent action, such as self-weight, the adverse or relieving partial load factor as applicable can generally be used throughout the entire structure when calculating each particular action effect. There can however be some exceptions, as stated in the Note to *2-1-1/clause 2.4.3(2)*. EN 1990 clause 6.4.3.1(4) states that 'where the results of a verification are very sensitive to variations of the magnitude of a permanent action from place to place in the structure, the unfavourable and the favourable parts of this action shall be considered as individual actions. **Note:** this applies in particular to the verification of static equilibrium and analogous limit states.' One such exception is intended to be the verification of uplift at bearings on continuous beams, where each span would be treated separately when applying adverse and relieving values of load. The same applies to holding down bolts. This is the basis for *2-1-1/clause 2.4.4*, which requires the reliability format for static equilibrium to be used in such situations to achieve this separation into adverse and relieving areas.

2-1-1/clause 2.4.3(2)

2-1-1/clause 2.4.4

2.5. Design assisted by testing

The characteristic resistances in EC2 have, in theory, been derived using Annex D of EN 1990. EN 1990 allows two alternative methods of calculating design values of resistance. Either the characteristic resistance R_k is first determined and the design resistance R_d determined from this using appropriate partial factors, or the design resistance is determined directly. R_k represents the lower 5% fractile for infinite tests. Where it is necessary to determine the characteristic resistance for products where this information is not available, one of these methods has to be used. Discussion on the use of EN 1990 is outside the scope of this guide and is not considered further here.

2.6. Supplementary requirements for foundations

Although 2-1-1/clause 2.6 refers specifically to foundations in its title, the effects of soil–structure interaction may need to be considered in the design of the whole bridge, as is the case with most integral bridges. This is stated in *2-1-1/clause 2.6(1)P*. Some further discussion on soil–structure interaction is given in Annex G of this guide. *2-1-1/clause 2.6(2)* recommends that the effects of differential settlement are checked where 'significant'. It is recommended here that the effects of differential settlement are always considered for bridges, as discussed under the comments to 2-1-1/clause 2.3.1.3.

2-1-1/clause 2.6(1)P
2-1-1/clause 2.6(2)

CHAPTER 3

Materials

This chapter discusses materials as covered in section 3 of EN 1992-2 in the following clauses:

- Concrete   *Clause 3.1*
- Reinforcing steel   *Clause 3.2*
- Prestressing steel   *Clause 3.3*
- Prestressing devices   *Clause 3.4*

3.1. Concrete
3.1.1. General
EC2 relies on EN 206-1 for the specification of concrete, including tests for confirming properties. 2-2/clause 3 does not cover lightweight concrete. Lightweight concrete is covered in 2-1-1/clause 11.

3.1.2. Strength
Compressive strength
EC2 classifies the compressive strength of normal concrete in relation to the cylinder strength (f_{ck}) and its equivalent cube strength ($f_{ck,cube}$) determined at 28 days. For example, the strength class C40/50 denotes normal concrete with cylinder strength of $40\,\text{N/mm}^2$ and cube strength of $50\,\text{N/mm}^2$. All formulae in EC2, however, use the cylinder strength. 2-1-1/Table 3.1, reproduced here as Table 3.1-1, provides material properties for normal concretes with typical cylinder strengths. The equivalent cube strengths are such that typically $f_{ck} \approx 0.8 f_{ck,cube}$. The characteristic compressive strength, f_{ck}, is defined as the value below which 5% of all strength test results would be expected to fall for the specified concrete.

It should be noted that EC2-1-1 covers significantly higher strength concrete than in BS 5400, but **2-2/clause 3.1.2(102)P** recommends limiting the range of strength classes that can be used to between C30/37 and C70/85. The National Annex can alter these limits. The UK has applied a more restrictive limit for use in calculation of the shear resistance. This is because testing carried out by Regan *et al.*[4] identified that $V_{Rd,c}$ (see 2-1-1/clause 6.2.2) could be significantly overestimated unless the value of f_{ck} was limited in calculation, particularly where limestone aggregate is to be used.

2-1-1/clause 3.1.2(6) gives an expression for estimating the mean compressive strength of concrete with time, assuming a mean temperature of 20°C and curing in accordance with EN 12390:

$$f_{cm}(t) = \beta_{cc}(t) f_{cm} \qquad \text{2-1-1/(3.1)}$$

2-2/clause 3.1.2(102)P

2-1-1/clause 3.1.2(6)

with

$$\beta_{cc}(t) = \exp\left\{s\left[1 - \left(\frac{28}{t}\right)^{0.5}\right]\right\}$$ 2-1-1/(3.2)

where:

$f_{cm}(t)$ is the mean compressive strength at an age of t days
f_{cm} is the mean compressive strength at 28 days given in 2-1-1/Table 3.1
t is the age of concrete in days
s is a coefficient which depends on cement type
   = 0.2 for rapid hardening high-strength cements
   = 0.25 for normal and rapid hardening cements
   = 0.38 for slow hardening cements.

2-1-1/clause 3.1.2(5)

The characteristic concrete compressive strength at time t can then similarly be estimated from **2-1-1/clause 3.1.2(5)**:

$$f_{ck}(t) = f_{cm}(t) - 8 \quad \text{for } 3 < t < 28 \text{ days}$$ (D3.1-1)

$$f_{ck}(t) = f_{ck} \quad \text{for } t \geq 28 \text{ days}$$ (D3.1-2)

Clauses 3.1.2(5) and 3.1.2(6) are useful for estimating the time required to achieve a particular strength (e.g. time to reach a specified strength to permit application of prestress or striking of formwork). It is still permissible to determine more precise values from tests and precasters may choose to do this to minimize waiting times. The clauses can also be used to predict 28-day strength from specimens tested earlier than 28 days, although it is desirable to have tests carried out at 28 days to be sure of final strength. 2-1-1/clause 3.1.2(6) makes it clear that they must not be used for justifying a non-conforming concrete tested at 28 days by re-testing at a later date.

Tensile strength

2-1-1/clause 3.1.2(7)P

2-1-1/clause 3.1.2(7)P defines concrete tensile strength as the highest stress reached under concentric tensile loading. Values for the mean axial tensile strength, f_{ctm}, and lower characteristic strength, $f_{ctk,0.05}$, are given in 2-1-1/Table 3.1 (reproduced below as Table 3.1-1). Tensile strengths are used in several places in EC2-2 where the effect of tension stiffening is considered to be important. These include:

- 2-2/clause 5.10.8(103) – calculation of prestress strain increases in external post-tensioned members (see section 5.10.8 of this guide);
- 2-2/clause 6.1(109) – prevention of brittle failure in prestressed members on cracking of the concrete;
- 2-1-1/clause 6.2.2(2) – shear tension resistance;
- 2-1-1/clause 6.2.5(1) – interface shear resistance at construction joints;
- 2-1-1/clause 7.3.2 – rules on minimum reinforcement;
- 2-1-1/clause 7.3.4 – rules on crack width calculation, which are influenced by tension stiffening between cracks;
- 2-1-1/clause 8.4 – rules on bond strength for reinforcement anchorage;
- 2-1-1/clause 8.7 – rules on laps for reinforcement;
- 2-1-1/clause 8.10.2 – transmission zones and bond lengths for pretensioned members.

Tensile strength is much more variable than compressive strength and is influenced a lot by the shape and texture of aggregate and environmental conditions than is the compressive strength. Great care should therefore be taken if the tensile strength is accounted for in design outside the application rules given.

2-1-1/clause 3.1.2(9)

2-1-1/clause 3.1.2(9) provides an expression, for estimating the mean tensile, $f_{ctm}(t)$, strength at time t:

$$f_{ctm}(t) = (\beta_{cc}(t))^{\alpha} f_{ctm}$$ 2-1-1/(3.4)

Table 3.1-1. Stress and deformation characteristics for concrete (2-1-1/Table 3.1)

	Strength classes for concrete														Formulae/notes
f_{ck} (MPa)	12	16	20	25	30	35	40	45	50	55	60	70	80	90	
$f_{ck,cube}$ (MPa)	15	20	25	30	37	45	50	55	60	67	75	85	95	105	
f_{cm} (MPa)	20	24	28	33	38	43	48	53	58	63	68	78	88	98	$f_{cm} = f_{ck} + 8$ (MPa)
f_{ctm} (MPa)	1.6	1.9	2.2	2.6	2.9	3.2	3.5	3.8	4.1	4.2	4.4	4.6	4.8	5.0	$f_{ctm} = 0.30 f_{ck}^{(2/3)} \leq C50/60$; $f_{ctm} = 2.12 \ln(1 + (f_{cm}/10)) > C50/60$
$f_{ctk,0.05}$ (MPa)	1.1	1.3	1.5	1.8	2.0	2.2	2.5	2.7	2.9	3.0	3.1	3.2	3.4	3.5	$f_{ctk,0.05} = 0.7 f_{ctm}$ (5% fractile)
$f_{ctk,0.95}$ (MPa)	2.0	2.5	2.9	3.3	3.8	4.2	4.6	4.9	5.3	5.5	5.7	6.0	6.3	6.6	$f_{ctk,0.95} = 1.3 f_{ctm}$ (95% fractile)
E_{cm} (GPa)	27	29	30	31	33	34	35	36	37	38	39	41	42	44	$E_{cm} = 22(f_{cm}/10)^{0.3}$ (f_{cm} in MPa)
ε_{c1} (‰)	1.8	1.9	2.0	2.1	2.2	2.25	2.3	2.4	2.45	2.5	2.6	2.7	2.8	2.8	ε_{c1} (‰) = $0.7 f_{cm}^{0.31} < 2.8$
ε_{cu1} (‰)	3.5 ←								→	3.2	3.0	2.8	2.8	2.8	ε_{cu1} (‰) = 3.5 for $f_{ck} < 50$ MPa ; ε_{cu1} (‰) = $2.8 + 27((98 - f_{cm})/100)^4$ for $f_{ck} \geq 50$ MPa
ε_{c2} (‰)	2.0 ←								→	2.2	2.3	2.4	2.5	2.6	ε_{c2} (‰) = 2.0 for $f_{ck} < 50$ MPa ; ε_{c2} (‰) = $2.0 + 0.085((f_{ck} - 50))^{0.53}$ for $f_{ck} \geq 50$ MPa
ε_{cu2} (‰)	3.5 ←								→	3.1	2.9	2.7	2.6	2.6	ε_{cu2} (‰) = 3.5 for $f_{ck} < 50$ MPa ; ε_{cu2} (‰) = $2.6 + 35((90 - f_{ck})/100)^4$ for $f_{ck} \geq 50$ MPa
n	2.0 ←								→	1.75	1.6	1.45	1.4	1.4	$n = 2.0$ for $f_{ck} < 50$ MPa ; $n = 1.4 + 23.4((90 - f_{ck})/100)^4$ for $f_{ck} \geq 50$ MPa
ε_{c3} (‰)	1.75 ←								→	1.8	1.9	2.0	2.2	2.3	ε_{c3} (‰) = 1.75 for $f_{ck} < 50$ MPa ; ε_{c3} (‰) = $1.75 + 0.55((f_{ck} - 50)/40)$ for $f_{ck} \geq 50$ MPa
ε_{cu3} (‰)	3.5 ←								→	3.1	2.9	2.7	2.6	2.6	ε_{cu3} (‰) = 3.5 for $f_{ck} < 50$ MPa ; ε_{cu3} (‰) = $2.6 + 35((90 - f_{ck})/100)^4$ for $f_{ck} \geq 50$ MPa

where:

$\beta_{cc}(t)$ is as defined in 2-1-1/Expression (3.2), reproduced above

$f_{ctm}(t)$ is the mean tensile strength at an age of t days (it should be noted that the tensile strength corresponding to a given early age compressive strength is less than that corresponding to the same 28-day compressive strength in 2-1-1/Table 3.1)

f_{ctm} is the mean tensile strength at 28 days given in 2-1-1/Table 3.1

t is the age of concrete in days

α = 1.0 for $t < 28$
  = 2/3 for $t \geq 28$

2-1-1/Expression (3.4) is only an approximation because of the large number of factors influencing the rate of strength gain. If a more accurate prediction is required, this should be obtained from tests which account for the actual exposure conditions and member dimensions, as identified in the Note to 2-1-1/clause 3.1.2(9).

3.1.3. Elastic deformation

The mean value of the modulus of elasticity, E_{cm}, can be obtained from 2-1-1/Table 3.1 which is based on the following relationship:

$$E_{cm} = 22\left(\frac{f_{ck} + 8}{10}\right)^{0.3} \quad \text{(D3.1-3)}$$

with f_{ck} in MPa. Limestone and sandstone aggregates typically lead to greater flexibility and the values derived from equation (D3.1-3) should be reduced by 10% and 30% respectively. For basalt aggregates the values should be increased by 20%.

Values of E_{cm} derived from equation (D3.1-3) are based on a secant stiffness for short-term loading up to a stress of $0.4f_{cm}$, as shown in Fig. 3.1-2. Consequently, for lower stresses, the response may be slightly stiffer and for higher stresses (which is unlikely under normal loading conditions) the response could be quite a lot more flexible. Given the inherent difficulty in predicting elastic moduli for concrete and the effects of creep on stiffness for sustained loading, the values obtained from equation (D3.1-3) will be satisfactory for elastic analysis in most normal bridge design applications. Where the differential stiffness between parts of the structure with different concretes or materials is unusually critical to the design, either testing could be carried out to determine a more accurate stiffnesses or a sensitivity analysis could be carried out. **2-1-1/clause 3.1.3(1)** relates to these considerations.

2-1-1/clause 3.1.3(1)

An approximation for estimating the variation of the modulus of elasticity with time (which should only be required for loading at very early age) is given in **2-1-1/clause 3.1.3(3)**:

2-1-1/clause 3.1.3(3)

$$E_{cm}(t) = \left(\frac{f_{cm}(t)}{f_{cm}}\right)^{0.3} E_{cm} \quad \text{2-1-1/(3.5)}$$

where $E_{cm}(t)$ and $f_{cm}(t)$ are the values at an age of t days and E_{cm} and f_{cm} are the values determined at an age of 28 days.

Other relevant properties of concrete defined in 2-1-1/clause 3.1.3 are:

- Poisson's ratio for uncracked concrete = 0.2
- Poisson's ratio for concrete cracked in tension = 0
- Coefficient of thermal expansion = $10 \times 10^{-6}/°C$.

3.1.4. Creep and shrinkage

3.1.4.1. Creep

Creep of concrete causes deformations under sustained forces to continue to grow beyond the initial elastic response or alternatively causes forces to reduce from the initial elastic values when a section is held at constant strain. Creep is particularly important in prestressed concrete as the continued long-term shortening of the concrete in compression leads to a reduction in prestressing force. Creep is also important for bridges built-up in stages as the long-term creep deformations cause changes in the internal actions derived solely from modelling the construction sequence. This is discussed in greater detail in Annex K of this guide. The creep parameters in this section only apply to normal density concrete. Section 11 gives supplementary requirements for lightweight concretes.

2-1-1/clause 3.1.4(1)P identifies that creep of the concrete depends on the ambient humidity, the dimensions of the element and the composition of the concrete. Creep is also influenced by the age of the concrete when the load is first applied and depends on the duration and magnitude of the loading.

2-1-1/clause 3.1.4(1)P

Creep deformation is normally related to the elastic deformation by way of a creep factor as given in 2-1-1/Expression (3.6) such that the total final creep deformation $\varepsilon_{cc}(\infty, t_0)$ at time $t = \infty$ for a constant compressive stress σ_c is:

$$\varepsilon_{cc}(\infty, t_0) = \phi(\infty, t_0)\frac{\sigma_c}{E_c} \qquad \text{2-1-1/(3.6)}$$

where E_c is the tangent modulus which, from *2-1-1/clause 3.1.4(2)*, may be taken equal to $1.05E_{cm}$ with E_{cm} according to 2-1-1/Table 3.1. The final creep coefficient $\phi(\infty, t_0)$ may be derived from 2-1-1/Fig. 3.1, provided that the concrete is not subjected to a compressive stress greater than $0.45f_{ck}(t_0)$ at an age t_0 at first loading, the ambient temperature is between $-40°C$ and $+40°C$ and the mean relative humidity is greater than 40%. The following definitions are used in EC2 for both creep and shrinkage calculations:

2-1-1/clause 3.1.4(2)

- t_0 is the age of the concrete at first loading in days
- h_0 is the notional size (or effective thickness) $= 2A_c/u$, where A_c is the concrete cross-sectional area and u is the perimeter(s) of the section which is exposed to drying, i.e. open to the atmosphere. Where h_0 varies along a member, an average value could be used as a simplification based on the sections which are most highly stressed
- S is for slow-hardening cements as identified in 2-1-1/clause 3.1.2(6)
- N is for normal and rapid hardening cements as identified in 2-1-1/clause 3.1.2(6)
- R is for rapid hardening high-strength cements as identified in 2-1-1/clause 3.1.2(6)

Where the humidity lies between 40% and 100%, the creep ratio should be determined by interpolation or extrapolation as relevant from 2-1-1/Figs 3.1a and 3.1b for 50% and 80% humidity respectively, or by direct calculation from 2-1-1/Annex B. For cases outside the humidity and temperature limits given, 2-1-1/Annex B may also be used. In the UK, it has been normal practice to use a relative humidity of 70% in design.

2-1-1/Annex B also gives information on how to determine the development of creep strain with time, which is needed in the analysis of bridges built by staged construction. The development of creep with time is determined by the parameter $\beta_c(t, t_0)$ which varies from 0.0 at first loading to 1.0 after infinite time such that $\phi_0 = \phi(\infty, t_0)$ and:

$$\phi(t, t_0) = \phi_0 \cdot \beta_c(t, t_0) = \phi(\infty, t_0)\beta_c(t, t_0) \qquad \text{2-1-1/(B.1)}$$

$$\beta_c(t, t_0) = \left[\frac{(t - t_0)}{(\beta_H + t - t_0)}\right]^{0.3} \qquad \text{2-1-1/(B.7)}$$

β_H is a coefficient depending on the relative humidity (RH in %), the notional member size (h_0 in mm) and the compressive strength as follows:

$$\beta_H = 1.5[1 + (0.012RH)^{18}]h_0 + 250 \leq 1500 \quad \text{for } f_{cm} \leq 35\,\text{MPa} \qquad \text{2-1-1/(B.8a)}$$

$$\beta_H = 1.5[1 + (0.012RH)^{18}]h_0 + 250\left[\frac{35}{f_{cm}}\right]^{0.5} \leq 1500\left[\frac{35}{f_{cm}}\right]^{0.5} \quad \text{for } f_{cm} \geq 35\,\text{MPa}$$

$$\text{2-1-1/(B.8b)}$$

2-1-1/Expression (B.7) can also be used in conjunction with the simple graphical method of 2-1-1/Fig. 3.1 to determine the development of creep with time, rather than calculating the creep factor directly from 2-1-1/Annex B.

The use of both 2-1-1/Fig. 3.1 and 2-1-1/Annex B to calculate $\phi(\infty, t_0)$ is illustrated in Worked example 3.1-1 where the relevant formulae in Annex B are reproduced. The creep factors produced in this way are average values. If the structure is particularly sensitive to creep then it would be prudent to allow for some variation in creep factor. The circumstances when this might be necessary are presented in 2-2/Annex B.105 and discussed in Annex B of this guide.

When the compressive stress of concrete at an age t_0 exceeds $0.45 f_{ck}(t_0)$, non-linear creep can give rise to greater creep deformations. Non-linear creep can often occur in pretensioned precast beams which are stressed at an early age and initially have only small dead load. A revised creep factor for use in 2-1-1/Expression (3.6) is given in **2-1-1/clause 3.1.4(4)** for this situation as follows:

2-1-1/clause 3.1.4(4)

$$\phi_k(\infty, t_0) = \phi(\infty, t_0) \exp(1.5(k_\sigma - 0.45)) \qquad \text{2-1-1/(3.7)}$$

where k_σ is the stress–strength ratio $\sigma_c/f_{cm}(t_0)$, σ_c is the compressive stress and $f_{cm}(t_0)$ is the mean concrete compressive strength at the time of loading. There is an anomaly in this equation as the criterion for its consideration is based on $f_{ck}(t_0)$, whereas the formula contains the mean strength $f_{cm}(t_0)$. This means that for a concrete stress of $0.45 f_{ck}(t_0)$, the formula actually reduces the creep factor, which is certainly not intended. A conservative approach is to redefine k_σ as $\sigma_c/f_{ck}(t_0)$. Arguably, as deformations are based on mean properties, it is the criterion for the start of non-linear creep which should be changed to $0.45 f_{cm}(t_0)$, rather than changing the formula, but this approach is not advocated here.

For high-strength concrete with grade greater than or equal to C55/67, 2-2/Annex B gives alternative rules for creep calculation which were considered by its drafters to be more accurate than those in EC2-1-1. This suggestion of 'greater accuracy' has not, however, been universally accepted. Concretes with and without silica fume are treated separately with significantly reduced creep strains possible in silica fume concretes. This is discussed further in Annex B of this guide.

Worked example 3.1-1: Calculation of $\phi(\infty, t_0)$ for bridge pier

A hollow rectangular pier with wall thickness 500 mm is first loaded by significant load from deck construction at an age of 30 days. The relative humidity is 80%, the concrete is C40/50 and the cement is Ordinary Portland. Calculate the creep factor $\phi(\infty, t_0)$ according to 2-1-1/Fig. 3.1 and 2-1-1/Annex B.

From 2-1-1/Fig. 3.1

From the construction shown in Fig. 3.1-1 using $h_0 = 500$ mm, $\phi(\infty, t_0)$ is found to be equal to 1.4.

(1) Construct horizontal line at $t = 30$ days to intersect 'N' curve.
(2) Construct line from the intersection point to the x-axis at $\phi = 0$.
(3) Construct vertical line at $h_0 = 500$ mm to intersect C40/50 curve.
(4) Construct horizontal line from the C40/50 curve to intersect line (2).
(5) Read off ϕ factor from horizontal axis at intersection point of lines (2) and (4).

From 2-1-1/Annex B

The creep coefficient $\phi(t, t_0)$ is calculated from 2-1-1/Expression (B.1):

$$\phi(t, t_0) = \phi_0 \cdot \beta_c(t, t_0) \qquad \text{2-1-1/(B.1)}$$

$\beta_c(t, t_0)$ is a factor that describes the amount of creep that occurs with time and at time $t = \infty$ it equals 1.0. The total creep is therefore given by ϕ_0 from 2-1-1/Expression (B.2):

$$\phi_0 = \phi_{RH} \cdot \beta(f_{cm}) \cdot \beta(t_0) \qquad \text{2-1-1/(B.2)}$$

ϕ_{RH} is a factor to allow for the effect of relative humidity on the notional creep coefficient. Two expressions are given depending on the size of f_{cm}.

From Table 3.1, $f_{cm} = f_{ck} + 8 = 40 + 8 = 48$ MPa.

For $f_{cm} > 35$ MPa, ϕ_{RH} is given by 2-1-1/Expression (B.3b):

$$\phi_{RH} = \left[1 + \frac{1 - RH/100}{0.1 \cdot \sqrt[3]{h_0}} \cdot \alpha_1\right] \cdot \alpha_2 \qquad \text{2-1-1/(B.3b)}$$

RH is the relative humidity of the ambient environment in % = 80% here.

$\alpha_{1/2}$ are coefficients to consider the influence of the concrete strength from 2-1-1/Expression (B.8c):

$$\alpha_1 = \left[\frac{35}{f_{cm}}\right]^{0.7} = \left[\frac{35}{48}\right]^{0.7} = 0.80$$

$$\alpha_2 = \left[\frac{35}{f_{cm}}\right]^{0.2} = \left[\frac{35}{48}\right]^{0.2} = 0.94 \qquad \text{2-1-1/(B.8c)}$$

From 2-1-1/Expression (B.3b):

$$\phi_{RH} = \left[1 + \frac{1 - 80/100}{0.1 \times \sqrt[3]{500}} \times 0.80\right] \times 0.94 = 1.13$$

$\beta(f_{cm})$ is a factor to allow for the effect of concrete strength on the notional creep coefficient from 2-1-1/Expression (B.4):

$$\beta(f_{cm}) = \frac{16.8}{\sqrt{f_{cm}}} = \frac{16.8}{\sqrt{48}} = 2.42$$

The effect of concrete age at loading on the notional creep coefficient is given by the factor $\beta(t_0)$ according to 2-1-1/Expression (B.5). For loading at 30 days this gives:

$$\beta(t_0) = \frac{1}{(0.1 + t_0^{0.20})} = \frac{1}{(0.1 + 30^{0.20})} = 0.48$$

This expression is only valid as written for normal or rapid-hardening cements, N. (For slow-hardening or high-strength cements, 2-1-1/Expression (B.5) needs to be modified by amending the age at loading (t_0) according to 2-1-1/Expressions (B.9) and (B.10).)

The final creep coefficient from 2-1-1/Expression (B.2) is then:

$$\phi(\infty, t_0) = \phi_0 = 1.13 \times 2.42 \times 0.48 = \mathbf{1.31}$$

compared to 1.4 from 2-1-1/Fig. 3.1b above.

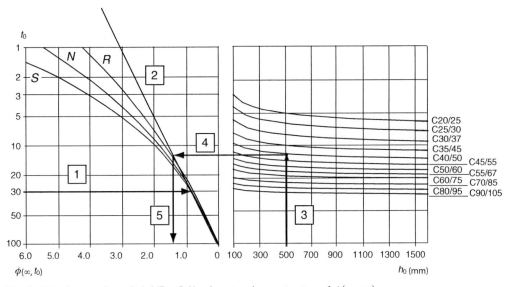

Fig. 3.1-1. Extract from 2-1-1/Fig. 3.1b, showing determination of $\phi(\infty, t_0)$

2-1-1/clause 3.1.4(6)

3.1.4.2. Shrinkage

2-1-1/clause 3.1.4(6) splits shrinkage into two components. Autogenous shrinkage occurs during hydration and hardening of the concrete without loss of moisture and the total strain depends only on concrete strength. The majority of this component therefore occurs relatively quickly and is substantially complete in a few months. Drying shrinkage is associated with movement of water through and out of the concrete section and therefore depends on relative humidity and effective section thickness as well as concrete composition. Drying shrinkage occurs more slowly taking several years to be substantially complete. The total shrinkage is the sum of these two components.

Shrinkage is particularly important in prestressed concrete as the continued long-term shortening of the concrete leads to a reduction in prestressing force, as discussed in section 5.10.6 of this guide. In calculating shrinkage losses, it is important to consider at what age of the concrete the prestressing force is to be applied. In composite sections, differential shrinkage between parts of the cross-section cast at different ages can also lead to locked in stresses in the cross-section and secondary moments and forces from restraint to the free deflections. This is discussed in Annex K4.2 of this guide. The shrinkage parameters in this section only apply to normal density concrete. Section 11 gives supplementary requirements for lightweight concretes.

Drying shrinkage

The drying shrinkage is given as:

$$\varepsilon_{cd}(t) = \beta_{ds}(t, t_s) \cdot k_h \cdot \varepsilon_{cd,0} \qquad \text{2-1-1/(3.9)}$$

$\varepsilon_{cd,0}$ is the nominal drying shrinkage taken from 2-1-1/Table 3.2 (reproduced below as Table 3.1-2) or can be calculated in accordance with 2-1-1/Annex B.

k_h is a coefficient depending on the notional size h_0 according to 2-1-1/Table 3.3 (reproduced below as Table 3.1-3).

$$\beta_{ds}(t, t_s) = \frac{(t - t_s)}{(t - t_s) + 0.04\sqrt{h_0^3}}$$

is a factor to calculate the rate of shrinkage with time (where t_s is the age of the concrete at the end of curing) which equals 1.0 for the final shrinkage value. All times are in days.

Autogenous shrinkage

The autogenous shrinkage strain is given as:

$$\varepsilon_{ca}(t) = \beta_{as}(t)\varepsilon_{ca}(\infty) \qquad \text{2-1-1/(3.11)}$$

where:

$$\varepsilon_{ca}(\infty) = 2.5(f_{ck} - 10) \times 10^{-6} \qquad \text{2-1-1/(3.12)}$$

$$\beta_{as}(t) = 1 - \exp(-0.2t^{0.5}) \qquad \text{2-1-1/(3.13)}$$

Table 3.1-2. Nominal unrestrained drying shrinkage values $\varepsilon_{cd,0}$ ($\times 10^6$)

$f_{ck}/f_{ck,cube}$ (MPa)	Relative humidity (%)					
	20	40	60	80	90	100
20/25	0.62	0.58	0.49	0.30	0.17	0
40/50	0.48	0.46	0.38	0.24	0.13	0
60/75	0.38	0.36	0.30	0.19	0.10	0
80/95	0.30	0.28	0.24	0.15	0.08	0
90/105	0.27	0.25	0.21	0.13	0.07	0

CHAPTER 3. MATERIALS

Table 3.1-3. Values of k_h in 2-1-1/(3.9)

h_0	k_h
100	1.0
200	0.85
300	0.75
≥500	0.70

Alternative rules for high-strength concrete (class C55/67 and above) are given in 2-2/Annex B.

3.1.5. Concrete stress–strain relation for non-linear structural analysis

When non-linear analysis is to be used according to 2-2/clause 5.7, the relation between short-term concrete stress and strain in 2-1-1/Fig. 3.2 may be used, which is reproduced here as Fig. 3.1-2. The stress–strain relationship is given as follows:

$$\frac{\sigma_c}{f_{cm}} = \frac{k\eta - \eta^2}{1 + (k-2)\eta} \qquad \text{2-1-1/(3.14)}$$

where:

$\eta = \varepsilon_c/\varepsilon_{c1}$
ε_{c1} is the strain at peak stress according to 2-1-1/Table 3.1
$k = 1.05 E_{cm} \times |\varepsilon_{c1}|/f_{cm}$.

2-2/clause 5.7 amends the above curve when carrying out non-linear analysis for bridges at the ultimate limit state, so that it is compatible with the reliability format proposed there. This effectively modifies the definition of f_{cm} in 2-1-1/Expression (3.14) from that given in 2-1-1/Table 3.1, as discussed in section 5.7 of this guide.

2-1-1/clause 3.1.5(2) states that other idealized stress–strain relations may be applied, if they 'adequately represent the behaviour of the concrete considered'. This could be interpreted as being adequate to produce a safe verification of the element being checked. This leaves scope to use design values of stress–strain response in analysis in circumstances where a reduced stiffness would be adverse. This method is discussed further in section 5.8.6 of this guide in the context of the design of slender piers and has the great advantage that a verification of the strength of cross-sections is then made directly in the analysis. 2-2/clause 5.7 also permits design values of material properties to be used in non-linear analysis, but adds a caveat that caution is required when there are imposed deformations to consider since more flexible properties will, in this case, underestimate the forces attracted. In all cases, creep may modify the stress–strain curve further. This is also discussed in section 5.8.6.

2-1-1/clause 3.1.5(2)

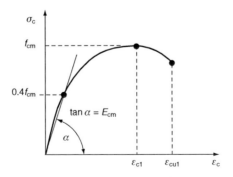

Fig. 3.1-2. Stress–strain relation for non-linear structural analysis

3.1.6. Design compressive and tensile strengths

Design strengths are obtained by combining partial safety factors for materials with their characteristic values. The design compressive strength for concrete is defined by **2-2/clause 3.1.6(101)P** as follows:

$$f_{cd} = \frac{\alpha_{cc} f_{ck}}{\gamma_c} \qquad \qquad 2\text{-}1\text{-}1/(3.15)$$

where:

γ_c is the partial safety factor for concrete

α_{cc} is a coefficient taking account of long-term effects on the compressive strength and of unfavourable effects resulting from the way the load is applied.

For persistent and transient design situations, the recommended value of γ_c for concrete is given as 1.5 – see section 2.4.2 of this guide.

The value of α_{cc} in 2-1-1/Expression (3.15) is recommended to be 0.85 for bridges, but this is predominantly intended to be applied to calculations on bending and axial load in 2-2/clause 6.1. α_{cc} may also in part be a correcting factor between the true stress–strain behaviour, where a peak stress of f_{ck} is reached but then this stress reduces up to the failure strain (similar in form to Fig. 3.1-2), and the idealized parabola-rectangle diagram (Fig. 3.1-3) which maintains the peak stress up to the failure strain. The factor α_{cc} therefore contributes to preventing flexural resistances from being overestimated by the neglect of the drop off in stress towards the failure strain. It was intended that a value of 1.0 should be used for shear calculations (as the formulae were produced from test results on this basis) and also in the membrane rules of 2-2/clause 6.9, which contain a 0.85 factor explicitly in the formulae for direct compressive strength. In other situations there is less clarity and this guide makes comment on appropriate values of α_{cc} to use in calculations throughout its sections. The requirements of National Annexes must be followed.

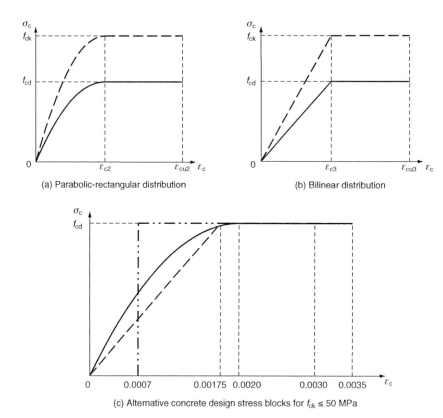

Fig. 3.1-3. Stress–strain relations for the design of concrete sections

CHAPTER 3. MATERIALS

The design tensile strength for concrete is defined in a similar manner in **2-2/clause 3.1.6(102)P**:

$$f_{ctd} = \frac{\alpha_{ct} f_{ctk,0.05}}{\gamma_c} \qquad \text{2-2/(3.16)}$$

2-2/clause 3.1.6(102)P

where:
- α_{ct} is a coefficient taking account of long-term effects on the tensile strength and of unfavourable effects resulting from the way the load is applied
- $f_{ctk,0.05}$ is the characteristic axial tensile strength below which 5% of all strength test results would be expected to fall for the specified concrete (see 2-1-1/Table 3.1)

The value for the α_{ct} factor is recommended to be taken as 1.0 for bridges. This value is appropriate for shear and is necessary for use in bond calculations to avoid longer bond and lap lengths being generated than in previous UK practice to BS 5400. It is, however, more appropriate to take it equal to 0.85 when using the tensile strength in the derivation of the resistance of compression struts to splitting, as discussed in section 6.7 of this guide.

3.1.7. Stress–strain relations for the design of sections

EC2 makes a distinction between the requirements for stress–strain relationships for use in global analysis (for non-linear analysis) and for use in the verification of cross-sections. The former is discussed in section 3.1.5 above which also references other parts of this guide. For cross-section design, **2-1-1/clause 3.1.7(1), (2) and (3)** provide three alternative stress–strain diagrams: parabolic rectangular, bilinear and simplified rectangular, as illustrated in Fig. 3.1-3. They are for ultimate limit state (ULS) design only, not for serviceability limit state (SLS).

2-1-1/clause 3.1.7(1), (2) and (3)

The stress–strain diagrams in Fig. 3.1-3 have been constructed using the following expressions given in EC2:

Parabolic-rectangular diagram

$$\sigma_c = f_{cd}\left[1 - \left(1 - \frac{\varepsilon_c}{\varepsilon_{c2}}\right)^n\right] \quad \text{for } 0 \leq \varepsilon_c \leq \varepsilon_{c2} \qquad \text{2-1-1/(3.17)}$$

$$\sigma_c = f_{cd} \quad \text{for } \varepsilon_{c2} \leq \varepsilon_c \leq \varepsilon_{cu2} \qquad \text{2-1-1/(3.18)}$$

where:
- n is the exponent defined in 2-1-1/Table 3.1
- ε_{c2} is the strain at reaching the maximum strength (defined in 2-1-1/Table 3.1)
- ε_{cu2} is the ultimate strain (defined in 2-1-1/Table 3.1)

Bilinear diagram
- ε_{c3} is the strain at reaching the maximum strength (defined in 2-1-1/Table 3.1), i.e. at f_{cd} when design values are used
- ε_{cu3} is the ultimate strain (defined in 2-1-1/Table 3.1)

Simplified rectangular diagram

λ and η factors are used to define the effective height of the compression zone and effective strength respectively, where:

$$\lambda = 0.8 \qquad \text{for } f_{ck} \leq 50\,\text{MPa} \qquad \text{2-1-1/(3.19)}$$

$$\lambda = 0.8 - \frac{(f_{ck} - 50)}{400} \qquad \text{for } 50 < f_{ck} \leq 90\,\text{MPa} \qquad \text{2-1-1/(3.20)}$$

$$\eta = 1.0 \qquad \text{for } f_{ck} \leq 50\,\text{MPa} \qquad \text{2-1-1/(3.21)}$$

$$\eta = 1.0 - \frac{(f_{ck} - 50)}{200} \qquad \text{for } 50 < f_{ck} \leq 90\,\text{MPa} \qquad \text{2-1-1/(3.22)}$$

Table 3.1-4 compares these three idealizations in terms of average stress over a rectangular compression zone (from extreme compression fibre to neutral axis) and the distance from the

Table 3.1-4. Comparison of stress block idealizations for $\alpha_{cc} = 0.85$

	Parabolic rectangular		Bilinear		Simplified rectangular	
Class	Average stress (MPa)	Centroid (as ratio of depth to n.a. depth)	Average stress (MPa)	Centroid (as ratio of depth to n.a. depth)	Average stress (MPa)	Centroid (as ratio of depth to n.a. depth)
C20	9.175	0.416	8.500	0.389	9.067	0.40
C25	11.468	0.416	10.625	0.389	11.333	0.40
C30	13.762	0.416	12.750	0.389	13.600	0.40
C35	16.056	0.416	14.875	0.389	15.867	0.40
C40	18.349	0.416	17.000	0.389	18.133	0.40
C45	20.643	0.416	19.125	0.389	20.400	0.40
C50	22.937	0.416	21.250	0.389	22.667	0.40
C55	23.194	0.393	22.098	0.374	23.930	0.39
C60	23.582	0.377	22.872	0.363	25.033	0.39

Note n.a. = neutral axis.

compression face of the section to the centre of compression. It is produced for $\alpha_{cc} = 0.85$. This table can be used for flexural design calculations as illustrated in section 6.1 of this guide. It will be seen in the worked examples presented there that the rectangular block generally gives the greater flexural resistance, which is obvious because the depth of the stress block required to provide a given force is smaller than for the other two alternatives. The bending resistances produced will, however, not vary significantly regardless of method chosen.

The average stresses, f_{av}, and centroid ratio, β (depth to the centroid of the compressive force over depth of compression zone), have been produced from the following expressions:

Parabolic rectangular

$$f_{av} = f_{cd}\left(1 - \frac{1}{n+1}\frac{\varepsilon_{c2}}{\varepsilon_{cu2}}\right) \tag{D3.1-4}$$

$$\beta = 1 - \frac{\dfrac{\varepsilon_{cu2}^2}{2} - \dfrac{\varepsilon_{c2}^2}{(n+1)(n+2)}}{\varepsilon_{cu2}^2 - \dfrac{\varepsilon_{cu2}\varepsilon_{c2}}{n+1}} \tag{D3.1-5}$$

Bilinear

$$f_{av} = f_{cd}\left(1 - 0.5\frac{\varepsilon_{c3}}{\varepsilon_{cu3}}\right) \tag{D3.1-6}$$

$$\beta = 1 - \frac{\dfrac{\varepsilon_{cu3}^2}{2} - \dfrac{\varepsilon_{c3}^2}{6}}{\varepsilon_{cu3}^2 - \dfrac{\varepsilon_{cu3}\varepsilon_{c3}}{2}} \tag{D3.1-7}$$

Simplified rectangular

$$f_{av} = \lambda\eta f_{cd} \tag{D3.1-8}$$

$$\beta = \lambda/2 \tag{D3.1-9}$$

3.1.8. Flexural tensile strength

2-1-1/*clause 3.1.8(1)* relates the mean flexural tensile strength of concrete to the mean axial tensile strength and the depth of the cross-section. The mean flexural tensile strength, $f_{ctm,fl}$, should be taken as the maximum of f_{ctm} taken from 2-1-1/Table 3.1 or $(1.6 - h/1000)f_{ctm}$ where h is the total member depth in millimetres. The increase in tensile strength for shallow beams or slabs (of depth less than 600 mm) arises because of the high stress gradient

2-1-1/clause 3.1.8(1)

reducing stress over the depth of a potential crack, thus elevating the peak stress at the surface needed to cause fracture. The flexural tensile strength is not explicitly used in the EC2 application rules, but it could be relevant in a non-linear analysis to determine tendon strain increases in external post-tensioning, as discussed in section 5.10.8 of this guide. The relationship above also holds for characteristic tensile strength values.

3.1.9. Confined concrete

In cases where a concrete element is under tri-axial stresses, 2-1-1/clause 3.1.9 allows enhancement of the characteristic compressive strength and ultimate strain limits. Such confinement may be provided by link reinforcement or prestressing but no guidance on detailing is given in the code. If benefit of confinement by links is taken, it is recommended that reference be made to test results demonstrating that the link geometry proposed can generate such constraint without premature failure of the concrete occurring. It was not intended that this rule be invoked for general calculations on bending and axial force. It was primarily intended for use with concentrated forces, and loosely forms the basis of the increased resistance in partially loaded areas, discussed in section 6.7, where the confinement is provided by the tensile strength of the surrounding concrete.

3.2. Reinforcing steel
3.2.1. General

2-1-1/clause 3.2.1(1)P allows the use of bars, de-coiled rods, welded fabric or lattice girders (made with ribbed bars) as suitable reinforcement. For bridge applications, reinforcing bars are most common and the rest of this guide deals only with reinforcing bars. EC2 relies on EN 10080 for the specification of reinforcement, including tests for confirming properties, classification details and methods of production. *2-1-1/clause 3.2.1(2)P* states that if site operations are such that the material properties can be altered, the properties need to be re-checked for conformance. If steels are supplied to standards other than EN 10080, then *2-1-1/clause 3.2.1(3)P* requires such steels to be checked to ensure they are in accordance with 2-1-1/clause 3.2.2 to 3.2.6 and 2-1-1/Annex C. As noted in 2-1-1/clause 1.1.2, plain round bars are not covered by EN 1992.

2-1-1/clause 3.2.1(1)P

2-1-1/clause 3.2.1(2)P

2-1-1/clause 3.2.1(3)P

3.2.2. Properties

The most important reinforcement property to the designer is usually the characteristic yield strength, f_{yk}. However, the specification of many other properties is necessary to fully characterize the reinforcement as noted in *2-1-1/clause 3.2.2(1)P*. These include tensile strength, ductility, bendability, bond characteristics, tolerances on section size and fatigue strength among others. 2-1-1/Annex C gives requirements for material properties and groups bars into three ductility classes A, B and C. The rules given in EC2 assume compliance with Annex C, which is clarified in *2-1-1/clause 3.2.2(2)P*. They are applicable only to ribbed and weldable reinforcement (and therefore cannot be used for plain round bars). *2-1-1/clause 3.2.2(3)P* states that the rules are also only valid for specified reinforcement yield strengths between 400 and 600 MPa, although this upper limit is a nationally determined parameter. Other paragraphs of clause 3.2.2 make further reference to 2-1-1/Annex C for property requirements.

2-1-1/clause 3.2.2(1)P

2-1-1/clause 3.2.2(2)P
2-1-1/clause 3.2.2(3)P

The commonest reinforcement grade has $f_{yk} = 500$ MPa. The yield strength and ductility class for weldable ribbed reinforcing steel have to be specified in accordance with EN 10080. For a bar with yield strength of 500 MPa and ductility class B, the bar is specified by 'B500B', where B500 refers to the yield strength of a 'bar' and the subsequent B is the ductility class.

3.2.3. Strength

The characteristic yield stress, f_{yk}, is obtained by dividing the characteristic yield load by the nominal cross-sectional area of the bar. Alternatively, for products without a pronounced

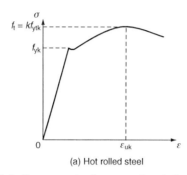 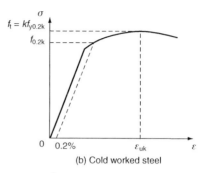

(a) Hot rolled steel (b) Cold worked steel

Fig. 3.2-1. Stress–strain diagrams of typical reinforcement steel

yield stress, the 0.2% proof stress, $f_{0.2k}$, may be used in place of the yield stress. 2-1-1/Fig. 3.7, reproduced here as Fig. 3.2-1, illustrates typical stress–strain curves for reinforcement.

2-1-1/clause C.2(1)P specifies that $f_{y,max}$ must not exceed $1.3 f_{yk}$. Since ductility generally decreases as yield stress increases, in applications where ductility is critical (such as seismic design or plastic methods of verification) it is important to ensure the actual yield strength does not excessively exceed the specified value. This may be the reason behind 2-1-1/clause C.2(1)P, although ductility is controlled by 2-2/clause 3.2.4. It is more likely to relate to aspects of design where consideration of over-strength is important, such as in seismic design where the formation of a plastic hinge is often assumed in elements visible above ground so as to limit the forces transmitted to uninspectable elements below ground. Excessive over-strength in the above-ground elements could therefore lead to greater forces developing in the below-ground parts and therefore greater damage.

3.2.4. Ductility

2-2/clause 3.2.4(101)P

2-2/clause 3.2.4(101)P specifies that reinforcement shall have adequate ductility as defined by the ratio of tensile strength to the yield stress, $(f_t/f_y)_k$, and the strain at maximum force, ε_{uk}. The use of strain at maximum force differs from previous UK practice, where the strain at fracture was used as a measure. The Eurocode's choice is the more rational as it relates to the stable plastic strain which can be developed without loss of load.

Three ductility classes (A, B and C) are defined in 2-1-1/Annex C, with ductility increasing from A to C. The ductility requirements are summarized in Table 3.2-1 and are discussed further in Annex C of this guide. The Note to 2-2/clause 3.2.4(101)P recommends that Class A reinforcement is not used for bridges, although this is subject to variation in the National Annex. This recommendation has been made for bridges because considerable reinforcement strain can be necessary in deep concrete beams in flexure (as are frequently used in bridge design) before the concrete compressive failure strain is reached. No explicit check on reinforcement strain is required in cross-section design if the idealization with horizontal plateau in 2-1-1/clause 3.2.7 is used. It was also considered that greater ductility should be provided in bridges to ensure that the usual assumptions made regarding adequacy of elastic global analysis could be made (see section 5.4 of this guide). Sections with Class A reinforcement can be prone to brittle flexural failures, proving very low rotation capacity. Typically only mesh reinforcement is Class A, so the restriction will have no practical consequence for bridge design.

Table 3.2-1. Ductility classes for reinforcement

Class	Characteristic strain at maximum force, ε_{uk}	Minimum value of $k = (f_t/f_y)_k$
A	$\geq 2.5\%$	≥ 1.05
B	$\geq 5\%$	≥ 1.08
C	$\geq 7.5\%$	$\geq 1.15, < 1.35$

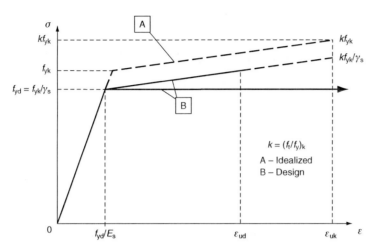

Fig. 3.2-2. Idealized and design stress–strain diagrams for reinforcement (tension and compression)

3.2.5. Welding
EN 1992 permits welding of reinforcement under certain circumstances as defined in 2-1-1/Table 3.4 (not reproduced here). Only bars with approximately the same diameters may be welded. The requirements are more restrictive for bars where the load is not 'predominantly static', i.e. there is fatigue loading, although there is less restriction on bars in compression. Welding processes for reinforcement are defined in EN 10080. The fatigue performance of welded reinforcement is much lower than that of non-welded reinforcement and so particular attention should be given to verifying performance in accordance with 2-2/clause 6.8, although all reinforcement should be checked in this way.

3.2.6. Fatigue
Fatigue strength of reinforcement has to be verified in accordance with EN 10080. 2-1-1/Annex C gives further information for such requirements.

3.2.7. Design assumptions
For cross-section design, *2-1-1/clause 3.2.7(2)* allows the use of two alternative stress–strain relations as indicated in 2-1-1/Fig. 3.8, reproduced here as Fig. 3.2-2. These are either:

- inclined top branch with a strain limit of ε_{ud} and a maximum stress of kf_{yk}/γ_s at ε_{uk} (which cannot be reached in design) or,
- horizontal top branch with stress equal to f_{yd} without strain limit.

2-1-1/clause 3.2.7(2)

The value of ε_{ud} may be found in the National Annex and is recommended to be taken as $0.9\varepsilon_{uk}$. ε_{uk} and k can be obtained from 2-1-1/Annex C. Material partial safety factors are discussed in section 2.4.2. For persistent and transient design situations, the recommended value of γ_s for reinforcing steel is 1.15. The design value of the modulus of elasticity, E_s, may be assumed to be 200 GPa in accordance with *2-1-1/clause 3.2.7(4)*.

2-1-1/clause 3.2.7(4)

3.3. Prestressing steel
3.3.1. General
EN 1992 relies on EN 10138 for the specification of prestressing tendons in concrete structures, including tests for confirming properties and methods of production. EC2 allows use of wires, bars and strands as prestressing tendons. Prestressing steel is usually specified by its strength, relaxation class and steel cross-sectional area.

Table 3.3-1. Relaxation classes for prestressing steel

Class	Prestressing steel type	ρ_{1000}, relaxation loss at 1000 hours at 20°C
1	Ordinary relaxation (wire or strand)	8.0%
2	Low relaxation (wire or strand)	2.5%
3	Hot rolled and processed bars	4%

3.3.2. Properties

2-1-1/clause 3.3.2(1)P refers to EN 101038 for general properties of prestressing steel. EN 10138 Part 1 gives general requirements, while Parts 2, 3 and 4 deal with wire, strand and bar respectively. These give standard designations to identify prestressing steel. For example, EN 10138-3 would refer to a 15.7 mm diameter, 7 wire strand (S7), with 1860 MPa ultimate tensile strength using the designation Y1860S7-15.7. Similar designations are used to describe bars and wires. Additionally, *2-1-1/clause 3.3.2(4)P* specifies three classes for prestressing tendons based on their relaxation behaviour. These classes, together with the relaxation losses to be assumed according to *2-1-1/clause 3.3.2(6)*, are indicated in Table 3.3-1. Despite the use of the words 'ordinary relaxation' to describe Class 1, most prestressing strand is Class 2.

The amount of relaxation of steel stress depends on time, temperature and level of stress. Standard tests for relaxation determine the value 1000 hours after tensioning at a temperature of 20°C. Values of the 1000-hour relaxation can be taken from Table 3.3-1 or alternatively from manufacturers' data or test certificates. The tabulated values from 2-1-1/clause 3.3.2(6) are based on an initial stress of 70% of the actual measured tensile strength of the prestressing steel.

The following three expressions are provided in *2-1-1/clause 3.3.2(7)* for determining relaxation losses:

Class 1: $\quad \dfrac{\Delta\sigma_{\mathrm{pr}}}{\sigma_{\mathrm{pi}}} = 5.39 \rho_{1000}\, \mathrm{e}^{6.7\mu} \left(\dfrac{t}{1000}\right)^{0.75(1-\mu)} \times 10^{-5}$ 2-1-1/(3.28)

Class 2: $\quad \dfrac{\Delta\sigma_{\mathrm{pr}}}{\sigma_{\mathrm{pi}}} = 0.66 \rho_{1000}\, \mathrm{e}^{9.1\mu} \left(\dfrac{t}{1000}\right)^{0.75(1-\mu)} \times 10^{-5}$ 2-1-1/(3.29)

Class 3: $\quad \dfrac{\Delta\sigma_{\mathrm{pr}}}{\sigma_{\mathrm{pi}}} = 1.98 \rho_{1000}\, \mathrm{e}^{8.0\mu} \left(\dfrac{t}{1000}\right)^{0.75(1-\mu)} \times 10^{-5}$ 2-1-1/(3.30)

where:

$\Delta\sigma_{\mathrm{pr}}$ is the absolute value of the relaxation losses of the prestress

σ_{pi} is the absolute value of the initial prestress taken as the prestress value applied to the concrete immediately after tensioning and anchoring (post-tensioning) or after prestressing (pre-tensioning) by subtracting the immediate losses – see sections 5.10.4 and 5.10.5

$\mu = \sigma_{\mathrm{pi}}/f_{\mathrm{pk}}$

f_{pk} is the characteristic value of the tensile strength of the tendon

ρ_{1000} is the value of relaxation loss (as a percentage) from Table 3.3-1

t is the time after tensioning (in hours)

2-1-1/clause 3.3.2(8) allows long-term (final) values of relaxation loss to be estimated using the above equations with a time of 500 000 hours. The authors are not aware of the origin of these equations. The equations produce curious results as a function of time:

- they do not give a loss equal to ρ_{1000} when evaluated at $t = 1000$ hours

CHAPTER 3. MATERIALS

- they are arbitrarily used to obtain long-term values at 500 000 hours even though the equations predict further losses after this time.

Nevertheless, the results so obtained appear to be conservative and can therefore be safely adopted for design.

Relaxation losses are sensitive to variations in stress levels over time and can therefore be reduced by taking account of other time-dependent losses occurring within the structure at the same time (such as creep). A method for determining reduced relaxation losses under such circumstances is given in 2-1-1/Annex D and discussed in Annex D of this guide. It has been previous UK practice to base design on the relaxation loss at 1000 hours without considering the interaction with creep and shrinkage and the example in Annex D suggests that this is generally a reasonable approximation.

Other losses to consider for prestressed structures are discussed fully in section 5.10.

> **Worked example 3.3-1: Relaxation loss of low relaxation prestressing tendon**
>
> A prestressing tendon has the following properties:
>
> - Type = 19 no. 15.7 mm diameter low relaxation strands (Class 2)
> - $A_p = 19 \times 150 = 2850 \, \text{mm}^2$ per tendon
> - Characteristic tensile strength of strand, $f_{pk} = 1860 \, \text{MPa}$.
>
> The long-term relaxation losses are found assuming that the tendon is tensioned to an initial stress of 1339.2 MPa (which includes the loss from friction, wedge slip and elastic deformation of the concrete during the stressing process).
>
> From Table 3.3-1: $\rho_{1000} = 2.5\%$ for low relaxation strands.
>
> For long-term relaxation losses, t should be taken as 500 000 hours (around 57 years) and $\mu = \sigma_{pi}/f_{pk} = 1339.2/1860 = 0.72$. Thus from 2-1-1/Expression (3.29) for low relaxation strands:
>
> $$\frac{\Delta\sigma_{pr}}{\sigma_{pi}} = 0.66\rho_{1000}\, e^{9.1\mu} \left(\frac{t}{1000}\right)^{0.75(1-\mu)} \times 10^{-5}$$
>
> $$= 0.66 \times 2.5 \times e^{(9.1 \times 0.72)} \times \left(\frac{500\,000}{1000}\right)^{0.75(1-0.72)} \times 10^{-5} = 0.043, \text{i.e. } \mathbf{4.3\%}$$
>
> This loss ignores any simultaneous reduction in tendon stress caused by creep and shrinkage of the concrete.

3.3.3. Strength

The high-strength steels used for prestressing do not exhibit a well-defined yield point and are therefore characterized by a proof stress rather than a yield stress. The '$x\%$ proof stress' is the stress for which there is a permanent strain deformation of $x\%$ when the load is removed. *2-1-1/clause 3.3.3(1)P* uses the 0.1% proof stress, $f_{p0.1k}$, defined as the characteristic value of the 0.1% proof load divided by the nominal cross-sectional area. Similarly, the specified value of tensile strength, f_{pk}, is obtained by dividing the characteristic maximum load in axial tension by the nominal cross-sectional area. For strand to EN 10138-3, $f_{p0.1k}$ is typically 86% of f_{pk}. The relationship is more varied for wires and bars. 2-1-1/Fig. 3.9, reproduced as Fig. 3.3-1, illustrates a typical stress–strain curve for prestressing steel where ε_{uk} is again defined as the strain at maximum force.

2-1-1/clause 3.3.3(1)P

3.3.4. Ductility characteristics

Prestressing tendons with ε_{uk} in accordance with EN 10138 and $f_{pk}/f_{p0.1k} \geq k$, where k is a constant which may be found in the National Annex, are deemed to have adequate ductility. The recommended value for k is 1.1.

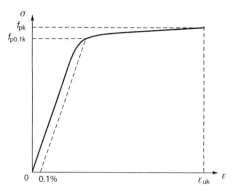

Fig. 3.3-1. Stress–strain diagram for typical prestressing steel

3.3.5. Fatigue
Fatigue stress ranges for prestressing steel must comply with EN 10138. See section 6.8 for full details of fatigue design requirements of prestressing steel to EC2.

3.3.6. Design assumptions

2-1-1/clause 3.3.6(2)
2-1-1/clause 3.3.6(3)

2-1-1/clause 3.3.6(2) and *2-1-1/clause 3.3.6(3)* allow the design value of the modulus of elasticity, E_p, to be assumed to be 205 GPa for wires and bars and 195 GPa for strands. There can typically be ±5% variation in these values and therefore it is usual to check site extensions during stressing on the basis of the test value on certificates accompanying the prestressing steel supplied.

2-1-1/clause 3.3.6(6)

The design proof stress for prestressing steel, f_{pd}, is defined in *2-1-1/clause 3.3.6(6)* as $f_{p0.1k}/\gamma_s$ where γ_s is the partial safety factor. For persistent and transient design situations, the recommended value of γ_s for prestressing steel is given as 1.15 – see section 2.4.2 of this guide.

2-1-1/clause 3.3.6(7)

For cross-section design, *2-1-1/clause 3.3.6(7)* allows the use of two alternative stress–strain relations as indicated in 2-1-1/Fig. 3.10, reproduced as Fig. 3.3-2. These are either:

- inclined top branch with a strain limit of ε_{ud} and a maximum stress of $k \times f_{pk}/\gamma_s$ at ε_{uk} (which cannot be reached). Alternatively, the design may be based on the actual stress–strain relationship if this is known; or,
- horizontal top branch with stress equal to $f_{p0.1k}/\gamma_s$ without strain limit.

The Note to 2-1-1/clause 3.3.6(7) allows the value of ε_{ud} to be given in the National Annex and is recommended to be taken as $0.9\varepsilon_{uk}$. Further recommendations of this Note allow ε_{ud}

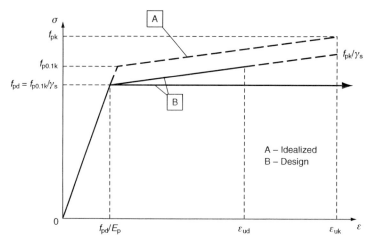

Fig. 3.3-2. Idealized and design stress–strain diagram for prestressing steel

and $f_{p0.1k}/f_{pk}$ to be taken as 0.02 and 0.9 respectively, where more accurate values are not readily available.

3.4. Prestressing devices
3.4.1. Anchorages and couplers
EC2 defines rules for the use of anchorages and couplers in post-tensioned construction. These are specifically related to ensuring that anchorage and coupler assemblies have sufficient strength, elongation and fatigue characteristics to meet the requirements of the design. Detailing of anchorage zones and couplers, as it affects the designer, is discussed in greater detail in section 8.10.

3.4.2. External non-bonded tendons
Detailing of anchorage zones is discussed in greater detail in section 8.10.

CHAPTER 4

Durability and cover to reinforcement

This chapter discusses durability and cover to reinforcement as covered in section 4 of EN 1992-2 in the following clauses:

- General . . . . Clause 4.1
- Environmental conditions . . . . Clause 4.2
- Requirements for durability . . . . Clause 4.3
- Methods of verification . . . . Clause 4.4

4.1. General

Bridges must be sufficiently durable so that they remain serviceable throughout their design life. EN 1990 section 2.4 gives the following general requirement:

'The structure shall be designed such that deterioration over its design working life does not impair the performance of the structure below that intended, having due regard to its environment and the anticipated level of maintenance.'

Durability is influenced by design and detailing, specification of materials used in construction and the quality of construction.

In recent years, durability problems have arisen in a number of concrete bridges in many countries and these problems have led to extensive research into the durability of concrete structures. The subject of durability is therefore treated extensively in Eurocode 2 in an effort to promote consideration of a lowest whole-life cost design philosophy, rather than minimizing the initial cost.

2-1-1/clause 4.1(1)P reiterates EN 1990 section 2.4. It requires that a concrete structure shall be designed, constructed and operated in such a way that, under the expected environmental conditions, the structure maintains its safety, serviceability, strength, stability and acceptable appearance throughout its intended working life, without requiring excessive unforeseen maintenance or repair.

2-1-1/clause 4.1(2)P requires the required protection of the structure to be established by considering its intended use, service life, maintenance programme and actions. For bridges, 'intended use' dictates the 'actions' that will be applied and hence, for example, the likely crack widths to be expected. Service life is relevant because deterioration, due to reinforcement corrosion for example, is a function of time. Maintenance programme is also important because regular routine maintenance provides the opportunity to intervene if deterioration is progressing at a rate greater than expected. *2-1-1/clause 4.1(3)P* reiterates the need to consider direct and indirect actions and also environmental conditions. Environmental

2-1-1/clause 4.1(1)P

2-1-1/clause 4.1(2)P

2-1-1/clause 4.1(3)P

conditions are very important because the onset of reinforcement corrosion, for example, can be vastly accelerated by an aggressive environment. These are discussed in section 4.2 below.

Concrete cover provides corrosion protection to steel reinforcement by both providing a physical barrier to contaminants and through its alkalinity, which inhibits the corrosion reaction. The density of the concrete, its quality and thickness are therefore relevant as noted by **2-1-1/clause 4.1(4)P**. Density and quality can be controlled through mix design in accordance with EN 206-1, together with specification of minimum strength classes. 2-2/clause 4.4 links the minimum concrete cover required to concrete strength, as discussed in the comments to that clause. Cracking is also relevant as it leads to local breaches in the concrete barrier. Cracks are controlled to acceptable sizes using design in accordance with 2-2/clause 7.3.

In exceptionally aggressive environments, consideration could also be given to epoxy-coated reinforcement, use of stainless steel reinforcement, applying surface coatings to inhibit the ingress of chlorides or carbon dioxide, or applying cathodic protection. Additionally, and importantly, concrete outlines should be detailed in such a way as to avoid ponding of water and promote free drainage. In post-tensioned construction, problems can arise due to incomplete grouting of ducts or poor detailing of anchorage zones, such that a path for water to reach the tendon exists. The Concrete Society Technical Report TR47[5] gives advice on these issues. Further discussions on durable detailing of post-tensioning systems is beyond the scope of this guide.

2-1-1/clause 4.1(6) reminds the designer that there may be some applications where the rules given in section 4 are insufficient and need to be supplemented by other requirements. Such additional considerations will rarely be needed for bridges.

4.2. Environmental conditions

Exposure conditions are defined in *2-1-1/clause 4.2(1)P* as the chemical and physical conditions to which the structure is exposed, in addition to the mechanical actions. The main mechanisms leading to the deterioration of concrete bridges that may need to be considered at the design stage are listed below. They all have the potential to lead to corrosion of the reinforcement or prestressing system:

- chloride ingress into the concrete
- carbonation of the concrete
- frost attack
- alkali–silica reactions
- attack from sulphates
- acid attack
- leaching
- abrasion.

The first three of the mechanisms above are covered by *2-1-1/clause 4.2(2)* while the others are covered by 2-1-1/clause 4.2(3) and are discussed below in the comments on that clause. Environmental conditions are classified by 'exposure classes' in 2-1-1/Table 4.1, based on the same classes as in EN 206-1. 2-1-1/Table 4.1 has been reproduced here as Table 4.2-1 for convenience and has been extended (in bold type) to incorporate the extra classification recommendations given in EC2-2 (which are subject to variation in the National Annex). The relevant clause references from EN 1992-2 are also given in the table. It should be noted that a detail may fall into more than one class, but only classes X0, XC, XD and XS lead to requirements for cover thickness.

The most common and serious cause of deterioration of reinforced concrete structures is the corrosion of reinforcement. The ingress of chlorides is particularly detrimental. In normal circumstances, the high alkaline nature of concrete protects any steelwork embedded within it from corrosion. Chlorides can destroy this passivating influence even if the alkalinity of the surrounding concrete remains high. This degradation usually occurs

CHAPTER 4. DURABILITY AND COVER TO REINFORCEMENT

locally, leading to local pitting corrosion of the steel components. Chlorides are contained in a variety of sources, including de-icing chemicals used on roads and sea water in marine environments. The rate at which chlorides penetrate the concrete depends mainly on the density and quality of the concrete.

The alkaline nature of the concrete may also be reduced due to carbonation. This is a reaction between carbon dioxide in the atmosphere and the alkalis in the cement matrix. The process starts at the concrete surface and over time gradually diffuses into the concrete, resulting in a reduction of alkalinity of the concrete. Natural protection of the reinforcement may be lost when the carbonation front reaches the level of the reinforcement. As with the ingress of chlorides, carbonation is less rapid with good quality concrete.

Once the passivity of steel has been eroded, corrosion will continue if there is sufficient moisture and oxygen present at the reinforcement. When concrete is wet, oxygen penetration is inhibited. In very dry conditions, where oxygen levels are sufficient, moisture levels are low. The greatest risk of corrosion is therefore in members subjected to cyclic wetting and drying and this is reflected in the high exposure class designations for these conditions in 2-1-1/Table 4.1, namely XC4 and XD3 for corrosion induced by carbonation and chlorides respectively.

Saturated concrete subjected to cyclic freezing and thawing is prone to the expansive effects of ice, leading to surface spalling. Frost damage of concrete is best avoided by either protecting the concrete from saturation, using air-entrained concrete, or by using higher-strength concretes.

In addition to the conditions detailed in Table 4.1-1, *2-1-1/clause 4.2(3)* requires the designer to give additional consideration to other forms of attack.

2-1-1/clause 4.2(3)

Alkali–silica reaction is a reaction between the alkalis in the cement matrix and certain forms of silica in the aggregate. This reaction leads to the formation of a hygroscopic silica gel that absorbs water, expands and causes cracking. The cracks resulting from the reactions can be several millimetres in width, albeit not extending excessively into the section – typically only 50 to 70 mm. As well as the potential risk for increasing reinforcement corrosion, the tensile and compressive strengths of the concrete are also reduced. Alkali–silica reaction can be avoided by using sources of aggregates that have previously been shown to perform satisfactorily in other structures with similar environmental conditions, by using cement with low alkali content, or by the inhibition of water ingress.

Sulphate attack may occur where, in the presence of water, sulphate ions react with the tricalcium aluminate component of the cement. This reaction again causes expansion, leading to cracking. Foundations are most susceptible to sulphate attack, due to the source of sulphates in the surrounding earth, but can be avoided by adopting sulphate resistant cements, such as Portland cement, with a low tricalcium aluminate content. Alternatively, carefully blended cements incorporating ground granulated blastfurnace slag or fly ash have been shown to improve concrete resistance to sulphate attack.

Acid attacks the calcium compounds in concrete, converting them to soluble salts which can be washed away. Thus the effect of acid on concrete is to weaken the surface and increase permeability. While large quantities of acid can seriously damage concrete, the relatively small amounts from acid rain, for example, will have little effect on a bridge over a typical design life.

A similar effect to acid attack is leaching by soft water. Calcium compounds are mildly soluble in soft water and over time can therefore be leached out if the concrete is constantly exposed to running soft water. This process is slow but can sometimes be observed at bridge abutments where the deck joint above has failed.

2-1-1/clause 4.2(3) also requires abrasion of the concrete to be considered. Abrasion of concrete in bridges may occur due to direct trafficking (uncommon in the UK) or from the effects of sand or gravel suspended in running water. Resistance to abrasion is best obtained by specifying higher-strength concrete and abrasion resistant aggregates or by adding a sacrificial thickness to the cover (as required by 2-2/clauses 4.4.1.2(114) and (115)).

2-2/clause 4.2(104) requires the possibility of water penetration into voided structures to be considered. It is often desirable to provide drainage points to the voids at low points in case there is water ingress, particularly if the deck drainage system passes inside the void.

2-2/clause 4.2(104)

Table 4.2-1. Exposure classes from 2-1-1/Table 4.1, incorporating EN 1992-2 recommendations

Class designation	Description of the environment	Informative examples where exposure classes may occur
1. No risk of corrosion or attack		
X0	For concrete without reinforcement or embedded metal: all exposures except where there is freeze/thaw, abrasion or chemical attack For concrete with reinforcement or embedded metal: very dry	Concrete inside buildings with very low air humidity
2. Corrosion induced by carbonation		
XC1	Dry or permanently wet	Concrete inside buildings with low air humidity Concrete permanently submerged in water
XC2	Wet, rarely dry	Concrete surfaces subject to long-term water contact Many foundations
XC3	Moderate humidity	Concrete inside buildings with moderate or high air humidity External concrete sheltered from rain **including the inside of voided bridge structures remote from water drainage or leakage from the carriageway** Surfaces protected by waterproofing (approved in accordance with national requirements) **including bridge decks. EN 1992-2 clause 4.2(105) refers**
XC4	Cyclic wet and dry	Concrete surfaces subject to water contact, not covered by exposure class XC2
3. Corrosion induced by chlorides		
XD1	Moderate humidity	Concrete surfaces exposed to airborne chlorides
XD2	Wet, rarely dry	Swimming pools Concrete components exposed to industrial waters containing chlorides
XD3	Cyclic wet and dry	Parts of bridges exposed to spray containing chlorides **including surfaces within 6 m (horizontally or vertically) of a carriageway where de-icing salts are used (e.g. parapets, walls and piers) and surfaces (such as the top of piers at expansion joints) likely to be exposed to water draining/leaking from the carriageway. EN 1992-2 clause 4.2(106) refers** Pavements Car park slabs
4. Corrosion induced by chlorides from sea water		
XS1	Exposed to airborne salt but not in direct contact with sea water	Structures near to or on the coast
XS2	Permanently submerged	Parts of marine structures
XS3	Tidal, splash and spray zones	Parts of marine structures
5. Freeze/thaw attack		
XF1	Moderate water saturation, without de-icing agent	Vertical concrete surfaces exposed to rain and freezing

Table 4.2-1. Continued

Class designation	Description of the environment	Informative examples where exposure classes may occur
XF2	Moderate water saturation, with de-icing agent	Vertical concrete surfaces of road structures exposed to freezing and airborne de-icing agents **including surfaces greater than 6 m from a carriageway where de-icing salts are used. EN 1992-2 clause 4.2(106) refers**
XF3	High water saturation, without de-icing agents	Horizontal concrete surfaces exposed to rain and freezing
XF4	High water saturation, with de-icing agents or sea water	Road and bridge decks directly exposed to de-icing agents Concrete surfaces exposed to direct spray containing de-icing agents and freezing **including surfaces within 6 m (horizontally or vertically) of a carriageway where de-icing salts are used (e.g. parapets, walls and piers) and surfaces (such as the top of piers at expansion joints) likely to be exposed to water draining/leaking from the carriageway. EN 1992-2 clause 4.2(106) refers.** Splash zone of marine structures exposed to freezing
6. Chemical attack		
XA1	Slightly aggressive chemical environment (in accordance with EN 206-1, Table 2)	Natural soils and ground water
XA2	Moderately aggressive chemical environment (in accordance with EN 206-1, Table 2)	Natural soils and ground water
XA3	Highly aggressive chemical environment (in accordance with EN 206-1, Table 2)	Natural soils and ground water

4.3. Requirements for durability

To adequately design members of concrete bridges for durability, the designer must first establish how aggressive the environment to which the member is exposed is, and second select suitable materials and design the structure to be able to resist the environment for its intended life. *2-1-1/clause 4.3(2)P* requires durability to be considered throughout all stages of the design including structural conception, material selection, construction details, construction, quality control, inspection, verification and potential uses of special measures such as cathodic protection or use of stainless steel.

2-2/clause 4.3(103) requires external prestressing tendons to comply with the requirements of National Authorities. This may include restrictions on the type of system permitted. The choice is typically between either:

- strands within a duct filled with either grout or grease, or
- strands individually sheathed and greased all within an outer grout-filled duct.

The latter affords the benefit of easier de-stressing of tendons and the possibility of removal and replacement of individual strands. Replacement of individual strands is, however, difficult in reality and it is usually only possible to replace a strand with one of a slightly smaller cross-section. The former system is sometimes preferred (when the duct is filled with grout) because of the passivating influence of the grout directly against the strand.

2-1-1/clause 4.3(2)P

2-2/clause 4.3(103)

4.4. Methods of verification
4.4.1. Concrete cover
4.4.1.1. General

2-1-1/*clause 4.4.1.1(1)P* defines cover as the distance from the concrete surface to the surface of the nearest reinforcement including links, stirrups and surface reinforcement where relevant. This definition will be familiar to UK designers. **2-1-1/*clause 4.4.1.1(2)P*** further defines the nominal cover, c_{nom}, as a minimum cover, c_{min} (see section 4.4.1.2), plus an allowance in the design for deviation, Δc_{dev} (see section 4.4.1.3):

$$c_{nom} = c_{min} + \Delta c_{dev} \qquad \text{2-1-1/(4.1)}$$

c_{nom} is the value of the nominal cover that should be stated on the drawings.

4.4.1.2. Minimum cover

The main durability provision in EC2 is the specification of concrete cover as a defence against corrosion of the reinforcement. In addition to the durability aspect, adequate concrete cover is also essential for the transmission of bond forces and for providing sufficient fire resistance (which is of less significance for bridge design) – **2-1-1/*clause 4.4.1.2(1)P***. The minimum cover, c_{min}, satisfying the durability and bond requirements is defined in **2-1-1/*clause 4.4.1.2(2)P*** by the following expression:

$$c_{min} = \max\{c_{min,b}; c_{min,dur} + \Delta c_{dur,\gamma} - \Delta c_{dur,st} - \Delta c_{dur,add}; 10\,\text{mm}\} \qquad \text{2-1-1/(4.2)}$$

where:

- $c_{min,b}$ is the minimum cover due to bond requirements and is defined in 2-1-1/Table 4.2 by way of **2-1-1/*clause 4.4.1.2(3)***. The anchorage and lap requirements in 2-1-1/clause 8 assume these minimum values are observed
- $c_{min,dur}$ is the minimum cover for durability requirements from 2-1-1/Table 4.4N or 4.5N for reinforcing or prestressing steel respectively. These tables can be modified in the National Annex
- $\Delta c_{dur,\gamma}$ is an additional safety element which **2-1-1/*clause 4.4.1.2(6)*** recommends to be 0 mm
- $\Delta c_{dur,st}$ is a reduction of minimum cover for the use of stainless steel, which, if adopted, should be applied to all design calculations, including bond. The recommended value in **2-1-1/*clause 4.4.1.2(7)*** is 0 mm
- $\Delta c_{dur,add}$ is a reduction of minimum cover for the use of additional protection. This could cover coatings to the concrete surface or reinforcement (such as epoxy coating). **2-1-1/*clause 4.4.1.2(8)*** recommends taking a value of 0 mm

The minimum cover according to **2-1-1/*clause 4.4.1.2(5)*** for durability requirements, $c_{min,dur}$, depends on the relevant exposure class taken from 2-1-1/Table 4.1 and the structural class from 2-1-1/Table 4.3N, which is subject to variation in a National Annex. 2-1-1/Annex E defines indicative strength classes for concrete which depend on the exposure class in Table 4.1-1 of the element under consideration. These indicative strengths are base strengths for each exposure class from which the necessary minimum covers are defined. For greater concrete strengths, these covers can be reduced. For the indicative minimum concrete strengths for different exposure classes given in 2-1-1/Annex E, EC2 recommends taking a structural class of S4 for structures with a design life of 50 years. 2-1-1/Table 4.3N (which can be modified in the National Annex and which is reproduced here as Table 4.4-1) contains recommended modifications to the structural class for other situations which include:

- a design life of 100 years (to cover bridges, although the UK National Annex to EN 1990 is likely to assign a design life of 120 years to bridges)
- provision of an increased strength class of concrete
- provision of special quality control on site (although requirements are not defined)
- 'members with slab geometry' where the placing of the particular reinforcement considered is not constrained by the construction sequence or the detailing of other

CHAPTER 4. DURABILITY AND COVER TO REINFORCEMENT

Table 4.4-1. Recommended structural classification

	Exposure class (from 2-1-1/Table 4.1)						
Criterion	X0	XC1	XC2/XC3	XC4	XD1	XD2/XS1	XD3/XS2/XS3
Service life of 100 years	Increase class by 2	Increase class by 2	Increase class by 2	Increase class by 2	Increase class by 2	Increase class by 2	Increase class by 2
Strength class (see notes 1 and 2)	≥ C30/37 reduce class by 1	≥ C30/37 reduce class by 1	≥ C35/45 reduce class by 1	≥ C40/50 reduce class by 1	≥ C40/50 reduce class by 1	≥ C40/50 reduce class by 1	≥ C45/55 reduce class by 1
Member with slab geometry (position of reinforcement not affected by construction process)	Reduce class by 1	Reduce class by 1	Reduce class by 1	Reduce class by 1	Reduce class by 1	Reduce class by 1	Reduce class by 1
Special quality control ensured	Reduce class by 1	Reduce class by 1	Reduce class by 1	Reduce class by 1	Reduce class by 1	Reduce class by 1	Reduce class by 1

Notes
1 The strength class and water/cement ratio are considered to be related values. The relationship is subject to a national code. A special composition (type of cement, w/c value, fine fillers) with the intent to produce low permeability may be considered.
2 The limit may be reduced by one strength class if air entrainment of more than 4% is applied.

reinforcement, as might occur where fixed length links are used which constrain the position of a layer of reinforcement in one face of the section with respect to that in the other face.

The recommended values for $c_{min,dur}$ are given in Tables 4.4N and 4.5N in EC2-1-1 for reinforcing and prestressing steel respectively, and are reproduced here as Tables 4.4-2 and 4.4-3. They can be amended in the National Annex.

Where in situ concrete is placed against other concrete elements, such as at construction joints, *2-2/clause 4.4.1.2(109)* allows the minimum concrete cover to reinforcement to be reduced. The recommended reduction is to the value required for bond, provided that the concrete class is at least C25/30, the exposure time of the temporary concrete surface to an outdoor environment is less than 28 days, and the interface is roughened. This recommendation can be modified in the National Annex.

2-2/clause 4.4.1.2(109)

Table 4.4-2. Minimum cover requirements, $c_{min,dur}$, for durability (reinforcing steel)

Environmental requirements for $c_{min,dur}$ (mm)

	Exposure class (from 2-1-1/Table 4.1)						
Structural class	X0	XC1	XC2/XC3	XC4	XD1/XS1	XD2/XS2	XD3/XS3
S1	10	10	10	15	20	25	30
S2	10	10	15	20	25	30	35
S3	10	10	20	25	30	35	40
S4	10	15	25	30	35	40	45
S5	15	20	30	35	40	45	50
S6	20	25	35	40	45	50	55

Table 4.4-3. Minimum cover requirements, $c_{min,dur}$, for durability (prestressing steel)

Environmental requirements for $c_{min,dur}$ (mm)

Structural class	Exposure class (from 2-1-1/Table 4.1)						
	X0	XC1	XC2/XC3	XC4	XD1/XS1	XD2/XS2	XD3/XS3
S1	10	15	20	25	30	35	40
S2	10	15	25	30	35	40	45
S3	10	20	30	35	40	45	50
S4	10	25	35	40	45	50	55
S5	15	30	40	45	50	55	60
S6	20	35	45	50	55	60	65

2-1-1/clause 4.4.1.2(11) requires further increases to the minimum covers for exposed aggregate finishes, while *2-1-1/clause 4.4.1.2(13)* gives requirements where the concrete is subject to abrasion. Specific requirements are given in *2-2/clause 4.4.1.2(114)* and *2-2/clause 4.4.1.2(115)* to cover bare concrete decks of road bridges and concrete exposed to abrasion by ice or solid transportation in running water.

2-1-1/clause 4.4.1.2(11)
2-1-1/clause 4.4.1.2(13)
2-2/clause 4.4.1.2(114)
2-2/clause 4.4.1.2(115)

4.4.1.3. Allowance in design for deviation

The actual cover to be specified on the drawings, c_{nom}, has to include a further allowance for deviation (Δc_{dev}) according to *2-1-1/clause 4.4.1.3(1)P* such that $c_{nom} = c_{min} + \Delta c_{dev}$. The value of Δc_{dev} for buildings and bridges may be defined in the National Annex and is recommended by EC2 to be taken as 10 mm. *2-1-1/clause 4.4.1.3(3)* allows the recommended value of Δc_{dev} to be reduced in situations where accurate measurements of cover achieved can be taken and non-conforming elements rejected, such as is the case for manufacture of precast units for example. Such modifications can again be given in the National Annex. *2-1-1/clause 4.4.1.3(4)* gives further requirements where concrete is cast against uneven surfaces, such as directly onto the ground, or where there are surface features which locally reduce cover (such as ribs). The former will typically cover bases cast on blinding or bored piles cast directly against soil.

2-1-1/clause 4.4.1.3(1)P
2-1-1/clause 4.4.1.3(3)
2-1-1/clause 4.4.1.3(4)

> **Worked example 4.4-1: Cover for deck slab**
> Determine the required nominal cover to 20 mm reinforcement for a C40/50 deck slab beneath waterproofing with a design life of 100 years.
>
> From Table 4.1 of EN 1992-1-1, exposure class is XC3. The starting structural class is S4. From Table 4.3N of EN 1992-1-1, the following additions to structural class apply:
>
> - for 100-year design life, add 2
> - for concrete grade in excess of C35/45, deduct 1
> - for member with slab geometry (as is the case for a deck slab without links), deduct 1.
>
> Therefore final structural class is S4. From Table 4.4N of EN 1992-1-1 for XC3 and structural class S4:
>
> $c_{min,dur} = 25 \text{ mm}$
>
> From bond considerations, $c_{min,b} = 20 \text{ mm}$ (the bar size):
>
> $c_{min} = \max\{c_{min,b}; c_{min,dur} + \Delta c_{dur,\gamma} - \Delta c_{dur,st} - \Delta c_{dur,add}; 10 \text{ mm}\}$
> $= \max\{20; 25 + 0 - 0 - 0; 10\} = 25 \text{ mm}$
>
> The recommended value of $\Delta c_{dev} = 10 \text{ mm}$ so:
>
> $c_{nom} = c_{min} + \Delta c_{dev} = 25 + 10 = \textbf{35 mm}$

CHAPTER 5

Structural analysis

This chapter discusses structural analysis as covered in section 5 of EN 1992-2 in the following clauses:

• General	*Clause 5.1*
• Geometric imperfections	*Clause 5.2*
• Idealization of the structure	*Clause 5.3*
• Linear elastic analysis	*Clause 5.4*
• Linear elastic analysis with limited redistribution	*Clause 5.5*
• Plastic analysis	*Clause 5.6*
• Non-linear analysis	*Clause 5.7*
• Analysis of second-order effects with axial load	*Clause 5.8*
• Lateral instability of slender beams	*Clause 5.9*
• Prestressed members and structures	*Clause 5.10*
• Analysis for some particular structural members	*Clause 5.11*

5.1. General

2-1-1/clause 5.1.1(1)P is a reminder that a global analysis may not cover all relevant structural effects or the true behaviour so that separate local analysis may also be necessary. A typical example of this situation includes the grillage analysis of a beam and slab deck where the longitudinal grillage members have been placed along the axis of the main beams only, thus not modelling the effects of local loads on the slab.

The Note to 2-1-1/clause 5.1.1(1)P makes an important observation regarding the use of shell finite element models: that the application rules given in EN 1992 generally relate to the resistance of entire cross-sections to internal forces and moments, but these stress resultants are not determined directly from shell finite element models. In such cases, either the stresses determined can be integrated over the cross-section to determine stress resultants for use with the member application rules or individual elements must be designed directly for their stress fields. *2-1-1/clause 5.1.1(3)* refers to Annex F for a method of designing elements subject to in-plane stress fields only. Annex LL provides a method for dealing with elements also subjected to out of plane forces and moments. The use of both these annexes can be conservative because they do not make allowance for redistribution across a cross-section as is implicit in many of the member rules in section 6 of EN 1992-2.

2-1-1/clause 5.1.1(2) gives other instances where local analysis may be needed. These relate to situations where the assumptions of beam-like behaviour are not valid and plane sections do not remain plane. Examples include those listed in the clause plus any situation where a discontinuity in geometry occurs, such as at holes in a cross-section. These local analyses can often be carried out using strut-and-tie analysis in accordance with section 6.5 of EN 1992-2. Some situations are covered fully or partially by the application rules

2-1-1/clause
5.1.1(1)P

2-1-1/clause
5.1.1(3)

2-1-1/clause
5.1.1(2)

elsewhere in EN 1992. For example, load application in the vicinity of bearings is partially covered by the shear enhancement rules in 2-1-1/clause 6.2, while the design of the bearing zone itself is partially covered by 2-1-1/clause 6.7. The design of post-tensioned anchorage zones is partially covered by 2-2/Annex J.

As a general principle, given in *2-1-1/clause 5.1.1(4)P*, 'appropriate' idealizations of both the geometry and the behaviour of the structure have to be made to suit the particular design verification being performed. Clauses 5.2 and 5.3.2 are relevant to geometry (covering imperfections and effective span). Behaviour relates to the choice of model and section properties. For example, a skew flat slab could be safely modelled for ultimate limit states using a torsionless grillage, but this would be inappropriate for serviceability limit state crack width checks; the model would fail to predict top cracking in the obtuse corners. Section properties, with respect to the choice of cracked or un-cracked behaviour, are discussed in section 5.4 of this guide. Clause 5.3.2 also covers shear lag, which affects section stiffness.

2-1-1/clause 5.1.1(5) and *2-1-1/clause 5.1.1(6)P* require each stage of construction to be considered in design as this may affect the final distribution of internal effects. Time-dependent effects, such as redistribution of moments due to creep in bridges built in stages, also need to be realistically modelled. 2-2/Annex KK addresses the specific case of creep redistribution, which provides some 'recognized design methods' as referred to in *2-2/clause 5.1.1(108)*.

2-1-1/clause 5.1.1(7) gives a general statement of the types of structural analysis that may be used, which include:

- linear elastic analysis with and without redistribution
- plastic analysis
- strut-and-tie modelling (a special case of plastic analysis)
- non-linear analysis.

Guidance on when and how to use these analysis methods is given in the sections of this guide corresponding to the relevant sections in EC2.

Soil–structure interaction is a special case of the application of 2-1-1/clause 5.1.1(4)P and is covered by *2-1-1/clause 5.1.2(1)P*. It should be considered where it significantly affects the analysis (as would usually be the case for integral bridge design). *2-1-1/clause 5.1.2(3) to (5)* also specifically mention the need to consider the interaction between piles in analysis where they are spaced centre to centre at less than three times the pile diameter.

2-2/clause 5.1.3(101)P essentially requires all possible combinations of actions and load positions to be considered, such that the most critical design situation is identified. This has been common practice in bridge design and often necessitates the use of influence surfaces. The Note to the clause allows a National Annex to specify simplified load arrangements to minimize the number of arrangements to consider. Its inclusion was driven by the buildings community, where such simplifications are made in current UK practice. The equivalent note in EC2-1-1 therefore makes recommendations for buildings but none are given for bridges in EC2-2. The comments made on load factors under clause 2.4.3 are also relevant to the determination of load combinations.

2-1-1/clause 5.1.4 requires that second-order effects should be considered in bridge design and these are much more formally addressed than previously was the case in UK practice. Detailed discussion on analysis for second-order effects and when they can be neglected is presented in section 5.8.

5.2. Geometric imperfections
5.2.1. General (additional sub-section)
The term 'geometric imperfections' is used to describe the departures from the exact centreline, setting out dimensions specified on drawings that occur during construction. This is inevitable as all construction work can only be executed to certain tolerances.

CHAPTER 5. STRUCTURAL ANALYSIS

2-1-1/clause 5.2(1)P requires these imperfections to be considered in analysis. The term does not apply to tolerances on cross-section dimensions, which are accounted for separately in the material factors, but does apply to load position. 2-1-1/clause 6.1(4) gives minimum requirements for the latter. Geometric imperfections can apply both to overall structure geometry and locally to members.

Geometric imperfections can give rise to additional moments from the eccentricity of axial loads generated. They are therefore particularly important to consider when a bridge or its elements are sensitive to second-order effects. *2-1-1/clause 5.2(2)P*, however, requires imperfections to be considered for ultimate limit states even when second-order effects can be ignored in accordance with 2-1-1/clause 5.8.2(6). For short bridge elements, the additional moments caused by imperfections will often be negligible and the effects of imperfections could then be ignored in such cases with experience. Imperfections need not be considered for serviceability limit states (*2-1-1/clause 5.2(3)*).

2-2/clause 5.2(104) states that the values of imperfections used within 2-2/clause 5.2 assume that workmanship is in accordance with deviation class 1 in EN 13670. If other levels of workmanship are to be used during construction, then the imperfections used in design should be modified accordingly.

In general, imperfections can be modelled as either bows or angular departures in members. EC2 generally uses angular departures as a simplification, but sinusoidal bows will often be slightly more critical and better reflect the elastic critical buckling mode shape. For this reason, sinusoidal imperfections have to be used in the design of arches to 2-2/clause 5.2(106). The type of imperfection relevant for member design will depend on the mode of buckling. An overall lean to a pier will suffice where buckling is in a sway mode, which EC2 describes as 'unbraced' conditions in section 5.8. This is because the moments generated by the imperfection will add to those from the additional deflections under load. An overall lean would not, however, suffice for buckling within a member when its ends are held in position, which EC2 describes as 'braced' conditions. In this latter case, a lean of the entire column alone would not induce any moments within the column length. It would, however, induce forces in the positional restraints, so an imperfection of this type would be relevant for the design of the restraints. A local eccentricity within the member is therefore required for buckling of braced members. This illustrates the need to choose the type of imperfection carefully depending on the effect being investigated. Further discussion on 'braced' and 'unbraced' conditions is given in section 5.8 of this guide and Fig. 5.2-1 illustrates the difference.

A basic lean imperfection, θ_1, is defined for bridges in *2-2/clause 5.2(105)* as follows:

$$\theta_1 = \theta_0 \alpha_h \qquad \text{2-2/(5.101)}$$

where:

- θ_0 is the basic value of angular departure
- α_h is the reduction factor for height with $\alpha_h = 2/\sqrt{l}$; $\alpha_h \leq 1$
- l is the length or height being considered in metres

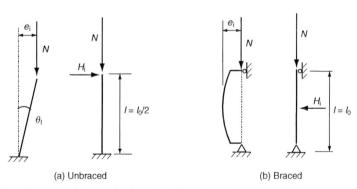

(a) Unbraced   (b) Braced

Fig. 5.2-1. Effect of geometric imperfections in isolated members

The lower limit for α_h given in EC2-1-1 was removed in EC2-2 to avoid excessive imperfections in tall bridge piers. The value of θ_0 is a nationally determined parameter but the recommended value is 1/200, which is the same as previously used in Model Code 90.[6]

2-1-1/clause 5.2(7)

2-1-1/clause 5.2(7) allows imperfections in isolated members to be taken into account either by modelling them directly in the structural system or by replacing them with equivalent forces. The latter is a useful alternative, as the same model can be used to apply different imperfections, but the disadvantage is that the axial forces in members must first be known before the equivalent forces can be calculated. This can become an iterative procedure. These alternatives are illustrated in Fig. 5.2-1 for the two simple cases of a pin-ended strut and a cantilever. They are:

(a) Application of an eccentricity, e_i

2-1-1/clause 5.2(7) gives the following formula for the imperfection eccentricity:

$$e_i = \theta_1 l_0 / 2 \qquad \qquad 2\text{-}1\text{-}1/(5.2)$$

where l_0 is the effective length.

For the unbraced cantilever in Fig. 5.2-1(a), the angle of lean from 2-2/(5.101) leads directly to the top eccentricity of $e_i = \theta_1 l = \theta_1 l_0/2$, when $l_0 = 2l$ (noting that $l_0 > 2l$ for cantilever piers with real foundations as discussed in section 5.8.3 of this guide).

For the braced pin-ended pier in Fig. 5.2-1(b), partially reproducing 2-1-1/Fig. 5.1(a2), the eccentricity is shown to be applied predominantly as an end eccentricity. This is not in keeping with the general philosophy of applying imperfections as angular deviations. An alternative, therefore, is to apply the imperfection for the pin-ended case as a kink over the half wavelength of buckling, based on two angular deviations, θ_1, as shown in Fig. 5.2-2. This is then consistent with the equivalent force system shown in Fig. 5.2-1(b). It is also the basis of the additional guidance given in EC2-2 for arched bridges where a deviation $a = \theta_1 l/2$ has to be attributed to the lowest symmetric modes as discussed below. This method of application is slightly less conservative.

2-1-1/Expression (5.2) can be misleading for effective lengths less than the height of the member, as the eccentricity e_i should really apply over the half wavelength of buckling, l_0. This interpretation is shown in Fig. 5.2-3 for a pier rigidly built in at each end. It leads to the same peak imperfection as for the pin-ended case, despite the fact the effective length for the built in case is half that of the pinned case. This illustrates the need to be guided by the buckling mode shape when choosing imperfections.

(b) Application of a transverse force, H_i, in the position that gives maximum moment

The following formulae for the imperfection forces to apply are given in 2-1-1/clause 5.2(7):

$$H_i = \theta_1 N \quad \text{for unbraced members (see Fig. 5.2-1(a))} \qquad 2\text{-}1\text{-}1/(5.3a)$$

$$H_i = 2\theta_1 N \quad \text{for braced members (see Fig. 5.2-1(b))} \qquad 2\text{-}1\text{-}1/(5.3b)$$

where N is the axial load.

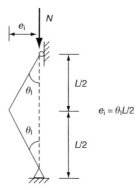

Fig. 5.2-2. Alternative imperfection for pin-ended strut as an angular deviation

CHAPTER 5. STRUCTURAL ANALYSIS

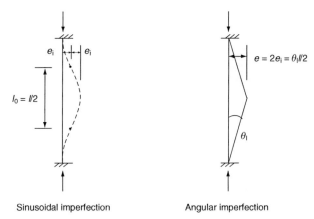

Fig. 5.2-3. Imperfection for pier built in at both ends

Where the imperfections are applied geometrically as a kink or lean as discussed above, these forces are directly equivalent to the imperfection.

5.2.2. Arches (additional sub-section)
2-2/clause 5.2(106) covers imperfections for arches for buckling in plane and out of plane.

2-2/clause 5.2(106)

In-plane buckling
For in-plane buckling cases where a symmetric buckling mode is critical, for example from arch spreading, a sinusoidal imperfection of $a = \theta_1 L/2$ has to be applied as shown in Fig. 5.2-4. This magnitude is derived by idealizing the actual buckling mode as a kink made up from angular deviations, θ_1, despite the clause's recommendation that the imperfection be distributed sinusoidally for arch cases as discussed above.

Where an arch does not spread significantly, the lowest mode of buckling is usually anti-symmetric, as shown in Fig. 5.2-5. In this case, the mode shape, and thus imperfection, can be idealized as a saw tooth using the same basic angular deviation in conjunction with the reduced length $L/2$ relevant to the buckling mode. The imperfection therefore becomes $a = \theta_1 L/4$. Once again, EC2 requires the imperfection to be distributed sinusoidally.

Out-of-plane buckling
For out-of-plane buckling, the same shape of imperfection as in Fig. 5.2-4 is suitable but in the horizontal plane.

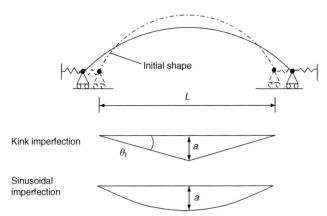

Fig. 5.2-4. Imperfection for in-plane buckling with spreading foundations

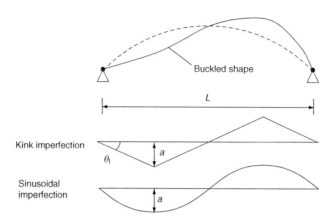

Fig. 5.2-5. Imperfection for in-plane buckling with 'rigid' foundations

5.3 Idealization of the structure

5.3.1 Structural models for overall analysis

2-1-1/clause 5.3.1(1)P

2-1-1/clause 5.3.1(1)P lists typical elements comprising a structure and states that rules are given in EC2 to cover the design of these various elements. While detailing rules are provided by element type in section 9 of EN 1992-2, rules for resistances are generally presented by resistance type rather than by element type in section 6. For example, 2-2/clause 6.1 covers the design of sections in general to combinations of bending and axial force. These rules apply equally to beams, slabs and columns.

2-1-1/clause 5.3.1(3), (4), (5) and (7)

2-1-1/clause 5.3.1(3), (4), (5) and (7) provide definitions of beams, deep beams, slabs and columns. The definitions given are self-explanatory and are often useful in defining the detailing and analysis requirements for the particular element. For example, the distinction made between a beam and deep beam is useful in determining the appropriate verification method and detailing rules. A beam can be checked for bending, shear and torsion using 2-2/clause 6.1 to 6.3, while deep beams are more appropriately treated using the strut-and-tie rules of 2-2/clause 6.5. No distinction is, however, made between beams with axial force and columns in EC2 in cross-section resistance design. A distinction, however, remains necessary when selecting the most appropriate detailing rules from section 9. Sometimes it may be appropriate to treat parts of a beam, such as a flange in a box girder, as a wall or column for detailing purposes, as discussed in section 9.5 of this guide.

5.3.2. Geometric data

5.3.2.1. Effective width of flanges (all limit states)

In wide flanges, in-plane shear flexibility leads to a non-uniform distribution of bending stress across the flange width. This effect is known as shear lag. The stress in the flange adjacent to the web is consequently found to be greater than expected from section analysis with gross cross-sections, while the stress in the flange remote from the web is lower than expected. This shear lag also leads to an apparent loss of stiffness of a section in bending. The determination of the actual distribution of stress is a complex problem which can, in theory, be determined by finite element analysis (with appropriate choice of elements) if realistic behaviour of reinforcement and concrete can be modelled. For un-cracked concrete, the behaviour is relatively simple but becomes considerably more complex with cracking of the concrete and yielding of the longitudinal reinforcement, which both help to redistribute the stress across the cross-section. The ability of the transverse reinforcement to distribute the forces is also relevant.

2-1-1/clause 5.3.2.1(1)P

2-1-1/clause 5.3.2.1(1)P accounts for both the loss of stiffness and localized increase in flange stresses by the use of an effective width of flange, which is less than the actual available flange width. The effective flange width concept is artificial but, when used with engineering bending theory, leads to uniform stresses across the whole reduced flange width that are equivalent to the peak values adjacent to the webs in the 'true' situation. It follows from the above that if finite element modelling of flanges is performed using appropriate analysis elements, shear lag will be taken into account automatically (the accuracy depending on the material properties specified in analysis as discussed above) so an effective flange need not be used.

The rules for effective width may be used for flanges in members other than just 'T' beams as suggested by 2-1-1/clause 5.3.2.1(1)P; box girders provide an obvious addition. The physical flange width is unlikely to be reduced for many typical bridges, such as precast beam and slab decks where the beams are placed close together. The effect of shear lag is greatest in locations of high shear where the force in the flanges is changing rapidly. Consequently, effective widths at pier sections will be smaller than those for the span regions.

2-1-1/clause 5.3.2.1(1)P notes that, in addition to the considerations discussed above, effective width is a function of type of loading and span (which affect the distribution of shear along the beam). These are characterized by the distance between points of zero bending moment. *2-1-1/clause 5.3.2.1(2) and (3)*, together with 2-1-1/Figs 5.2 and 5.3 (not reproduced), allow effective widths to be calculated as a function of the actual flange width and the distance, l_0, between points of zero bending moment in the main beam adjacent to the location considered. This length actually depends on the load case being considered and the approximations given are intended to save the designer from having to determine actual values of l_0 for each load case. The total effective width acting with a web, b_{eff}, is given as follows:

2-1-1/clause 5.3.2.1(2) and (3)

$$b_{eff} = \sum b_{eff,i} + b_w \leq b \qquad \text{2-1-1/(5.7)}$$

$$b_{eff,i} = 0.2b_i + 0.1l_0 \leq b_i \quad \text{and} \quad \leq 0.2l_0 \qquad \text{2-1-1/(5.7a)}$$

where b is the total flange width available for the particular web and b_i is the width available to one side of the web, measured from its face.

2-1-1/Expression (5.7) and 2-1-1/Expression (5.7a) differ from similar ones in EN 1994-2, as a minimum of 20% of the actual flange width may always be taken to act each side of the web where the span is short. The reference in 2-1-1/clause 5.3.2.1(3) to 'T' or 'L' beams is only intended to describe webs with flanges to either one or both sides of a web. It is not intended to limit use to main beams of these shapes.

The limitations on span length ratios for use of 2-1-1/Fig. 5.2, given in the Note to 2-1-1/clause 5.3.2.1(2), are made so that the bending moment distribution within a span conforms with the assumption that spans have hog moments at supports and sag moments in the span. The simple rules do not cater for other cases, such as entire spans that are permanently hogging. If spans or moment distributions do not comply with the above, then the distance between points of zero bending moment, l_0, should be calculated for the actual moment distribution. This is iterative because analysis will have to be done first with cross-section properties based on the full flange width to determine the likely distribution of moment.

The same effective width for shear lag applies to both SLS and ULS. This is unlike previous design to BS 5400, where it was permissible to neglect shear lag at ULS on the basis that the effects of concrete cracking and reinforcement yielding discussed above allow stresses to redistribute across a flange. EC2, however, bases effective widths at ULS on widths approximating more closely to the elastic values, thus avoiding the complexity of providing rules to calculate effective widths which allow for these redistribution effects. This differs from the approach in EN 1993-1-5 for steel flanges, where consideration of plasticity is allowed at ULS and greater effective widths can be achieved. The practical significance of using the same effective width at ULS and SLS will not usually be great for concrete bridges and often the full width will be available.

For global analysis, *2-1-1/clause 5.3.2.1(4)* permits section properties to be based on the mid-span value throughout that entire span. Quite often this will lead to the full flange width being used. It can, however, be advantageous to use the actual distribution of effective widths near to supports and at mid-span in continuous beams to reduce the stiffness at supports and hence also the support hogging moment. A more accurate prediction of stiffness, and hence effective width distribution throughout the span, may also be necessary where prediction of

2-1-1/clause 5.3.2.1(4)

deflections is important to the construction process, such as in balanced cantilever construction. For section design, the actual effective width at the location being checked must always be used.

Where it is necessary to determine a more realistic distribution of longitudinal stress across the width of the flange, as may be required in a check of combined local and global effects in a flange, the formula in EN 1993-1-5 clause 3.2.2 could be used to estimate stresses. This is explicitly permitted for slabs in steel–concrete composite construction in EN 1994-2. A typical location where this might be necessary would be at a transverse diaphragm between main beams at a support where the deck slab is in tension under global bending and also subjected to a local hogging moment from wheel loads. The use of the formula in EN 1993-1-5 can be beneficial here as often the greatest local effects in a slab occur in the middle of the slab between webs where the global longitudinal stresses are lowest.

The effective flange widths in 2-1-1/clause 5.3.2.1 do not apply to the introduction of axial loads, such as those from prestressing or anchorages in cable-stayed bridges. The phenomenon of shear lag still applies to local concentrated axial loads, but stresses spread out across the section at a rate un-connected to the bending moment profile. Where concentrated axial loads are applied to a section, separate assessment must be made of the area over which this force acts at each cross-section through the span – see 2-1-1/clause 8.10.3.

Worked example 5.3-1: Effective flange width for a box girder

A box girder bridge has the span layout and cross-section shown in Fig. 5.3-1. Determine the effective width of top flange acting with an outer web at mid-span and supports for the main span.

Considering mid-span first:

From 2-1-1/Fig. 5.2, $l_0 = 0.7 l_2 = 0.7 \times 40\,000 = 28\,000$ mm.

From 2-1-1/(5.7a), the cantilever portion has effective width given by:

$$b_{\text{eff},i} = 0.2 b_i + 0.1 l_0 = 0.2 \times 4000 + 0.1 \times 28\,000 = 3600 \text{ mm} < 0.2 l_0 = 5600 \text{ mm}$$

Similarly, the internal flange associated with the web has effective width:

$$b_{\text{eff},i} = 0.2 b_i + 0.1 l_0 = 0.2 \times \frac{5700}{2} + 0.1 \times 28\,000 = 3370 \text{ mm} < 0.2 l_0 = 5600 \text{ mm}$$

but this is greater than the available width of $5700/2 = 2850$ mm so the effective width is taken as 2850 mm.

Finally, from 2-1-1/(5.7) the total width of flange acting with an outer web is:

$$b_{\text{eff}} = \sum b_{\text{eff},i} + b_w = 3600 + 2850 + 300 = \mathbf{6750 \text{ mm}}$$

This is almost the whole available width.

Considering the supports:

From 2-1-1/Fig. 5.2, $l_0 = 0.15(l_1 + l_2) = 0.15 \times (30\,000 + 40\,000) = 10\,500$ mm.

From 2-1-1/(5.7a), the cantilever portion has effective width given by:

$$b_{\text{eff},i} = 0.2 b_i + 0.1 l_0 = 0.2 \times 4000 + 0.1 \times 10\,500 = 1850 \text{ mm} < 0.2 l_0 = 2100 \text{ mm}$$

Similarly, the internal flange associated with the web has effective width:

$$b_{\text{eff},i} = 0.2 b_i + 0.1 l_0 = 0.2 \times \frac{5700}{2} + 0.1 \times 10\,500 = 1620 \text{ mm} < 0.2 l_0 = 2100 \text{ mm}$$

Finally, from 2-1-1/(5.7) the total width of flange acting with an outer web is:

$$b_{\text{eff},i} = \sum b_{\text{eff},i} + b_w = 1850 + 1620 + 300 = \mathbf{3770 \text{ mm}}$$

This represents only 53% of the available width of 7150 mm, illustrating that shear lag can be significant at supports for wide flanges and short spans.

CHAPTER 5. STRUCTURAL ANALYSIS

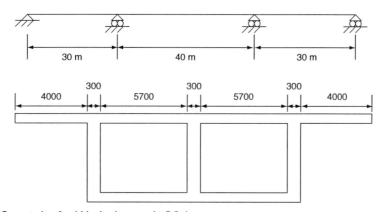

Fig. 5.3-1. Box girder for Worked example 5.3-1

5.3.2.2. Effective spans of beams and slabs
2-2/clause 5.3.2.2 gives requirements for the effective span, l_{eff}, of beams and slabs. Examples are given in 2-1-1/Fig. 5.4. The main cases of interest in bridge design are:

(i) Beams monolithic with supports
Where a horizontal member is built in monolithically to another vertical member, the effective span is taken to a point within the vertical member which is the minimum of half the vertical member thickness, t, and half the horizontal member depth, h, from the edge of the vertical member according to Figs 5.3-2 (a) and (b). The limitation to $h/2$ is intended to keep the centre of reaction in a realistic position where the supporting member is very thick. Cases (a) and (b) could apply, for example, to integral bridges and to the transverse cantilever design of box girder bridges respectively. In this situation, *2-1-1/clause 5.3.2.2(3)* and its Note state that the design moment should be taken as that at the face of the support but not less than 65% of the actual maximum end moment.

2-1-1/clause 5.3.2.2(3)

(ii) Beams on bearings
2-1-1/clause 5.3.2.2(2) permits rotational restraint from bearings to generally be neglected. This is the usual assumption made for most mechanical bearings but the designer needs to exercise judgement. Clearly, for example, the torsional restraint provided by linear rocker bearings should not be ignored.

2-1-1/clause 5.3.2.2(2)

The effective span is measured between centres of bearings as shown in Fig. 5.3-2(c). The moment so obtained may, however, be rounded off over each bearing in accordance with *2-2/clause 5.3.2.2(104)*. This has the effect of reducing the moment by $F_{\text{Ed,sup}}t/8$, where t can be defined in the National Annex. The recommended value in the Note to 2-2/clause 5.3.2.2(104) is the 'breadth of the bearing'. This is intended to be the dimension in the direction of the span of the bearing contact surface with the deck. $F_{\text{Ed,sup}}$ is the support reaction coexisting with the moment case considered.

2-2/clause 5.3.2.2(104)

The definition of t above is intended to be used for a rectangular contact area. For a bearing with circular contact area, using the diameter for t is slightly unconservative and, for consistency, the moment should be reduced by $F_{\text{Ed,sup}}D/3\pi$, based on the centroid of half of a circular area, where D is the contact diameter. The contact dimensions should be taken as the lesser of the physical top plate size or the dimension obtained by spreading from the edges of the stiff part of the bearing through the top plate. A spread of 1:1 is suggested, in keeping with that given in EN 1993-1-5 clause 3.2.3, although a greater spread at 60° to the vertical was allowed by BS 5400 Part 3.[7]

The wording of 2-2/clause 5.3.2.2(104) differs from that in 2-1-1/clause 5.3.2.2(4) in order to clarify that the rounding can only be made if the analysis assumes point support. The wording in EC2-1-1, which starts with 'Regardless of method of analysis...', was considered undesirable as the support width could be included in a more detailed analysis model.

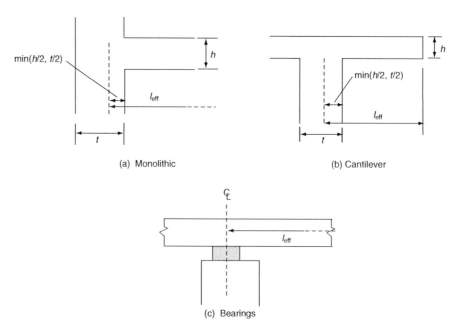

Fig. 5.3-2. Examples of effective span

5.4. Linear elastic analysis

Linear elastic analysis is the most commonly used technique for bridge design and may be used to calculate action effects at all limit states in accordance with *2-1-1/clause 5.4(1)*.

2-1-1/clause 5.4(2) allows linear elastic analysis to assume un-cracked cross-sections and linear stress–strain relationships with mean values of the elastic modulus. The use of un-cracked linear elastic analysis, despite the apparent anomaly that sections will behave in a non-linear manner at the ultimate limit state, is justified by the lower-bound theorem of plasticity. This states that providing equilibrium is maintained everywhere and the yield strength of the material is nowhere exceeded, a safe lower-bound estimate of resistance is obtained. Some ductility is necessary for the lower-bound theorem to be applied, so that peak resistance at a section is maintained during redistribution of moments, but this has not been found to be a problem in real structures. The margin of safety in this respect may, however, not be as high in Eurocode 2 as it was in previous UK practice, since there is no requirement to prevent over-reinforced flexural behaviour in reinforced and prestressed concrete section design (which may lead to sudden concrete failure before reinforcement yield).

Un-cracked elastic analysis has many advantages over other calculation methods, including the fact that the reinforcement does not need to be known prior to performing the global analysis and the principle of superposition may be applied to the results of individual load cases. Additionally, the alternatives of plastic analysis and elastic analysis with redistribution can only be used for ultimate limit states so an elastic analysis must then additionally be used for serviceability limit states. Non-linear analysis can be used for both SLS and ULS as an alternative, but superposition of load cases cannot be performed.

An un-cracked elastic analysis leads to only one solution of a possible infinite number of solutions that can be justified by the lower-bound theorem as above. The modelled behaviour must, however, be realistic (particularly for SLS), as required by 2-2/clause 5.1, and therefore it may sometimes be more appropriate to consider some cracked sections. An example is in the calculation of transverse stiffness for a bridge with prestressing in the longitudinal direction only. In this case, the concrete in the transverse direction is liable to be cracked while that in the longitudinal direction remains un-cracked. A non-linear analysis is most realistic in this situation (and in general) as discussed in section 5.7 of this guide, as the analysis can model cracking and other aspects of material non-linear response.

Where there are significant second-order effects (see section 5.8 of this guide), these must also be taken into account and linear elastic analysis must then only be performed in conjunction with further magnification of moments and reduced stiffness properties accounting for cracking and creep, as described in section 5.8.7 of this guide.

At the ultimate limit state, *2-1-1/clause 5.4(3)* allows an analysis using fully cracked section properties to be used for load cases where there are applied deformations from temperature, settlement and shrinkage. Creep should also be included by reducing the effective modulus for these load cases, as discussed in sections 5.8.7 and 3.1.4 of this guide. However, it should be noted that it will usually be possible to neglect these effects altogether at the ultimate limit state, providing there is sufficient ductility, as discussed in sections 2.3 and 5.6 of this guide.

2-1-1/clause 5.4(3)

At the serviceability limit state, 2-1-1/clause 5.4(3) requires a 'gradual evolution of cracking' to be considered. This appears to cover the situation where cracking occurs only in some parts of the structure, which increases the moments at other sections compared to the values which would be obtained were all sections fully cracked (or fully un-cracked). To investigate this would require a non-linear analysis, modelling the effects of tension stiffening. While non-linear analysis is covered in section 5.7, no guidance is given on modelling tension stiffening in global analysis. The use of either mean or lower characteristic values of the tensile strength can be justified depending on whether the effect is favourable or unfavourable. Tension stiffening is generally favourable as, for example, in second-order analysis of piers. Where this is the case, it would be conservative to use lower characteristic values of the tensile strength. However, for the non-linear analysis of externally post-tensioned bridges, where cracking at sections other than the critical section is beneficial in increasing the tendon strain, mean values of tensile strength would be safer, as discussed in section 5.10.8 of this guide. Since 2-1-1/clause 5.7(4) requires 'realistic' properties to be used, arguably mean tensile properties for concrete could be justified in all cases.

Since a fully un-cracked elastic analysis will generally be conservative (as it does not lead to redistribution of moment away from the most highly stressed, and therefore cracked, areas), an un-cracked elastic global analysis will be adequate to comply with 2-1-1/clause 5.4(3) for the serviceability limit state. This avoids the need to consider tension stiffening. Where significant imposed deformations are present, an un-cracked analysis provides a stiffer response and will therefore also be conservative. A fully cracked analysis is not conservative.

5.5. Linear elastic analysis with limited redistribution

Elastic analysis results in both a set of internal actions that are in equilibrium with the applied loads and geometric compatibility in the structure after the predicted elastic deflections have taken place. The 'real' moments in the structure may be significantly different to these predicted ones because the stiffness of the bridge is unlikely to conform accurately to that predicted by the use of un-cracked gross-section properties and mean elastic concrete properties. Factors affecting real stiffness are the variability in material properties, cracking of concrete and material non-linearity with load level. In practice, therefore, some redistribution of moment from that assumed in un-cracked elastic analysis usually occurs in real bridges as the ultimate limit state is approached, even if redistribution was not assumed in the design. The neglect of this redistribution in elastic analysis has traditionally been justified as being safe on the basis of the lower-bound theorem of plasticity discussed in section 5.4.

Designed redistribution of the moments from an elastic analysis is permitted for bridges at the ultimate limit state according to *2-1-1/clause 5.5(2)*, providing equilibrium is still maintained between the applied loads and the resulting distribution of moments and shears – *2-1-1/clause 5.5(3)* refers. This again follows from the lower-bound theorem of plasticity, but there must also be sufficient ductility (or rotation capacity) to allow this to occur.

2-1-1/clause 5.5(2)
2-1-1/clause 5.5(3)

In the limit, an elastic analysis with unlimited redistribution of moment effectively becomes a lower-bound plastic analysis, as discussed in section 5.6. However, the amount of redistribution possible is limited by the rotation capacity of the section, which is itself limited by the compressive failure strain of the concrete and the tensile failure strain of the reinforcement. Simplified rules for determining the amount of redistribution permitted are given in *2-2/clause 5.5(104)*. They depend on the relative depth of the compression zone. Members with a relatively small depth of concrete in compression can produce a greater strain in the reinforcement at concrete failure and hence a greater curvature and rotation capacity. If the simplified rules are used, no explicit check of rotation capacity, as discussed in section 5.6.3, is necessary. Moments should not be redistributed for SLS verifications.

2-2/clause 5.5(104)

The simplified rules in 2-2/clause 5.5(104) limit the amount of redistribution to 15% (the recommended value by way of the parameter k_5 below) for members with class B or C reinforcement. This is half the maximum redistribution recommended for buildings. This reduction in limit was made for bridges due to the lack of test results for rotation capacity for the deep flanged beams typically encountered in bridge design. Note 2 to clause 5.5(104) therefore permits the greater amount of redistribution allowed in EN 1992-1-1 to be applied to solid slabs. The following limits are given for the ratio, δ, of the redistributed moment to that from the elastic analysis:

$$\delta \geq k_1 + k_2 x_u/d \geq k_5 \quad \text{for} \quad f_{ck} \leq 50\,\text{MPa} \qquad \text{2-2/(5.10a)}$$

$$\delta \geq k_3 + k_4 x_u/d \geq k_5 \quad \text{for} \quad f_{ck} > 50\,\text{MPa} \qquad \text{2-2/(5.10b)}$$

where:

- x_u is the depth of the compression zone at the ultimate limit state under the total moment after redistribution (which will usually be the ultimate moment resistance of the section)
- d is the effective depth of the section
- k_1 to k_5 are nationally determined parameters whose recommended values are as follows:
  $k_1 = 0.44$
  $k_2 = 1.25(0.6 + 0.0014/\varepsilon_{cu2})$
  $k_3 = 0.54$
  $k_4 = 1.25(0.6 + 0.0014/\varepsilon_{cu2})$
  $k_5 = 0.85$
- ε_{cu2} is the ultimate concrete compressive strain which is 0.0035 for concretes with $f_{ck} \leq 50\,\text{MPa}$.

For concretes with strength $f_{ck} \leq 50\,\text{MPa}$, the limit on $x_u/d = 0.328$ for 15% redistribution while no redistribution is permitted when $x_u/d = 0.448$. The limit on amount of redistribution also helps to provide satisfactory performance at the serviceability limit state, where the behaviour may be close to that of the un-cracked elastic analysis. Further checks must be done in any case at serviceability, based on elastic analysis without redistribution. Worked example 5.5-1 illustrates the use of 2-2/clause 5.5(104).

Redistribution without explicit check of rotation capacity in accordance with 2-2/clause 5.5(104) is not permitted in the following situations:

- Members where the reinforcement is Class A (which are deemed to have inadequate ductility). It is additionally recommended in 2-2/clause 3.2.4 that Class A reinforcement should not be used at all for bridges.
- Bridges where the ratio of adjacent span lengths exceeds 2.
- Bridges with elements subject to significant compression.

For the common case of multiple beam and slab decks, redistribution of longitudinal beam moments implies a change in the transverse moments arising from the additional deflections in the main beams. Redistribution cannot therefore be considered for a beam in isolation. One possibility is to force the desired redistribution in such decks by applying imposed settlements to the beam intermediate supports so as to achieve the amount of redistribution required in the main beams. If the deck became overstressed transversely under this redistribution, in principle it would be possible to check rotation capacity in the transverse members as discussed in section 5.6.3 of this guide, so that this additional transverse moment could itself be redistributed. This would become iterative and would involve introduction of plastic hinges into the model.

2-1-1/clause 5.5(1)P requires the 'influence of any redistribution of the moments on all aspects of the design' to be considered. No application rules are given for this principle but it is recommended here that the design shear force at a section should be taken as the greater of that before or after redistribution. This is because shear failure may not be a sufficiently ductile failure mechanism to permit the assumed redistribution of moment. Similarly, reactions used in the design of the substructure should be the greater of those before or after redistribution. Caution with moment redistribution would also be required in integral bridges with soil–structure interaction as the implied deflections from redistribution in the frame could also cause a change in the soil forces attracted.

Redistribution of moments is not permitted in some other specific cases. These are:

- Bridges where the actual rotation capacity could not be calculated with confidence. Examples include curved bridges (where redistribution of flexural moment can lead to an increase of torsion and sudden brittle failure) and similarly skew bridges – *2-2/clause 5.5(105)* refers.
- The design of columns. *2-1-1/clause 5.5(6)* requires columns to be designed for the elastic moments from frame action without redistribution.

2-1-1/clause 5.5(1)P

2-2/clause 5.5(105)

2-1-1/clause 5.5(6)

Worked example 5.5-1: Two-span beam with uniform load
A two-span box girder bridge with concrete cylinder strength $f_{ck} = 40$ MPa is subjected to uniform load W per span and has moments from elastic analysis, as shown in Fig. 5.5-1. In the hog zone, the reinforcement to be provided gives a moment resistance of $0.113\,WL$ which results in $x_u/d = 0.35$. In the sag zone the moment resistance is $0.080\,WL$. Check that the bridge can be shown to be adequate at the ultimate limit state using moment redistribution.

The maximum redistribution possible from the hogging zone is given by 2-2/(5.10a):

$$\delta \geq k_1 + k_2 x_u/d = 0.44 + 1.25(0.6 + 0.0014/0.0035) \times 0.35 = 0.88$$

The ratio of hog moment after redistribution to that before redistribution needed to give adequate bending resistance $= 0.113/0.125 = 0.90 > 0.88$, so sufficient redistribution is available. Redistribution of moment must be achieved by a linear change of the moments between supports, as shown in Fig. 5.5-1, and the increased span moment must be checked. The required reduction in moment at the support is $0.012\,WL$ and the maximum moment in the span will therefore increase by approximately half this moment so the final check is:

Moment at support after redistribution $= 0.113\,WL$ which equals the bending resistance

Moment in span after redistribution $\approx 0.070\,WL + 0.006\,WL = 0.076\,WL < 0.080\,WL$ resistance. The bridge is therefore acceptable at the ultimate limit state. Shear should also be checked both before and after redistribution, as discussed above.

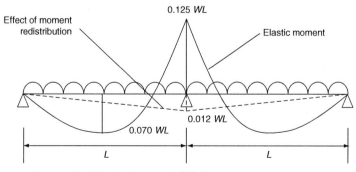

Fig. 5.5-1. Moment diagram for Worked example 5.5-1

5.6. Plastic analysis

5.6.1. General

2-2/clause 5.6.1(101)P permits plastic analysis for verifications at ULS only. Plastic analysis is an extreme case of moment redistribution, where the moments are distributed in accordance with the structure's ability to resist them. This requires a high level of rotation capacity to allow the bridge to deform sufficiently to develop the assumed moments. *2-1-1/clause 5.6.1(2)P* relates to this need to check rotation capacity.

As with the method of moment redistribution, it is necessary to maintain equilibrium between internal action effects and external actions to automatically arrive at a safe solution. This is achieved if a lower-bound (static) analysis is used. However, *2-1-1/clause 5.6.1(103)P* also permits the use of an upper-bound (kinematic) approach, based on balancing internal and external work in assumed collapse mechanisms. Many standard texts are available on this. A commonly used example of this method is yield line analysis for the assessment of slabs. When using an upper-bound approach, it is therefore necessary either to verify that equilibrium is satisfied and that the plastic resistance moment is nowhere exceeded in the assumed mechanism (in which case the 'actual' load resistance has been found), or to consider sufficient collapse mechanisms, such that a very close approximation to the real collapse load is found from the lowest collapse load obtained. The former is impractical for slabs while the latter requires experience to achieve.

By way of 2-2/clause 5.6.1(101)P, national authorities may further restrict the use of plastic methods. It is likely that most will limit the use of plastic analysis to accidental situations only. Some countries might permit its use for deck slab design, although the large number of load cases necessary to consider in bridge design is likely to make plastic analysis less attractive. Consideration of compressive membrane action is often a better way to improve efficiency in deck slabs and it should also be borne in mind that serviceability requirements are likely to become critical.

5.6.2. Plastic analysis for beams, frames and slabs

Two methods are available for ensuring that the bridge possesses the necessary rotation capacity for the use of plastic analysis. These are:

(1) Method without direct check of rotation capacity (permitted by *2-1-1/clause 5.6.2(1)P*).
(2) Method with direct check of rotation capacity.

The first method is discussed here while the second is discussed in section 5.6.3 below.

Method without direct check of rotation capacity

Adequate rotation capacity for plastic analysis is deemed to be achieved in accordance with *2-2/clause 5.6.2(102)* if the depth of the compression zone is everywhere restricted to a depth as follows:

$$x_u/d \leq 0.15 \quad \text{for} \quad f_{ck} \leq 50\,\text{MPa} \tag{D5.6-1}$$

$$x_u/d \leq 0.10 \quad \text{for} \quad f_{ck} \geq 55\,\text{MPa} \tag{D5.6-2}$$

The ratio of moments at intermediate supports to those in adjacent spans must also lie between 0.5 and 2.0 and, for bridges, Class A reinforcement must not be used due to its low ductility. If the above criteria are met, the effects of imposed deformations, such as those from settlement and creep, may be neglected at the ultimate limit state, as discussed in section 2.3 of this guide.

The restrictions on compression depth are more onerous than the corresponding ones for buildings in EC2-1-1 because of the greater depth of typical bridge members and the lack of test results for them. However, for solid slabs, the limits in EC2-1-1 can be used, which are as follows:

$$x_u/d \leq 0.25 \quad \text{for} \quad f_{ck} \leq 50\,\text{MPa} \tag{D5.6-3}$$

$$x_u/d \leq 0.15 \quad \text{for} \quad f_{ck} \geq 55\,\text{MPa} \tag{D5.6-4}$$

5.6.3. Rotation capacity

5.6.3.1. Method with direct check of rotation capacity (additional sub-section)

If the criteria in 2-2/clause 5.6.2 are not fulfilled, it is necessary to verify the plastic rotation capacity of the bridge against the actual rotation implicit in the analysis. The rules in 2-2/clause 5.6.3 apply to continuous beams and one-way spanning slabs. They cannot be applied to yield line analyses of two-way spanning slabs.

The plastic rotation capacity can be derived by integration of the plastic curvature along the length, L_p, of the beam where the reinforcement strain exceeds that at first yield. This plastic length is determined by the difference in section moment from first yield to final rupture (as shown in Fig. 5.6-1) and also by any shift in the tensile force caused by shear truss action, as discussed in section 6.2 of this guide. (The latter is not shown in Fig. 5.6-1, but, from the shift method, would give a length of beam where the steel is yielding at least equal to the effective depth of the beam.) The plastic rotation capacity then follows from:

$$\theta_{\text{pl,d}} = \int_{-L_p/2}^{L_p/2} \frac{\Delta\varepsilon(a)}{d - x(a)} \, da \qquad \text{(D5.6-5)}$$

where:

$\Delta\varepsilon(a)$ is the mean reinforcement strain in excess of that at first yield
d is the effective depth
$x(a)$ is the depth of the compression zone
a represents the longitudinal position within the length L_p

The mean reinforcement strain at each position, $\Delta\varepsilon(a)$, should include the effects of tension stiffening as this reduces the rotation capacity. A method of accounting for the effects of tension stiffening in this way is given in reference 6.

It is not, however, necessary to perform the integration in (D5.6-5) since 2-1-1/Fig. 5.6N (reproduced as Fig. 5.6-2) gives simplified values for $\theta_{\text{pl,d}}$ which link the plastic rotation capacity to x_u/d for different reinforcement ductility and concrete strength classes – **2-1-1/ clause 5.6.3(4)** refers. It is conservatively based on a length deforming plastically of approximately 1.2 times the depth of the section and is produced for a 'shear slenderness', $\lambda = 3.0$.

The shear slenderness itself gives a measure of the length of the plastic zone. It is defined as the distance between the points of maximum and zero moment after redistribution, divided by the effective depth. The distance between the points of maximum and zero moment is typically 15% of the span, L. With this assumption, a shear slenderness of 3.0 equates to a span to depth ratio of 20, which is typical for continuous construction. The assumed plastic length used in 2-1-1/Fig. 5.6N relates to this shear slenderness. For other values of λ, the plastic rotation capacity from 2-1-1/Fig. 5.6N thus needs to be corrected by multiplying by a factor $k_\lambda = \sqrt{\lambda/3}$. For intermediate concrete strengths, the plastic rotation capacity can be obtained by interpolation.

The maximum curvature in the section, and hence rotation capacity, occurs when the section is balanced such that both reinforcement and concrete reach their failure strain simultaneously. This accounts for the maxima in the values of rotation capacity in

2-1-1/clause 5.6.3(4)

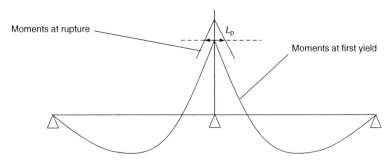

Fig. 5.6-1. Length of beam undergoing plastic deformation

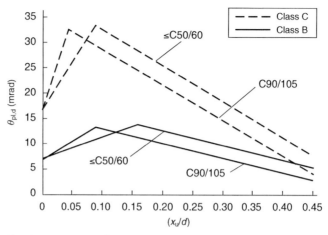

Fig. 5.6-2. Allowable plastic rotation, $\theta_{pl,d}$, of reinforced concrete sections for Class B and C reinforcement ($\lambda = 3.0$)

Fig. 5.6-2. At high x_u/d, failure of the concrete may occur shortly after yielding of the reinforcement so the class of reinforcement has little effect on rotation capacity. At low values of x_u/d, the reinforcement fails before the concrete reaches its failure strain. These situations are illustrated in Fig. 5.6-3.

A further absolute limit is placed on the depth of the compression block at plastic hinge locations by *2-2/clause 5.6.3(102)* as follows:

2-2/clause
5.6.3(102)

$$x_u/d \leq 0.30 \quad \text{for} \quad f_{ck} \leq 50\,\text{MPa} \tag{D5.6-6}$$

$$x_u/d \leq 0.23 \quad \text{for} \quad f_{ck} \geq 55\,\text{MPa} \tag{D5.6-7}$$

These are again more onerous than corresponding limits in EC2-1-1 due to the greater potential depth of bridge beams.

The plastic rotation capacity has to be compared with the actual rotation at the hinge implied from the global analysis. *2-1-1/clause 5.6.3(3)* states only that the rotation should be calculated using design values for materials. One possibility is to use elastic analysis to determine the fraction of the load, α, at which the first plastic hinge forms (that is when the moment resistance is just exceeded at the hinge location) and then to apply the remaining load increments to a model with the plastic hinge modelled as a hinge without any moment rigidity. This determination of the plastic rotation, θ_s, is shown in Fig. 5.6-4.

2-1-1/clause
5.6.3(3)

The question then arises as to what stiffness to use for the as yet unyielded parts of the bridge in Fig. 5.6-4 for a uniformly distributed load. If gross elastic cross-section properties are used, based on the design value of the concrete, Young's modulus $E_{cd} = E_{cm}/\gamma_{cE}$ with $\gamma_{cE} = 1.2$ (from 2-1-1/clause 5.8.6(3)), this will overestimate the real stiffness and therefore underestimate rotation at the hinge. A reasonable approximation might be to use fully cracked properties, again based on the design value of the concrete, Young's modulus

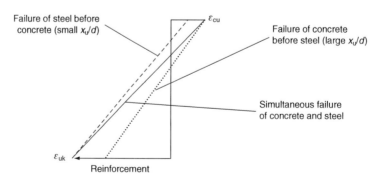

Fig. 5.6-3. Possible strain diagrams at failure

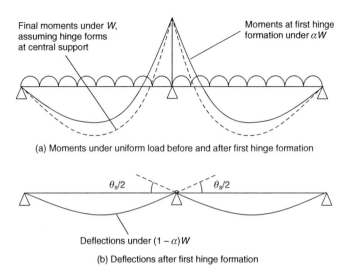

Fig. 5.6-4. Rotation at plastic hinge for two-span bridge beam with uniformly distributed load, W

$E_{cd} = E_{cm}/\gamma_{cE}$ and steel modulus E_s. This could still overestimate stiffness in the most highly stressed areas where another hinge was about to form, but would underestimate stiffness everywhere else. To minimize any un-conservatism in this respect, it is desirable to base the section bending resistance here on the reinforcement diagram with the flat yield plateau in 2-1-1/Fig. 3.8. The 'real' behaviour can only be obtained through non-linear analysis considering the 'real' material behaviour.

5.6.3.2. Check of rotation capacity when neglecting imposed deformations at the ultimate limit state (additional sub-section)

A similar calculation of rotation capacity and plastic hinge rotation has to be made when elastic analysis is used, but the effects of imposed deformations are to be ignored at the ultimate limit state as discussed in section 2.3 of this guide. The rotation caused by, for example, settlement can be checked on the basis of the angular change produced if the settlement is applied to a model with a hinge at the location where rotation capacity is being checked in a similar way to that in Fig. 5.6-4. In general (for differential temperature or differential shrinkage for example), the plastic rotations can be obtained from the 'free' displacements as shown in Fig. 5.6-5, or by again applying the free curvatures to a model with hinged supports.

In verifying the rotation capacity in this way, it should be recognized that even the use of elastic global analysis may place some demands on rotation capacity – see section 5.4 of this guide. The entire plastic rotation capacity may not therefore be available for the above check, so it is advisable to leave some margin between plastic rotation capacity and plastic rotation due to the imposed deformation. Generally, the proportion of total rotation capacity required for this check will be small and adequate by inspection. It has, in any case, been common practice in the UK to ignore imposed deformations at the ultimate limit state without an explicit check of rotation capacity. EC2, however, demands more caution, particularly as there are no restrictions on designing reinforced concrete beams with over-reinforced behaviour.

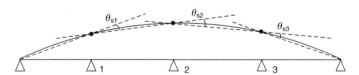

Fig. 5.6-5. Plastic rotations caused by imposed curvatures such as from differential temperature

If the limits in (D5.6-1) or (D5.6-2) are met, the effect of imposed deformations could automatically be neglected. Alternatively, 2-2/Expression (5.10a) or 2-2/Expression (5.10b) could be used where it was only necessary to shed 15% of the moment (or other limit as specified in the National Annex).

5.6.4. Strut-and-tie models

Analysis with strut-and-tie models is a special case of the application of the lower-bound theorem of plasticity. Strut-and-tie models have not been commonly used by UK engineers, mainly because of the lack of codified guidance. When such models have been used, as for example in the design of diaphragms in box girder bridges, there has not been a consistent approach used for the check of the strength of compression struts and nodes. EC2 now provides guidance on these limits, but it is far from a complete guide in itself and reference can usefully be made to texts such as reference 8 for more background. There will often be difficulties in applying specific rules from EC2 for compression limits in nodes and struts, and engineering judgement will still be needed. However, the use of strut-and-tie modelling is still very valuable, even when used simply to determine the locations and quantities of reinforcement.

2-1-1/clause 5.6.4(1)

2-1-1/clause 5.6.4 gives general guidance on the use of strut-and-tie models. Strut-and-tie models are intended to be used in areas of non-linear strain distribution unless rules are given elsewhere in EC2. Such exceptions include beams with short shear span which are covered in section 6.2. In this particular case, the use of the strut-and-tie rules in preference to the test-based shear rules would lead to a very conservative shear resistance based on concrete crushing.

Typical examples where non-linear strain distribution occurs are areas where there are concentrated loads, corners, openings or other discontinuities. These areas are often called 'D-regions', where 'D' stands for discontinuity, detail or disturbance. Outside these areas, where the strain distribution again becomes linear, stresses may be derived from traditional beam or truss theory depending on whether the concrete is cracked or uncracked respectively. These areas are known as 'B-regions', where 'B' stands for Bernoulli or beam. EN 1992 also refers to them as 'continuity' regions. Some typical D-regions, together with their approximate extent, are shown in Fig. 5.6-6. Strut-and-tie models are best developed by following the flow of elastic force from the B-region boundaries as shown in Fig. 5.6-6, or from other boundary conditions, such as support reactions, when the entire system is a D-region (such as a deep beam). *2-1-1/clause 5.6.4(1)* indicates that strut-and-tie modelling can be used for both continuity and discontinuity regions.

2-1-1/clause 5.6.4(3)

Strut-and-tie modelling makes use of the lower-bound theorem of plasticity which states that any distribution of stresses used to resist a given applied loading is safe, as long as equilibrium is satisfied throughout and all parts of the structure have stresses less than 'yield'. Equilibrium is a fundamental requirement of *2-1-1/clause 5.6.4(3)*. In reality, concrete has limited ductility so it will not always be safe to design any arbitrary force system using this philosophy. Since concrete permits limited plastic deformations, the force system has to be chosen in such a way as to not exceed the deformation limit anywhere before the assumed state of stress is reached in the rest of the structure. In practice, this is best achieved by aligning struts and ties to follow the internal forces predicted by an un-cracked elastic analysis. To this end, it may sometimes be advisable to first model the region with finite elements to establish the flow of elastic forces before constructing the strut-and-tie model. This is the basis of *2-1-1/clause 5.6.4(5)*.

2-1-1/clause 5.6.4(5)
2-1-1/clause 5.6.4(2)

The advantage of closely following the elastic behaviour in choosing a model is that the same analysis can then be used for both ultimate and serviceability limit states. *2-1-1/clause 5.6.4(2)* requires orientation of the struts in accordance with elastic theory. The stress limit to use at the serviceability limit state for crack control may be chosen according to bar size or bar spacing, as discussed in section 7 of this guide.

Failure to follow the elastic flow of force and overly relying on the lower-bound theorem can result in resistance being overestimated. This can occur for example in plain concrete

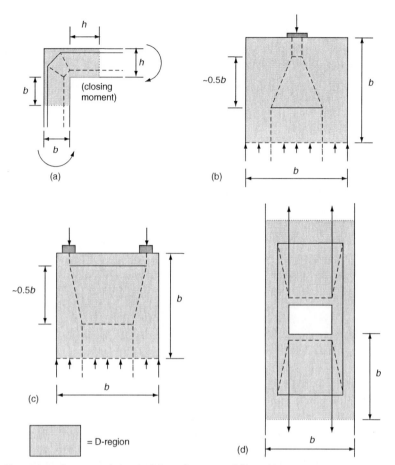

Fig. 5.6-6. Examples of strut-and-tie models and extent of D-regions

under a concentrated vertical load if the load is applied to a small width. Neglecting the transverse tensions generated, as shown in Fig. 6.5-1, by assuming that the load does not spread across the section, can actually lead to an overestimate of resistance. This particular case is discussed at length in section 6.7 of this guide. Similar problems can arise if the assumed strut angles depart significantly from the elastic trajectories, as discussed in section 6.5.2.

Experience, however, shows that it is not always necessary to rigidly follow the elastic flow of force at the ultimate limit state. The most obvious example is the truss model for reinforced concrete shear design, which permits considerable departure of both reinforcement and compression strut directions from the principal stress directions of 45° to the vertical at the neutral axis. Generally, it is desirable to follow the lines of force from elastic analysis, unless experience shows that it is unnecessary to do so.

Often there appears to be a choice of model even when the elastic load paths have been followed. In selecting the best solution, it should be borne in mind that the loads in the real structure will try to follow the paths involving least force and deformation. Since reinforcement is much more deformable than the stiff concrete struts, the best model will minimize the number and length of the ties. An optimization criterion is given in reference 8. This requires the minimization of the internal strain energy, $\sum F_i L_i \varepsilon_{mi}$, where:

F_i = force in strut or tie i
L_i = length of strut or tie i
ε_{mi} = mean strain in strut or tie i

The terms for the concrete struts can usually be ignored as their strains are usually much smaller than those of the ties. As a guide to constructing models, in highly stressed node regions (e.g. near concentrated loads) the compression struts and ties should form an angle of about 60° and not less than 45°.

Detailing of nodes is also important and guidance is given in section 6.5.4. This guidance also applies to the design of local areas subject to concentrated loads even if the design is not performed using strut-and-tie analysis.

5.7. Non-linear analysis

2-1-1/clause 5.7(1)

Non-linear analysis of concrete bridges, in the context of *2-1-1/clause 5.7(1)*, accounts for the non-linear nature of the material properties. This includes the effects of cracking in concrete and non-linearity in the material stress–strain curves. The analysis then ensures both equilibrium and compatibility of deflections using these material properties. This represents the most realistic representation of structural behaviour, provided that the material properties assumed are realistic. Such an analysis may then be used at SLS and ULS. Non-linear analysis may also model non-linearity in structural response due to the changing geometry caused by deflections. This is important in the design of elements where second-order effects are significant, as discussed in section 5.8 of this guide. It is for second-order calculation for slender members, such as bridge piers, that non-linear analysis is often particularly beneficial, as the simplified alternatives are usually quite conservative. This is discussed in section 5.8.6.

2-1-1/clause 5.7(4)P

2-1-1/clause 5.7(4)P requires the analysis to be performed using 'realistic' values of structural stiffnesses and a method which takes account of uncertainties in the resistance model. The most 'realistic' values are the mean properties as these are the properties expected to be found in the real structure. If mean strengths and stiffnesses are used in analysis, rather than design values, there is an apparent incompatibility between local section design and overall analysis. The rationale often given for this is that material factors are required to account for bad workmanship. It is unlikely that such a severe drop in quality would affect the materials in large parts of the structure. It is more likely that this would be localized and, as such, would not significantly affect global behaviour, but would affect the ability of local sections to resist the internal effects derived from the global behaviour. This is noted by *2-1-1/clause 5.7(2)*, which requires local critical sections to be checked for inelastic behaviour.

2-1-1/clause 5.7(2)

For the buckling of columns, however, it could be argued that even a relatively small area of 'design material' at the critical section could significantly increase deflections and hence the moments at the critical section. Care and experience is therefore needed in selecting appropriate material properties in different situations.

2-1-1/clause 5.7(3)

2-1-1/clause 5.7(3) generally permits load histories to be ignored for structures subjected predominantly to static load and all the actions in a combination may then be applied by increasing their values simultaneously.

5.7.1. Method for ultimate limit states (additional sub-section)

2-2/clause 5.7(105)

2-2/clause 5.7(105) makes a proposal for the properties to use in non-linear analysis for ultimate limit states and provides a safety format. The proposed method, which may be amended in the National Annex, essentially uses mean properties for steel and a reference strength for concrete of $0.84f_{ck}$. The material stress–strain responses are derived by using the strengths given below in conjunction with the non-linear concrete stress–strain relationship given in 2-1-1/clause 3.1.5 (Fig. 3.2), the reinforcement stress–strain relationship for curve A in 2-1-1/clause 3.2.7 (Fig. 3.8) and the prestressing steel stress–strain relationship for curve A in 2-1-1/clause 3.3.6 (Fig. 3.10):

2-1-1/Fig. 3.2 for concrete: f_{cm} is replaced by $1.1\dfrac{\gamma_s}{\gamma_c}f_{ck}$

2-1-1/Fig. 3.8 for reinforcement: f_{yk} is replaced by $1.1f_{yk}$

kf_{yk} is replaced by $1.1kf_{yk}$

2-1-1/Fig. 3.10 for prestressing steel: f_{pk} is replaced by $1.1f_{pk}$

CHAPTER 5. STRUCTURAL ANALYSIS

These modifications are necessary to make the material characteristics compatible with the verification format given, which uses a single value of material safety factor, $\gamma_{O'}$, to cover concrete, reinforcement and prestressing steel. This can be seen as follows.

For reinforcement failure, the strength used above corresponds approximately to a mean value and is taken as $f_{ym} = 1.1 f_{yk}$, and since the *design* ultimate strength is $f_{yd} = f_{yk}/1.15$ the equivalent material factor to use with $f_{ym} = 1.1 f_{yk}$ is $\gamma_{O'} = 1.1 \times 1.15 = \mathbf{1.27}$.

For concrete failure, the reference strength for analysis is taken as $f_c = 1.1 \times (1.15/1.5) f_{ck} = 0.843 f_{ck}$, and since the *design* ultimate strength is $f_{cd} = f_{ck}/1.5$ the equivalent material factor to use with $f_c = 0.843 f_{ck}$ is $\gamma_{O'} = 0.843 \times 1.5 = \mathbf{1.27}$. The concrete reference strength is therefore not a mean strength but one that is necessary to give the same material factor for concrete and steel.

The above concrete strengths do not include the factor α_{cc}, which is needed for the resistance. To adopt the same global safety factor as above and include α_{cc} in the resistance, it must also be included in the analysis. This implies that f_{cm} should be replaced by $1.1(\gamma_s/\gamma_c)\alpha_{cc} f_{ck}$ in 2-1-1/Fig. 3.2, contrary to the above. This change appeared to have been agreed by the EC2-2 Project Team on several occasions but failed to be made in the final text.

Although not stated, further modification is required to the stress–strain curves where there is significant creep. The stress–strain curve of 2-1-1/Fig. 3.2 for concrete (and Figs 3.3 and 3.4) is for short-term loading. Creep will therefore have the effect of making the response to long-term actions more flexible. This can conservatively be accounted for by multiplying all strain values in the concrete stress–strain diagram by a factor $(1 + \phi_{ef})$, where ϕ_{ef} is the effective creep ratio discussed in section 5.8.4 of this guide. This has the effect of stretching the stress–strain curve along the strain axis.

If the analysis is performed until the ultimate strength is reached at one location (based on the above material properties) such that the maximum combination of actions reached is $q_{ud} = \alpha_{Ud}(\gamma_G G + \gamma_Q Q)$, where $\gamma_G G + \gamma_Q Q$ is the applied design combination of actions and α_{Ud} is the load factor on these design actions reached in the analysis, then the safety verification on load would be:

$$\gamma_G G + \gamma_Q Q \leq \frac{\alpha_{Ud}(\gamma_G G + \gamma_Q Q)}{\gamma_{O'}} \tag{D5.7-1}$$

The above verification based on applied load is modified in EC2-2 so as to be based on internal actions and resistances. This is done in order to:

(1) Distinguish between over-proportional, linear and under-proportional behaviour which a global safety factor applied to the maximum load cannot do as it takes no account of the path by which the ultimate load was reached.
(2) Introduce the effects of model uncertainties separately on both internal actions and resistances.

In respect of the second point, the material factor $\gamma_{O'}$ includes a partial factor γ_{Rd} which accounts for uncertainties in the resistance model used together with uncertainties in the geometric imperfections modelled, such that $\gamma_{O'} = \gamma_{Rd} \gamma_O$. As a result, EC2-2 provides the following safety verification:

$$\gamma_{Rd} E(\gamma_G G + \gamma_Q Q) \leq R\left(\frac{q_{ud}}{\gamma_O}\right) \qquad \text{2-2/(5.102 aN)}$$

where $R(q_{ud}/\gamma_O)$ is the material resistance corresponding to the combination of actions q_{ud}/γ_O and $E(\gamma_G G + \gamma_Q Q)$ are the internal actions under the design combination of actions $\gamma_G G + \gamma_Q Q$. The suggested values of γ_O and γ_{Rd} are 1.20 and 1.06 respectively such that $\gamma_{O'} = \gamma_{Rd} \gamma_O = 1.06 \times 1.20 = 1.27$ as discussed above, thus giving the same material factor for steel and concrete.

At this point it should be noted that the load factors γ_Q and γ_G themselves contain a partial factor γ_{Sd}, which represents error due to structural modelling uncertainties, such that $\gamma_Q = \gamma_{Sd} \gamma_q$ and $\gamma_G = \gamma_{Sd} \gamma_g$. This then gives the possibility in the non-linear analysis of applying the partial factor γ_{Sd} to either the action itself or to the action effect it produces. The former possibility is catered for in 2-2/Expression (5.102 aN), but a further inequality is provided to cater for the second possibility:

$$\gamma_{Rd}\gamma_{Sd}E(\gamma_g G + \gamma_q Q) \leq R\left(\frac{q_{ud}}{\gamma_O}\right) \qquad \text{2-2/(5.102 cN)}$$

The value of γ_{Sd} is suggested to be taken as 1.15.

2-2/clause 5.7(105) allows either 2-2/Expression (5.102 aN) or 2-2/Expression (5.102 cN) to be used (or a third formulation in 2-2/Expression (5.102 bN) below). Clause 6.3.2 of EN 1990 effectively says the factor γ_{Sd} should be applied to the loading where the resulting action effect increases at a faster rate than the loading, and should be applied to the action effect where this increases at a lower rate than the loading. EN 1990 would therefore appear to suggest that both inequalities should be verified, although this seems overly onerous.

The basic procedure for carrying out the non-linear analysis and verifying the structure is then as follows:

(1) Determine the maximum value of action combination q_{ud} reached in the non-linear analysis, which corresponds to the attainment of the ultimate strength $R(q_{ud})$ in one region of the structure (based on the analysis material properties) or instability for second-order calculations.
(2) Apply a global safety factor γ_O to the ultimate structural strength $R(q_{ud})$ to get $R(q_{ud}/\gamma_O)$.
(3) Apply 2-2/Expression (5.102 aN) or 2-2/Expression (5.102 cN) for the global safety verification.

It is not obvious how to apply these inequalities in all situations so 2-2/Annex PP attempts to illustrate their use. The method is simplest where there is only one action effect contributing to failure (scalar combination – typically a beam in bending). The alternative is that several action effects contribute to failure (vector combination – typically a column under bending and axial force). These are both explained further below where it is noted that it will often not be clear which part of the structure is the critical element with respect to attaining the ultimate strength $R(q_{ud})$ in one region of the structure. This tends to favour the use of simpler, more tried and tested methods of analysis.

A late addition to the drafts of EN 1992-2 was the verification format of 2-2/Expression (5.102bN):

$$E(\gamma_G G + \gamma_Q Q) \leq R\left(\frac{q_{ud}}{\gamma_{O'}}\right) \qquad \text{2-2/(5.102 bN)}$$

This follows the simpler format of (D5.7-1), having only one safety factor on the resistance side, and is simpler to interpret and use. It does not, however, follow item 2 above, taken from 2-2/clause 5.7(105) itself.

It should be noted that other methods of non-linear analysis are possible, including analysis with mean properties and subsequent section by section resistance checks against the internal actions from the global analysis, or use of design properties throughout. The latter is discussed at the end of this section. The former is often appropriate for serviceability analysis.

5.7.2. Scalar combinations (additional sub-section)

2-2/clause PP.1(101)

Scalar combinations of internal actions are covered by **2-2/clause PP.1(101)**. The method of application of the inequalities (except for 2-2/Expression (5.102 bN) which is simpler) is illustrated in Fig. 5.7-1 for under-proportional behaviour. The case of over-proportional behaviour is similar. The ultimate strength in one area of the structure is reached at point A, which corresponds to the action combination at point B. This action combination is reduced by the global safety factor γ_O to give the reduced combination of actions at point C. This corresponds to a new point on the internal action path at point D which defines a

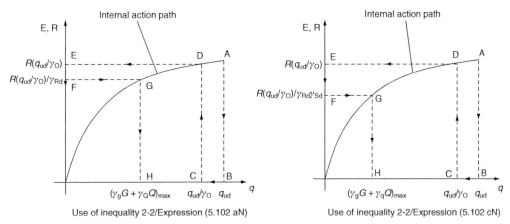

Fig. 5.7-1. Safety format for scalar under-proportional behaviour

reduced resistance given by point E. For inequality 2-2/Expression (5.102 aN), the resistance at point E is reduced further by γ_{Rd} to give point F, which corresponds to point G on the internal action path. Point G corresponds to the final maximum permissible value of the combination of actions at point H, which must be greater than or equal to the actual value of the load combination $\gamma_G G + \gamma_Q Q$. For inequality 2-2/Expression (5.102 cN), the resistance at point E is reduced further by $\gamma_{Rd}\gamma_{Sd}$ to give point F, which corresponds to point G on the internal action path. Point G corresponds to the final maximum allowable value of the load combination at point H and this must be greater than or equal to the actual value of the combination of actions $\gamma_g G + \gamma_q Q$.

5.7.3. Vector combinations (additional sub-section)

Vector combinations of internal actions are covered by **2-2/clause PP.1(102)**. The method of application of the inequalities is illustrated in Fig. 5.7-2 for over-proportional behaviour and the inequality of 2-2/Expression (5.102 aN). The case of under-proportional behaviour is similar. The ultimate strength in one area of the structure is reached at point A. For simple beam-like structures, it is possible, by section analysis, using the same material properties as in the non-linear analysis, to construct the failure surface a, which defines local failure for all combinations of M_{Ed} and N_{Ed}. The *applied* action combination is reduced by the global safety factor γ_O so that the internal actions reduce following the internal action path to give the reduced internal actions at point B.

The effect of uncertainties in the material resistance model is next taken into account by reducing both M_{Ed} and N_{Ed} at B by γ_{Rd} to give point C. (The reduction is made by reducing the length of the vector $\overline{OB}$ by γ_{Rd}.) Point C will not generally lie on the internal action path, whereas the point corresponding to actual design failure must do so. To determine point D, corresponding to the maximum allowable design combination of actions on the internal action path, it is necessary to construct part of the design failure surface, b. This surface has exactly the same shape as surface a but is scaled down radially everywhere by the ratio $\overline{OC}/\overline{OC'}$. Point D lies on the intersection of this surface and the internal action path. The final verification is then performed by ensuring that the final maximum value of the load combination $\gamma_G G + \gamma_Q Q$ lies inside the failure surface b, i.e. point D is not reached under the application of $\gamma_G G + \gamma_Q Q$. The same procedure is used for inequality 2-2/Expression (5.102 cN), except γ_{Rd} is replaced by $\gamma_{Rd}\gamma_{Sd}$ and $\gamma_G G + \gamma_Q Q$ is replaced by $\gamma_g G + \gamma_q Q$.

A major criticism of this method is that it is not clear how it should be applied when the critical section, defining the attainment of the ultimate strength $R(q_{ud})$ in one region of the structure, is not readily identifiable. This will not usually apply to analysis of a simple element, such as a single column, but it usually will to the analysis of a slab. It will also be noted that this procedure is rather lengthy. It will therefore often be much simpler to use design values directly in the structural analysis so that the bridge resistance is verified

2-2/clause PP.1(102)

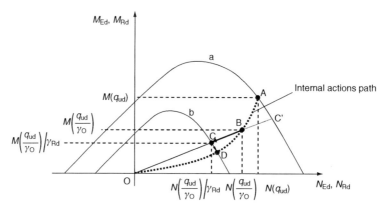

Fig. 5.7-2. Safety format for vector over-proportional behaviour using inequality 2-2/Expression (5.102 aN)

directly in the analysis. Where second-order effects are important, as in the design of slender piers at the ultimate limit state, it will also be more conservative to use design values directly in this way. It will not, however, always be conservative to do this, particularly at SLS when there are imposed displacements as the structural response is then artificially flexible. This is noted in Note 2 to 2-2/clause 5.7(105). The use of design properties in global analysis is discussed further in section 5.8.6 of this guide.

5.7.4. Method for serviceability limit states (additional sub-section)

For global analysis at the serviceability limit state, the concrete stress–strain curve of 2-1-1/Fig. 3.2 could be used directly without the modification proposed in 2-2/clause 5.7(105). Creep should be accounted for using ϕ_{ef} according to section 5.8.4, except that M_{0Ed} should be taken applicable to the serviceability limit state.

5.8. Analysis of second-order effects with axial load

5.8.1. Definitions and introduction to second-order effects

Second-order effects are additional action effects caused by the interaction of axial forces and deflections under load. First-order deflections lead to additional moments caused by the axial forces and these in turn lead to further increases in deflection. Such effects are also sometimes called $P - \Delta$ effects because additional moments are generated by the product of the axial force and element or system deflections. The simplest case is a cantilevering pier with axial and horizontal forces applied at the top, as in Fig. 5.8-1. Second-order effects can be calculated by second-order analysis, which takes into account this additional deformation.

Second-order effects apply to both 'isolated' members (for example, as above or in Fig. 5.8-2(a)) and to overall bridges which can sway involving several members (Fig. 5.8-2(b)). EC2 refers to two types of isolated member. These are:

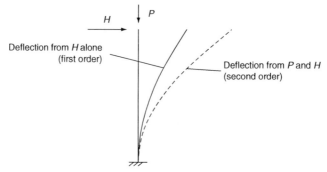

Fig. 5.8-1. Deflections for an initially straight pier with transverse load

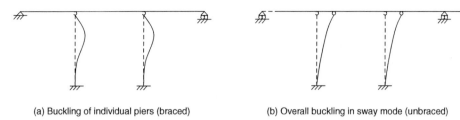

(a) Buckling of individual piers (braced) (b) Overall buckling in sway mode (unbraced)

Fig. 5.8-2. Buckling modes for braced and unbraced members

Braced members – members that are held in position at both ends and which may or may not have restraining rotational stiffness at the ends. An example is a pin-ended strut. The effective length for buckling will always be less than or equal to the actual member length.

Unbraced members – members where one end of the member can translate with respect to the other and which have restraining rotational stiffness at one or both ends. An example is the cantilevering pier above. The effective length for buckling will always be greater than or equal to the actual member length.

The compression members of complete bridges can often be broken down into equivalent isolated members that are either 'braced' or 'unbraced' by using an effective length and appropriate boundary conditions. This is discussed in section 5.8.3.

Some engineers may be unfamiliar with performing second-order analysis, which is the default analysis in the Eurocodes. A significant disadvantage of second-order analysis is that the principle of superposition is no longer valid and all actions must be applied to the bridge together with all their respective load and combination factors. Fortunately, there will mostly be no need to do second-order analysis, as alternative methods are given in this chapter and frequently second-order effects are small in any case and may therefore be neglected. Some rules for when second-order effects may be neglected are given in clauses 5.8.2 and 5.8.3 and are discussed below.

Slender compression elements are most susceptible to second-order effects. (The definition of slenderness is discussed in section 5.8.3.) The degree of slenderness and magnitude of second-order effects are related to well-known elastic buckling theory. Although elastic buckling in itself has little direct relevance to real design, it gives a good indication of susceptibility to second-order effects and can be used as a parameter in determining second-order effects from the results of a first-order analysis; section 5.8.7 refers.

Second-order analysis can be performed with most commercially available structural software. In addition to the problem of lack of validity of the principle of superposition, the flexural rigidity of reinforced concrete structures, EI, is not constant. For a given axial load, EI reduces with increasing moment due to cracking of the concrete and inherent non-linearity in the concrete stress–strain response. This means that second-order analysis for reinforced concrete elements is non-linear with respect to both geometry and material behaviour. This non-linearity has to be taken into account in whatever method is chosen to address second-order effects. This is dealt with in subsequent sections.

For bridges, it is slender piers that will most commonly be affected by considerations of second-order effects. Consequently, all the Worked examples in this section relate to slender piers. The provisions, however, apply equally to other slender members with significant axial load, such as pylons and decks of cable-stayed bridges.

5.8.2. General

When second-order calculations are performed, it is important that stiffnesses are accurately determined as discussed above. *2-1-1/clause 5.8.2(2)P* therefore requires the analysis to consider the effects of cracking, non-linear material properties and creep. This can be done either through a materially non-linear analysis (as discussed in section 5.8.6) or by using linear material properties based on a reduced secant stiffness (as discussed in section 5.8.7). Imperfections must be included as described in section 5.2 as these lead to additional

2-1-1/clause 5.8.2(2)P

first-order moments and consequently additional second-order moments in the presence of axial compression.

2-1-1/clause 5.8.2(3)P

Soil–structure interaction must also be taken into account (*2-1-1/clause 5.8.2(3)P*), as it should be in a first-order analysis. There are few rules specific to the analysis of integral bridges, but the slenderness of integral bridge piers can be determined using the general procedures discussed in section 5.8.3 of this guide.

2-1-1/clause 5.8.2(4)P

2-1-1/clause 5.8.2(4)P requires the structural behaviour to be considered 'in the direction in which deformation can occur, and biaxial bending shall be taken into account when necessary'. Often, deformation in two orthogonal directions needs to be considered in bridge design under a given combination of actions, although the moments in one direction may be negligible compared to the effect of moments in the other. A related clause is *2-1-1/clause 5.8.2(5)P* which requires geometric imperfections to be considered in accordance with clause 5.2. 2-1-1/clause 5.8.9(2) states that imperfections need only be considered in one direction (the one that has the most unfavourable effect), so biaxial bending conditions need not always be produced simply due to considerations of imperfections.

2-1-1/clause 5.8.2(5)P

2-1-1/clause 5.8.2(6)

It will not always be necessary to consider second-order effects. *2-1-1/clause 5.8.2(6)* permits second-order effects to be ignored if they are less than 10% of the corresponding first-order effects. This is not a very useful criterion as it is first necessary to perform a second-order analysis to check compliance. As a result, 2-1-1/clause 5.8.3 provides a simplified criterion for isolated members based on a limiting slenderness. This is discussed below.

5.8.3. Simplified criteria for second-order effects
5.8.3.1. Slenderness criterion for isolated members

Where simplified methods are used to determine second-order effects, rather than a second-order non-linear computer analysis, the concept of effective length can be used to determine slenderness. This slenderness can then be used to determine whether or not second-order effects need to be considered. According to 2-1-1/clause 5.8.3.2(1), the slenderness ratio is defined as follows:

$$\lambda = l_0/i \qquad \text{2-1-1/(5.14)}$$

where l_0 is the effective length and i is the radius of gyration of the uncracked concrete cross-section.

A simplified criterion for determining when second-order analysis is not required is given in *2-1-1/clause 5.8.3.1(1)*, based on a recommended limiting value of the slenderness λ as follows:

2-1-1/clause 5.8.3.1(1)

$$\lambda \leq \lambda_{\text{lim}} = 20A \cdot B \cdot C/\sqrt{n} \qquad \text{2-1-1/(5.13N)}$$

This limiting slenderness may be modified in the National Annex. $n = N_{\text{Ed}}/(A_c f_{\text{cd}})$ is the relative normal force. The greater the axial force and thus n become, the more the section will be susceptible to the development of second-order effects and, consequently, the lower the limiting slenderness becomes. Higher limiting slenderness can be achieved where:

- there is little creep (because the stiffness of the concrete part of the member in compression is then higher);
- there is a high percentage of reinforcement (because the total member stiffness is then less affected by the cracking of the concrete);
- the location of the peak first-order is not the same as the location of the peak second-order moment.

These effects are accounted for by the terms A, B and C respectively where:

$$A = 1/(1 + 0.2\phi_{\text{ef}}) \qquad \text{(D5.8-1)}$$

where ϕ_{ef} is the effective creep ratio according to 2-1-1/clause 5.8.4. If ϕ_{ef} is not known, A may be taken as 0.7. This corresponds approximately to $\phi_{ef} = 2.0$ which would be typical of a concrete loaded at relatively young age, such that $\phi_\infty = 2.0$, and with a loading that is entirely quasi-permanent. This is therefore reasonably conservative. A is not, in any case, very sensitive to realistic variations in ϕ_{ef}, so using the default value of 0.7 is reasonable.

$$B = \sqrt{1 + 2\omega} \qquad (D5.8\text{-}2)$$

where $\omega = A_s f_{yd}/(A_c f_{cd})$ is the mechanical reinforcement ratio. If ω is not known, B may be taken as 1.1 which is equivalent to $\omega = 0.1$. This value would usually be achieved in a slender column, but is slightly generous compared to the actual minimum reinforcement required by 2-1-1/clause 9.5.2(2):

$$C = 1.7 - r_m \qquad (\,(D5.8\text{-}3)$$

where r_m is the moment ratio M_{01}/M_{02}, where M_{01} and M_{02} are the first-order end moments such that $|M_{02}| \geq |M_{01}|$. If r_m is not known, C may be taken as 0.7 which corresponds to uniform moment throughout the member. C should also be taken as 0.7 where there is transverse loading, where first-order moments are predominantly due to imperfections or where the member is unbraced. The reasons for this are explained in section 5.8.7 of this guide.

Before carrying out a non-linear analysis or using the simplified methods of 5.8.7 and 5.8.8, it would be usual to check whether such effects can be ignored using 2-1-1/Expression (5.13N). The use of this formula is illustrated in the Worked example 5.8-2.

5.8.3.2. Slenderness and effective length of isolated members

2-1-1/clause 5.8.3.2 gives methods of calculating effective lengths for isolated members. Typical examples of isolated members and their corresponding effective lengths are given in 2-1-1/Fig. 5.7, reproduced here as Fig. 5.8-3. Examples of application include piers with free sliding bearings at their tops (case (b)), piers with fixed bearings at their tops but where the deck (through its connection to other elements) provides no positional restraint (case (b) again), and piers with fixed (pinned) bearings at their tops which are restrained in position by connection by way of the deck to a rigid abutment or other stocky pier (case (c)).

The effective lengths given in cases (a) to (e) assume that the foundations (or other restraints) providing rotational restraint are infinitely stiff. In practice, this will never be the case and the effective length will always be somewhat greater than the theoretical value for rigid restraints. For example, BS 5400 Part 4[9] required the effective length for case (b) to be taken as $2.3l$ instead of $2l$. **2-1-1/clause 5.8.3.2(3)** gives a method of accounting for this rotational flexibility in the effective length using equations 2-1-1/Expression (5.15)

2-1-1/clause 5.8.3.2(3)

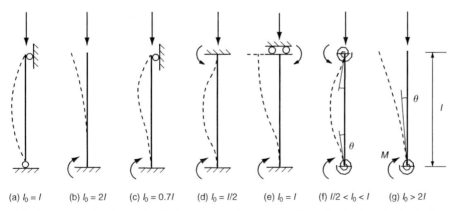

Fig. 5.8-3. Examples of different buckling modes and corresponding effective lengths for isolated members

for braced members and 2-1-1/Expression (5.16) for unbraced members:

$$l_o = 0.5l\sqrt{\left(1 + \frac{k_1}{0.45 + k_1}\right) \cdot \left(1 + \frac{k_2}{0.45 + k_2}\right)} \qquad \text{2-1-1/(5.15)}$$

$$l_o = l \max\left\{\sqrt{1 + 10 \cdot \frac{k_1 \cdot k_2}{k_1 + k_2}}; \left(1 + \frac{k_1}{1 + k_1}\right) \cdot \left(1 + \frac{k_2}{1 + k_2}\right)\right\} \qquad \text{2-1-1/(5.16)}$$

where k_1 and k_2 are the flexibilities of the rotational restraints at ends 1 and 2 respectively relative to the flexural stiffness of the member itself such that:

$$k = (\theta/M) \cdot (EI/l)$$

where:

θ is the rotation of the restraint for bending moment M
EI is the bending stiffness of compression member – see discussion below
l is the clear height of compression member between end restraints

2-1-1/Expression (5.16) can be used for unbraced members with rotational restraint at both ends. Quick inspection of 2-1-1/Expression (5.16) shows that the theoretical case of a member with ends built in rigidly for moment ($k_1 = k_2 = 0$), but free to sway in the absence of positional restraint at one end, gives an effective length $l_o = l$ as expected.

It is the relative rigidity of restraint to flexural stiffness of compression member that is important in determining the effective length. Consequently, using the un-cracked value of EI for the pier itself will be conservative as the restraint will have to be relatively stiffer to reduce the buckling length to a given value. It is also compatible with the definition of radius of gyration, i, in 2-1-1/clause 5.8.3.2(1) which is based on the gross cross-section. **2-1-1/clause 5.8.3.2(5)**, however, requires cracking to be considered in determining the stiffness of a restraint, such as a reinforced concrete pier base, if it significantly affects the overall stiffness of restraint offered to the pier. For the pier example, however, often the overall stiffness is dominated by the soil stiffness rather than that of the reinforced concrete element.

2-1-1/clause 5.8.3.2(5)

The Note to 2-1-1/clause 5.8.3.2(3) recommends that no value of k is taken less than 0.1. For integral bridges, or other bridges where restraint is provided at the top of the pier by its connection to the deck, cracking of the deck must also be considered in producing the stiffness. The value of end stiffness to use for piers in integral construction can be determined from a plane frame model by deflecting the pier to give the deflection relevant to the mode of buckling and determining the moment and rotation produced in the deck at the connection to the pier. Alternatively, the elastic critical buckling method described below could be used to determine the effective length more directly.

It should be noted that the cases in Fig. 5.8-3 do not allow for any rigidity of positional restraint in the sway cases. If significant lateral restraint is available, as might be the case in an integral bridge where one pier is very much stiffer than the others, ignoring this restraint will be very conservative as the more flexible piers may actually be 'braced' by the stiffer one. In this situation, a computer elastic critical buckling analysis will give a reduced value of effective length. (In many cases, however, it will be possible to see by inspection that a pier is braced.)

Where cases are not covered by 2-1-1/clause 5.8.3.2, effective lengths can be calculated from first principles according to **2-1-1/clause 5.8.3.2(6)**. This might be required, for example, for a member with varying section, and hence EI, along its length. The procedure is to calculate the buckling load, N_B, from a computer elastic critical buckling analysis, using the actual varying geometry and loading. It will be conservative to assume un-cracked concrete section for the member of interest and cracked concrete for the others (unless they can be shown to be un-cracked). An effective length is then calculated from:

2-1-1/clause 5.8.3.2(6)

$$l_0 = \pi\sqrt{EI/N_B} \qquad \text{2-1-1/(5.17)}$$

(a) Buckling of individual piers (braced)   (b) Overall buckling in sway mode (unbraced)

Fig. 5.8-4. Typical braced and unbraced situations

where EI may be freely chosen but a compatible value of radius of gyration, i, and therefore concrete cross-sectional area, A_c, must be used in calculating the slenderness according to 2-1-1/Expression (5.14). A 'sensible' choice of EI would be to base it on the actual cross-section in the middle third of the buckling half wave.

Effective lengths can also be derived for piers in integral bridges and other bridges where groups of piers of varying stiffness are connected to a common deck. In this instance, the buckling load, and hence effective length, of any one pier depends on the load and geometry of the other piers also. All piers may sway in sympathy and act as unbraced (Fig. 5.8-4(b)) or a single stiffer pier or abutment might prevent sway and give braced behaviour for the other piers (Fig. 5.8-4(a)). The analytical method above could also be used in this situation to produce an accurate effective length by applying coexisting loads to all columns and increasing all loads proportionately until a buckling mode involving the pier of interest is found. N_B is then taken as the axial load in the member of interest at buckling.

Finally, effective lengths can be taken from tables of approximate values such as was provided in Table 11 of BS 5400 Part 4.[9]

Worked example 5.8-1: Effective length of cantilevering pier
A bridge pier with free-sliding bearing at the top is 27.03 m tall and has cross-section dimensions and reinforcement layout (required for later examples), as shown in Fig. 5.8-5 and Fig. 5.8-6. The pier base has a foundation flexibility (representing the rotational flexibility of the pile group and pile cap) of 6.976×10^{-9} rad/kNm. The short term E for the concrete is $E_{cm} = 35 \times 10^3$ MPa. Calculate the effective length about the minor axis.

The inertia of the cross-section about the minor axis $= 3.1774 \, \text{m}^4$ so:

$$\frac{EI}{l} = \frac{35 \times 10^6 \times 3.1774}{27.03} = 4.114 \times 10^6 \, \text{kNm/rad}$$

At the base of the pier, $k_1 = (\theta/M) \cdot (EI/l) = 6.976 \times 10^{-9} \times 4.114 \times 10^6 = 28.7 \times 10^{-3}$. This is less than the lowest recommended value of 0.1 given in 2-1-1/clause 5.8.3.2(3). However, as the stiffness above was derived using lower-bound soil properties and pile cap stiffness, the stiffer calculated value of k will be used.

At the top of the pier, there is no restraint so $k_2 = \infty$.
From 2-1-1/Expression (5.16), the effective length is then:

$$l_0 = l \times \max\left[\sqrt{1 + 10 \times \frac{0.0287 \times \infty}{0.0287 + \infty}}; \left(1 + \frac{0.0287}{1 + 0.0287}\right) \times \left(1 + \frac{\infty}{1 + \infty}\right)\right]$$

$$= l \times \max[1.13; 2.06]$$

$$= 2.06l$$

The effective length is therefore close to the value of $2l$ for a completely rigid support.

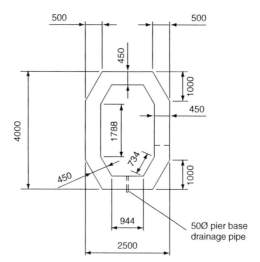

Fig. 5.8-5. Pier dimensions for Worked example 5.8-1

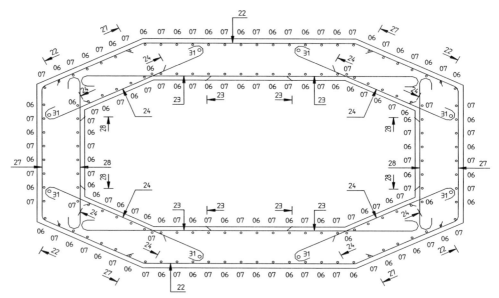

Fig. 5.8-6. Pier reinforcement layout for Worked example 5.8-1

Worked example 5.8-2: Check limiting slenderness of bridge pier
The bridge pier in Worked example 5.8-1 has concrete with cylinder strength 40 MPa and carries an axial load of 31 867 kN. Calculate the slenderness about the minor axis and determine whether second-order effects may be ignored. Take the effective length as 2.1 times the height.

The inertia of the cross-section about the minor axis = 3.1774 m^4. The area of the cross-section = 4.47 m^2. The limiting slenderness is determined from 2-1-1/Expression (5.13N) as follows:

$$\lambda \leq \lambda_{\text{lim}} = 20 A \cdot B \cdot C / \sqrt{n}$$

Since the split of axial load into short-term and long-term is not given, the recommended value of $A = 0.7$ will be conservatively used as discussed in the main text. The reinforcement ratio is also not known at this stage, so the recommended value of $B = 1.1$. Since the pier is free to sway, this is an unbraced member and benefit cannot be taken from the moment ratio at each end of the pier. Hence $C = 0.7$ (which also corresponds to equal moments at each end of a pier that is held in position at both ends).

The relative normal force is given by:

$$n = N_{Ed}/(A_c f_{cd}) = \frac{31\,867 \times 10^3}{4.47 \times 10^6 \times 22.67} = 0.314$$

where $f_{cd} = \alpha_{cc} f_{ck}/\gamma_c = 0.85 \times 40/1.5 = 22.67\,\text{MPa}$.

Hence from 2-1-1/Expression (5.13N), $\lambda_{lim} = 20 \times 0.7 \times 1.1 \times 0.7/\sqrt{0.31} = 19.4$. The radius of gyration of the pier cross-section $i = 0.84\,\text{m}$ and the effective length $l_0 = 2.1 \times 27.03 = 56.763\,\text{m}$ so the slenderness $\lambda = 56\,763/843 = 67.3 \gg 19.4$.

Second-order effects cannot therefore be ignored for this pier.

5.8.4. Creep

Creep tends to increase deflections from those predicted using short-term properties as discussed in section 3.1.4, so **2-1-1/clause 5.8.4(1)P** requires creep effects to be included in second-order analysis. To perform this calculation rigorously, different stress–strain relationships would be required for different load applications. To overcome this problem, a simplified relationship is given using an effective creep ratio, ϕ_{ef}, which, used together with the total design load, gives a creep deformation corresponding to that from the quasi-permanent load only, as is required. The effective creep factor is given in **2-1-1/clause 5.8.4(2)** as follows:

$$\phi_{ef} = \phi(\infty, t_0) M_{0Eqp}/M_{0Ed} \qquad \text{2-1-1/(5.19)}$$

where:

$\phi(\infty, t_0)$ is the final creep coefficient according to 2-1-1/clause 3.1.4
M_{0Eqp} is the first-order bending moment in the quasi-permanent load combination (SLS)
M_{0Ed} is the first-order bending moment in the design load combination (ULS)

The use of an SLS value for quasi-permanent loads and a ULS value for the design combination does not appear logical and it is recommended here that either SLS or ULS values are used to calculate both moments. By way of illustration, if all the moments were due to permanent load then 2-1-1/Expression (5.19) as written would lead to $\phi_{ef} < \phi(\infty, t_0)$ which is incorrect.

With appropriate modification as suggested, the use of 2-1-1/Expression (5.19) in second-order calculations will generally be conservative. This is because short-term increments of axial load from live load will lead proportionately to a greater increase in second-order moment than a similar increment of dead load as the increment is occurring at higher axial load. The first-order moment ratio in 2-1-1/Expression (5.19) therefore overestimates the proportion of dead load moment in the lead up to failure. To avoid this conservatism, the Note to 2-1-1/clause 5.8.4(2) allows ϕ_{ef} to be based on the moment ratio M_{Eqp}/M_{Ed} including second-order effects, but this would become iterative.

2-1-1/clause 5.8.4(4) allows creep to be ignored and thus short-term concrete properties to be used if all three of the following are satisfied:

$\phi(\infty, t_0) \leq 2$
$\lambda \leq 75$
$M_{0Ed}/N_{Ed} \geq h$

where h is the cross-section depth in the plane of bending.

The latter criterion is unlikely to apply very often in bridge pier design, so creep will normally need to be considered as above. A warning is also made in the Note to 2-1-1/clause 5.8.4(4) that creep should not be neglected, as well as second-order effects if the mechanical reinforcement ratio, ω, from 2-1-1/clause 5.8.3.1(1) is less than 0.25.

2-2/clause 5.8.4(105) permits a more accurate method of accounting for creep to be used, allowing for the creep deformations from individual load cases, rather than by the use of one effective creep factor to apply to the total combination of actions. Reference is made to 2-2/Annex KK. The additional effort involved in such an approach is generally not warranted for the analysis of second-order effects with axial load and has the disadvantage of requiring the analysis to be split into several stages.

5.8.5. Methods of analysis

2-1-1/clause 5.8.5(1)

Three methods for taking second-order effects into account are given in *2-1-1/clause 5.8.5(1)*. These are:

(1) Non-linear analysis according to clause 5.8.6.
(2) Method based on magnification of first-order moments according to clause 5.8.7.
(3) Method based on maximum predicted curvatures according to clause 5.8.8.

Method (1) will give the lowest total moments while method (3) is the quickest to perform. Of methods (2) and (3), method (2) typically gives lower moments for small creep ratios ($\phi_{ef} < 0.5$, say) or for high reinforcement content, but for higher creep ratios and normal reinforcement content, method (3) gives lower moments.

5.8.6. General method – second-order non-linear analysis

For the reasons explained in section 5.8.1 of this guide, it is essential that a second-order analysis with axial load for reinforced concrete sections realistically models material non-linearity as well as the geometric non-linearity. *2-1-1/clause 5.8.6(1)P* provides a general method based on non-linear analysis, which allows for both these sources of non-linearity.

2-1-1/clause 5.8.6(1)P

To illustrate the effects of non-linearity on the resistance of a system, it is simplest to consider a cantilevering pier such as the one in Fig. 5.8-1. The moment–curvature relationship for a section under a specified axial load, N, can be determined and used to produce a moment–deflection curve for the member by relating curvature directly to displacement. For a cantilevering pier, an approximate relationship can be obtained by assuming the maximum moment at the base of the pier acts over the full member height. This leads to a tip deflection $\Delta = kL^2/2$ where k is the curvature at the base of the pier. From equilibrium, the total applied moment $= M_0 + N\Delta$ where M_0 is the first-order moment including moment from initial imperfections. This can be plotted on the pier resistance moment–deflection curve as shown in Fig. 5.8-7, and used to find the equilibrium deflection such that there is equilibrium and compatibility. Equilibrium is achieved at the stable equilibrium position, as shown in Fig. 5.8-7.

Structural analysis packages perform a similar process to the above within the lengths of members, so approximate relationships between moments and deflections as above are not required. Such approximate methods based on critical cross-sections are, however, permitted by *2-1-1/clause 5.8.6(6)*.

2-1-1/clause 5.8.6(6)
2-1-1/clause 5.8.6(2)P

While the above illustrates the structural behaviour in general terms, it does not address the issue of what material properties should be used in the analysis. *2-1-1/clause 5.8.6(2)P* states only that stress–strain curves for concrete and steel should be 'suitable for overall analysis' and should take creep into account. The resistances of local sections are governed

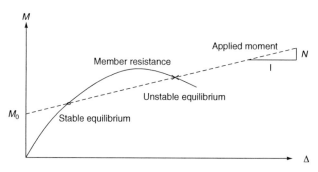

Fig. 5.8-7. Moment–deflection relationship for a cantilever pier

by design values of the material strengths while, arguably, the overall behaviour will be most similar to that produced with mean material strengths. EC2 generally requires 'realistic' stiffnesses to be used in the analysis, as described in section 5.7, and this leads to the lengthy verification format described therein, which can be used for second-order analysis. An alternative allowed by both 2-2/clause 5.7 and **2-1-1/clause 5.8.6(3)** is to use design values of material properties throughout the analysis so that, if equilibrium and compatibility are attained in the analysis, no further local design checks are required. This is conservative where all applied actions are external forces as the resulting deflections (and hence $P - \Delta$ effects) will be greater because of the uniformly reduced stiffnesses implicit in the method. In this case, it will also be conservative to neglect the effects of tension stiffening as noted in **2-1-1/clause 5.8.6(5)**. The clause permits the inclusion of the effects of tension stiffening, but no method of its consideration is given.

2-1-1/clause 5.8.6(3)

2-1-1/clause 5.8.6(5)

It should be noted that ignoring tension stiffening is not always conservative, despite the Note to 2-1-1/clause 5.8.6(5). One example is in the analysis of externally post-tensioned beams, where greater force increase can be obtained in the tendons if all sections are assumed not to be tension stiffened.

If design properties are used, the stress–strain relationships given in 2-1-1/Fig. 3.2 for concrete (modified as specified in 2-1-1/clause 5.8.6(3) for design values) and 2-1-1/Fig. 3.8 for reinforcing steel can be used. Creep may be accounted for through **2-1-1/clause 5.8.6(4)** by multiplying all strain values in the above concrete stress–strain diagram by a factor $(1 + \phi_{ef})$, where ϕ_{ef} is the effective creep ratio discussed in section 5.8.4. The analysis would be performed using the design combination of actions relevant to the ultimate limit state. When this procedure is followed, no further checks of local sections are required, as strength and stability are verified directly by the analysis. Care is needed, however, where there are indirect actions (imposed displacements) as a stiffer overall system may attract more load to the critical design section, despite the reduction in $P - \Delta$ effects. A sensitivity analysis could be tried in such cases.

2-1-1/clause 5.8.6(4)

Analysis at SLS should, however, be performed using material stress–strain diagrams based on realistic stiffnesses, particularly where imposed deformations can generate internal effects. The modelling should therefore follow that suggested for SLS non-linear analysis in section 5.7 of this guide.

5.8.7. Second-order analysis based on nominal stiffness

Although the elastic critical buckling load or moment itself has little direct relevance to real design of reinforced concrete, it gives a good indication of susceptibility to second-order effects and can also be used as a parameter in determining second-order effects from the results of a first-order analysis. The method of 2-1-1/clause 5.8.7 is based on the elastic theory that total moments in a pin-ended strut, including second-order effects, can be derived by multiplying first-order moments (including moments arising from initial imperfections) by a magnifier that depends on the axial force and the Euler buckling load of the pier. The simplest example of this is a pin-ended column, length L, under axial load only with an initial sinusoidal bow imperfection of maximum displacement a_0. The Euler buckling load is given by:

$$N_B = \pi^2 EI/L^2 \tag{D5.8-4}$$

(Determination of EI for reinforced concrete columns is covered by 2-1-1/clause 5.8.7.2 and is discussed later in this section.)

If the axial load is N_{Ed} then the final deflection is given by:

$$a = a_0 \cdot \left[\frac{1}{1 - (N_{Ed}/N_B)}\right] \tag{D5.8-5}$$

(This is obtained from simple elastic theory by solving $EI[d^2(v - v_0)/dx^2] + N_{Ed}v = 0$, where v is the lateral displacement as a function of height x up the column and $v_0 = a_0 \sin \pi x/L$.)

The corresponding final maximum moment including second-order effects, $M_{Ed} = N_{Ed}a$, is then given by:

$$M_{Ed} = N_{Ed} \cdot \left[\frac{a_0}{1 - (N_{Ed}/N_B)}\right] = M_{0Ed}\left[\frac{1}{1 - (N_{Ed}/N_B)}\right] \tag{D5.8-6}$$

where $M_{0Ed} = N_{Ed}a_0$ is the first-order moment. The magnifier here is $1/(1 - N_{Ed}/N_B)$, which assumes that the initial imperfection is sinusoidal. Similar results are produced for the magnification of moments in pin-ended struts with applied end moments or transverse load, but the magnifier varies slightly depending on the distribution of the first-order moment. For uniform moment, the amplifier above is slightly unconservative but it will generally suffice with sufficient accuracy.

2-1-1/clause 5.8.7.3(1)

The above illustrates the basis of *2-1-1/clause 5.8.7.3(1)*, which allows total moments in bridges and bridge components, including second-order effects, to be found by increasing the first-order moments (including the effects of all imperfections) as follows:

$$M_{Ed} = M_{0Ed} \cdot \left[1 + \frac{\beta}{(N_B/N_{Ed}) - 1}\right] \qquad \text{2-1-1/(5.28)}$$

2-1-1/clause 5.8.7.3(2)

where $\beta = \pi^2/c_0$ from *2-1-1/clause 5.8.7.3(2)* and $N_B = \pi^2 EI/l_0^2$ where l_0 is the effective length for buckling determined in accordance with 2-1-1/clause 5.8.3.2. M_{0Ed} is the design moment from a first-order analysis but this must include the moment from initial imperfections (M_{0Ed} and N_{Ed} are design values and must include all load factors). c_0 depends on the distribution of moment and hence curvature in the column. For uniform curvature, $c_0 = 8$. For sinusoidal curvature, $c_0 = \pi^2$ and the expression for moment simplifies to that of (D5.8-6) as given in *2-1-1/clause 5.8.7.3(4)*:

2-1-1/clause 5.8.7.3(4)

$$M_{Ed} = \left[\frac{M_{0Ed}}{1 - (N_{Ed}/N_B)}\right] \qquad \text{2-1-1/(5.30)}$$

2-1-1/clause 5.8.7.3(4) recommends the general use of this magnifier as a reasonable approximation for other moment diagram shapes. It is not, however, conservative for cases of uniform or near uniform moment where $\beta = \pi^2/8$. $\beta = \pi^2/8$ should also be used where an equivalent uniform moment has been assumed when there are differing first-order end moments as discussed below.

The above expressions assume that the peak first-order moment occurs at the same section as the peak moment from the $P - \Delta$ effect. Considering first braced columns, this is true for a pin-ended strut without end moments or with equal end moments but is not necessarily true when end moments are present and are not equal. It would therefore be conservative to magnify the first-order moments throughout the height of the pier by the above magnification in this latter case. *2-1-1/clause 5.8.7.3(3)* (as did BS 5400 Part 4[9]) partly overcomes this conservatism by allowing an equivalent first-order moment to be used only where there is *no transverse load* applied in the height of the column *and the column is braced*. Differing first-order end moments M_{01} and M_{02}, giving rise to a linear variation of moment along the height of the pier, are replaced by an equivalent first-order end moment M_{0e} according to 2-1-1/clause 5.8.8.2(2):

2-1-1/clause 5.8.7.3(3)

$$M_{0e} = 0.6M_{02} + 0.4M_{01} \geq 0.4M_{02} \quad \text{where} \quad |M_{02}| > |M_{01}| \qquad \text{2-1-1/(5.32)}$$

Where moments give tension on the same side of the member, they have the same sign, otherwise moments should be given opposite signs. This gives an equivalent first-order moment in the middle of the column which can be magnified using 2-1-1/Expression (5.28) and used in design of the middle of the column. The total moment to design for at the column end should also not be taken as less than M_{02} as discussed below, although this is not explicitly stated. Similarly, where reinforcement varies, each end of the column should be designed for at least the first-order end moments.

A linearly varying moment is only obtained when imperfections are ignored and 2-1-1/ Expression (5.32) only applies to the linearly varying part of the moment. Imperfections must, however, be considered and the use of 2-1-1/Expression (5.28) becomes slightly confusing in this case as it includes the first-order term from imperfections in M_{0Ed}. To overcome this problem, it is recommended that the first-order effects from linearly varying moment and imperfection are kept separate and the reduction according to 2-1-1/Expression (5.32) is made only to the linearly varying part, such that the total effective first-order moment is as follows:

$$M_{0Ed} = M_{0Ed,l} + M_{0Ed,i} \qquad \text{(D5.8-7)}$$

where:

$M_{0Ed,l}$ is the effective first-order uniform moment determined in accordance with 2-1-1/Expression (5.32) from the linearly varying moment, but excluding imperfections

$M_{0Ed,i}$ is the maximum first-order moment from imperfections in the middle of the pier

The moment according to equation (D5.8-7) is then used in 2-1-1/Expression (5.28) to derive the total design effects including second-order effects at the middle part of the member.

In reality, for braced columns, the moments from linearly varying first-order moment, first-order moments from imperfections and additional second-order moments from $P - \Delta$ effects add as shown in Fig. 5.8-8. (The second-order end moments shown do not develop if there is no end restraint rotational stiffness.) It is then necessary to consider three design locations:

(1) $P - \Delta$ effects add to first-order moments in the middle of the member. This total moment is obtained approximately when the equivalent first-order moment from equation (D5.8-7) is magnified according to 2-1-1/Expression (5.28).
(2) $P - \Delta$ effects reduce the first-order moments at the end with the larger first-order moment so the initial moment M_{02} only needs to be checked at this end, assuming no $P - \Delta$ effects or imperfections.
(3) $P - \Delta$ effects increase the first-order moments at the end with the smaller first-order moment so the moment M_{01} plus the first-order effects from imperfections should be increased. EC2, however, gives no requirement for this check and neither did BS 5400 Part 4.[9] In practice, this check is unlikely to be critical. Work by Cranston[10] indicated that this moment would be less than those in (1) and (2). If there were specific concerns that this check might be critical, computer second-order non-linear analysis would have to be carried out. Care is particularly needed if reinforcement is curtailed.

For unbraced columns (that are therefore able to sway), the above reduction according to 2-1-1/Expression (5.32) should not be made, although this is not made clear in EC2. This is again best illustrated through the simplest case of a cantilevering pier where the peak first-order moment at the base obviously coincides with the peak moment from the $P - \Delta$ effect. In these cases the first-order moments throughout the height of the pier should be magnified according to 2-1-1/Expression (5.28).

This method of addressing second-order effects is straightforward for structural steel members, where EI can be taken as constant up to yield. For concrete structures, the situation is more complicated as there is significant non-linearity involved in cracking of the concrete and inherent non-linearity in the concrete stress–strain response. The result is that 'EI' for a given axial force reduces with increasing moment and is not unique. **2-1-1/ clause 5.8.7.2(1)** overcomes this difficulty by providing a constant 'nominal stiffness' EI for a given cross-section, which depends on all the relevant parameters, i.e. reinforcement content, axial force, concrete strength, creep and slenderness:

$$EI = K_c E_{cd} I_c + K_s E_s I_s \qquad \text{2-1-1/(5.21)}$$

2-1-1/clause 5.8.7.2(1)

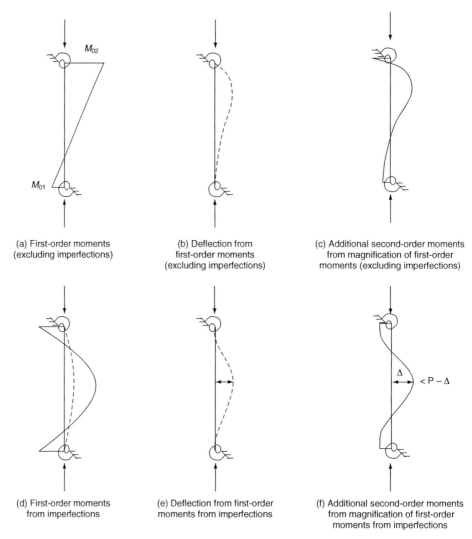

Total moments including second-order effects are the sum of (a), (c), (d) and (f)

Fig. 5.8-8. Braced columns with end moments and imperfections

where:

E_{cd} = design value of Young's modulus for concrete, E_{cm}/γ_{cE} from 2-1-1/clause 5.8.6(3). γ_{cE} is recommended to be equal to 1.20 but is a nationally determined parameter
I_c = inertia of gross concrete section
E_s = design value of Young's modulus for reinforcement
I_s = inertia of reinforcement about the concrete section centroid
K_c = factor allowing for cracking, concrete material non-linearity and creep
K_s = factor for contribution of reinforcement $= 1.0$

The calculation of these parameters is illustrated in Worked example 5.8-3.

Only one EI value can be used throughout the member height in this calculation method. Care has to be taken with this method when reinforcement is curtailed in the member as this might affect the distribution of curvature assumed when using the above β parameter. Where reinforcement is curtailed continuously to match the moment capacity envelope, it will be more appropriate to use the β value of $\pi^2/8$ for constant moment. Since the second-order moments depend on the stiffness of the section which is itself influenced by the reinforcement, the calculation according to the method of 2-1-1/clause 5.8.7 becomes iterative if the reinforcement is to be reduced from that assumed in an initial calculation.

CHAPTER 5. STRUCTURAL ANALYSIS

Worked example 5.8-3: Cantilevering pier check to 2-1-1/clause 5.8.7
The pier of Worked example 5.8-1 is subjected to a 31 867 kN axial load and a 1366 kN lateral load about the minor axis at the ultimate limit state, as shown in Fig. 5.8-9. The main vertical reinforcement is 136 no. 32 mm diameter bars. The effective creep ratio ϕ_{ef} for this particular load case is 1.0 and $E_{cm} = 35 \times 10^3$ MPa. Calculate the final moment at the base of the pier.

The inertias of the concrete cross-section and of the reinforcement about the concrete cross-section centroid are first calculated. These are as follows:

$I_c = 3.177 \, \text{m}^4$

$I_s = 7.81 \times 10^{-2} \, \text{m}^4$

(Compression reinforcement is included in the above – see the discussion on compression reinforcement in section 5.8.8 below.)

The factor K_c is then calculated:

$$k_1 = \sqrt{\frac{f_{ck}}{20}} = \sqrt{\frac{40}{20}} = 1.41 \qquad \text{2-1-1/(5.23)}$$

$$k_2 = n \cdot \frac{\lambda}{170} = 0.314 \times \frac{67.3}{170} = 0.125 \leq 0.2 \text{ as required} \qquad \text{2-1-1/(5.24)}$$

(n and λ are taken from the Worked example 5.8-2)

$$K_c = \frac{k_1 \cdot k_2}{1 + \phi_{ef}} = \frac{1.41 \times 0.125}{1 + 1.0} = 0.088 \text{ (for } \rho > 0.002\text{)} \qquad \text{2-1-1/(5.22)}$$

The concrete modulus

$$E_{cd} = E_{cm}/\gamma_{cE} = 35 \times 10^3/1.2 = 29.2 \times 10^3 \, \text{MPa} \qquad \text{2-1-1/(5.20)}$$

$K_s = 1.0$ for reinforcement (for $\rho > 0.002$) \qquad 2-1-1/(5.22)

The nominal flexural rigidity of the pier is then given by 2-1-1/Expression (5.21) as:

$$EI = K_c E_{cd} I_c + K_s E_s I_s$$

$$= 0.088 \times 29.2 \times 10^3 \times 3.177 \times 10^{12} + 1.0 \times 200 \times 10^3 \times 7.81 \times 10^{10}$$

$$= 2.38 \times 10^{16} \, \text{Nmm}^2$$

From Worked example 5.8-2, the effective length for buckling, $l_0 = 2.1 \times 27.03 = 56.763$ m, so the buckling load

$$N_B = \frac{\pi^2 EI}{l_0^2} = \frac{\pi^2 \times 2.38 \times 10^{16}}{56\,763^2} = 72\,903 \, \text{kN}$$

The initial imperfection displacement at the pier top is obtained from 2-2/Expression (5.101) as $l \cdot \theta_i = l \cdot \theta_0 \cdot \alpha_h$ where $\alpha_h = 2/\sqrt{l} = 2/\sqrt{27.03} = 0.38$ and θ_0 is a nationally determined parameter whose recommended value is 1/200. In this example, the lower limit of $\alpha_h \geq 2/3$ in EN 1992-1-1 was applied to allow greater tolerance on site, but this restriction does not exist in EN 1992-2. Therefore:

$$l \cdot \theta_i = 27\,030 \cdot \frac{1}{200} \cdot \frac{2}{3} = 90 \, \text{mm}$$

The first-order moment at the base $M_{0Ed} = 1366 \times 27.030 + 31\,867 \times 0.09 = 39\,791$ kNm.

The final moment, including second-order effects, is given by 2-1-1/Expression (5.28):

$$M_{Ed} = M_{0Ed} \cdot \left[1 + \frac{\beta}{(N_B/N_{Ed}) - 1}\right]$$

With β taken equal to 1.0 for approximately sinusoidal curvature this simplifies to the expression of 2-1-1/Expression (5.30) which is then used:

$$M_{Ed} = \left[\frac{M_{0Ed}}{1-(N_{Ed}/N_B)}\right] = \frac{39\,791}{1-31\,867/72\,903} = 70\,691\,\text{kNm}$$

As there is no transverse load giving moments about the major axis in this example, the pier should separately be checked about the major axis for the moments arising from initial imperfections alone. By inspection, the pier would be adequate in that direction if it were adequate about the minor axis. It is not necessary to check biaxial bending in this example, as the moments from imperfections need only be considered in one direction in accordance with 2-1-1/clause 5.8.9(2); the major axis moment caused by imperfections would not therefore coexist with the minor axis moment calculated above. If first-order moments were applied about both axes together (as would be the case in most real situations), biaxial bending would need to be addressed in accordance with 2-1-1/clause 5.8.9.

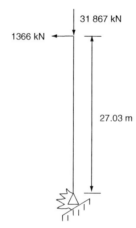

Fig. 5.8-9. Pier loading for Worked example 5.8-3

5.8.8. Method based on nominal curvature

The method of 2-1-1/clause 5.8.8 is based on similar theory to the slender column method in BS 5400 Part 4[9] in that an estimate of the maximum possible curvature is used to calculate the second-order moment. *2-1-1/clause 5.8.8.1(1)* notes that the method is primarily intended for use with members that can be isolated from the rest of the bridge, whose boundary conditions can be represented by an effective length applied to the member. The first-order moment, including that from initial imperfections, is added to the moment from the additional maximum deflection according to the expression in *2-1-1/clause 5.8.8.2(1)*. (This differs from the method in BS 5400 where initial imperfections are not considered.)

2-1-1/clause 5.8.8.1(1)

2-1-1/clause 5.8.8.2(1)

$$M_{Ed} = M_{0Ed} + M_2 \qquad \text{2-1-1/(5.31)}$$

where:

M_{0Ed} is the first-order moment, including the effect of imperfections
M_2 is the estimated (nominal) second-order moment

The additional second-order moment is given as follows:

$$M_2 = N_{Ed}e_2 \qquad \text{2-1-1/(5.33)}$$

M_2 is determined by calculating e_2 from the estimated curvature at failure, $1/r$, according to the formula, $e_2 = (1/r)l_0^2/c$. c depends on the distribution of curvature in the column. The definition of c differs from c_0 used in 2-1-1/clause 5.8.7 as it depends on the shape of the

CHAPTER 5. STRUCTURAL ANALYSIS

total curvature, not just the curvature from first-order moment. For sinusoidal curvature, $c = \pi^2$ and for constant curvature, $c = 8$ as discussed in section 5.8.7.

The latter value of c is best illustrated by considering a free-standing pier of length L with rigid foundations and hence $l_0 = 2L$. For constant curvature, $1/r$, the deflection is obtained by integration of the curvature as follows:

$$\Delta = \int_0^L \int_0^x \left(\frac{1}{r}\right) dx\, dx = \left(\frac{1}{r}\right) L^2/2 \tag{D5.8-8}$$

From the above formula for e_2, with $c = 8$ and $l_0 = 2L$, the deflection is:

$$e_2 = \left(\frac{1}{r}\right) l_0^2/c = \left(\frac{1}{r}\right) 4L^2/8 = \left(\frac{1}{r}\right) L^2/2$$

which is the same result as that in (D5.8-8).

The value of $c = \pi^2$ is recommended in *2-1-1/clause 5.8.8.2(4)*, but care should again be taken when reinforcement is curtailed continuously to match the moment capacity envelope. In that situation, it will be more appropriate to use $c = 8$ as for constant moment.

The value of curvature $1/r$ depends on creep and the magnitude of the applied axial load. For members with constant symmetrical cross-section (including reinforcement) it can be determined according to *2-1-1/clause 5.8.8.3(1)*:

$$\frac{1}{r} = K_r K_\phi \frac{1}{r_0} \qquad \text{2-1-1/(5.34)}$$

2-1-1/clause 5.8.8.2(4)

2-1-1/clause 5.8.8.3(1)

where:

$1/r_0$ is the basic value of curvature, discussed below
K_r is a correction factor depending on axial load, discussed below
K_ϕ is a factor for taking account of creep, discussed below

2-1-1/Expression (5.34) is only applicable to constant symmetric sections with symmetric reinforcement. The latter implies that the reinforcement in compression is considered in the stiffness calculation. Reinforcement in compression is also considered in the stiffness calculation in 2-1-1/clause 5.8.7.2(1). No criteria are given for the detailing of reinforcement in compression to enable its contribution to stiffness to be considered. Criteria for the detailing of compression bars to enable their use in the cross-section resistance calculation are, however, given in the following clauses:

- Beams   *2-1-1/clause 9.2.1.2(3)*
- Columns *2-1-1/clause 9.5.3(6)*
- Walls   *2-1-1/clause 9.6.3(1)*

These requirements are discussed under the relevant clauses. The rules for columns, in particular, require compression bars in an outer layer to be held by links if they are to be included in the resistance check. It is, however, not considered necessary here to provide such links in order to consider the contribution of reinforcement in compression to the stiffness calculation. This apparent incompatibility is justified by the conservative nature of the methods of clauses 5.8.7 and 5.8.8 compared to a general non-linear analysis and the similar approach taken in EN 1994-2 clause 6.7 for composite columns. If there is specific concern over the adequacy of the restraint to compression bars, the suggested curvature in (D5.8-10) below could be used as a more conservative value.

The curvature $1/r_0$ is based on a rectangular beam with symmetrical reinforcement and strains of yield in reinforcement at each fibre separated by a lever arm $z = 0.9d$, where d is the effective depth (the compression and tension reinforcement thus being considered to reach yield). Hence the curvature is given by:

$$1/r_0 = \frac{\varepsilon_{yd}}{0.45d} \tag{D5.8-9}$$

77

This differs from the method in BS 5400 Part 4,[9] where curvature was based on steel yield strain in tension and concrete crushing strain at the other fibre. Despite the apparent reliance on compression reinforcement to reduce the final concrete strain, the results produced will still be similar to those from BS 5400 Part 4 because:

(1) The moment from imperfections has to be added in EC2.
(2) The strain difference across the section is less in BS 5400 Part 4, but it occurs over a smaller depth (not the whole cross-section depth) – thus producing proportionally more curvature.

For situations where the reinforcement is not just in opposite faces of the section, d is taken as $h/2 + i_s$ in accordance with **2-1-1/clause 5.8.8.3(2)**, where i_s is the radius of gyration of the total reinforcement area. This expression is again only applicable to uniform symmetric sections with symmetric reinforcement.

No rule is given where the reinforcement is not symmetrical. One possibility would be to determine the curvature from similar assumptions to those used in reference 9. These are that the tension steel yields at ε_{yd} and the extreme fibre in compression reaches its failure strain ε_c, so the curvature $1/r_0$ would be given approximately by:

$$1/r_0 = \frac{\varepsilon_{yd} + \varepsilon_c}{h} \tag{D5.8-10}$$

where h is the depth of the section in the direction of bending (used as an approximation to the depth to the outer reinforcement layer). The concrete strain can conservatively be taken as $\varepsilon_c = \varepsilon_{cu2}$. If equation (D5.8-10) is used, the factor K_r below should be taken as 1.0.

K_r is a factor which accounts for the reduction in curvature with increasing axial load and is given as $(n_u - n)/(n_u - n_{bal}) \leq 1.0$. n_u is the ultimate capacity of the section under axial load only, N_u, divided by $A_c f_{cd}$. N_u implicitly includes all the reinforcement area, A_s, in calculating the compression resistance such that $N_u = A_c f_{cd} + A_s f_{yd}$ so that

$$n_u = \frac{(A_c f_{cd} + A_s f_{yd})}{A_c f_{cd}} = 1 + \frac{A_s f_{yd}}{A_c f_{cd}}$$

as given in **2-1-1/clause 5.8.8.3(3)**. n_{bal} is the value of design axial load, divided by $A_c f_{cd}$, which would maximize the moment resistance of the section – see Fig. 5.8-10.

The clause allows a value of 0.4 to be used for n_{bal} for all symmetric sections. In other cases, the value can be obtained from a section analysis. K_r may always be conservatively taken as 1.0 (even though for $n < n_{bal}$ it is calculated to be greater than 1.0), and this approximation will usually not result in any great loss of economy for bridge piers unless the compressive load is unusually high.

K_ϕ is a factor which allows for creep and is given by **2-1-1/clause 5.8.8.3(4)** as follows:

$$K_\phi = 1 + \beta \phi_{ef} \geq 1.0 \qquad \text{2-1-1/(5.37)}$$

where:

ϕ_{ef} is the effective creep ratio, discussed in section 5.8.4
β = $0.35 + f_{ck}/200 - \lambda/150$ and λ is the slenderness ratio discussed in section 5.8.3

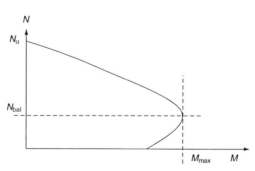

Fig. 5.8-10. Axial force for a 'balanced' section

CHAPTER 5. STRUCTURAL ANALYSIS

For braced members (held in position at both ends) which do not have transverse loading, an equivalent first-order moment for the linearly varying part of the moment may be used according to *2-1-1/clause 5.8.8.2(2)*. This, together with the other checks required, is discussed at length in section 5.8.7 above. It is discussed there that the final first-order moment M_{0Ed} should comprise the reduced equivalent moment from 2-1-1/Expression (5.32) added to the full first-order moment from imperfections.

2-1-1/clause 5.8.8.2(2)

Worked example 5.8-4: Cantilevering pier check to 2-1-1/clause 5.8.8

The pier of Worked example 5.8-1 is subjected to a 31 867 kN axial load and a 1366 kN lateral load about the minor axis at the ultimate limit state, as shown in Fig. 5.8-9. The main vertical reinforcement is 136 no. 32 mm diameter bars with yield strength 460 MPa (less than the standard 500 MPa). The effective creep ratio ϕ_{ef} for this particular load case is 1.0 and $E_{cm} = 35 \times 10^3$ MPa. Calculate the final moment at the base of the pier.

Since the section is symmetrical (with respect to cross-section and reinforcement), the method of 2-1-1/clause 5.8.8 can be used without modification. The radius of gyration, i_s, of the reinforcement was found to be 845 mm so the effective depth, d, is found from 2-1-1/Expression (5.35):

$$d = h/2 + i_s = 2500/2 + 845 = 2095 \text{ mm}$$

(Compression reinforcement is included in the above – see the discussion on compression reinforcement in section 5.8.8 above.)

The curvature $1/r_0$ is then calculated from 2-1-1/Expression (5.34):

$$1/r_0 = \frac{\varepsilon_{yd}}{0.45d} = \frac{400/200 \times 10^3}{0.45 \times 2095} = 2.12 \times 10^{-6} \text{ mm}^{-1}$$

Since the relative axial force $n = 0.314$ (from Worked example 5.8-2), which is less than n_{bal} which may be taken as 0.4, K_r will be greater than 1.0 according to the formula 2-1-1/Expression (5.36) and should therefore be taken equal to 1.0.

In order to calculate K_ϕ, the parameter β must first be calculated taking the slenderness $\lambda = 67.3$ from Worked example 5.8-2:

$$\beta = 0.35 + \frac{f_{ck}}{200} - \frac{\lambda}{150} = 0.35 + \frac{40}{200} - \frac{67.3}{150} = 0.101$$

$K_\phi = 1 + \beta \cdot \phi_{ef} = 1 + 0.101 \times 1.0 = 1.101$ from 2-1-1/Expression (5.37).

The nominal curvature according to 2-1-1/Expression (5.34) is then:

$$1/r = K_r \cdot K_\phi \cdot 1/r_0 = 1.0 \times 1.101 \times 2.12 \times 10^{-6} = 2.33 \times 10^{-6} \text{ mm}^{-1}$$

The effective length for buckling $l_0 = 2.1 \times 27.03 = 56.763$ m and $c = \pi^2$ for a sinusoidal distribution of curvature. From 2-1-1/Expression (5.33):

$$e_2 = (1/r)l_0^2/c = 2.33 \times 10^{-6} \times 56763^2/\pi^2 = 761 \text{ mm}$$

$$M_2 = N_{Ed} \cdot e_2 = 31867 \times 0.761 = 24251 \text{ kNm}$$

The initial imperfection displacement at the pier top is obtained from 2-2/Expression (5.101) as $l \cdot \theta_i = l \cdot \theta_0 \cdot \alpha_h$, where $\alpha_h = 2/\sqrt{l} = 2/\sqrt{27.03} = 0.38$ and θ_0 is a nationally determined parameter whose recommended value is 1/200. In this example, the lower limit of $\alpha_h \geq 2/3$ in EN 1992-1-1 was applied to allow greater tolerance on site, but this restriction does not exist in EN 1992-2. Therefore:

$$l \cdot \theta_i = 27030 \cdot \frac{1}{200} \cdot \frac{2}{3} = 90 \text{ mm}$$

The first-order moment at the base $M_{0Ed} = 1366 \times 27.030 + 31867 \times 0.09 = 39791$ kNm.

> The final moment including second-order effects is given by 2-1-1/Expression (5.31):
>
> $M_{Ed} = M_{0Ed} + M_2 = 39\,791 + 24\,251 = \mathbf{64\,042\,kNm}$
>
> The pier should also be checked about the major axis for the moments arising from initial imperfections and the nominal second-order moment. The check of biaxial bending should be carried out as discussed in section 5.8.9, but imperfections should only be considered in one direction.

5.8.9. Biaxial bending

The effects of slenderness for columns bent biaxially are most accurately determined using non-linear analysis, as discussed in section 5.8.6 of this guide. The provisions of 2-1-1/clause 5.8.9 apply when simplified methods have been used.

The approximate methods described in sections 5.8.7 and 5.8.8 can also be used for the case of biaxial bending. The second-order moment is first determined separately in each direction following either of the above methods, including imperfections. *2-1-1/clause 5.8.9(2)* states that it is only necessary to consider imperfections in one direction, but the direction should be chosen to determine the most unfavourable overall effect. *2-1-1/clause 5.8.9(3)* allows the interaction between the moments to be neglected (i.e. consider bending in each direction separately) if the slenderness ratios in the two principle directions do not differ by more than a factor of 2 *and* the 'relative eccentricities' satisfy one of the criteria in 2-1-1/Expression (5.38b) (not reproduced here). Where this is not satisfied, the moments in the two directions (including second-order effects) must be combined, but imperfections only need to be considered in one direction such as to produce the most unfavourable conditions overall. Section design under the biaxial moments and axial force may be done either by a rigorous cross-section analysis using the strain compatibility method discussed in section 6.1.4.4 of this guide, or the simple interaction provided in *2-1-1/clause 5.8.9(4)* may be used.

Where the method of 2-1-1/clause 5.8.8 is used, it is not explicitly stated whether a nominal second-order moment, M_2, should be considered in both orthogonal directions simultaneously, given that the section can only 'fail' in one plane of bending. If the method of section 5.8.7 is used, the first-order moments in both directions would be amplified, but the resulting moments in a given direction would be small if the first-order moments (including those from imperfections) were small. M_2, however, can be significant in both directions.

From above, a case could be made for considering M_2 only in the direction that gives the most unfavourable verification. For circular columns, it is possible to take the vector resultant of moments in two orthogonal directions, thus transforming the problem into a uniaxial bending problem with M_2 considered only in the direction of the resultant moment. In general, however, it is recommended here that M_2 conservatively be calculated for both directions, as was practice to BS 5400 Part 4.[9] Bending should then be checked in each direction independently, and then biaxial bending should be considered (with M_2 applied in both directions together unless second-order effects can be neglected in one or both directions in accordance with 2-1-1/clause 5.8.2(6) or 2-1-1/clause 5.8.3) if 2-1-1/clause 5.8.9(3) is not fulfilled. Imperfections should only be considered in one direction. In many cases, M_2 will not be very significant for bending about the major axis, as the curvature from equation (D5.8-9), and hence nominal second-order moment, is smaller for a wider section.

5.9. Lateral instability of slender beams

2-1-1/clause 5.9(1)P requires the designer to consider lateral instability of slender concrete beams. The instability referred to involves both lateral and torsional displacement of the beam when subjected to bending about the major axis. Such instability needs to be considered for both erection and finished conditions, but is only likely to be a potential problem for concrete bridge beams during transportation or erection before they are sufficiently braced (by deck slab and diaphragms, for example) within the final structure.

2-1-1/clause 5.9(3) defines geometric conditions to be satisfied so that second-order effects from the above mode of buckling can be ignored. These limits are not applicable where there is axial force (such as due to external prestressing), as the axial force leads to additional second-order effects as discussed in section 5.8. It is recommended that sections are generally designed to be within these limits to avoid the complexity of verifying the beam through second-order analysis. The limits should be met for most practical beam geometries used in bridge design with the possible exception of edge beams with continuous integral concrete parapets. Where such upstands are outside the geometric limits but have been ignored in the ultimate limit state checks of the edge beam, engineering judgement may often be used to conclude that the upstand is adequate. (Some care would still be required in the verification of cracking in the upstand.)

If the simple requirements of 2-1-1/clause 5.9(3) are not met, then second-order analysis needs to be carried out to determine the additional transverse bending and torsional moments developed. Geometric imperfections must be taken into account and *2-1-1/ clause 5.9(2)* requires a lateral deflection of $l/300$ to be assumed as a geometric imperfection, where l is the total length of the beam. It is not necessary to include an additional torsional imperfection as well. Any bracing, whether continuous from a deck slab or discrete from diaphragms, should be included in the model. Such analysis is complex as it must allow for both the non-linear behaviour of the materials and the geometric non-linearity of the instability type, for which finite element modelling (with shell elements) would be required.

No further guidance is offered in EN 1992 and further discussion of a suitable non-linear analysis is beyond the scope of this guide. The commentary to clause 6.6.3.3.4 of Model Code 90[6] gave a simplified method of designing slender beams and reference could be made to this if required. Regardless of the method used, the supporting structures and restraints must be designed for the resulting torsion – *2-1-1/clause 5.9(4)*.

5.10. Prestressed members and structures
5.10.1. General
2-2/clause 5.10 covers specific rules for prestressed concrete members and structures and covers both pre-tensioned as well as post-tensioned bridges. It deals with maximum permissible prestressing forces, prestress losses and the treatment of prestress in section design and global analysis. It does not cover the design of anchorage zones, which is covered by 2-2/ clause 8.10. The rules in this section are very much geared towards post-tensioning and some interpretation is needed for pre-tensioned beams made composite with a deck slab, as noted in the text and examples below.

2-1-1/clause 5.10.1(2) allows the effects of prestressing to be considered as an action or as part of the resistance, but individual clauses usually make it clear as to which approach is to be used so there is little real choice. In general, prestress is treated as an action (*2-1-1/clause 5.10.1(3)*) and is included in the combinations in EN 1990 as such. For example, prestress is treated as an applied force in the design of end blocks (see section 8.10) and, where elastic analysis is used, in member serviceability design and in the flexural design of unbonded or externally post-tensioned members. The effects of prestress are usually split into axial force and moment components.

For the bending resistance of members with bonded prestressing, the prestressing is most conveniently treated as part of the resistance. *2-1-1/clause 5.10.1(4)* requires that the contribution of the prestressing tendons to the section resistance should be limited to their additional strength beyond prestressing. This is intended to prevent double-counting of the design prestressing force, which would occur if it was included both on the loading side (as a primary prestress moment and axial force) and in the section resistance calculation. This requirement is most simply achieved by treating the secondary effects of prestress as an applied action on the loading side, but omitting the primary effects of the design prestressing force. The design prestressing force is then taken into account in the section bending resistance by shifting the origin of the design stress–strain diagram for the prestressing tendons by an amount corresponding to the design prestress. The initial strain in the prestress corresponding to this prestressing force is called the prestrain. This method is discussed further in section 6.1 of this guide.

2-1-1/clause 5.10.1(5)P requires that brittle failure of a prestressed member, caused by sudden failure of prestressing tendons upon cracking of the concrete in flexure, is avoided. *2-2/clause 5.10.1(106)* requires this to be achieved using one of the methods in 2-2/clause 6.1(109), which are discussed in section 6.1 of this guide. It is a new codified check for UK designers. The requirement is analogous to that for minimum reinforcement in reinforced concrete elements.

2-1-1/clause 5.10.1(5)P
2-2/clause 5.10.1(106)

5.10.2. Prestressing force during tensioning
5.10.2.1. Maximum stressing force
EC2 defines maximum permissible limits to the stressing force both during and after tensioning in order to reduce the risk of tendon failure, to avoid stressing into the non-linear portion of the prestressing cable's stress–strain curve and to ensure that excessive relaxation of the stress in the tendons will not occur. The limit during tensioning is covered in this clause while the limit after tensioning is covered in 2-1-1/clause 5.10.3.

The maximum force applied to a tendon, $P_{\max}$, should not exceed the value given in *2-1-1/ clause 5.10.2.1(1)P* as follows:

2-1-1/clause 5.10.2.1(1)P

$$P_{\max} = A_p \sigma_{p,\max} \qquad \text{2-1-1/(5.41)}$$

where A_p is the cross-sectional area of the tendon and $\sigma_{p,\max}$ is the maximum allowable stress for the tendon, defined as the minimum of $k_1 f_{pk}$ or $k_2 f_{p0.1k}$ where f_{pk} and $f_{p0.1k}$ are the characteristic stress and 0.1% proof stress respectively, as discussed in section 3.3. The values of k_1 and k_2 may be given in the National Annex and are recommended by EC2 to be taken as 0.8 and 0.9 respectively. It is worth noting that, for these recommended values (and assuming a minimum $f_{pk} = 1.1 f_{p0.1k}$, as discussed in section 3.3.4), the maximum permissible tendon jacking stress is slightly greater than the design yield strength ($0.88 f_{p0.1k}$ or $0.90 f_{p0.1k}$ compared to design yield of $0.87 f_{p0.1k}$ if $\gamma_s = 1.15$). For prestressing strand to EN 10138-3, typically $f_{p0.1k} = 0.86 f_{pk}$ whereupon the first limit equates to the even higher value of $0.93 f_{p0.1k}$. A similar situation arose in BS 5400 Part 4.[9]

2-1-1/clause 5.10.2.1(2) allows higher stressing, to $k_3 f_{pk}$, if the force in the jack can be measured to an accuracy of ±5% of the final value of prestressing force. The value of k_3 may be given in the National Annex and is recommended by EC2 to be taken as 0.95. This value should never be assumed during design. Its use is intended for overcoming shortfalls in prestressing force caused by unforeseen problems during construction, such as unexpectedly high friction and wobble losses in ducted tendons. (2-1-1/clause 5.10.3 requires that prestress force should be checked on site by measuring both force and tendon extensions as is normal good practice.) The decision to prestress to this elevated stress must be made in conjunction with the prestressing supplier as it carries an increased risk of strand failure during stressing.

2-1-1/clause 5.10.2.1(2)

5.10.2.2. Limitation of concrete stress
2-1-1/clause 5.10.2.2 defines several rules for prestressed concrete members to ensure that crushing or splitting of the concrete is avoided during the prestresssing operations and throughout the life of the structure. At anchorages, such effects are generally critical at initial application of the prestress due to the long-term reductions in prestress force and the strength gain of concrete with time, although the prestress force in unbonded and externally prestressed members can potentially increase to a higher value under ultimate load conditions – see section 5.10.8.

In the design of post-tensioned members, prestressing forces are applied directly to the member ends as concentrated forces from relatively small anchorages. These forces must then spread out over the cross-section of the member resulting in high local concrete stresses in this zone. The design and detailing of such end blocks is discussed further in section 8.10 of this guide. Transmission lengths in pre-tensioned members are also discussed in section 8.10.

Although rules for checks on bursting of concrete around anchorages are covered in EC2-2, no explicit check is given on crushing and splitting of the concrete directly in front of the anchor plate in post-tensioned beams. **2-1-1/clause 5.10.2.2(1)P** requires such local damage to be avoided. As in previous UK practice, it is therefore necessary to ensure that the concrete has achieved the required minimum strength quoted in the prestressing suppliers' system specification, referred to in **2-1-1/clause 5.10.2.2(2)** as the 'European Technical Approval' (ETA).

Where a tendon is loaded in stepped increments or is not stressed to the maximum force assumed in the ETA, **2-1-1/clause 5.10.2.2(4)** allows the required minimum concrete strength at transfer to be reduced, subject to an absolute minimum value. This value can be specified in the National Annex and is recommended by EC2 to be taken as 50% of the required minimum concrete strength for full prestressing according to the ETA. 2-1-1/clause 5.10.2.2(4) recommends that the required concrete strength can be obtained by linear interpolation such that a prestressing force of 30% of the maximum requires 50% of the concrete strength for full prestress and full prestress requires the full concrete strength specified in the ETA.

The compressive stress in the concrete away from anchorages should also be limited to prevent longitudinal cracking, which is undesirable from durability considerations. **2-1-1/clause 5.10.2.2(5)** defines a limit of $0.6f_{ck}(t)$, where $f_{ck}(t)$ is the characteristic compressive strength of the concrete at the time of prestressing. This limit corresponds to that provided in 2-2/clause 7.2, which is used to prevent longitudinal cracking where the element is in an aggressive environment. This limit can be increased (to a recommended value of $0.7f_{ck}(t)$, which may be varied in the National Annex) for pre-tensioned elements where tests or experience show that longitudinal cracking will not occur. If the compressive stress in the concrete exceeds $0.45f_{ck}(t)$ under the quasi-permanent combination of actions, 2-1-1/clause 5.10.2.2(5) requires non-linear creep to be considered as discussed in section 3.1.4.

No limits are given for concrete tensile stresses at transfer so it must be assumed that the serviceability limit state crack width limits of 2-2/clause 7.3 apply. The decompression check required by 2-2/Table 7.101N need only be applied at 100 mm from the strands so is inappropriate for the beam top fibre, where this is remote from the strands. Crack widths could, however, be checked and limited to 0.2 mm in accordance with 2-2/Table 7.101N. Alternatively, the National Annex may modify 2-2/Table 7.101N to give further guidance. Possibilities would be to redefine the decompression check so that it applies to the extreme fibres for checks at transfer, or to specify a limit of 1 MPa of tension as was permitted in BS 5400 Part 4.[9]

5.10.3. Prestress force

At any given time, t, and distance, x, from the stressing end of the tendon, the mean prestress force, $P_{m,t}(x)$, is equal to the maximum force applied at the jacking end (P_{max}) minus the immediate losses, $\Delta P_i(x)$, and time-dependent losses, ΔP_{c+s+r}:

$$P_{m,t}(x) = P_{max} - \Delta P_i(x) - \Delta P_{c+s+r} \qquad (D5.10\text{-}1)$$

This definition is provided in both **2-1-1/clause 5.10.3(1)P** and **2-1-1/clause 5.10.3(4)**. Care is needed in applying the above, as EC2 presents equations for short-term losses, $\Delta P_i(x)$, for a single tendon, whereas the long-term losses, ΔP_{c+s+r}, are presented for the whole group of tendons.

Whatever the initial jacking load, 2-1-1/clause 5.10.3 defines maximum permissible stresses in the tendon immediately after anchoring (post-tensioning) or transfer (pre-tensioning) including the effects of immediate losses only (i.e. at a time of $t = t_0$). The prestressing force after stressing and immediate losses is given by:

$$P_{m0}(x) = P_{max} - \Delta P_i(x) \qquad (D5.10\text{-}2)$$

This must nowhere exceed the limit given in **2-1-1/clause 5.10.3(2)**:

$$P_{m0}(x) = A_p \sigma_{pm0}(x) \qquad 2\text{-}1\text{-}1/(5.43)$$

where $\sigma_{pm0}(x)$ is the stress in the tendon at point x immediately after tensioning or transfer. 2-1-1/clause 5.10.3(2) limits $\sigma_{pm0}(x)$ to the minimum of $k_7 f_{pk}$ or $k_8 f_{p0.1k}$, where f_{pk} and $f_{p0.1k}$ are the characteristic stress and 0.1% proof stress respectively. The values of k_7 and k_8 may be given in the National Annex and are recommended by EC2 to be taken as 0.75 and 0.85 respectively. These are typically a little higher than used in previous UK bridge design practice. For prestressing strand to EN 10138-3, typically $f_{p0.1k} = 0.86 f_{pk}$ so the second limit governs, giving an allowable force of 73.1% of the characteristic tensile strength. The limit on force after tensioning was 70% of the characteristic tensile strength of the tendon in BS 5400 Part 4.[9]

2-1-1/clause 5.10.3(3)

2-1-1/clause 5.10.3(3) requires the following to be considered in determining the immediate losses, $\Delta P_i(x)$:

- losses due to the elastic deformation of concrete, ΔP_{el};
- losses due to short-term relaxation, ΔP_r (only affecting pre-tensioned members where there is a delay between stressing and transfer to the concrete);
- losses due to friction, $\Delta P_\mu(x)$;
- losses due to anchorage slip (or wedge draw-in), ΔP_{sl}.

These losses are discussed in sections 5.10.4 and 5.10.5. The time-dependent losses of prestress are designated ΔP_{c+s+r} and result from creep and shrinkage of the concrete and the long-term relaxation of the prestressing steel. Time-dependent losses are discussed in section 5.10.6.

5.10.4. Immediate losses of prestress for pre-tensioning

2-1-1/clause 5.10.4(1)

2-1-1/clause 5.10.4(1) requires the following losses to be considered for pre-tensioned members:

(1) Loss due to friction at the bends (for curved wires or strands) during the stressing process – calculation of friction losses is analogous to that for externally post-tensioned bridges discussed in section 5.10.5.
(2) Loss due to wedge draw-in of the anchorage devices. This depends on the construction process and it is not normally considered by the designer, although the losses can be calculated in the same way as for post-tensioned members discussed in section 5.10.5, if the draw-in is known.
(3) Loss due to the relaxation of the pre-tensioning tendons during the period which elapses between the tensioning of the tendons and the prestressing of the concrete. This is calculated according to 2-1-1/clause 3.3.2.
(4) Loss due to the elastic deformation of the concrete as the result of the action of the pre-tensioned tendons when they are released from the anchorages. The loss of force in each tendon of area A_p varies along its length and can be approximated from:

$$\Delta P_{el}(x) = A_p \frac{E_p}{E_{cm}(t)} \sigma_c(x) \tag{D5.10-3}$$

where $\sigma_c(x)$ is the stress in the concrete adjacent to the tendon at transfer. $E_p/E_{cm}(t)$ is the modular ratio, with the modulus for concrete based on its age at transfer. This loss will typically be a greater percentage than that for post-tensioned members for the reasons discussed in the next section. It is possible to refine this equation to allow for the change in concrete stress during transfer by adding a denominator similar to that in 2-1-1/Expression (5.46), but with a zero creep factor, ϕ, as follows:

$$\Delta P_{el}(x) = \frac{A_p \dfrac{E_p}{E_{cm}(t)} \sigma_c(x)}{1 + \dfrac{E_p}{E_{cm}(t)} \dfrac{A_p}{A_c}\left(1 + \dfrac{A_c}{I_c} z_{cp}^2\right)} \tag{D5.10-4}$$

Definitions of A_c and z_{cp} are given with the comments on 2-1-1/Expression (5.46). In equation (D5.10-4), A_p can be based on either one tendon or on a group of tendons as an approximation. In the latter case, $\sigma_c(x)$ then relates to the cable group centroid. Care is needed to be consistent with definitions.

5.10.5. Immediate losses of prestress for post-tensioning
5.10.5.1. Losses due to elastic deformation of the concrete
To perform a rigorous calculation of elastic loss in a sequentially stressed tendon group, each tendon has to be considered individually and the loss in each determined from the progressive stressing of each subsequent tendon. The change in stress induced in each tendon is determined from the change in strain induced in the adjacent concrete, averaged over the tendon's length, from stressing of each subsequent tendon. The need to use an average concrete strain arises because, where individual tendons are unbonded prior to stressing subsequent tendons, the strain changes in tendon and adjacent concrete are not constrained to be equal and the loss of force in the tendon will be uniform along its length (neglecting the effects of friction). It is therefore usual to calculate an average elastic loss for the entire length of tendon. If an individual tendon is bonded prior to stressing subsequent tendons, the loss of force in it will vary throughout its length as the change in steel strain is constrained to be the same as the change in strain of the concrete immediately adjacent. (This is the case for pre-tensioned beams as discussed in section 5.10.4.)

As a simpler alternative to separate consideration of tendons, *2-1-1/clause 5.10.5.1(2)* allows an entire group of tendons to be treated together (acting at their centroid) and the stress from tensioning the complete group used to determine an average loss. This is a common approximation, albeit sometimes slightly unconservative, which has been used in previous UK practice and which is also used in the calculation of long-term losses in 2-1-1/clause 5.10.6. An approximate formula is provided for this average loss when tendons are 'identical', by which it is meant that they are the same size and have the same initial stressing force. 2-1-1/Expression (5.44) gives the mean loss per tendon, wherein A_p is the area of one tendon. The progressive loss of prestress with sequential stressing is accounted for by way of the factor, j:

2-1-1/clause 5.10.5.1(2)

$$\Delta P_{el} = A_p E_p \sum \left[\frac{j \Delta \sigma_c(t)}{E_{cm}(t)} \right] \qquad \text{2-1-1/(5.44)}$$

where:

A_p is the cross-sectional area of a tendon
E_p is the modulus of elasticity of prestressing steel
$E_{cm}(t)$ is the modulus of elasticity of concrete at time t (see section 3.1.3)
$\Delta \sigma_c(t)$ is the variation of stress at the centroid of the tendon group applied at time t. It will include contributions from the prestress force together with any simultaneous change in other permanent actions, such as the gradual development of self-weight forces developed as a beam lifts from its formwork during stressing. As discussed above, $\Delta \sigma_c(t)$ will be an average value along the tendon group centroid where all tendons are unbonded prior to completion of stressing. It will also include stresses from variations of permanent actions applied after prestressing (e.g. removal of supports, a jacking operation or addition of superimposed dead load), but these need to be considered separately as they will have a different 'j' value

$j = (n-1)/2n$, where n is the number of 'identical' tendons successively prestressed. As an approximation, it can be taken as 0.5. Where the stress varies in the tendons due to variations of permanent actions applied after prestressing, $j = 1$ as all tendons are affected similarly. This is the reason for the '$\sum$' sign in 2-1-1/Expression (5.44). 2-1-1/Expression (5.44) can be applied to determine the total loss in a group of tendons directly if A_p is adjusted accordingly (as is done in Worked examples 5.10-1 and 5.10-3).

One final point to note is that if already installed bonded tendons are included in the beam section properties for subsequent stressing operations, the 'elastic loss' in these bonded tendons from stressing subsequent tendons will be included directly in the calculation of concrete cross-sectional stresses. It is then not necessary to apply 2-1-1/Expression (5.44) to these tendons. This approach is followed by some software.

5.10.5.2. Losses due to friction

In post-tensioned systems, prestress losses occur due to friction in the duct as a cable turns through an angle. These losses are caused both by the intended angular deviations forming the cable profile and by unintentional variations in the tendon profile, often referred to as wobble, which arise from tolerances in setting out, from sag in the ducts between duct supports and from movement of ducts during concreting. With external prestressing, friction is concentrated at the points of angular deviation.

2-1-1/clause 5.10.5.2(1) gives the following expression, from which the loss due to friction, $\Delta P_\mu(x)$, in post-tensioned tendons may be estimated:

$$\Delta P_\mu(x) = P_{\max}(1 - e^{-\mu(\theta + kx)}) \qquad \text{2-1-1/(5.45)}$$

where:

- θ is the sum of the angular deviations over a distance x (irrespective of direction or sign)
- μ is the coefficient of friction between the tendon and its duct. Values for μ depend on both tendon and duct type and are best determined from manufacturer's data for the particular prestressing system to be used. In the absence of such data, 2-1-1/Table 5.1 defines values which may be assumed for μ, reproduced here as Table 5.10-1 for convenience
- k is the 'wobble' factor to account for unintentional angular deviation of the tendon (per unit length). Again, values are best determined from manufacturer's data. *2-1-1/clause 5.10.5.2(3)* recommends values within the range $0.005 < k < 0.01$ (per metre) if no such data is available. For external prestressing tendons, *2-1-1/clause 5.10.5.2(4)* allows the wobble loss due to unintentional angles to be ignored, although strictly some allowance should be made if it is possible to have any significant angular error in setting out deviation angles on site. Usually such unintentional angles are a small fraction of the intended ones and can therefore be ignored
- x is the distance along the length of the tendon from the stressing end (i.e. the point where the stress in the tendon is equal to $P_{\max}$)

2-1-1/Expression (5.45) can be rearranged such that the tendon force at x, $P(x)$ is given by:

$$P(x)/P_{\max} = e^{-\mu(\theta + kx)} \qquad (D5.10\text{-}5)$$

For small values of $\mu(\theta + kx)$, equation (D5.10-5) may be written as:

$$P(x)/P_{\max} = 1 - \mu\theta - \mu kx \qquad (D5.10\text{-}6)$$

It can be seen that equation (D5.10-6) gives a linear reduction in force along the tendon where there is either no intentional deviation of the tendon (i.e. wobble loss only), or where the angular deviation per metre is linear (parabolic profile).

Table 5.10-1. Coefficients of friction recommended for post-tensioned tendons

	Internal tendons (for tendons filling about half of the duct)	External unbonded tendons			
		Steel duct (non-lubricated)	HDPE duct (non-lubricated)	Steel duct (lubricated)	HDPE duct (lubricated)
Cold drawn wire	0.17	0.25	0.14	0.18	0.12
Strand	0.19	0.24	0.12	0.16	0.10
Deformed bar	0.65	–	–	–	–
Smooth round bar	0.33	–	–	–	–

Note HDPE = high-density polyethylene.

It should be noted that the definition and value of the wobble factor, k, used above is not the same as is used in previous UK practice, due to the format of the loss equation. (The EC2 values of k are equivalent to those used previously within BS 5400 Part 4,[9] divided by the coefficient of friction, μ.)

5.10.5.3. Losses at anchorage

Account has to be taken of the losses due to wedge draw-in of the anchorages, during the anchoring operation after tensioning, and due to the deformation of the anchorage itself. Appropriate values of wedge draw-in are given in manufacturers' data sheets for their systems. Current systems used in the UK have a design draw-in of between 6 mm and 12 mm, although actual site draw-in values are usually lower than the quoted values for the particular system. For relatively short tendons, this loss can be particularly significant and can affect the entire length of the tendon. For long tendons, the loss from draw-in usually does not affect the whole length of the tendon due to the interaction with friction losses. Calculation of loss from anchoring is illustrated in Worked example 5.10-1.

Worked example 5.10-1: Immediate loss of prestress in a concrete box girder
Calculate the immediate loss of prestress in the three-span continuous concrete box girder illustrated in Fig. 5.10-1, assuming the section properties and material properties below and vertical prestressing profile illustrated in Fig. 5.10-2. The tendons are straight in plan.

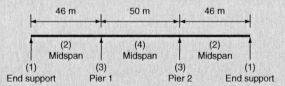

Fig. 5.10-1. Schematic elevation of three-span concrete box girder

Section properties:	Area, A_c	Inertia, I_c	Height to centroid, z_c
(1) End supports	12.337 m²	4.245 m⁴	1.0122 m
(2) Midspan 1 & 3	10.882 m²	4.370 m⁴	1.0420 m
(3) Pier 1/Pier 2	10.445 m²	4.001 m⁴	0.9706 m
(4) Midspan 2	10.882 m²	4.370 m⁴	1.0420 m

Material properties:
Deck concrete class is C40/50, thus $f_{ck} = 40$ MPa and, from 2-1-1/Table 3.1, $E_{cm} = 35$ GPa. Prestressing comprises low-relaxation strand with $f_{pk} = 1820$ MPa, coefficient of friction, $\mu = 0.19$ and wobble factor, $k = 0.005/\text{m}$. The tendons comprise 28 strands with a total area of 4620 mm² and each of the 24 tendons in the length of the deck is stressed to $0.70 f_{pk}$ from both ends. The stressing force per tendon is therefore $0.70 \times 4620 \times 1820 \times 10^{-3} = 5886$ kN. The average E_p for the particular batch of prestressing strand is 200 GPa. (195 GPa would usually be assumed in the absence of such information in accordance with 2-1-1/clause 3.3.6.) The draw-in at anchorages is 6 mm from the supplier's data.

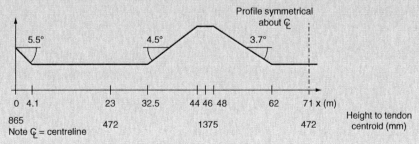

Fig. 5.10-2. Idealized elevation of vertical tendon profiles

First, the elastic shortening losses are found using the simplified formula for average loss for the tendon group centroid given in 2-1-1/clause 5.10.5.1. The stresses in the concrete at the level of the tendon centroids is found using the expression $\sigma_c = P_0/A_c + P_0 e^2/I_c$. The stresses due to beam self-weight and prestress secondary moments, which develop during stressing, are conservatively ignored here as they reduce the concrete stress. The tendon force ignoring friction losses is conservatively used:

(1) End supports:

$$\sigma_c = \frac{24 \times 5886 \times 10^3}{12.337 \times 10^6} + \frac{24 \times 5886 \times 10^3 \times (1012.2 - 865)^2}{4.245 \times 10^{12}}$$

$$= 11.45 + 0.72 = 12.17 \, \text{MPa}$$

(2) Midspan:

$$\sigma_c = \frac{24 \times 5886 \times 10^3}{10.882 \times 10^6} + \frac{24 \times 5886 \times 10^3 \times (1042 - 472)^2}{4.370 \times 10^{12}}$$

$$= 12.98 + 10.50 = 23.48 \, \text{MPa}$$

(3) Piers:

$$\sigma_c = \frac{24 \times 5886 \times 10^3}{10.445 \times 10^6} + \frac{24 \times 5886 \times 10^3 \times (970.6 - 1375)^2}{4.001 \times 10^{12}}$$

$$= 13.52 + 5.77 = 19.30 \, \text{MPa}$$

The average stress in each constituent straight length needs to be estimated:

$0 < x \le 4.1$: Average stress $= (12.17 + 23.48)/2 = 17.8 \, \text{MPa}$
$4.1 < x \le 32.5$: Average stress $= 23.5 \, \text{MPa}$
$32.5 < x \le 44$: Average stress $= 17 \, \text{MPa}$ approximately (as the stress goes from $23.48 \, \text{MPa}$ where the tendon is at the soffit, to approximately $13 \, \text{MPa}$ as it reaches the section neutral axis and back up to $19.3 \, \text{MPa}$ at the top of its profile)
$44 < x \le 48$: Average stress $= 19.3 \, \text{MPa}$
$48 < x \le 62$: Average stress $= 17 \, \text{MPa}$ assumed as above
$62 < x \le 71$: Average stress $= 23.5 \, \text{MPa}$

The average stress in the concrete adjacent to the tendon over its length is therefore:

$$\sigma_c = \frac{4.1 \times 17.8 + 28.4 \times 23.5 + 11.5 \times 17 + 4.0 \times 19.3 + 14.0 \times 17 + 9.0 \times 23.5}{71}$$

$$= 20.6 \, \text{MPa}$$

(conservative, given the neglect of self-weight stresses).

The elastic loss is calculated from 2-1-1/Expression (5.44). $j = (n-1)/2n = (24-1)/48 = 0.479$. It is assumed that the tendons would be stressed when the concrete reached an age approaching 28 days, such that from 2-1-1/clause 3.1.3, $E_{cm}(t) \approx E_{cm}$.

From 2-1-1/Expression (5.44):

$$\Delta P_{el} = A_p E_p \sum \left[\frac{j \Delta \sigma_c(t)}{E_{cm}(t)} \right] = 4620 \times 200 \times 10^3 \times 0.479 \times \frac{20.6}{35 \times 10^3} \times 10^{-3} = 260 \, \text{kN}$$

which represents a loss of $260/5886 = 0.044$, i.e. **4.4%** – a loss factor of 0.956. This value would be reduced by considering the self-weight and secondary prestress stresses that come on to the beam during stressing. They would need to be considered in order to meet the allowable compressive stress limit of $0.6 f_{ck}(t) = 0.6 \times 40 = 24 \, \text{MPa}$ from 2-1-1/clause 5.10.2.2(5) which would otherwise be exceeded at an extreme fibre in this example.

Second, the losses due to friction can be calculated by considering the sum of deviated angles at various sections along the deck from the stressing end. In this example, all the friction losses have been assumed to occur at the sharp changes in the idealized prestressing profile. In reality, for internal post-tensioned structures, the friction losses will occur gradually over the length of the tendon.

A diagram can be constructed illustrating the friction (and anchorage draw-in) loss as a factor of the initial prestressing force. Thus, the following values have been used to construct Fig. 5.10-3:

At $x = 0$ m:

$\Delta\theta = 0°$, from equation (D5.10-5), $P(x)/P_{max} = 1.0$

At $x = 4.1$ m:

$\Delta\theta = 0°$ giving $P(x)/P_{max} = e^{-0.19(0.0 + 0.005 \times 4.1)} = 0.996$

$\Delta\theta = 5.5°$ (0.096 rad) giving $P(x)/P_{max} = e^{-0.19(0.096 + 0.005 \times 4.1)} = 0.978$

At $x = 23$ m:

$\Delta\theta = 5.5°$ (0.096 rad) giving $P(x)/P_{max} = 0.961$

At $x = 32.5$ m:

$\Delta\theta = 5.5°$ (0.096 rad) giving $P(x)/P_{max} = 0.952$

$\Delta\theta = 10°$ (0.175 rad) giving $P(x)/P_{max} = 0.938$

At $x = 44$ m:

$\Delta\theta = 10°$ (0.175 rad) giving $P(x)/P_{max} = 0.928$

$\Delta\theta = 14.5°$ (0.253 rad) giving $P(x)/P_{max} = 0.914$

At $x = 46$ m:

$\Delta\theta = 14.5°$ (0.253 rad) giving $P(x)/P_{max} = 0.912$

At $x = 48$ m:

$\Delta\theta = 14.5°$ (0.253 rad) giving $P(x)/P_{max} = 0.911$

$\Delta\theta = 18.2°$ (0.318 rad) giving $P(x)/P_{max} = 0.899$

At $x = 62$ m:

$\Delta\theta = 18.2°$ (0.318 rad) giving $P(x)/P_{max} = 0.888$

$\Delta\theta = 21.9°$ (0.382 rad) giving $P(x)/P_{max} = 0.877$

At $x = 71$ m:

$\Delta\theta = 21.9°$ (0.382 rad) giving $P(x)/P_{max} = 0.869$

For double-end stressing and symmetrical spans, the friction losses are symmetrical about the centre of the second span.

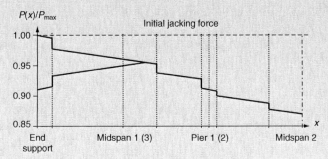

Fig. 5.10-3. Loss of prestress from friction and anchorage draw-in

Finally, the loss in prestress due to the anchorage draw-in is obtained by considering the loss diagram illustrated in Fig. 5.10-4. The total initial extension e of the cable over its length, L, is given by $e = \int_0^L \varepsilon(x)\,dx$, where $\varepsilon(x)$ is the tendon strain at distance x so that $\varepsilon(x) = P(x)/A_p E_p$. The extension then becomes $e = (1/A_p E_p) \int_0^L P(x)\,dx$. This can be seen to be $1/A_p E_p$ times the area under the cable force–distance curve. When a draw-in δ_{ad} occurs, the force–distance curve changes, as shown in Fig. 5.10-4, with the friction reversing at the end near the anchorage. The new curve is approximately a mirror image of the original over the length from the stressing face to the point where the curves meet. (This symmetry follows from the linearized version of the friction equation in (D5.10-6).) Since the area under the new curve times $1/A_p E_p$ is the new cable extension, and this must be an amount δ_{ad} less than the original extension, the shaded area A_{ad} between the two curves times $1/A_p E_p$ must equal the draw in δ_{ad} so that $A_{ad} = \delta_{ad} E_p A_p$.

Thus, for this example, consider the 6 mm draw-in for one tendon:

$$A_{ad} = 6 \times 200 \times 10^3 \times 4620 \times 10^{-6} = 5544\,\text{kNm}$$

and the force–distance diagram can be constructed as shown in Fig. 5.10-4.

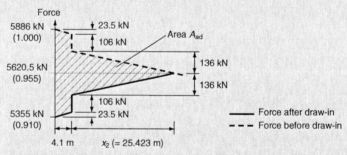

Fig. 5.10-4. Anchorage draw-in loss

The shaded area in Fig. 5.10-4 with $x_2 = 25.423$ m is: $(23.5 + 2 \times 106 + 2 \times 136) \times 4.1 + 136 \times 25.423 = 2081 + 3458 = 5539\,\text{kNm} \approx 5544\,\text{kNm}$ as required. The loss at the stressing end is therefore seen to be 9%. The draw-in loss does not extend beyond the mid-point of the cable so there is no interaction with the draw-in loss from anchoring at the other end. The total final losses along the length of the tendon are those shown in Fig. 5.10-3 multiplied by a further elastic loss factor of 0.956 as calculated above.

5.10.6. Time-dependent losses

Over time, further losses of prestressing force occur due to the reduction of steel strain caused by the deformation of the concrete due to creep and shrinkage and the reduction of stress in the steel due to its relaxation under tension. Relaxation losses are sensitive to variations in stress levels and can therefore be reduced by taking account of other time-dependent losses occurring within the structure at the same time. The Note to *2-1-1/clause 5.10.6(1)* notes that the relaxation of steel depends also on the reduction of steel strain caused by creep and shrinkage of the concrete. This reduces the steel force and hence the relaxation loss. This can approximately be accounted for by using a reduction factor of 0.8 applied to the relaxation loss calculated, based on the initial stress in the prestressing after anchoring according to 2-1-1/clause 3.3.2.

A method for determining reduced relaxation losses under such circumstances is given in 2-1-1/Annex D and discussed in Annex D of this guide. It has been previous UK practice to base design on the relaxation loss at 1000 hours without considering the interaction with creep and shrinkage, and the example in Annex D suggests that this is generally a reasonable approximation. For low relaxation strand, it is generally not worth the additional calculation effort of using Annex D.

2-1-1/clause 5.10.6(1)

CHAPTER 5. STRUCTURAL ANALYSIS

To rigorously account for creep, shrinkage and relaxation losses usually requires a computer program because the losses produced in an interval of time affects the state of stress and therefore also the creep and relaxation losses over the next interval of time. This is discussed in more detail in Annex K of this guide. *2-1-1/clause 5.10.6(2)* therefore gives the following simplified expression to evaluate the time-dependent losses at a point, x, under the permanent loads for the group of tendons:

2-1-1/clause 5.10.6(2)

$$\Delta P_{c+s+r} = A_p \Delta \sigma_{p,c+s+r} = A_p \frac{\varepsilon_{cs} E_p + 0.8 \Delta \sigma_{pr} + \dfrac{E_p}{E_{cm}} \phi(t,t_0) \sigma_{c,QP}}{1 + \dfrac{E_p}{E_{cm}} \dfrac{A_p}{A_c} \left(1 + \dfrac{A_c}{I_c} z_{cp}^2 \right)(1 + 0.8\phi(t,t_0))} \qquad \text{2-1-1/(5.46)}$$

where all compressive stresses and corresponding strains should be taken as positive and:

$\Delta \sigma_{p,c+s+r}$ is the absolute value of the variation of stress in the tendons due to creep, shrinkage and relaxation at location x, at time t

ε_{cs} is the estimated shrinkage strain in accordance with 2-1-1/clause 3.1.4(6)

E_p is the modulus of elasticity for the prestressing steel

E_{cm} is the short-term modulus of elasticity for the concrete from 2-1-1/Table 3.1

$\Delta \sigma_{pr}$ is the absolute value of the variation of stress in the tendons at location x, at time t, due to the relaxation of the prestressing steel. This is defined as being determined for a stress of $\sigma_p(G + P_{m0} + \psi_2 Q)$, i.e. the initial stress in the tendons due to the initial prestress and quasi-permanent actions

$\phi(t, t_0)$ is the creep coefficient at a time t for initial load application at time t_0

$\sigma_{c,QP}$ is the stress in the concrete adjacent to the tendons due to self-weight, initial prestress and all other quasi-permanent actions where relevant

A_p is the area of all the prestressing tendons at the section being considered – note that elsewhere it is defined as the area of one tendon. Either definition can be used providing it is used consistently

A_c is the area of the concrete section

I_c is the second moment of area of the concrete section

z_{cp} is the eccentricity of the tendons, i.e. the distance between the centroid of the tendons and the centroid of the concrete section

The geometric terms above are shown in Fig. 5.10-5 for a typical beam. A more rigorous analysis would need to consider the change of concrete stresses caused by the losses throughout and the effect of this on the creep losses still to take place. If the above formula is applied to unbonded tendons, then mean stresses averaged along the tendon must be used as identified by *2-1-1/clause 5.10.3(3)*.

2-1-1/clause 5.10.3(3)

The derivation of 2-1-1/Expression (5.46) can be illustrated for bonded tendons. For the beam in Fig. 5.10-5, a change in prestress stress $\Delta \sigma_p$ leads to a change in concrete stress given by:

$$\Delta \sigma_c = \frac{\Delta \sigma_p A_p}{A_c}\left(1 + \frac{A_c z_{cp}^2}{I_c}\right) \qquad \text{(D5.10-7)}$$

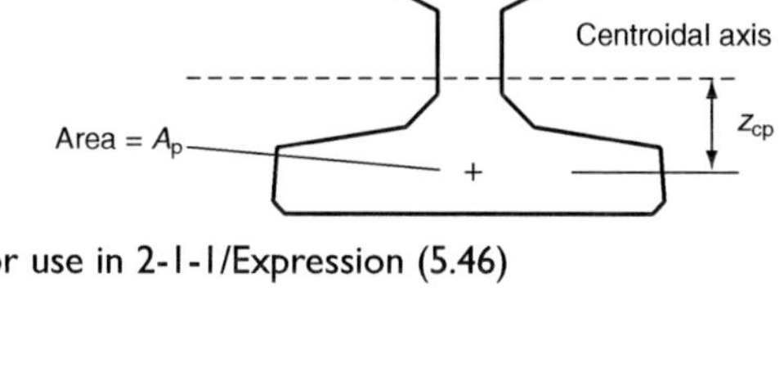

Fig. 5.10-5. Properties for use in 2-1-1/Expression (5.46)

The losses for shrinkage, relaxation and creep are then derived individually using the above equilibrium equation and the following strain compatibility relationships.

Shrinkage loss

The free shrinkage strain is ε_{cs}. However, as the concrete shrinks, the prestressing steel must compress by the same strain which causes a loss of prestress and a change in the concrete stress. The net concrete strain is therefore $\varepsilon_{cs} - \Delta\sigma_c/E_{ce}$, where E_{ce} is the effective concrete modulus which must allow for creep as the change in stress occurs over some time. This strain equals the change in prestress strain, thus:

$$\varepsilon_{cs} - \frac{\Delta\sigma_c}{E_{ce}} = \frac{\Delta\sigma_p}{E_p} \quad \text{(D5.10-8)}$$

From equation (D5.10-7) and (D5.10-8), the prestress loss due to concrete shrinkage is:

$$\Delta\sigma_{p,s} = \frac{\varepsilon_{cs}E_p}{1 + \frac{E_p}{E_{ce}}\frac{A_p}{A_c}\left(1 + \frac{A_c}{I_c}z_{cp}^2\right)} \quad \text{(D5.10-9)}$$

Relaxation loss

The unrestrained relaxation of stress at constant strain in the prestressing is $\Delta\sigma_{pr}$. However, as this loss of prestress leads to a change of concrete stress $\Delta\sigma_c$ and hence a change in concrete strain, it must also lead to a change of strain in the tendon so that the actual loss of prestress $\Delta\sigma_p$ is as follows:

$$\frac{\Delta\sigma_{pr} - \Delta\sigma_p}{E_p} = \frac{\Delta\sigma_c}{E_{ce}} \quad \text{(D5.10-10)}$$

From equation (D5.10-7) and (D5.10-10), the prestress loss due to steel relaxation is:

$$\Delta\sigma_{p,r} = \frac{\Delta\sigma_{pr}}{1 + \frac{E_p}{E_{ce}}\frac{A_p}{A_c}\left(1 + \frac{A_c}{I_c}z_{cp}^2\right)} \quad \text{(D5.10-11)}$$

Creep loss

The free creep strain is $\sigma_c\phi(t,t_0)/E_c$, where σ_c is the initial concrete stress adjacent to the tendons. However, as the concrete creeps, the prestressing steel must change strain by the same amount, which causes a change in prestress and a change in the concrete stress. The net concrete strain is therefore $\sigma_c\phi(t,t_0)/E_c - \Delta\sigma_c/E_{ce}$. This strain equals the change in prestress strain, thus:

$$\frac{\sigma_c}{E_c}\phi(t,t_0) - \frac{\Delta\sigma_c}{E_{ce}} = \frac{\Delta\sigma_p}{E_p} \quad \text{(D5.10-12)}$$

This equation ignores the fact that losses in prestress will alter the value of σ_c and hence the creep strain and iteration is necessary to get the correct answer. From (D5.10-7) and (D5.10-12), the prestress loss due to creep is:

$$\Delta\sigma_{p,c} = \frac{\frac{E_p}{E_c}\phi(t,t_0)\sigma_c}{1 + \frac{E_p}{E_{ce}}\frac{A_p}{A_c}\left(1 + \frac{A_c}{I_c}z_{cp}^2\right)} \quad \text{(D5.10-13)}$$

In all the losses above, the effective concrete modulus E_{ce} must allow for creep. For a constant stress applied at time t_0, the relevant modulus would be $E_{cm}/(1 + \phi(t,t_0))$ (taking $E_c = E_{cm}$ rather than $E_c = 1.05E_{cm}$ as specified in 2-1-1/clause 3.1.4, but there is only 5% difference). However, as the modulus is in each case needed to calculate the stress from a concrete strain that occurs slowly with time, a more appropriate modulus is $E_{ce} = E_{cm}/(1 + 0.8\phi(t,t_0))$. The 0.8 multiplier on the creep factor is the equivalent of the ageing coefficient discussed in Annex K. If this modulus is substituted in the above equations, the relaxation loss is multiplied by the factor 0.8, discussed at the start of this section, and the three losses are added, the total loss in 2-1-1/Expression (5.46) is obtained.

The denominator of 2-1-1/Expression (5.46) effectively makes allowance for the resistance provided by the prestressing steel in resisting the shortening of the concrete, thus reducing the concrete strain change and hence the prestress loss. Often the denominator approximates to 1.0 and may conservatively be taken as such so that the equation simplifies to a form which will be familiar to most engineers. A more rigorous analysis would need to consider the change of concrete stress caused by the losses throughout and the effect of this on the creep and relaxation losses still to take place. A general method as in Annex K2 of this guide could be used which would require a computer program. If 2-1-1/Annex D is used to calculate a reduced relaxation loss, a further reduction should not be obtained by using this loss in 2-1-1/Expression (5.46) in conjunction with the 0.8 factor; the 0.8 factor should be omitted. 2-1-1/Expression (5.46) can be applied in one go, with some simplifications, at $t = \infty$ to give long-term losses, as illustrated in Worked example 5.10-3.

Staged construction
If staged construction is used, the losses will need to be calculated over several time intervals and summed. Where dead load and prestress effects accumulate with time, the creep factor, in principle, needs to be obtained for each additional loading depending on the age of the concrete and the loss for each determined to time infinity and summed. When considering each additional loading, only the increment of stress should be considered in calculating the creep loss from this particular loading. For simplicity, as additional stages of construction often reduce the concrete stress at the level of the tendons, the total loss can often conservatively be based on the initial loading where this produces the greatest stress adjacent to the tendons. (Alternatively, if little creep occurs before the concrete stress is reduced by a further stage of construction, the full creep loss could be based on this lower stress, although this would be slightly unconservative.) Loss calculation in staged construction is complex and simplifications like the above usually have to be made if a computer program is not used.

For indeterminate structures built by staged construction, the deformations associated with creep lead to the development of restraint moments and a further change of stress. Methods of calculating this creep redistribution are discussed in Annex K.

Composite construction
Where prestressed beams are used which are subsequently made composite with a deck slab, the losses from 2-1-1/Expression (5.46) need to be calculated and applied in two phases. First, the loss occurring prior to casting the deck slab should be calculated. The effects of this loss of prestress are determined by proportionately reducing the prestress stress calculated on the prestressed beam section alone. Second, the remaining loss after casting the deck slab needs to be calculated. Since this loss of force is applied to the composite section, the effects of this loss are best represented by applying the loss of force as a tensile load to the composite section at the level of the centroid of the tendons. In determining the loss of force before and after casting the top slab, it is simplest in using 2-1-1/Expression (5.46) to take $\sigma_{c,QP}$ as the stress due to beam self-weight and prestress only in both phases. For simplicity, it is also possible to take the denominator as unity to avoid the problem of the section properties differing in the two phases. With these two assumptions, 2-1-1/Expression (5.46) need only be calculated once (based on the beam only case). The loss of force occurring before and after casting the slab should still be applied to the beam only section and composite section in the appropriate ratios (based on lapsed time) when determining the change in concrete stress from this loss.

For composite beams, additional stresses are set up by differential creep and differential shrinkage, and these are discussed in Annex K4 of this guide. Creep also redistributes moments for beams that are made continuous, as discussed in Annex K. Stress checks in a pre-tensioned beam made composite are illustrated in Worked example 5.10-3.

Worked example 5.10-2: Time-dependent loss of prestress in a concrete box girder

Find the long-term loss of prestress at midspan 1 for the box girder in Worked example 5.10-1, assuming the combined total bending moment from secondary prestress and permanent load to be $M = 47\,654$ kNm. The entire bridge is to be cast on falsework and it is assumed that the tendons will be stressed when the concrete is an average of 30 days old. The average effective thickness of the cross-section, h_0, is 300 mm and the relative humidity is 70%. Low relaxation strand is used.

The shrinkage, creep and relaxation parameters are first determined.

2-1-1/clause 3.1.4

Shrinkage – from *2-1-1/clause 3.1.4*
The total autogenous shrinkage strain is, from 2-1-1/Expression (3.12):

$$\varepsilon_{ca}(\infty) = 2.5(f_{ck} - 10) \times 10^{-6} = 2.5 \times (40 - 10) \times 10^{-6} = 75 \times 10^{-6}$$

but after 30 days, a fraction $\beta_{as}(t) = 1 - \exp(-0.2t^{0.5}) = 1 - \exp(-0.2 \times 30^{0.5}) = 0.666$ has occurred so the remaining autogenous shrinkage after stressing is $(1 - 0.666) \times 75 \times 10^{-6} = 25.1 \times 10^{-6}$.

The total drying shrinkage is, by interpolation in 2-1-1/Table 3.2, $\varepsilon_{cd,0} = 310 \times 10^{-6}$. From 2-1-1/Table 3.3 with $h_0 = 300$ mm, this is modified by $k_h = 0.75$ to be $0.75 \times 310 \times 10^{-6} = 232.5 \times 10^{-6}$. After 30 days and assuming three days curing period, a fraction

$$\beta_{ds}(t, t_s) = \frac{(t - t_s)}{(t - t_s) + 0.04\sqrt{h_0^3}} = \frac{(30 - 3)}{(30 - 3) + 0.04\sqrt{300^3}} = 0.115$$

has occurred so the remaining drying shrinkage after stressing is $(1 - 0.115) \times 232.5 \times 10^{-6} = 206 \times 10^{-6}$. The total shrinkage to consider for loss calculation is therefore $25 \times 10^{-6} + 206 \times 10^{-6} = \mathbf{231 \times 10^{-6}}$.

2-1-1/clause 3.1.4

Creep coefficient – from *2-1-1/clause 3.1.4* or Annex B
Prestress and self-weight are applied when the concrete is 30 days old. $\phi(\infty, t_0)$ for $t_0 = 30$ days is found to be equal to **1.5**. (For a calculation of a typical creep factor, see section 3.1.4 of this guide.) This creep factor is used as an approximation for the superimposed dead load (SDL) contribution also, as it was assumed to be applied shortly after. In general, SDL can be ignored in this calculation as it reduces the concrete stress adjacent to the tendons.

2-1-1/clause 3.3.2

Relaxation – from *2-1-1/clause 3.3.2*
From Worked example 3.3-1, the long-term relaxation is typically of the order of 4% for low relaxation strand, therefore $\Delta\sigma_{pr} = 0.04 \times 0.70 \times 1820 = 51$ MPa. This is conservatively based on the jacking force, when the force after lock-off could have been used.

The stress in the concrete at the tendon level under quasi-permanent loads is next calculated, taking into account the immediate losses calculated in the previous example.

The concrete stress at the tendon centroid is:

$$\sigma_{c,QP} = 0.945 \times 0.956 \times \left(\frac{24 \times 5886 \times 10^3}{10.882 \times 10^6} + \frac{24 \times 5886 \times 10^3 \times (1042 - 472)^2}{4.370 \times 10^{12}}\right)$$

$$- \frac{47\,654 \times 10^6 \times (1042 - 472)}{4.370 \times 10^{12}} = 11.727 + 9.487 - 6.216 = 14.998 \text{ MPa}$$

The concrete stress at the extreme fibre is:

$$\sigma_{bot} = 0.945 \times 0.956 \times \left(\frac{24 \times 5886 \times 10^3}{10.882 \times 10^6} + \frac{24 \times 5886 \times 10^3 \times (1042 - 472) \times 1042}{4.370 \times 10^{12}}\right)$$

$$- \frac{47\,654 \times 10^6 \times (1042)}{4.370 \times 10^{12}} = 11.727 + 17.345 - 11.363 = 17.709 \text{ MPa}$$

CHAPTER 5. STRUCTURAL ANALYSIS

This meets the allowable compressive stress limit of $0.6 \times 40 = 24$ MPa from 2-1-1/clause 5.10.2.2(5) and the similar limit given in 2-2/clause 7.2. The stress also is below the limit of $0.45 \times 40 = 18.0$ MPa at which non-linear creep can occur according to the same clauses. If the stress had exceeded this limit, the creep factor would have needed adjustment according to 2-1-1/clause 3.1.4(4), basing the calculation on the stress at the extreme fibre.

The loss due to long-term creep, shrinkage and relaxation is given by 2-1-1/Expression (5.46).

Thus, for midspan (1):

$$\Delta\sigma_{p,c+s+r}$$

$$= \frac{231 \times 10^{-6} \times 200 \times 10^3 + 0.8 \times 51 + \frac{200}{35} \times 1.5 \times 14.998}{1 + \frac{200}{35} \times \frac{24 \times 4620}{10.882 \times 10^6} \times \left(1 + \frac{10.882 \times 10^6}{4.370 \times 10^{12}} \times (1042 - 472)^2\right) \times (1 + 0.8 \times 1.5)}$$

$$= \frac{215.6}{1.232} = 175.0 \text{ MPa and } \Delta P_{c+s+r} = 24 \times 4620 \times 175.0 \times 10^{-3}$$

$$= 19\,404 \text{ kN } (13.7\%)$$

The total loss from elastic loss, friction and draw-in and creep, shrinkage and relaxation $= 4.4 + 5.5 + 13.7 = 23.6\%$.

If the spans were built by staged construction, an additional redistribution of moment due to creep would need to be considered in working out long-term stresses – see Annex K.

5.10.7. Consideration of prestress in the analysis

2-1-1/clause 5.10.7 gives miscellaneous requirements for the treatment of prestress in analysis, in addition to the general statements in 2-1-1/clause 5.10.1. The main issues raised are:

- the difference in behaviour between external and internal post-tensioning;
- the difference in behaviour between bonded and unbonded prestress;
- treatment of primary and secondary effects of prestress.

External versus internal post-tensioning

There are two main differences between external and internal post-tensioning. First, second-order effects can arise from prestressing with external tendons – *2-1-1/clause 5.10.7(1)*. This occurs because of the lack of continuous contact between tendon and concrete so that the tendon does not everywhere deflect by the same amount as the concrete member. This can lead to a loss of prestress moment under applied load, as shown in Fig. 5.10-6 for an extreme case, or can similarly lead to an increase in prestressing moment where the first-order

2-1-1/clause 5.10.7(1)

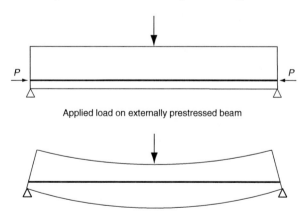

Fig. 5.10-6. Second-order effects in externally post-tensioned beam with no intermediate deviators

deflections are dominated by the pretressing. This effect must be considered in the member design. The impact is greatly reduced by the provision of intermediate deviators, constraining the tendon to move with the concrete. **2-1-1/clause 5.10.7(6)** allows external tendons to be considered to be straight between deviators, i.e. the effect of sag under their self-weight can be ignored.

Second, by comparison with bonded internal post-tensioning only, a different approach to calculation of the ultimate flexural strength for externally post-tensioned members is required, as discussed in section 5.10.8. This is because the strain in external tendons does not increase at the same rate as the strain in the surrounding concrete.

Bonded and unbonded prestress

Where the prestress is unbonded, the strain in tendon and surrounding concrete are not equal. This leads to the different treatment of ultimate flexural strength discussed in section 5.10.8. Similarly, losses of prestress depend on the strain in the adjacent concrete averaged along the tendon's length rather than varying continuously along the tendon length with the local concrete strain. This is discussed in section 5.10.5 above.

Primary and secondary effects of prestress

For statically determinate members, the prestress moment at any section is given by Pe; the axial force of the prestress at the section multiplied by the eccentricity of the tendon centroid to the cross-section centroid. This is known as the primary prestress moment. Secondary or 'parasitic' moments may be introduced due to prestressing of statically indeterminate structures. These arise due to the restraint by the supports of the deflections caused by the prestressing. These secondary moments are often very significant and, unlike the name might suggest, should never be neglected.

To illustrate the source of these secondary prestress moments, consider the two-span continuous bridge deck in Fig. 5.10-7(a), which has a constant axial prestressing force, P,

2-1-1/clause 5.10.7(6)

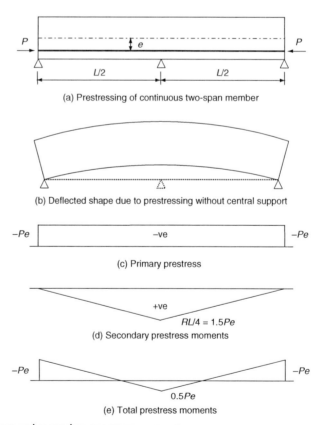

Fig. 5.10-7. Primary and secondary prestress moments

acting at a constant eccentricity, e. First, assuming the central support is unable to resist any vertical upward movement, the deflected shape of the beam due to prestressing would be as shown in Fig. 5.10-7(b). This structure is now statically determinate and the prestress moment at any section would be the primary moment, Pe, as shown in Fig. 5.10-7(c). In practice, however, the continuous beam is restrained at the central support and a downward reaction, R, must be applied in order to maintain the compatibility of zero deflection there. This applied force induces secondary moments in the beam, which are illustrated in Fig. 5.10-7(d). For this case, the force R is $6Pe/L$ and the maximum secondary moment is $1.5Pe$. Note that the secondary moment distribution varies linearly between supports since the secondary moments are produced only by the support reactions induced by the prestressing. Figure 5.10-7(e) shows the final total distribution of primary and secondary moments along the beam.

If elastic global analysis is used, the secondary moments are assumed to remain locked in at ULS and are considered on the loading side when checking the bending resistance. *2-1-1/clause 5.10.7(3)* allows redistribution of moments to be carried out in accordance with 2-1-1/clause 5.5 providing the primary and secondary prestressing moments are applied before the redistribution is carried out. If plastic global analysis is carried out, *2-1-1/clause 5.10.7(4)* suggests that the secondary moments are treated as additional plastic rotations at the supports and included in the check of rotation capacity. The plastic rotations can be determined as discussed in section 5.6.3.2 of this guide.

2-1-1/clause 5.10.7(3)

2-1-1/clause 5.10.7(4)

5.10.8. Effects of prestressing at the ultimate limit state

The design value of the prestressing force at the ultimate limit state is defined in *2-1-1/clause 5.10.8(1)* as $P_{d,t}(x) = \gamma_P P_{m,t}(x)$, where $P_{m,t}(x)$ is the mean value of the prestress force at time t and distance x, discussed in section 5.10.3 above.

2-1-1/clause 5.10.8(1)

The ultimate bending resistance of beams with bonded tendons is discussed in section 5.10.1 of this guide and Worked examples are presented in section 6.1. Un-bonded tendons, however, do not undergo strain increases at the same rate as in the adjacent concrete section. Any increase in tendon strain arises only from overall deformation of the structure. This can often lead to only relatively small increases in tendon force (which can conservatively be ignored), making the ultimate limit state critical. *2-1-1/clause 5.10.8(2)* allows an increase of stress from the effective prestress to the stress at ultimate limit state to be assumed without any calculation. This assumed increase, $\Delta\sigma_{p,ULS}$, may be given in the National Annex and is recommended by EC2 to be taken as 100 MPa. Generally, this value will be suitably conservative. Some caution is, however, required. The strain increase in a tendon that does not follow the bending moment profile and passes through areas where the concrete is in compression could be less than this value. Such arrangements are not uncommon. The top flange cantilevering tendons used in balanced cantilever designs, for example, frequently extend from hogging zones over the piers into sagging zones at midspan while remaining in the top flange. Caution should also be exercised where the tendons have very low eccentricity, as, again, the 100 MPa might not be achieved.

2-1-1/clause 5.10.8(2)

It might also appear optimistic to assume a stress increase of 100 MPa, where a tendon is initially stressed to its maximum permissible limit from 2-1-1/clause 5.10.2.1 and little loss has occurred at the time considered. In such a case, the assumed stress increase might take the tendon beyond its design proof stress, as illustrated in Fig. 5.10-8. If 'design' material were present throughout the tendon, the strain increase and hence overall extension needed to generate the 100 MPa stress might not be achievable from the overall structural deformation. It can be argued, however, that 'design' tendon material would only be present locally at the critical section and would not therefore significantly alter the overall stiffness of the tendon (other than for very short tendons). A tendon stress increase of 100 MPa would not be sufficient to reach the inclined branch of the 'mean' or even characteristic prestress stress–strain curve, given the restrictions on allowable force after lock-off in 2-1-1/clause 5.10.3, and would not therefore require significant tendon extension.

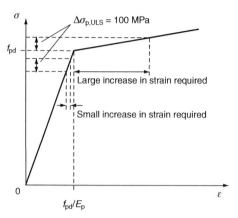

Fig. 5.10-8. Strain increase for external tendons with high initial stress

Where the overall deformation of the bridge is to be considered in deriving the increase in prestressing force, *2-2/clause 5.10.8(103)* requires a non-linear analysis to be used in accordance with 2-2/clause 5.7. Although not explicitly stated, this analysis should also take into account the effects of tension stiffening of the concrete away from the critical section. This is to ensure the strain increase is not overestimated due to overestimation of the total beam deformation. Away from the critical section, the stiffening effects from tension stiffening between cracks and from entire sections remaining uncracked in flexure should be considered. Mean values of concrete tensile strength are most representative of the values to be found in the real structure but, for safety, it would be appropriate to use upper characteristic values in determining the adverse effects of tension stiffening.

2-2/clause 5.7 does not provide unique requirements for other material properties and allows the use of either something close to mean properties or the use of design properties, as discussed in section 5.7 of this guide. In either case, the adverse effects of tension stiffening discussed above should be included away from the critical section as otherwise structure deformations and tendon strain increases may be overestimated. The former method gives a verification of the member rather than an explicit calculation of stress increase. Analysis with design properties has the advantage that the non-linear analysis is then itself the verification of the structure and the actual strain increase achieved is not itself important; if convergence is achieved in the analysis under the ultimate limit state applied loads, the bridge is adequate under that load case. Analysis with design properties has been commonly used in the UK for this type of analysis in the past because of its greater convenience. It may be slightly less conservative than the other analysis method proposed in 2-2/clause 5.7, due to the slightly greater deformation away from the critical section due to the greater extent of design material assumed.

An alternative to using a non-linear analysis is to calculate the strain increase from a linear elastic model using uncracked section properties. This will give some stress increase, but probably less than that permitted in 2-1-1/clause 5.10.8(2) without calculation.

Non-linear analysis can sometimes illustrate a problem with short tendons in highly stressed areas. It is possible for short tendons to reach stresses in excess of yield at the ultimate limit state (for example, those for the shortest tendons in balanced cantilever construction) and therefore care must be taken in designing the bursting zones accordingly for the full characteristic breaking load of the tendon. Section 8.10 of this guide gives further commentary on the design of bursting zones for prestressed concrete anchorages. A greater potential problem with short yielding tendons is that the tendons themselves might fail at the anchorage wedges at a stress just below their characteristic tensile strength if the strain were sufficiently high.

5.10.9. Effects of prestressing at the serviceability and fatigue limit states

For serviceability limit state verifications, *2-1-1/clause 5.10.9(1)P* requires allowance to be made in the design for possible variations in prestress. Consideration of this variation is

CHAPTER 5. STRUCTURAL ANALYSIS

achieved through the definition of an upper and a lower characteristic value of the prestressing force:

$$P_{k,\sup} = r_{\sup}P_{m,t}(x) \qquad 2\text{-}1\text{-}1/(5.47)$$

$$P_{k,\inf} = r_{\inf}P_{m,t}(x) \qquad 2\text{-}1\text{-}1/(5.48)$$

The values of $r_{\sup}$ and $r_{\inf}$ are defined in the National Annex but the recommended values are 1.05 and 0.95 respectively for pre-tensioning or un-bonded tendons, and 1.10 and 0.90 respectively for post-tensioning with bonded tendons. The formal use of favourable and unfavourable values of prestress at the serviceability limit state was not found in BS 5400 Part 4[9] and does not seem to be warranted given the good performance of the UK's pre-stressed concrete bridges. The UK National Annex sets both $r_{\sup}$ and $r_{\inf}$ to 1.0, although a relevant consideration was that the verification criteria are not equivalent in EN 1992-2 and BS 5400 Part 4 in terms of either the loading considered or the acceptance criteria for stresses and cracks.

Worked example 5.10-3: Simply supported pre-tensioned beam with straight, fully bonded tendons

A simply-supported M beam has the tendon geometry and section properties shown in Fig. 5.10-11. The exposure class for the beam soffit is XD3. The prestress is made up of 29 no. 15.2 mm strands, each with area = 139 mm² (i.e. total area = 4031 mm²), characteristic tensile strength (CTS) of 232 kN and E_p of 195 GPa. The tendons are placed in several layers, with the centroid as shown in the figure. Initial stressing is to 75% CTS. The SLS moments due to beam self-weight, slab weight and SDL at midspan are 465 kNm, 288 kNm and 275 kNm respectively. The live load moments at midspan are 1120 kNm (frequent value) and 1500 kNm (characteristic value). The concrete grade is C40/50 for both beam and slab, but initial stressing is done at a characteristic cylinder strength of 32 MPa, whereupon $E_{cm} = 35$ GPa and $E_{cm}(t) = [(32+8)/(40+8)]^{0.3} \times 35 = 33$ GPa from 2-1-1/Expression (3.5). The stresses at transfer and in the final condition are checked. In the example, it is assumed that the National Annex will set the values of $r_{\sup}$ and $r_{\inf}$ in 2-1-1/clause 5.10.9 to unity. If they have values other than unity, the resulting variation in prestressing force would need to be considered in the stress checks.

(i) Stresses at transfer at beam end (critical location)
Short-term losses:
The initial prestressing force $= 29 \times 0.75 \times 232 = 5046$ kN. The position at the end of the transmission zone is critical at transfer as the dead load bending moment is approximately zero there. Assuming 1% of steel relaxation to take place before transfer, the stress at the level of the tendons is:

$$\sigma_{cp} = 0.99 \times \left(\frac{5046 \times 10^3}{3.87 \times 10^5} + \frac{5046 \times 10^3 \times 208^2}{4.76 \times 10^{10}}\right) = 17.449 \text{ MPa}$$

Elastic loss of prestress force from equation (D5.10-3):

$$\Delta P_{el} = A_p \frac{E_p}{E_{cm}(t)} \sigma_c = 4031 \times \frac{195}{33} \times 17.499 = 417 \text{ kN for the whole group of tendons}$$

More accurately, elastic loss of prestress force from equation (D5.10-4):

$$\Delta P_{el} = \frac{A_p \dfrac{E_p}{E_{cm}(t)} \sigma_c}{1 + \dfrac{E_p}{E_{cm}(t)} \dfrac{A_p}{A_c}\left(1 + \dfrac{A_c}{I_c} z_{cp}^2\right)} = \frac{417 \times 10^3}{1 + \dfrac{195}{33} \dfrac{4031}{3.87 \times 10^5}\left(1 + \dfrac{3.87 \times 10^5}{4.76 \times 10^{10}} \times 208^2\right)} = 385 \text{ kN}$$

Prestress force at transfer is therefore:
$P_{mo} = 0.99 \times 5046 - 385 = 4610$ kN, which can be shown to satisfy the maximum allowable force from 2-1-1/Expression (5.43).

Assuming the dead load stress to be approximately zero at the end of the transmission zone:

Bottom fibre stress

$$= P_{mo}(1/A + e/W_{p,1}) = 4610 \times 10^3(1/3.87 \times 10^5 + 208/116.2 \times 10^6) = 20.16\,\text{MPa}$$
$$< \text{compressive limit of } k_6 f_{ck}(t) = 0.7 f_{ck}(t) = 0.7 \times 32 = 22.4\,\text{MPa}$$

from 2-1-1/clause 5.10.2.2(5) for a pre-tensioned beam, so adequate.

Top fibre stress

$$= P_{mo}(1/A - e/W_{p,2}) = 4610 \times 10^3(1/3.87 \times 10^5 - 208/75.4 \times 10^6) = -0.81\,\text{MPa}$$

The relevant tensile limit is not explicitly stated in EN 1992, as discussed in the main text under the comments on clause 5.10.2.2. Of the possibilities discussed there, a limit of 1 MPa of tension, as was permitted in BS 5400 Part 4,[9] is used here simply because the strand pattern of the beam in this example was designed to BS 5400. With this assumption, the stresses at transfer are just adequate. It would still be necessary to check stresses or crack widths after all losses for the finished structure at this location in accordance with EN 1992-2 Table 7.101N, as discussed in the main text. This is not done here.

(ii) Final serviceability stresses at midspan
Short-term losses:
The short-term losses are conservatively taken from above, although the elastic loss could be reduced slightly at midspan because the moment from beam self-weight reduces the compressive stress at the level of the tendons.

Long-term losses:
(a) The total steel relaxation loss from 2-1-1/Expression (3.29) is found to be 3.6% (corresponding to $\mu = 0.68$ for this case). Therefore, assuming 1% loss before transfer and 2.6% after:

$$\Delta \sigma_{pr} = 0.026 \times 4610/(29 \times 139) = 29.7\,\text{MPa}$$

(b) The concrete shrinkage strain remaining after transfer is found to be 300×10^{-6} (from 2-1-1/clause 3.1.4 with appropriate assumptions made for age at transfer).

(c) The creep strain must be calculated from the concrete stress adjacent to the tendons. As midspan is to be checked, the stresses are calculated there. Midspan bending moment due to beam self-weight only = 465 kNm. This moment, excluding moments from slab weight and SDL, is conservatively used here. The other moments reduce the concrete compressive stress adjacent to the tendons and hence the creep. They could be included but, since they are applied at later stages, they would individually require different (lower) creep factors.

Stresses due to beam self-weight alone are:
Bottom fibre stress $= -465 \times 10^6/116.2 \times 10^6 = -4.00\,\text{MPa}$
Top fibre stress $= 465 \times 10^6/75.4 \times 10^6 = 6.17\,\text{MPa}$
Therefore total stresses at midspan at transfer are shown in Fig. 5.10-9.

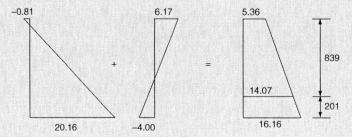

Fig. 5.10-9. Stresses at midspan at transfer

The maximum concrete stress is 16.16 MPa, which exceeds $0.45f_{ck}(t) = 0.45 \times 32 = 14.4$ MPa, so the basic creep factor needs to be increased to allow for non-linear creep in accordance with 2-1-1/Expression (3.7). There is an anomaly in this equation, as the criterion for its consideration is based on $f_{ck}(t)$, whereas the formula contains the mean strength $f_{cm}(t)$. This means that for a concrete stress of $0.45f_{ck}(t)$, the formula actually reduces the creep factor, which is not intended. In this example, therefore, $f_{cm}(t)$ is replaced by $f_{ck}(t)$ in the formula and the increase factor is $e^{1.5(16.16/32 - 0.45)} = 1.09$. The basic creep factor for load applied at transfer (concrete age t_0) was found to be $\phi(\infty, t_0) = 2.0$. The final creep factor is therefore $\phi_k(\infty, t_0) = 2.0 \times 1.09 = 2.18$.

From 2-1-1/Expression (5.46), but conservatively taking the denominator as unity for simplicity, the long-term loss of force is:

$$\Delta P_{c+s+r} = A_p \Delta \sigma_{p,c+s+r} = A_p E_p \left(\varepsilon_{cs} + \frac{0.8 \Delta \sigma_{pr}}{E_p} + \frac{\phi(\infty, t_0) \sigma_{c,qp}}{E_{cm}} \right)$$

$$= 29 \times 139 \times 195 \times 10^3$$

$$\times (300 \times 10^{-6} + 0.8 \times 29.7/195 \times 10^3 + 2.18 \times 14.07/35 \times 10^3)$$

$$= 1020 \text{ kN}$$

Final prestress force after all losses = $4610 - 1020 = 3590$ kN (29% loss).

Final stresses due to prestress and losses at midspan:

It is common to apply the above long-term losses to the beam only section, whereupon the final stresses from prestress after all losses are:

Bottom fibre stress = $20.16 \times 3590/4610 = 15.70$ MPa

Top fibre stress = $-0.81 \times 3590/4610 = -0.63$ MPa

However, it is more realistic to apply part of the loss to the precast beam and partly to the composite as some of the loss occurs after the beam is made composite. It is assumed here that one-third of the long-term loss occurs before making the beam composite with the slab. Greater accuracy is not warranted due to the uncertainty in the true time interval. The stresses due to prestress after all losses then become:

Bottom fibre stress

$$= 20.16 - 1020/3 \times 10^3 (1/3.87 \times 10^5 + 208/116.2 \times 10^6)$$

$$- 2 \times 1020/3 \times 10^3 (1/6.27 \times 10^5 + 469/174.3 \times 10^6)$$

$$= 15.76 \text{ MPa cf. } 15.70 \text{ MPa based on the precast beam alone above}$$

Top of precast beam fibre stress

$$= -0.81 - 1020/3 \times 10^3 (1/3.87 \times 10^5 - 208/75.4 \times 10^6)$$

$$- 2 \times 1020/3 \times 10^3 (1/6.27 \times 10^5 - 469/315.7 \times 10^6)$$

$$= -0.82 \text{ MPa cf. } -0.63 \text{ MPa based on the precast beam alone above}$$

This illustrates that there is little difference in the two approaches, with the former being slightly conservative with respect to the bottom flange.

Before checking total stresses at midspan, stresses from differential shrinkage, differential creep and differential temperature need to be calculated.

Differential shrinkage:

From equation (DK.4) in Annex K of this guide and assuming a differential shrinkage strain between slab and precast unit of 200×10^{-6} and a creep factor of 2.0 to cover the behaviour of both precast beam and slab as discussed in Annex K:

$$F_{sh} = \varepsilon_{diff} E A_{slab} \frac{(1-e^{-\phi})}{\phi}$$

$$= 200 \times 10^{-6} \times 35 \times 10^3 \times 1500 \times 160 \times \frac{(1-e^{-2})}{2} = 726\,\text{kN}$$

(tensile) at a stress of $[(1-e^{-2})/2] \times 200 \times 10^{-6} \times 35 \times 10^3 = 3.03\,\text{MPa}$.

The average restrained differential shrinkage axial stress $= -726/627 = -1.16\,\text{MPa}$.

The restrained differential shrinkage bending stress at the bottom of the composite section $= 726 \times 0.42/174.3 = 1.75\,\text{MPa}$.

The restrained differential shrinkage bending stress at the top of the composite section $= -726 \times 0.42/233.6 = -1.31\,\text{MPa}$.

These and the locked in self-equilibriating stresses are shown in Fig. 5.10-10. All but the self-equilibriating stresses are released.

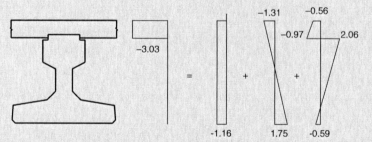

Fig. 5.10-10. Differential shrinkage stresses

Differential creep:
Differential creep stresses were here found to be favourable at the precast beam extreme fibres and were therefore ignored. For the deck slab, an additional 0.86 MPa of compressive stress was produced. The calculation is not performed here but Worked example K-2 in Annex K of this guide illustrates the calculation on a slightly different beam geometry.

Differential temperature:
Self-equilibriating stresses from temperature difference are included in the check of final stresses below. There calculation is similar to that for differential shrinkage and is not shown here.

Final serviceability stress check for beams:

(a) Cracking and decompression (beam soffit):
The exposure class for the beam soffit is XD3, so, from Table 7.101N of EN 1992-2, decompression should be checked in the frequent load combination. As some of the strands are within 100mm of the concrete free surface, decompression must be checked at the extreme fibre – the Note below the Table refers. The self-equilibriating stresses from temperature difference are included in this check below but have their quasi-permanent values in the frequent combination. The critical combination here has traffic as the leading action.

Bottom fibre stress for precast beam

$$= \underset{\text{(prestress)}}{15.76} - \underset{\substack{116.2 \times 10^6 \\ \text{(precast only)}}}{\frac{(465+288) \times 10^6}{}} - \underset{\substack{174.3 \times 10^6 \\ \text{(composite)}}}{\frac{(275+1120) \times 10^6}{}} - \underset{\text{(shrinkage)}}{0.59} - \underset{\text{(temperature)}}{0.70}$$

$$= 0.01\,\text{MPa} > 0\,\text{MPa}$$

The decompression check is therefore satisfied.

(b) Compression limits (top of precast beam and top of deck slab):
If the top of the precast beam is also treated as being in an XD class, then from 2-2/clause 7.2(102) the maximum allowable compressive stress is $k_1 f_{ck} = 0.6 f_{ck}(t) = 0.6 \times 40 = 24\,\text{MPa}$. This must be checked under the characteristic combination of actions.

Top fibre stress for precast beam

$$= \underset{\text{(prestress)}}{0.82} - \underset{\substack{75.4 \times 10^6 \\ \text{(precast only)}}}{\frac{(465 + 288) \times 10^6}{}} - \underset{\substack{315.7 \times 10^6 \\ \text{(composite)}}}{\frac{(275 + 1500) \times 10^6}{}} - \underset{\text{(shrinkage)}}{2.06} - \underset{\text{(temperature)}}{0.10}$$

$$= 16.95\,\text{MPa}$$

This compressive stress is satisfactory as it is less than the limit of 24 MPa from 2-2/clause 7.2(102) for exposure class XD.

Top fibre stress for deck slab

$$= \underset{\substack{315.7 \times 10^6 \\ \text{(composite)}}}{\frac{(275 + 1500) \times 10^6}{}} + \underset{\text{(creep redistribution)}}{0.86} + \underset{\text{(temperature)}}{1.30} = 9.76\,\text{MPa} < 24\,\text{MPa}$$

Shrinkage was ignored for the deck slab as it relieves compression. The deck slab extreme fibre is exposure class XC3 so this check on compression stress limit is not strictly necessary. Where a beam is found to be decompressed under the characteristic combination, strictly, cracking should be considered in the section analysis when verifying compression limits.

A check would also generally be required of crack width in the deck slab, but this is not done here as the deck slab is in global compression (although local moments from wheel load could cause overall tension in the reinforcement).

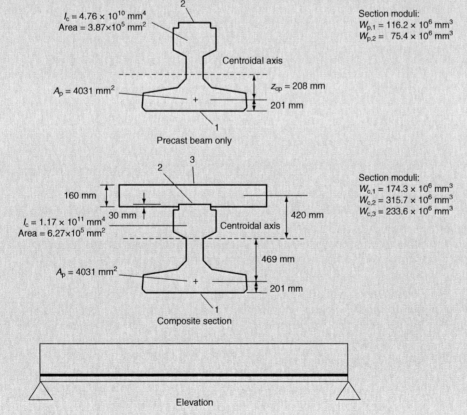

Fig. 5.10-11. Beam for Worked example 5.10-3

5.11. Analysis for some particular structural members

EC2 gives additional information on the analysis of flat slabs and shear walls in Annex I. Most of this annex has little relevance to bridge design and is not discussed further in this guide.

CHAPTER 6

Ultimate limit states

This chapter discusses ultimate limit states (ULS) as covered in section 6 of EN 1992-2 in the following clauses:

• Bending with or without axial force	*Clause 6.1*
• Shear	*Clause 6.2*
• Torsion	*Clause 6.3*
• Punching	*Clause 6.4*
• Design with strut-and-tie models	*Clause 6.5*
• Anchorage and laps	*Clause 6.6*
• Partially loaded areas	*Clause 6.7*
• Fatigue	*Clause 6.8*
• Membrane elements	*Clause 6.9*

6.1. ULS bending with or without axial force
6.1.1. General (additional sub-section)
This section of the guide deals with the design at ultimate limit state of members subject to bending with or without axial force. It is split into the following additional sub-sections for convenience:

• Reinforced concrete beams	*Section 6.1.2*
• Prestressed concrete beams	*Section 6.1.3*
• Reinforced concrete columns	*Section 6.1.4*
• Brittle failure of members with prestress	*Section 6.1.5*

6.1.2. Reinforced concrete beams (additional sub-section)
6.1.2.1. Assumptions
2-1-1/clause 6.1(2)P makes standard assumptions for the calculation of ultimate moments of resistance as follows:

2-1-1/clause 6.1(2)P

(1) Plane sections remain plane.
(2) Strain in bonded reinforcement, whether in tension or compression, is the same as the strain in the concrete at the same level.
(3) Tensile strength of the concrete is ignored.
(4) The stresses in the concrete in compression are given by the design stress–strain relationships discussed in section 3.1.7.
(5) The stresses in the reinforcing steel are given by the design stress–strain relationships discussed in section 3.2.7.
(6) The initial strain in prestressing is taken into account.

Assumption (1), relating to linear strains, is only appropriate for 'beam-like' behaviour and does not apply to deep beams (see definition in 2-1-1/clause 5.3.1) or to local load introduction, such as in prestressed end blocks or in the vicinity of bearings. In these situations the stresses and strains vary in a complex manner and *2-1-1/clause 6.1(1)P* notes that strut-and-tie analysis is more appropriate, as discussed in section 6.5.

Local bond slip also means that the strains in reinforcement will not always exactly match those in the surrounding concrete, but the assumption of equal strains in (2) above is adequate for design.

2-1-1/clause 6.1(3)P describes the stress–strain curves to be used for concrete, reinforcement and prestressing steel. The design stress–strain curves for reinforcement include one with an inclined branch (representing strain hardening) with a limit on the ultimate strain. While its use can give rise to a small saving in reinforcement in under-reinforced beams, the calculations involved are more time-consuming and are not suited to hand calculations. Computer software can be used to automate the process. For the purposes of developing design equations and worked examples, the rest of this chapter considers only the reinforcement stress–strain curve with a horizontal top branch and no strain limit. The same principles, however, apply to the use of the inclined branch. The curves for prestressing also include one with an inclined branch and its use is illustrated in Worked example 6.1-5 to illustrate the improvement over the use of the curve with the horizontal plateau. Whereas use of the curve with horizontal plateau for reinforcement leads to reinforced concrete bending resistances similar to those from BS 5400 Part 4,[9] use of the curve with horizontal plateau for prestressing leads to a lower bending resistance.

For concrete, EC2 allows three different stress–strain relationships, as discussed in section 3.1 of this guide and illustrated in Fig. 3.1-3. As was shown in section 3.1.7, the differences between the three alternatives are very small and, in fact, the simple rectangular stress block generally gives the greatest moment of resistance. This is unlike both BS 5400 Part 4 and *Model Code 90*,[6] where the peak stress used in the rectangular block was lower than that in the parabola-rectangle block. (*Model Code 90* reduces the peak allowable stress in the rectangular block by a factor equivalent to ν' as used in 2-1-1/clause 6.5.) The design equations developed in the following sections apply to all three concrete stress blocks but it is simplest and most economic to use the simpler rectangular stress block, as illustrated in Worked example 6.1-1. The worked examples, however, generally use the parabolic-rectangular stress block as it is more general.

6.1.2.2. Strain compatibility
The ultimate moment resistance of a section can be determined by using the strain compatibility method, by either algebraic or iterative approaches. An iterative approach is possible using the following steps:

(1) Guess a neutral axis depth and calculate the strains in the tension and compression reinforcement by assuming a linear strain distribution and a strain of ε_{cu2} (or ε_{cu3} if not using the parabolic-rectangular stress–strain idealization) at the extreme fibre of the concrete in compression.
(2) Calculate from the stress–strain idealizations the steel stresses appropriate to the calculated steel strains.
(3) Calculate from the stress–strain idealizations the concrete stresses appropriate to the strains associated with the assumed neutral axis depth.
(4) Calculate the net tensile and compressive forces at the section. If these are not equal, adjust the neutral axis depth and return to step (1).
(5) When the net tensile force is equal to the net compressive force, take moments about a common point in the section to determine the ultimate moment of resistance.

The strain compatibility method described above is tedious for hand analysis, but must be used for non-uniform sections (or at least for sections which are non-uniform in the compression zone). This method is illustrated in Worked example 6.1-4 for a flanged beam. A further difficulty, in this case for flanged beams, stems from the provisions of *2-1-1/clause 6.1(5)* and

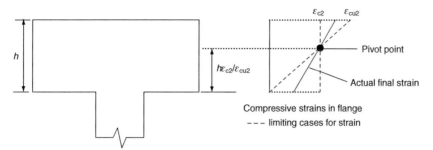

Fig. 6.1-1. Allowable maximum strain in flanges depending on strain distribution (ε_{c3} and ε_{cu3} used for bilinear and rectangular stress block)

2-1-1/clause 6.1(6). The former requires the mean strain in parts of the section which are fully in compression with approximately concentric loading (such as the flanges of box girders where the neutral axis is in the webs) to be limited to ε_{c2} or ε_{c3} (as appropriate). 'Approximately concentric' is defined as $e/h < 0.1$, which is equivalent to a neutral axis at a depth below top of flange greater than $1.33h$ for a flange of depth h. This statement is a simplification of the range of limiting strain distributions in 2-1-1/Fig. 6.1, referenced by 2-1-1/clause 6.1(6), which is of general application. (The simplification of 2-1-1/clause 6.1(5) does not, however, simplify the resistance calculation itself.) Where a part has zero compressive strain at one face, the limiting strain can still be taken as ε_{cu2} or ε_{cu3} as appropriate for the stress block used. Where the part has equal compressive strains at both faces, a reduced limit of ε_{c2} or ε_{c3} applies as appropriate, depending on the stress–strain idealization used.

The reduction in limiting strain for pure compression arises because the real concrete behaviour is such that the peak stress is reached at a strain approximating to ε_{c2} (or ε_{c3}) and then drops off before the final failure strain is obtained. Thus, for pure compression, peak load is obtained at approximately ε_{c2}, but for pure flexure the resistance continues to increase beyond the attainment of this strain. For intermediate cases of strain diagram, the limiting strain needs to be obtained by interpolation between these cases and this can be done by rotating the strain diagram about the fixed pivot point, shown in Fig. 6.1-1. The same applies for entire sections which are wholly in compression. For a flange of thickness h wholly in compression, this means limiting the strain to ε_{c2} at a height of $h\varepsilon_{c2}/\varepsilon_{cu2}$ in the flange. The simplified rule in 2-1-1/clause 6.1(5) limits the strain to ε_{c2} at mid-height.

The need for this additional level of complexity for bridges is partly mitigated by the use of the recommended value of $\alpha_{cc} = 0.85$ which compensates for the drop-off in strength at increasing strain, as discussed in section 3.1.6. While the theoretical need for this additional complexity can be explained as above, the practical need seems dubious and was not required in BS 5400 Part 4.[9] The method is illustrated in Worked example 6.1-4. For beams with steel that yields with the usual assumption of concrete limiting stress of ε_{cu2} or ε_{cu3}, the effect of this modification is typically negligible. Where the steel does not yield, the effect can be a little more significant, but still usually relatively small. Where such calculation is required, it is considerably simpler to perform with the rectangular stress block.

For uniform sections (or at least uniform in the compression zone), it is possible to use the simplified design equations which are developed in the following sections.

6.1.2.3. Singly reinforced beams and slabs

Consider the singly reinforced rectangular beam illustrated in Fig. 6.1-2, with $F_s = (f_{yk}/\gamma_s)A_s$ and $F_c = f_{av}bx$ acting at a lever arm βx from the compression fibre, where f_{av} is the average stress in the concrete above the neutral axis at the ultimate limit state. f_{av} and β are parameters relating to the geometry of the concrete stress block being used. Formulae for these are given in section 3.1.7 along with a tabulation of their values for varying concrete strengths and stress–strain idealizations. The failure strain shown in Fig. 6.1-2 of ε_{cu2} is only appropriate for the parabolic-rectangular block and should be replaced by ε_{cu3} for the

2-1-1/clause 6.1(6)

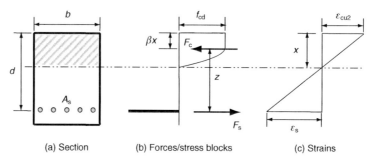

Fig. 6.1-2. Singly reinforced rectangular beam at failure (using parabolic-rectangular concrete stress block)

bilinear and rectangular blocks. The equivalent diagram for the rectangular stress block is shown in Fig. 6.1-3.

Equations are now developed with reference to Fig. 6.1-2. From moment equilibrium (assuming the steel yields), taking moments about the centroid of the compressive force:

$$M = F_s z = \frac{f_{yk}}{\gamma_s} A_s z \tag{D6.1-1}$$

Alternatively, taking moments about the centroid of the tensile force:

$$M = F_c z = f_{av} bxz \tag{D6.1-2}$$

For equilibrium:

$$F_c = F_s \Rightarrow f_{av} bx = \frac{f_{yk}}{\gamma_s} A_s$$

which can be rewritten as:

$$\frac{x}{d} = \frac{f_{yk}}{f_{av}\gamma_s} \rho \tag{D6.1-3}$$

with

$$\rho = \frac{A_s}{bd} \tag{D6.1-4}$$

From the geometry of the stress–strain diagrams:

$$z = d - \beta x \tag{D6.1-5}$$

Substituting this into equation (D6.1-2) gives:

$$M = f_{av} bx(d - \beta x) = f_{av} bx\left(1 - \beta \frac{x}{d}\right) d$$

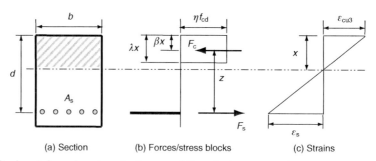

Fig. 6.1-3. Singly reinforced rectangular beam at failure (using rectangular concrete stress block)

or

$$\frac{M}{bd^2} = f_{av}\frac{x}{d}\left(1 - \beta\frac{x}{d}\right) \qquad (D6.1\text{-}6)$$

(D6.1-1) and (D6.1-6) can be used as design equations for the section's ultimate moment resistance. A check must, however, be made to ensure the strain in the reinforcement is sufficient to cause yielding as assumed.

From the reinforcing steel stress–strain idealization (refer to section 3.2):

$$\varepsilon_{s,\text{yield}} = \frac{f_{yk}}{\gamma_s E_s}$$

and from the strain diagram in Fig. 6.1-2:

$$\varepsilon_s = \frac{\varepsilon_{cu2}}{x}(d - x) = \varepsilon_{cu2}\left(\frac{d}{x} - 1\right) \qquad (D6.1\text{-}7)$$

To ensure yielding:

$$\varepsilon_s \geq \varepsilon_{s,\text{yield}} \Rightarrow \varepsilon_{cu2}\left(\frac{d}{x} - 1\right) \geq \frac{f_{yk}}{\gamma_s E_s}$$

which can be expressed as:

$$\frac{x}{d} \leq \frac{1}{\left(\dfrac{f_{yk}}{\gamma_s E_s \varepsilon_{cu2}} + 1\right)} \qquad (D6.1\text{-}8)$$

(Note that ε_{cu2} should be replaced by ε_{cu3} in equation (D6.1-8) if the rectangular or bilinear concrete stress blocks are used.)

If x/d does not satisfy equation (D6.1-8) then the reinforcement does not yield and the expressions in equations (D6.1-1) and (D6.1-3) are not valid, so neither is the resistance in equation (D6.1-6). In this case there are the following options:

(1) increase the section size to comply with equation (D6.1-8);
(2) add compression reinforcement (see section 6.1.2.4 on doubly reinforced beams) to comply with equation (D6.1-8);
(3) use the strain-compatibility method to establish actual reinforcement force and hence moment resistance;
(4) conservatively take the reinforcement strain from equation (D6.1-7) and the lever arm from equation (D6.1-5) using the depth of compression zone, x, from equation (D6.1-3) (determined assuming the steel yields), so that from equation (D6.1-1), $M = A_s E_s \varepsilon_s z$.

For the simple rectangular block, the above can be simplified by substituting for $f_{av} = \lambda \eta f_{cd}$ and $\beta = \lambda/2$ (from the expressions in section 3.1.7) in equations (D6.1-3) and (D6.1-5) above so that:

$$M = A_s f_{yd} z \quad \text{with} \quad z = d\left(1 - \frac{f_{yd} A_s}{2\eta f_{cd} bd}\right) \qquad (D6.1\text{-}9)$$

and

$$x = \frac{A_s f_{yd}}{\lambda b \eta f_{cd}} \qquad (D6.1\text{-}10)$$

It is still necessary to check that equation (D6.1-8) is satisfied (but ε_{cu2} should be replaced by ε_{cu3}) in order to use equation (D6.1-9), as it assumes that the steel yields. The modified equation is:

$$\frac{x}{d} \leq \frac{1}{\left(\dfrac{f_{yk}}{\gamma_s E_s \varepsilon_{cu3}} + 1\right)}$$

If the steel does not yield, one of the above options needs to be considered. The last option leads to conservatively taking the moment resistance as:

$$M = E_s A_s \varepsilon_s z \quad \text{with} \quad z = d\left(1 - \frac{f_{yd} A_s}{2\eta f_{cd} bd}\right) \tag{D6.1-11}$$

In equation (D6.1-11), ε_s is determined from equation (D6.1-7) but substituting ε_{cu3} for ε_{cu2}.

The equations above are obviously more suited to analysing given sections rather than designing a section to resist a given moment. For design purposes it is therefore convenient to rearrange the equations. If $K_{av} = M/bd^2 f_{av}$ is defined and substituted into equation (D6.1-6) then:

$$K_{av} = \frac{x}{d}\left(1 - \beta\frac{x}{d}\right) = \frac{x}{d} - \beta\left(\frac{x}{d}\right)^2 \tag{D6.1-12}$$

which can be written as a quadratic equation $\beta(x/d)^2 - (x/d) + K_{av} = 0$ with solutions:

$$\left(\frac{x}{d}\right) = \frac{1 \pm \sqrt{1 - 4\beta K_{av}}}{2\beta} \tag{D6.1-13}$$

The lower root of equation (D6.1-13) is the relevant one. Again, the ratio of x/d should be checked against the limit in equation (D6.1-8) and the section designed from rearranging equation (D6.1-1) to give:

$$A_s \geq \frac{M\gamma_s}{f_{yk} z} \tag{D6.1-14}$$

with $z = d - \beta x$ as before from equation (D6.1-5) and x taken from equation (D6.1-13).

Worked example 6.1-1: Reinforced concrete deck slab

A 250 mm thick deck slab has a maximum applied ULS sagging moment of 150 kNm/m. Assuming 40 mm cover to the main reinforcement and C35/45 concrete, find the required reinforcement area for adequate ultimate moment resistance using both the parabolic-rectangular and simplified rectangular stress blocks. Reinforcement is B500B.

$f_{ck} = 35$ MPa. Assuming $\alpha_{cc} = 0.85$ and $\gamma_c = 1.5$, from section 3.1.7 for the parabolic-rectangular concrete stress block, $f_{av} = 16.056$ MPa and $\beta = 0.416$.

From equation (D6.1-12):

$$K_{av} = \frac{150 \times 10^6}{1000 \times 200^2 \times 16.056} = 0.234$$

From equation (D6.1-13):

$$\frac{x}{d} = \frac{1 \pm \sqrt{1 - 4 \times 0.416 \times 0.234}}{2 \times 0.416} = 0.262 \text{ (and an irrelevant root of 2.14)}$$

Check limit from equation (D6.1-8) to ensure reinforcement is yielding ($f_{yk} = 500$ MPa, $E_s = 200$ GPa and $\gamma_s = 1.15$):

$$\frac{1}{\left(\frac{f_{yk}}{\gamma_s E_s \varepsilon_{cu2}} + 1\right)} = \frac{1}{\left(\frac{500}{1.15 \times 200 \times 10^3 \times 0.0035} + 1\right)} = 0.6169 > \frac{x}{d} \text{ therefore okay}$$

Thus, $x = 0.262 \times 200 = 52.4$ mm and, from equation (D6.1-5), $z = 200 - 0.416 \times 52.4 = 178.2$ mm.

Thus, from equation (D6.1-14):

$$A_s \geq \frac{150 \times 10^6 \times 1.15}{500 \times 178.2} = 1936 \text{ mm}^2/\text{m}$$

Using 20ϕ bars at 150 mm centres gives $A_s = 2094$ mm^2/m > 1936 mm^2/m required. The calculation is now repeated with the rectangular stress block for comparison:

From section 3.1.7 for the rectangular concrete stress block, $f_{av} = 15.867$ MPa and $\beta = 0.400$.

From equation (D6.1-12):

$$K_{av} = \frac{150 \times 10^6}{1000 \times 200^2 \times 15.867} = 0.236$$

From equation (D6.1-13):

$$\frac{x}{d} = \frac{1 - \sqrt{1 - 4 \times 0.400 \times 0.236}}{2 \times 0.400} = 0.264$$

Reinforcement will yield by inspection, thus $x = 0.264 \times 200 = 52.9$ mm and, from equation (D6.1-5), $z = 200 - 0.400 \times 52.9 = 178.9$ mm.

Thus, from equation (D6.1-14):

$$A_s \geq \frac{150 \times 10^6 \times 1.15}{500 \times 178.9} = 1929 \text{ mm}^2/\text{m}$$

which is slightly less reinforcement than required for the parabolic-rectangular block.

Equation (D6.1-9) can be used as a check on the moment resistance with this steel area:

$$M = A_s f_{yd} d \left(1 - \frac{f_{yd} A_s}{2\eta f_{cd} b d}\right)$$

$$= 1929 \times \frac{500}{1.15} \times 200 \times \left(1 - \frac{500/1.15 \times 1929}{2 \times 1.0 \times \frac{0.85 \times 35}{1.5} \times 1000 \times 200}\right)$$

$$= 150.0 \text{ kNm/m as required}$$

Worked example 6.1-2: Voided reinforced concrete slab

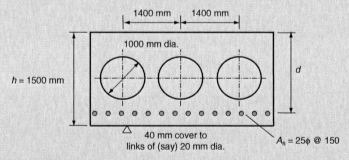

Fig. 6.1-4. Voided reinforced concrete slab

Consider the voided reinforced concrete slab section shown in Fig. 6.1-4. Assuming C35/45 concrete, $f_{yk} = 500$ MPa, $\gamma_s = 1.15$, $\gamma_c = 1.5$ and $\alpha_{cc} = 0.85$, find the ultimate sagging moment resistance. If the ultimate applied moment is increased to 3000 kNm/m, find the additional reinforcement required to provide sufficient ultimate resistance.

(1) To find the moment resistance, consider the voided slab as a flanged beam of width, $b = 1400$ mm, slab depth above hole = slab depth below hole = 250 mm and effective depth, $d = h - \text{cover} - \text{link } \phi - \text{bar } \phi/2$; therefore $d = 1500 - 40 - 20 - 12.5 \approx 1425$ mm:

$$A_s = \pi \times 12.5^2 \times \frac{1400}{150} = 4581.5 \text{ mm}^2$$

From section 3.1.7 for the parabolic-rectangular concrete stress block using $\alpha_{cc} = 0.85$ and $\gamma_c = 1.5$, $f_{av} = 16.056$ MPa and $\beta = 0.416$.

From equation (D6.1-4):

$$\rho = \frac{4581.5}{1400 \times 1425} = 0.002296$$

From equation (D6.1-3):

$$\frac{x}{d} = \frac{500}{16.056 \times 1.15} \times 0.002296 = 0.0622$$

Check against limit from equation (D6.1-8) to ensure reinforcement is yielding:

$$\frac{1}{\left(\frac{f_{yk}}{\gamma_s E_s \varepsilon_{cu2}} + 1\right)} = \frac{1}{\left(\frac{500}{1.15 \times 200 \times 10^3 \times 0.0035} + 1\right)} = 0.6169 > \frac{x}{d} \text{ therefore okay}$$

Thus $x = 0.0622 \times 1425 = 88.6$ mm < 250 mm; therefore neutral axis lies in the flange and the above equations are still valid.

From equation (D6.1-5), $z = 1425 - 0.416 \times 88.6 = 1388.1$ mm, and from equation (D6.1-1):

$$M_{Rd} = \frac{500}{1.15} \times 4581.5 \times 1388.1 \times 10^{-6} = \mathbf{2765\,kNm} \text{ for a 1.4 m width}$$

(2) To find required increase in A_s to resist $M = 3000$ kNm/m:

For 1400 mm wide section, $M = 1.4 \times 3000 = 4200$ kNm.

From equation (D6.1-12):

$$K_{av} = \frac{4200 \times 10^6}{1400 \times 1425^2 \times 16.056} = 0.0920$$

From equation (D6.1-13):

$$\frac{x}{d} = \frac{1 - \sqrt{1 - 4 \times 0.416 \times 0.0920}}{2 \times 0.416} = 0.0958$$

This is less than the limit from equation (D6.1-8); therefore the reinforcement will yield.

Thus $x = 0.0958 \times 1425 = 136.6$ mm < 250 mm; therefore neutral axis remains in the flange and, from equation (D6.1-5), $z = 1425 - 0.416 \times 136.6 = 1368$ mm.

Thus from equation (D6.1-14):

$$A_s \geq \frac{4200 \times 10^6 \times 1.15}{500 \times 1368} = 7060 \text{ mm}^2 (= 5043 \text{ mm}^2/\text{m})$$

Thus, additional

$$A_s = \frac{7060 - 4581.5}{1.4} = \frac{2479}{1.4} = 1770 \text{ mm}^2/\text{m}$$

i.e. add 20ϕ at 150 mm centres (additional $A_s = 2094$ mm²/m) or **use 32ϕ at 150 mm centres** in place of 25ϕ bars (giving total $A_s = 5361$ mm²/m).

6.1.2.4. Doubly reinforced rectangular beams

Where the tension zone is very heavily reinforced, for efficiency it can become necessary to add compression reinforcement to reduce the depth of the concrete compression zone and thereby allow the tensile reinforcement to yield. This situation arises where the neutral axis depth exceeds the limit in (D6.1-8). It may also be necessary to analyse sections with known compression reinforcement for their ultimate flexural resistance. Of note is the fact that EC2 uses the same stress–strain relationship for reinforcement in tension and compression, unlike the relationships in BS 5400 Part 4.[9]

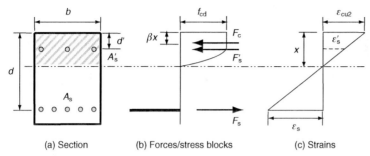

Fig. 6.1-5. Doubly reinforced rectangular beam at failure with parabolic-rectangular stress block

Consider the doubly reinforced rectangular beam illustrated in Fig. 6.1-5 with associated strain and stress diagrams.

For equilibrium, assuming that all reinforcement yields:

$$F_c + F'_s = F_s \Rightarrow f_{av}bx + \frac{f_{yk}}{\gamma_s}A'_s = \frac{f_{yk}}{\gamma_s}A_s$$

therefore:

$$x = \frac{f_{yk}}{\gamma_s}\frac{(A_s - A'_s)}{f_{av}b} \qquad (D6.1\text{-}15)$$

or alternatively:

$$A'_s = A_s - \frac{f_{av}bx\gamma_s}{f_{yk}} \qquad (D6.1\text{-}16)$$

Equation (D6.1-16) can be used to determine the required compression reinforcement to allow the tensile reinforcement to yield, when x/d for A_s alone exceeds the limit given in equation (D6.1-8). In using equation (D6.1-16), the value of x should first be reduced to comply with the limit in equation (D6.1-8).

Equation (D6.1-15) can be used to analyse sections with known reinforcement. x should be checked against the limit given in equation (D6.1-8), as before, to ensure the reinforcement is yielding. A further check is required to ensure that the compression reinforcement is also yielding:

$$\varepsilon'_s \geq \varepsilon_{s,yield} \Rightarrow \varepsilon_{cu2}\left(1 - \frac{d'}{x}\right) \geq \frac{f_{yk}}{\gamma_s E_s}$$

so that:

$$\frac{x}{d'} \geq \frac{1}{(1-C)} \qquad (D6.1\text{-}17)$$

where:

$$C = \frac{f_{yk}}{\gamma_s E_s \varepsilon_{cu2}} \qquad (D6.1\text{-}18)$$

If the reinforcement does not yield, then the ultimate resistance must be determined using the strain compatibility method. An example is given in section 6.1.2.5.

Worked example 6.1-3: Doubly reinforced concrete slab

Consider a 1000 mm wide, class C35/45 concrete slab with effective depth of 275 mm to 40ϕ reinforcement at 150 mm centres. Assuming $f_{yk} = 500$ MPa, $\gamma_s = 1.15$, $\gamma_c = 1.5$, $\alpha_{cc} = 0.85$ and a parabolic-rectangular concrete stress distribution, find the area of compression reinforcement (at a depth, $d' = 50$ mm) required to fully utilize the tension reinforcement and determine the ultimate moment resistance.

From section 3.1.7 with $f_{ck} = 35$ MPa, $f_{av} = 16.056$ MPa and $\beta = 0.416$.

From equation (D6.1-4):

$$\rho = \frac{8377}{1000 \times 275} = 0.0305 \ (3.05\%)$$

From equation (D6.1-3):

$$\frac{x}{d} = \frac{500}{1.15 \times 16.056} \times 0.0305 = 0.825$$

Check against limit in equation (D6.1-8):

$$\frac{1}{\left(\dfrac{500}{1.15 \times 200 \times 10^3 \times 0.0035} + 1\right)} = 0.6169$$

but this is not greater than the actual x/d above; therefore it is necessary to add compression reinforcement until $x/d = 0.6169$.

$$\Rightarrow x_{\max} = 0.6169 \times 275 = 169.6 \, \text{mm}$$

For this depth of concrete in compression, check yielding of compression reinforcement at depth d'.

From equation (D6.1-18):

$$C = \frac{500}{1.15 \times 200 \times 10^3 \times 0.0035} = 0.6211$$

From equation (D6.1-17) for compression reinforcement to yield:

$$x > \frac{d'}{(1-C)} = \frac{50}{(1-0.6211)} = 132 \, \text{mm}$$

and as $x = 169.6$ mm, the steel yields.

Equation (D6.1-16) can therefore be used to find the required area:

$$A'_s = 8377 - \frac{16.056 \times 1000 \times 169.6 \times 1.15}{500} = 2113.1 \, \text{mm}^2/\text{m}$$

Therefore adopt 25ϕ compression reinforcement at 225 mm centres (2182 mm^2/m). The moment of resistance is now found.

From equation (D6.1-15):

$$x = \frac{500}{1.15} \times \frac{(8377 - 2182)}{16.056 \times 1000} = 167.8 \, \text{mm} < x_{\max}$$

limit above for tensile yield. This depth also exceeds the critical depth of 132 mm above for compression reinforcement yield. Therefore, the forces in concrete and steel are as follows:

$$F_c = f_{av} b x = 16.056 \times 1000 \times 167.8 \times 10^{-3} = 2693.9 \, \text{kN}$$

$$F'_s = \frac{f_{yk}}{\gamma_s} A'_s = \frac{500}{1.15} \times 2182 \times 10^{-3} = 948.5 \, \text{kN}$$

$$F_s = \frac{f_{yk}}{\gamma_s} A_s = \frac{500}{1.15} \times 8377 \times 10^{-3} = 3642.4 \, \text{kN}$$

$(F_c + F'_s - F_s = 2693.9 + 948.5 - 3642.4 = 0 \, \text{kN}$ therefore section balances)

Taking moments about top fibre to find moment of resistance:

$$M_{Rd} = F_s d - F_c \beta x - F'_s d'$$

$$= (3642.4 \times 275 - 2693.9 \times 0.416 \times 167.8 - 948.5 \times 50) \times 10^{-3} = \mathbf{764.1 \, kNm}$$

6.1.2.5. Flanged beams

Flanged sections may be treated as a rectangular section using the equations derived above, providing that the neutral axis at the ultimate limit state remains within the flange. If the simplified rectangular concrete stress block is assumed, a flanged section could be considered as rectangular for neutral axis depths up to $1/\lambda$ times the flange thickness, which is 1.25 times the flange thickness for $f_{ck} < 50$ MPa. This is because the stress block remains in the flange and 2-1-1/clause 6.1(5) only applies for approximately 'concentric' loading which is equivalent to a neutral axis at a depth of $1.33h$ for a flange of depth h, as discussed in 6.1.2.2 above. Strictly, 2-1-1/clause 6.1(6) could be considered to require adjustment to the limiting concrete strain for a neutral axis depth of $1.25h$, but this will generally have minimal effect.

Where the neutral axis lies within the web, the section should be analysed using the strain compatibility method discussed in section 6.1.2.2 above and illustrated in Worked example 6.1-4. A further complexity for flanged beams is that the variable concrete strain limit, discussed in section 6.1.2.2, should theoretically be applied to the flange regions, although some judgement can be applied here to small flange projections – EC2-1-1 specifically refers to box girder flanges which are usually wide. This adds to the iteration necessary and gives greater need for the analysis by computer.

Worked example 6.1-4: Flanged reinforced concrete beam

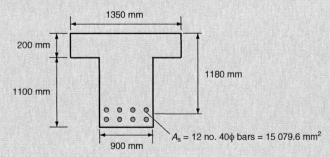

Fig. 6.1-6. Flanged reinforced concrete section

Consider the flanged section shown in Fig. 6.1-6. Assuming C35/45 concrete, $f_{yk} = 500$ MPa, $\gamma_s = 1.15$, $\gamma_c = 1.5$ and $\alpha_{cc} = 0.85$, find the ultimate sagging moment resistance using the parabolic-rectangular stress block.

By trial and error, guess a depth to neutral axis of 331 mm and consider the concrete stress block and strains illustrated in Fig. 6.1-7.

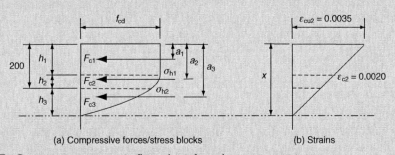

(a) Compressive forces/stress blocks   (b) Strains

Fig. 6.1-7. Compressive stresses in flanged reinforced concrete section

From geometry of parabolic-rectangular concrete stress block:

$$h_1 = \left(1 - \frac{0.0020}{0.0035}x\right) = \frac{3}{7} \times 331 = 141.9 \text{ mm}$$

$$h_2 = 200 - h_1 = 200 - 141.9 = 58.1 \text{ mm}$$

and

$$h_3 = 331 - 200 = 131 \text{ mm}$$

Thus

$$a_1 = \frac{h_1}{2} = \frac{141.9}{2} = 70.95 \text{ mm} \qquad a_2 = h_1 + \frac{3}{8}h_2 = 141.9 + \frac{3 \times 58.1}{8} = 163.69 \text{ mm}$$

and

$$a_3 = h_1 + h_2 + 0.358 h_3 = 141.9 + 58.1 + 0.358 \times 131 = 246.9 \text{ mm}$$

(Note that the 0.358 factor above, for calculating the centroid location, must be derived for the specific geometry of the remaining part of the parabola for each neutral axis depth considered in the iteration. This is particularly tedious for hand calculations and therefore using the simplified rectangular stress block is vastly easier – as illustrated at the end of this example.)

$$f_{cd} = \frac{0.85 \times 35}{1.5} = 19.833 \text{ MPa} \quad \text{and therefore} \quad \sigma_{h1} = f_{cd} = 19.833 \text{ MPa}$$

$$\sigma_{h2} = f_{cd}\left\{1 - \left[1 - \left(\frac{\left(1 - \frac{200}{331}\right) \times 0.0035}{0.0020}\right)\right]^2\right\} = 19.833 \times 0.9055 = 17.959 \text{ MPa}$$

Therefore, $F_{c1} = f_{cd} b_1 h_1 = 19.833 \times 1350 \times 141.9 \times 10^{-3} = 3799 \text{ kN}$:

$$F_{c2} = (\tfrac{2}{3}(\sigma_{h1} - \sigma_{h2}) + \sigma_{h2}) b_1 h_2 = (\tfrac{2}{3}(19.833 - 17.959) + 17.959) \times 1350 \times 58.1 \times 10^{-3}$$
$$= 1507 \text{ kN}$$

$$F_{c3} = [\tfrac{2}{3}\sigma_{h1}(h_2 + h_3) - (\tfrac{2}{3}(\sigma_{h1} - \sigma_{h2}) + \sigma_{h2}) h_2] b_2$$
$$= [\tfrac{2}{3} \times 19.833 \times (58.1 + 131) - (\tfrac{2}{3}(19.833 - 17.959) + 17.959) \times 58.1] \times 900 \times 10^{-3}$$
$$= 1246 \text{ kN}$$

and $\sum F_c = 3799 + 1507 + 1246 = 6552 \text{ kN}$.

The reinforcement strain is $(1180 - 331)/331 \times 0.0035 = 0.009$ so it will yield. Reinforcement force

$$F_s = \frac{f_{yk}}{\gamma_s} A_s = \frac{500}{1.15} \times 15\,079.6 \times 10^{-3} = 6556.3 \text{ kN} \approx F_c$$

so assumed neutral axis depth is correct and taking moments about top fibre gives moment of resistance:

$$M = F_s d - F_{c1} a_1 - F_{c2} a_2 - F_{c3} a_3$$

$$M_{Rd} = (6556.3 \times 1180 - 3799 \times 70.95 - 1507 \times 163.69 - 1246 \times 246.9) \times 10^{-3}$$

$$= \mathbf{6912 \text{ kNm}}$$

Strictly, since the neutral axis is below the flange, a new limiting strain diagram should be used in accordance with 2-1-1/clause 6.1(5) or (6), as discussed in section 6.1.2.2 of this guide. The latter would require a strain limit of $\varepsilon_{c2} = 2.0 \times 10^{-3}$ to be maintained at $h\varepsilon_{c2}/\varepsilon_{cu2} = 200 \times 2.0/3.5 = 114.3$ mm from the underside of the flange, i.e. 85.7 mm from the top. The neutral axis would consequently have to be deeper and therefore a new guess might be, say, 364 mm (the correct value) with corresponding stress and strain diagram, as shown in Fig. 6.1-8.

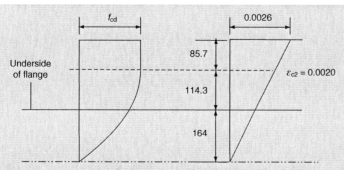

Fig. 6.1-8. Revised stress and strain distribution for further iteration

Repeating the above calculation process with the new strain diagram (the correct one) leads to a new moment resistance of 6880 kNm, which is less than 0.5% reduction and therefore not really worth the extra effort. In this case, the designer might decide that the flange outstands are too small to be subject to this modification on limiting strain. In general, where the steel is predicted to yield on the basis of a limiting concrete strain of ε_{cu2}, the change in resistance caused by this extra complexity will usually be minimal.

As discussed above, this worked example is much simpler if the rectangular stress block is used. Assuming a depth to the neutral axis of 334 mm (obtained by trial and error) gives a rectangular stress block distribution, as shown in Fig. 6.1-9.

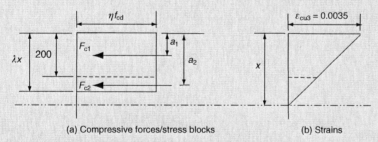

Fig. 6.1-9. Stress and strain distribution for rectangular stress block

From geometry of stress block:

$$a_1 = \frac{200}{2} = 100 \text{ mm}$$

and

$$a_2 = \frac{(\lambda x - 200)}{2} + 200 = \frac{(0.8 \times 334 - 200)}{2} + 200 = 233.6 \text{ mm}$$

Thus $F_{c1} = \eta f_{cd} b_1 h_1 = 1.0 \times 19.833 \times 1350 \times 200 \times 10^{-3} = 5355$ kN, $F_{c2} = \eta f_{cd} b_2 h_2 = 1.0 \times 19.833 \times 900 \times (0.8 \times 334 - 200) \times 10^{-3} = 1200$ kN and $\sum F_c = 5355 + 1200 = 6555$ kN.

The reinforcement strain is $(1180 - 334)/334 \times 0.0035 = 0.009$ so will yield, giving the same reinforcement force as calculated above, $F_s = 6556.3$ kN $\approx F_c$, so assumed neutral axis depth is correct. Taking moments about top fibre gives moment of resistance:

$$M_{Rd} = F_s d - F_{c1} a_1 - F_{c2} a_2 = (6556.3 \times 1180 - 5355 \times 100 - 1200 \times 233.6) \times 10^{-3}$$
$$= \mathbf{6921 \text{ kNm}}$$

Again, since the neutral axis is below the flange, a new limiting strain diagram strictly should be used in accordance with 2-1-1/clause 6.1(5) or (6) as discussed for the parabolic-rectangular stress block example above.

6.1.3. Prestressed concrete beams (additional sub-section)

6.1.3.1. Assumptions

The general assumptions for the design of prestressed concrete sections are the same as those for reinforced concrete. In addition, 2-1-1/clause 6.1(2)P requires that the initial strain in prestressing tendons is taken into account when assessing the ultimate resistance. This 'prestrain' is the strain corresponding to the design prestress force after all losses, $P_{d,t}(x) = \gamma_P P_{m,t}(x)$ from 2-1-1/clause 5.10.8(1). The prestrain is then taken into account in the section bending resistance by shifting the origin of the design stress–strain diagrams for the prestressing tendons in section 3.3 by an amount corresponding to the prestrain. For bonded prestressing, the change in strain in the prestressing steel is assumed to be the same as the change in strain in the adjacent concrete. This assumption is obviously not valid for un-bonded tendons, which are discussed separately in section 6.1.3.4.

6.1.3.2. Strain compatibility

The strain compatibility method described for reinforced concrete in section 6.1.2.2 can also be applied to prestressed concrete, but the prestrain in the tendons should be added to the strain calculated from the strain diagram at failure to give a total strain. This is then used to calculate the stress in the prestressing steel from the stress–strain design curve.

Worked example 6.1-5: Prestressed concrete 'M' beam

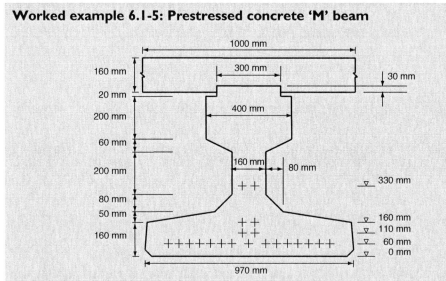

Fig. 6.1-10. Prestressed concrete 'M' beam

Consider an M3 prestressed concrete beam with 160 mm deep in situ deck slab, as illustrated in Fig. 6.1-10. Calculate the ultimate moment of resistance (i) using a prestressing steel stress–strain relationship with a horizontal top branch and (ii) using a prestressing steel stress–strain relationship with an inclined top branch, assuming the following properties:

Class C30/37 slab concrete, $f_{ck} = 30$ MPa
Class C40/50 beam concrete, $f_{ck} = 40$ MPa
Parabolic concrete stress distribution therefore $\varepsilon_{cu2} = 0.0035$ and $\varepsilon_{c2} = 0.0020$

Prestressing strands (using properties from EN 10138-3, Table 4):

21 no. 15 mm strands of type Y1670S7
Nominal diameter = 15.2 mm
Nominal cross-sectional area = 139 mm^2
Characteristic tensile strength, $f_{pk} = 1670$ MPa
Characteristic value of maximum force = $139 \times 1670 \times 10^{-3} = 232$ kN

Characteristic value of 0.1% proof force = 204 kN
$f_{p,0.1k} = (204/232) \times 1670 = 0.879 \times 1670 = 1468$ MPa
$E_p = 195$ GPa
$\gamma_s = 1.15$

Stressing and losses:

Assume initial stressing to 75% f_{pk}.

Allow 10% losses at transfer and a further 20% long-term losses. (For detailed loss calculations, see section 5.10 of this guide.)

Prestrain = long-term strand stress/E_p; therefore prestrain = $0.75 \times 0.9 \times 0.8 \times 1670/(195 \times 10^3) = 0.0046$. The values in 2-1-1/Fig. 3.10 first need to be defined as:

$$f_{pd} = \frac{1468}{1.15} = 1276.5 \text{ MPa}$$

$$\frac{f_{pk}}{\gamma_s} = \frac{1670}{1.15} = 1452.2 \text{ MPa}$$

$$\frac{f_{pd}}{E_p} = \frac{1276.5}{195 \times 10^3} = 0.00655$$

(1) Consider horizontal top branch and a neutral axis depth, obtained by trial and error, of 335 mm.

The strain profile is shown in Fig. 6.1-11.

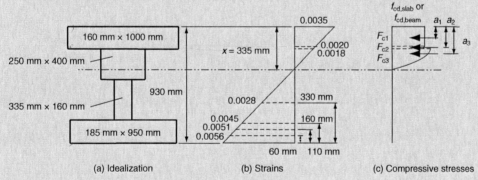

(a) Idealization   (b) Strains   (c) Compressive stresses

Fig. 6.1-11. Stress–strain profile for prestressed 'M' beam

Therefore the total strains at ULS in the four layers of strands, including prestrain, are:

$\varepsilon_{s1} = 0.0056 + 0.0046 = 0.0102$

$\varepsilon_{s2} = 0.0051 + 0.0046 = 0.0097$

$\varepsilon_{s3} = 0.0045 + 0.0046 = 0.0091$

$\varepsilon_{s4} = 0.0028 + 0.0046 = 0.0074$

All strains are greater than $f_{pd}/E_p (= 0.00655)$; therefore all stresses can be taken as $f_{pd} = 1276.5$ MPa. Thus, total steel force, $F_s = 21 \times 139 \times 1276.5 \times 10^{-3} = 3726.2$ kN.

The neutral axis is in the top flange of the beam; therefore treat concrete in a similar manner to the flanged beam calculations splitting the compression zone into the following three sections and taking account of the different concrete strengths:

(i) Rectangular part of stress block in top slab (top 143.6 mm)
(ii) Parabolic part of stress block in top slab (bottom 16.4 mm)
(iii) Parabolic part of stress block in top flange to neutral axis (175 mm deep)

Using geometry of the parabolic–rectangular stress–strain distribution, it can be shown that:

(i) $F_{c1} = 2440.7$ kN with depth to centroid, $a_1 = 71.8$ mm
(ii) $F_{c2} = 278.6$ kN with depth to centroid, $a_2 = 149.7$ mm
(iii) $F_{c3} = 1008.5$ kN with depth to centroid, $a_3 = 225.6$ mm

Thus $F_c = 2440.7 + 278.6 + 1008.5 = 3727.8$ kN $\approx F_s$; therefore the section balances and neutral axis is at the correct level. (Note that using the rectangular stress block would be much easier here.)

Taking moments about the neutral axis level gives:

$$M = \begin{bmatrix} 2440.7 \times (335 - 71.8) + 278.6 \times (335 - 149.7) + 1008.5 \times (335 - 225.6) \\ + 354.9 \times (3 \times 595 - 330 - 160 - 110) + 2661.5 \times (595 - 60) \end{bmatrix}$$
$$\times 10^{-3} = \mathbf{2648.8\,kNm}$$

Strictly, since the neutral axis is in the web, a new limiting strain diagram should be used in compliance with 2-1-1/clause 6.1(6) as discussed in section 6.1.2.2 of this guide. This would require a strain limit of $\varepsilon_{c2} = 2.0 \times 10^{-3}$ to be maintained at $h\varepsilon_{c2}/\varepsilon_{cu2} = 160 \times 2.0/3.5 = 91.4$ mm from the underside of the deck slab, i.e. 68.6 mm from the top. The analysis was conservatively repeated using a strain limit of $\varepsilon_{c2} = 2.0 \times 10^{-3}$ at the top of the deck slab which gave a resistance moment of 2585 kNm. This result reiterates the conclusions of Worked example 6.1-4.

(2) Consider inclined top branch and a neutral axis depth, obtained by trial and error, of 348 mm.

For the purposes of this example (in the absence of the National Annex), the recommended value of ε_{ud} of 0.02 from 2-1-1/clause 3.3.6(7) and:

$$\varepsilon_{uk} = \frac{\varepsilon_{ud}}{0.9} = \frac{0.020}{0.9} = 0.022$$

will be used to define the stress–strain inclined branch. EN 10138-3 Table 5 actually gives $\varepsilon_{uk} = 0.035$, which could be used together with $\varepsilon_{ud} = 0.9\varepsilon_{uk}$ to define the inclined branch. The smaller strain limit of $\varepsilon_{ud} = 0.02$ used below gives a slightly greater resistance than would be obtained from a calculation using the greater ε_{ud} determined from EN 10138-3 Table 5, as the inclined branch of the stress–strain diagram is steeper for a smaller ε_{ud}. The use of $\varepsilon_{ud} = 0.02$, however, produces results more similar to those from BS 5400 Part 4.[9] (The resistance is quite sensitive to changes in the slope of the branch assumed, but is still not vastly greater than that obtained by ignoring the inclination.)

The following strains are obtained in the prestressing strands (including prestrain):

$\varepsilon_{s1} = 0.0053 + 0.0046 = 0.0099$

$\varepsilon_{s2} = 0.0047 + 0.0046 = 0.0093$

$\varepsilon_{s3} = 0.0042 + 0.0046 = 0.0088$

$\varepsilon_{s4} = 0.0025 + 0.0046 = 0.0071$

and corresponding stresses of:

$$\sigma_{s1} = 1276.5 + \left(\frac{1670}{1.15} - 1276.5\right) \times \frac{(0.0099 - 0.00655)}{(0.022 - 0.00655)} = 1314\,\text{MPa}$$

$$\sigma_{s2} = 1276.5 + \left(\frac{1670}{1.15} - 1276.5\right) \times \frac{(0.0093 - 0.00655)}{(0.022 - 0.00655)} = 1308\,\text{MPa}$$

$$\sigma_{s3} = 1276.5 + \left(\frac{1670}{1.15} - 1276.5\right) \times \frac{(0.0088 - 0.00655)}{(0.022 - 0.00655)} = 1302\,\text{MPa}$$

CHAPTER 6. ULTIMATE LIMIT STATES

$$\sigma_{s4} = 1276.5 + \left(\frac{1670}{1.15} - 1276.5\right) \times \frac{(0.0071 - 0.00655)}{(0.022 - 0.00655)} = 1283\,\text{MPa}$$

therefore forces are:

$F_{s1} = 15 \times 139 \times 1314 \times 10^{-3} = 2739.7\,\text{kN}$

$F_{s2} = 2 \times 139 \times 1308 \times 10^{-3} = 363.6\,\text{kN}$

$F_{s3} = 2 \times 139 \times 1302 \times 10^{-3} = 362.0\,\text{kN}$

$F_{s4} = 2 \times 139 \times 1283 \times 10^{-3} = 356.7\,\text{kN}$

Thus total steel force, $F_s = 2739.7 + 363.6 + 362.0 + 356.7 = 3822\,\text{kN}$.

Again, the neutral axis is in the top flange of the beam and it can therefore be shown that:

(i) $F_{c1} = 2535.4\,\text{kN}$ with depth to centroid, $a_1 = 74.6\,\text{mm}$
(ii) $F_{c2} = 184.4\,\text{kN}$ with depth to centroid, $a_2 = 153.2\,\text{mm}$
(iii) $F_{c3} = 1103.6\,\text{kN}$ with depth to centroid, $a_3 = 230.5\,\text{mm}$

Thus, $F_c = 2535.4 + 184.4 + 1103.6 = 3823.5\,\text{kN} \approx F_s$; therefore section balances and neutral axis is at the correct level.

Taking moments about the neutral axis level gives

$$M = \begin{bmatrix} 2535.4 \times (348 - 74.6) + 184.4 \times (348 - 153.2) + 1103.6 \times (348 - 230.5) \\ + 356.7 \times (582 - 330) + 362 \times (582 - 160) + 363.6 \times (582 - 110) \\ + 2739.7 \times (582 - 60) \end{bmatrix}$$

$$\times 10^{-3}$$

$= \mathbf{2702.8\,kNm}$ (which is 2.0% higher than in (1) above).

The comments in (1) above on limiting concrete strain apply here also. Their consideration would not significantly affect the resistance.

6.1.3.3. Simplified concrete stress block
Design of simple sections can be made easier by considering the simplified rectangular concrete stress block in a similar manner to reinforced beams.

6.1.3.4. Un-bonded tendons
2-1-1/clause 5.10.1(3) requires that, generally, prestress should be included in the action combinations defined in EN 1990 as part of the loading, and its effects included in the applied internal moment and axial force.

The design of prestressed concrete sections, where the prestressing tendons are not bonded to the concrete section, cannot be treated using the general rules above for bonded tendons since the tendon strains do not increase at the same rate as the strain in the concrete at the same level. *2-2/clause 6.1(108)* allows increase in strain in external tendons to be considered by assuming that the strain between fixed points is constant and considering the increase in strain resulting from structural deformation between these fixed points. It should be noted that deviators are not usually fixed points, especially as they often have small deviation angles which provide little frictional restraint. External post-tensioning is discussed in section 5.10.7 of this guide.

2-2/clause 6.1(108)

6.1.4. Reinforced concrete columns (additional sub-section)
6.1.4.1. Assumptions
EC2 covers the design of reinforced concrete columns under the same clauses as those for bending design, since the same basic assumptions discussed in section 6.1.2.1 apply. EC2

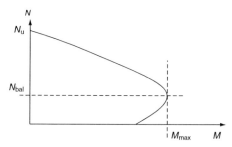

Fig. 6.1-12. Interaction diagram for typical reinforced column

defines the failure of concrete in compression by means of a compressive strain limit, as seen previously, of ε_{cu2} (or ε_{cu3}). 2-1-1/clause 6.1(5), however, requires this limit to be adjusted from ε_{cu2} (typically equal to 0.0035 for Class 50/60 concrete or lower), where the section is purely in flexure or the neutral axis remains within the section under combined bending and axial load, to ε_{c2} (or ε_{c3}), which is typically 0.0020, for sections subjected to loading where the whole section is in uniform compression. For intermediate strain distributions, the limiting strain diagram is determined from the construction in Fig. 6.1-1. This rule also applies to *parts* of sections in compression under axial loads (such as compression flanges of box girders, for example) as discussed in 6.1.2.2.

From the limited calculations undertaken by the authors, the calculated section strengths for circular and rectangular sections are usually relatively insensitive to variations in the assumptions of ultimate concrete strain. Some caution is needed in assuming that the effect is negligible where sections are heavily reinforced in tension and the reinforcement is not expected to yield.

2-1-1/clause 6.1(4)

2-1-1/clause 6.1(4) defines minimum moments, which should be considered for column design. These are defined by applying the axial loads at minimum eccentricities given by the depth of section, h, divided by 30, but not less than 20 mm. The minimum bending moments should be considered about any axis, but often a nominal moment about a major axis will have little effect on the resistance to a bending moment about the minor axis and can be ignored. Where columns are slender (see section 5.8.3), the additional second-order moments developed must be allowed for in accordance with 2-1-1/clauses 5.8.6, 5.8.7 or 5.8.8. In all cases, moments from imperfections should be considered in accordance with 2-2/clause 5.2.

6.1.4.2. Strain compatibility

A strain compatibility approach (see section 6.1.2.2) can be adopted for any cross-section. First, an area of reinforcement must be assumed and a neutral axis depth estimated. The extreme fibre compressive strain is set to ε_{cu2} (or ε_{cu3}) and therefore the strains throughout the section can be calculated. From these strains, the stresses in the various levels of reinforcement can be obtained, and hence the axial load and moment that the section can resist. This procedure will simply give one solution of the coexisting axial load and moment that can be carried simultaneously, as shown in Fig. 6.1-12.

Since it is usually desired to verify a particular combination of axial load and moment, further iteration is needed to do this to tailor the internal stresses to produce the desired combination of resistances. This is usually done so that either:

(1) the moment resistance is determined for a given applied axial force and it is verified that this moment resistance exceeds the coexisting applied moment, or
(2) the applied moment and axial force are increased pro rata together and it is verified that the load factor exceeds unity.

These methods are illustrated in Worked example 6.1-6.

Where, after this first iteration, parts of the section are shown to be wholly in compression, the concrete strain limit should be adjusted, as described above, and the iteration process

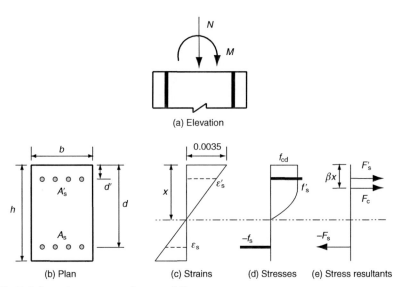

Fig. 6.1-13. Reinforced concrete column at failure

repeated to obtain new stresses and coexisting resistances to combinations of moment and axial force.

If the solution is deficient compared to the design values, the area of reinforcement originally assumed must be modified and the procedure (including iteration of the neutral axis depth) repeated. This procedure is obviously tedious and it can be seen that a computer is required for efficient design of reinforced concrete columns subjected to axial forces and bending moments.

6.1.4.3. Axial load plus uniaxial bending
To illustrate the principles, for simplicity, the following section develops expressions only for rectangular concrete columns. As with beam design above, the equations still apply for flanged beams where the neutral axis remains in the flange. In practice, computer programs will be required to automate the iterative strain-compatibility procedures involved for the design of general sections.

Only the case where the neutral axis remains within the section is considered here. The strains, stresses and stress resultants for failure of a rectangular reinforced concrete column are illustrated in Fig. 6.1-13.

For equilibrium, $N = F_c + F'_s + F_s$, thus:

$$N = f_{av}bx + f'_s A'_s + f_s A_s \tag{D6.1-19}$$

and taking moments about the application of N (the column centreline) gives:

$$M = f_{av}bx\left(\frac{h}{2} - \beta x\right) + f'_s A'_s\left(\frac{h}{2} - d'\right) + f_s A_s\left(\frac{h}{2} - d\right) \tag{D6.1-20}$$

In the above expressions, tensile forces and stresses should be taken as negative. f'_s and f_s should be calculated from the strains implied by the neutral axis depth x, and, where the reinforcement strains are sufficient to cause yielding, f'_s and/or f_s can be taken as f_{yd} but again should observe the sign convention. f_{av} and β are defined in section 3.1.7.

These equations are difficult to apply for the design of sections subjected to known axial loads and bending moments, since f_s and x are unknown. Thus, one design procedure is to assume values for x (and hence f'_s and f_s) and A'_s, then calculate A_s from equation (D6.1-19) for a given N. Equation (D6.1-20) can then be used to check that the value of M is greater than the applied moment. If M is less than the applied moment, the procedure should be repeated by altering the assumed values of x and A'_s. This procedure is illustrated in Worked example 6.1-6.

Worked example 6.1-6: Reinforced concrete pier

A 1200 mm by 600 mm rectangular reinforced concrete column is to be used as a bridge pier subjected to an axial compressive force of 2600 kN and a coexistent moment about the minor axis of 2300 kNm (which can be applied in either direction) at the ultimate limit state. Assuming Class 50/60 concrete and high-yield reinforcement with $f_{yk} = 500$ MPa, find the reinforcement required for an adequate design.

Try a neutral axis depth, x, of 125 mm.

Assuming $d' = 60$ mm and $d = 540$ mm (to give suitable cover), the strains and corresponding stresses are:

$$\varepsilon'_s = 0.0035 \times \left(1 - \frac{60}{125}\right) = 0.00182 < \frac{f_{yd}}{E_s} \, (= 0.002174)$$

therefore

$$f'_s = 0.00182 \times 200 \times 10^3 = 364 \text{ MPa (compression)}$$

and

$$\varepsilon_s = 0.0035 \times \left(\frac{540}{125} - 1\right) = 0.01162 > \frac{f_{yd}}{E_s}$$

therefore

$$f_s = f_{yd} = 434.8 \text{ MPa (tension)}$$

From section 3.1.7, for a parabolic-rectangular concrete stress block distribution with $f_{ck} = 50$ MPa, $f_{av} = 22.94$ MPa and $\beta = 0.416$. $N = 2600$ kN can be substituted into equation (D6.1-19) and, for a given value of A_s, the equation can be solved to find the corresponding value of A'_s to satisfy equilibrium. Finally, equation (D6.1-20) can be solved to find the moment resistance for the given reinforcement. The procedure must be repeated with different reinforcement quantities until the value of M obtained from equation (D6.1-20) is sufficient to resist the applied moment. For this example the following values are obtained by way of this iterative procedure:

A_s (mm²)	A'_s (mm²)	M (kNm)
11 no. 16ϕ = 2211.7	331.5	1113.1
11 no. 20ϕ = 3455.8	1817.5	1372.8
11 no. 25ϕ = 5399.6	4139.4	1778.4
11 no. 32ϕ = 8846.7	**8257.0**	**2497.9**
11 no. 40ϕ = 13 823.0	14 201.2	3536.4

Thus providing 22 no. 32ϕ bars (11 no. at 100 mm centres in each face, $A_s = A'_s = 8846.7$ mm²) will give adequate resistance for the column for the reversible moment. With this reinforcement now check:

(1) the maximum moment that can accompany the design axial force of 2600 kN and
(2) the load factor that can be applied to the design axial load and moment in combination.

(1) Maximum moment coexisting with $N = 2600$ kN

As compression steel has been added compared to the solution with $A'_s = 8257.0$ mm² and $x = 125$ mm above, try reducing neutral axis depth to 121 mm, to keep the axial force resistance at 2600 kN, and substitute $A_s = A'_s = 8846.7$ mm² into equation (D6.1-19) and equation (D6.1-20):

$$\varepsilon'_s = 0.0035 \times \left(1 - \frac{60}{121}\right) = 0.001764 < \frac{f_{yd}}{E_s} \quad \text{therefore} \quad f'_s = 353 \text{ MPa (compression)}$$

$$\varepsilon_s = 0.0035 \times \left(\frac{540}{121} - 1\right) = 0.011225 > \frac{f_{yd}}{E_s} \quad \text{therefore} \quad f_s = 434.8 \text{ MPa (tension)}$$

$$N = (22.94 \times 121 \times 1200 + 353 \times 8846.7 - 434.8 \times 8846.7) \times 10^{-3} = 2607\,\text{kN}$$

compared to a target of 2600 kN, so neutral axis guess was acceptable. The coexisting moment is therefore:

$$M = \begin{pmatrix} 22.94 \times 121 \times 1200 \times (300 - 0.416 \times 121) \\ + 353 \times 8846.7 \times (300 - 60) - 434.8 \times 8846.7 \times (300 - 540) \end{pmatrix} \times 10^{-6}$$

$$= 2504\,\text{kNm}$$

(2) Load factor on $N = 2600\,\text{kN}$ and $M = 2300\,\text{kNm}$

The load factor can be checked by substituting $A_s = A'_s = 8846.7\,\text{mm}^2$ into equations (D6.1-19) and (D6.1-20) and adjusting the depth of the neutral axis to 127 mm:

$$\varepsilon'_s = 0.0035 \times \left(1 - \frac{60}{127}\right) = 0.001846 < \frac{f_{yd}}{E_s} \text{ therefore } f'_s = 369.3\,\text{MPa (compression)}$$

$$\varepsilon_s = 0.0035 \times \left(\frac{540}{127} - 1\right) = 0.011382 > \frac{f_{yd}}{E_s} \text{ therefore } f_s = 434.8\,\text{MPa (tension)}$$

$$N = (22.94 \times 127 \times 1200 + 369.3 \times 8846.7 - 434.8 \times 8846.7) \times 10^{-3} = 2917\,\text{kN}$$

with a load factor of $2917/2600 = 1.12$ and

$$M = \begin{pmatrix} 22.94 \times 127 \times 1200 \times (300 - 0.416 \times 127) \\ + 369.3 \times 8846.7 \times (300 - 60) - 434.8 \times 8846.7 \times (300 - 540) \end{pmatrix} \times 10^{-6}$$

$$= 2571\,\text{kNm}$$

with load factor of $2571/2300 = 1.12$, the same as for axial force, so the guessed neutral axis depth was correct.

The load factor on the above design loads needed to cause failure is 1.12.

6.1.4.4. Axial load plus biaxial bending

If a column of known geometry and reinforcement content is analysed rigorously using the strain compatibility approach, for example, it is possible to construct interaction diagrams relating failure values of design axial force (N_{Ed}) and moments (M_{Edz} and M_{Edy}) about the major and minor axes respectively. The shape of the diagram varies according to the value of N_{Ed} in relation to the full axial load carrying resistance, N_{Rd}. The shape of this diagram is represented approximately in EC2 for bisymmetric sections by the following expression given in 2-1-1/clause 5.8.9(4):

$$\left(\frac{M_{Edz}}{M_{Rdz}}\right)^a + \left(\frac{M_{Edy}}{M_{Rdy}}\right)^a \leq 1.0 \qquad\qquad \text{2-1-1/(5.39)}$$

where:

M_{Edz}, M_{Edy} are the design moments about the respective axes (including second-order moments)

M_{Rdz}, M_{Rdy} is the moment resistance about the respective axes

a is the exponent:
for circular and elliptical cross-sections $a = 2.0$
for rectangular cross-sections:

N_{Ed}/N_{Rd}	0.1	0.7	1.0
$a =$	1.0	1.5	2.0

with linear interpolation for intermediate values. These stem from observations that the interaction is linear near the balance point ($N_{Ed}/N_{Rd} \sim 0.1$) and circular near the squash load ($N_{Ed}/N_{Rd} \sim 1.0$)

N_{Ed} is the design value of axial force
N_{Rd} is the design axial resistance of the section given by:
$$N_{Rd} = A_c f_{cd} + A_s f_{yd}$$

where:

A_c is the gross area of the concrete section
A_s is the area of longitudinal reinforcement

This method is not suited to design as N_{Rd} depends on the reinforcement. It therefore has to be used to check the adequacy of sections with guessed reinforcement and the reinforcement then modified accordingly.

Section 5.8.9 of this guide provides further guidance on appropriate imperfections to consider for biaxial bending, and details the methods which can be used to determine the effects of biaxial bending on slender elements. In particular, 2-1-1/clause 5.8.9(2) requires imperfections to be considered only in the direction in which they will have the most unfavourable effect.

6.1.5. Brittle failure of members with prestress (additional sub-section)
Prestressed beams generally

2-1-1/clause 5.10.1(5)P requires that prestressed beams should not fail in a brittle manner due to corrosion or failure of individual tendons. It is desirable for a beam to first exhibit cracking, as a warning that there is corrosion occurring. A potential problem arises where tendons are corroding but the concrete remains uncracked. No sign of distress may be apparent when the concrete compensates for the loss of prestress through acting in tension. However, if the concrete suddenly cracks, this tensile strength is permanently lost and the structure may fail suddenly if there is insufficient bending reserve in the remaining tendons and reinforcement. For prestressed beams, this is a new requirement to UK designers, which will only influence the design of members where the cracking moment of the unprestressed section is a significant proportion of the applied moment being checked. This may occur where sections are designed with relatively little prestress compared with the beam cross-section. Protection against this brittle failure can be achieved in one of three ways according to *2-2/clause 6.1(109) (a)* to *(c)*:

2-2/clause 6.1(109) (a) to (c)

(a) Ensure that the remaining cables, after corrosion or failures have led to cracking, are adequate to carry the frequent combination design moment

To do this check, the tendon area is hypothetically reduced so that the calculated cracking moment with the reduced area of tendons is less than or equal to that from the frequent combination of actions, as defined in EN 1990 for serviceability limit states. The cracking moment should be based on the tensile stress, f_{ctm}. For each extreme tension fibre of the beam with section modulus Z, the moment at cracking, M, and the prestress force, P, are related by:

$$P/A + Pe/Z - M/Z = -f_{ctm} \tag{D6.1-21}$$

where e is the eccentricity of the prestress and A is the section cross-sectional area.

The prestressing force required to make the beam crack at moment, M, is thus:

$$P = (M/Z - f_{ctm})/(1/A + e/Z) \tag{D6.1-22}$$

The ultimate strength of the beam with the reduced tendon area, but using the material factors for accidental situations, is then again compared against the applied moment from the frequent combination of actions. This is illustrated in Worked example 6.1-7. When calculating the ultimate moment resistance with the hypothetically reduced tendon area, the resistance of any reinforcement present may also be used. It is also permissible to allow for moment redistribution to adjacent areas in indeterminate bridges when checking the moment resistance of the section with the reduced tendon area. Redistribution

CHAPTER 6. ULTIMATE LIMIT STATES

of moment and checks of rotation capacity are discussed in sections 5.5 and 5.6 respectively.

The requirement to reduce the number of tendons until cracking results under the frequent combination of actions is intended to ensure that cracking would become visible under normal traffic loading conditions, and investigations and repair could then be instigated quickly. It is slightly odd that the frequent combination is used both to determine the reduced number of tendons and then to check the ultimate strength after cracking, as the loading might increase in the intervening period between cracking and subsequent detection and remedial action. However, the check is new to UK practice and the required loading means it should rarely govern.

It is not made clear whether the requirement has to be met both before and after long-term losses. Meeting the requirement before long-term losses would be more onerous, as a greater area of prestress would have to be discounted, but it is unlikely that corrosion would occur before long-term losses were substantially complete. It is therefore recommended here that a check only be made after long-term losses.

If equation (D6.1-22) is used to determine the prestress force to cause cracking at the moment, M, then the reduced prestress area, $A_{p,Red}$, for cracking can be determined accordingly as:

$$A_{p,Red} = \frac{(M/Z - f_{ctm})}{\sigma_p(1/A + e/Z)} \tag{D6.1-23}$$

where σ_p is the stress in the tendons just prior to cracking. If there is no reinforcement considered (minimum reinforcement would also have to be present) then the ultimate resistance must satisfy:

$$A_{p,Red} \times z_s \times f_{p0.1k} \geq M \tag{D6.1-24}$$

where z_s is the lever arm for the prestress at the ultimate limit state.

From equations (D6.1-23) and (D6.1-24), the moment from the frequent combination of actions, M, must exceed a minimum value to avoid brittle fracture in the absence of reinforcement, thus:

$$M \geq \frac{f_{ctm} z_s}{\dfrac{z_s}{Z} - \dfrac{\sigma_p}{f_{p0.1k}}(1/A + e/Z)} \tag{D6.1-25}$$

This can be seen to be related mainly to properties of the concrete section, the stress in the remaining tendons just prior to cracking and the prestress eccentricity. The total amount of prestress does not affect the behaviour other than through z_s, which will have a limited range.

(b) Minimum reinforcement

Brittle fracture may also be avoided by ensuring that there is sufficient longitudinal reinforcement provided to compensate for the loss of resistance when the tensile strength of the concrete is lost. This is achieved by providing a minimum area of reinforcement according to 2-2/Expression (6.101a). This reinforcement is not additional to requirements for other effects and may be used in ultimate bending checks. This check will always produce a requirement for some reinforcement (unlike that in (a)), but avoids the need to directly quantify the change in bending strength due to the loss of tendons from corrosion:

$$A_{s,min} = \frac{M_{rep}}{z_s f_{yk}} \qquad \text{2-2/(6.101a)}$$

where:

M_{rep} is the cracking bending moment calculated assuming a recommended tensile strength equal to f_{ctm} at the extreme tension fibre, ignoring the prestressing. This tensile strength is a nationally determined parameter.

z_s is the lever arm at the ultimate limit state related to the reinforcing steel

It is not clear whether, in calculating the lever arm, the centre of force of the compression zone for flexure should be derived ignoring the prestressing or including it. The latter is the most conservative, as a smaller lever arm will result, and is consistent with the idea of reinforcement compensating for lost prestress. Either way, method (b) tends to be more conservative than method (a).

The clause states that within joints of segmental precast elements, f_{ctm} should be taken as zero. This leads to the conclusion that no minimum reinforcement is required for brittle fracture in precast segmental construction. This was intended because such reinforcement could not, in any case, be placed across the joints. It was envisaged that no reinforcement should be required since, if corrosion occurs, the joints will open up and give warning of the problem. However, this will not be true if the strength of the glue is significant; the glue may be stronger than the parent concrete. If a designer feels strongly that a check of brittle failure should be made, method (a) could be used assuming the glue to have the same tensile strength as the concrete. However, if the check is not satisfied, the only measure that can be taken is to change the concrete cross-section!

2-2/clause 6.1(110)

Item (i) of **2-2/clause 6.1(110)** requires the reinforcement area from 2-2/Expression (6.101a) to be provided in all areas where tension occurs under the characteristic combination of actions, including the secondary effects of prestress but ignoring the primary (isostatic) effects. It is assumed that local loss of prestress at one section does not alter the distribution of secondary prestress effects.

To ensure adequate ductility in the bridge when a section suddenly cracks, 2-2/clause 6.1(110) (iii) requires the minimum reinforcing steel area, $A_{s,min}$, in the spans of continuous beams to extend to the supports of the span considered. The reasons for this are unclear, but it is suggested here that it is intended to provide compression reinforcement to the bottom flange in the adjacent section so that the flange does not fail in compression when moment is shed from the span when cracking occurs. In principle, therefore, the same should apply for reinforcement provided in the top flange at supports; that is, this reinforcement should be taken into the span to prevent compressive failure in the top flange. Typically, in box girders or beam and slab decks this will not be necessary as the top flange is relatively very large. However, this would not be the case for half-through construction, where the top flange may be small or non-existent and it would then be logical to take this reinforcement into the span. Additional care would be required with haunched sections where more reinforcement would be needed in the span as the moment shed from the supports gives a bigger flange force on a smaller section.

This continuation of reinforcement may, in any case, be avoided if, at the ultimate limit state, the resisting tensile force provided by the reinforcement and prestressing steel at the section at the supports, calculated with characteristic strengths f_{yk} and $f_{p0.1k}$ respectively, is less than the resisting compressive force of the bottom flange, according to 2-2/Expression (6.102):

$$A_s f_{yk} + k_p A_p f_{p0.1k} < t_{inf} b_0 \alpha_{cc} f_{ck} \qquad \text{2-2/(6.102)}$$

where:

t_{inf}, b_0 are, respectively, the thickness and the width of the bottom flange of the box girder section

A_s, A_p denote, respectively, the area of resisting and prestressing steel in the tensile zone at the ultimate limit state

k_p is a nationally determined parameter with recommended value of 1.0

The above comment on similarly checking the span for redistribution from the supports also applies.

(c) Provide proven monitoring facility

If external post-tensioning is used, the cables can readily be inspected for signs of corrosion or failure so no further contingency in the design is required under this clause. However, the requirement of 2-1-1/clause 9.2.1.1(4) must still be met.

Worked example 6.1-7: Post-tensioned concrete box girder
Consider the following example of a concrete box girder with bonded prestressing tendons, as shown in Fig. 6.1-14.

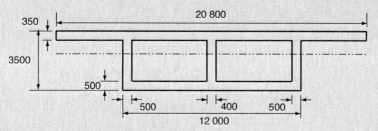

Fig. 6.1-14. Section through concrete box girder

The box girder has the following section properties: $A = 17\,\text{m}^2$, $I_{xx} = 33.5\,\text{m}^4$, $Z_{top} = 21.1\,\text{m}^3$, $Z_{bot} = 17.5\,\text{m}^3$. The prestressing tendons comprise 35 no. tendons each of area $= 2850\,\text{mm}^2$, with $f_{pk} = 1860\,\text{MPa}$ and $f_{p0.1k} = 1600\,\text{MPa}$. The neutral axis is 1.914 m from the bottom of the section. The tendons are stressed to 70% of their characteristic tensile strength and there are total losses (long term plus short term) of 25%. The depth to the centroid of the prestressing force $= 3000$ mm. (In this example it is assumed that all the prestressing cables are actually at this position.)

Check the beam for brittle fracture using method (a) under sagging moments from the frequent combination of actions of (i) 240 MNm and (ii) 100 MNm. Also, (iii) determine the reinforcement to prevent brittle fracture using method (b).

Case (i): The frequent combination sagging moment = 240 MNm
The reduced hypothetical prestress to make the beam crack at the frequent combination sagging moment is from equation (D6.1-22):

$$P = (M/Z_{bot} - f_{ctm})/(1/A + e/Z_{bot})$$

For concrete with cylinder strength $= 40\,\text{MPa}$, $f_{ctm} = 3.5\,\text{MPa}$ from 2-1-1/Table 3.1.
Therefore:

$$P = (240 \times 10^9/17.5 \times 10^9 - 3.5)/(1/(17 \times 10^6) + 1086/(17.5 \times 10^9)) = 84.5\,\text{MN}$$

If the above check is done after losses, then the tendon area required is:

$$A_{p,\text{Red}} = \frac{84.5 \times 10^6}{0.7 \times 0.75 \times 1860} = 86\,532\,\text{mm}^2$$

which corresponds to 30.3 tendons. (This could have been derived directly from equation (D6.1-23).) For the ultimate limit state check with hypothetically reduced tendons, the material factors for accidental situations may be used from 2-1-1/Table 2.1. Thus, $\gamma_c = 1.2$ and $\gamma_s = 1.0$. For simplicity, the tendon stress–strain curve with the horizontal plateau at $f_{p0.1k}$ will be used. Assuming that the tendons yield, the force in the tendons

$$F_s = 86\,532 \times 1600/1.0 = 138.5\,\text{MN}$$

If the simplified rectangular stress block of 2-1-1/Fig. 3.5 is used for the concrete, the area of the compression zone is thus:

$$= F_s/\eta f_{cd} = \frac{138.5 \times 10^6}{1.0 \times 0.85 \times 40/1.2} = 4.89 \times 10^6\,\text{mm}^2$$

The depth of the block $\lambda x = 4.89 \times 10^6/20\,800 = 235\,\text{mm}$, which is in the top flange so the calculation of depth is consistent. From 2-1-1/Fig. 3.5 the neutral axis is at a depth $x = 235/\lambda = 235/0.8 = 294\,\text{mm}$. Since the whole flange is not in compression, a limiting compressive strain of $\varepsilon_{cu3} = 0.0035$ may be used rather than the lower ε_{c3}, which is applicable where the whole flange is in compression – see section 6.1.2.2. The strain in the tendons is therefore $(3000 - 294)/294 \times 0.0035 = 0.032$ so the tendons will clearly reach yield and the above assumption was justified. The moment resistance is therefore $M_{Rd} = 138.5 \times 10^6 \times (3000 - 235/2) = 399.2 \text{ MNm} \gg 240 \text{ MNm}$ applied frequent combination moment, so there is no problem with brittle fracture.

(This check could also have been done directly with equation (D6.1-24) using the lever arm of 2883 mm above, thus, for adequacy, the applied moment must exceed:

$$\frac{f_{ctm}z_s}{\frac{z_s}{Z} - \frac{\sigma_p}{f_{p0.1k}}(1/A + e/Z)} = \frac{3.5 \times 2883}{\frac{2883}{17.5 \times 10^9} - \frac{976.5}{1600}\left(\frac{1}{17 \times 10^6} + \frac{1086}{17.5 \times 10^9}\right)}$$

$$= 111 \text{ MNm} < 240 \text{ MNm so OK}$$

This also immediately shows that case (ii) below, where the moment is 100 MNm, will not work without adding reinforcement.)

It is quite likely that a bridge of this sort would be cracked in flexure under the frequent combination of moments, even with all the tendons intact, so no further checks would be required. The amount of prestress was chosen to be artificially high here simply so a hypothetical reduction would need to be made.

Case (ii): The example above is now repeated assuming a frequent combination sagging moment = 100 MNm

The reduced hypothetical prestress to make the beam crack at the frequent combination sagging moment is from equation (D6.1-22):

$$P = (M/Z_{bot} - f_{ctm})/(1/A + e/Z_{bot})$$

For concrete with cylinder strength $= 40 \text{ MPa}$, $f_{ctm} = 3.5 \text{ MPa}$, from 2-1-1/Table 3.1.
Therefore:

$$P = (100 \times 10^9/(17.5 \times 10^9) - 3.5)/(1/(17 \times 10^6) + 1086/(17.5 \times 10^9)) = 18.3 \text{ MN}$$

If the above check is done after losses, then the tendon area required is

$$A_{P,red} = \frac{18.3 \times 10^6}{0.7 \times 0.75 \times 1860} = 18\,759 \text{ mm}^2$$

which corresponds to 6.6 tendons.

Assuming that the tendons yield, the force in the tendons, $F_s = 18\,579 \times 1600/1.0 = 30.0 \text{ MN}$. If the simplified rectangular stress block of 2-1-1/Fig. 3.5 is used for the concrete, the area of the compression zone is thus:

$$= F_s/\eta f_{cd} = \frac{30.0 \times 10^6}{1.0 \times 0.85 \times 40/1.2} = 1.059 \times 10^6 \text{ mm}^2$$

The depth of the block $\lambda x = 1.059 \times 10^6/20\,800 = 50.9 \text{ mm}$ which is in the top flange, so the calculation of depth is consistent. From 2-1-1/Fig. 3.5 the neutral axis is at depth $x = 50.9/\lambda = 50.9/0.8 = 64 \text{ mm}$. The strain in the tendons is therefore $(3000 - 64)/64 \times 0.0035 = 0.16$, so the tendons will clearly reach yield and the above assumption was justified. The moment resistance is therefore $M_{Rd} = 30.0 \times 10^6 \times (3000 - 51/2) = 89 \text{ MNm} < 100 \text{ MNm}$ applied frequent moment, so there is now a problem with brittle fracture. Reinforcement present in the bottom flange (that part not utilized for other effects, such as torsion) could be included in the global flexural resistance to make this check work or additional reinforcement provided.

Case (iii): Reinforcement with $f_{yk} = 500 \text{ MPa}$ is provided at $d = 3250 \text{ mm}$
From 2-2/Expression (6.101a):

$$A_{s,min} = \frac{M_{rep}}{z_s f_{yk}}$$

CHAPTER 6. ULTIMATE LIMIT STATES

$$M_{\text{rep}} = Z_{\text{bot}} \cdot f_{\text{ctm}} = 17.5 \times 10^9 \times 3.5 = 61.25\,\text{MNm}$$

z_s is the lever arm relating to this reinforcement at the ultimate limit state. From the above calculations in case (i), where most of the prestress is present and the lever arm is $0.96d$, it can be seen that it would be reasonable to use $0.9d$ as an approximation. Therefore:

$$A_{s,\text{min}} = \frac{M_{\text{rep}}}{z_s f_{yk}} = \frac{61.25 \times 10^9}{0.9 \times 3250 \times 500} = 41\,880\,\text{mm}^2$$

which is a little less than 20ϕ bars at 175 mm centres in the top and bottom faces of the bottom flange. This is a reasonably large amount of reinforcement compared to the usual in a prestressed post-tensioned box girder and would generally not be desirable economically.

Pretensioned beams
For pretensioned beams, methods (a) and (b) above can be used, but method (b) may be applied without providing additional reinforcement if the check is modified. Since pretensioned strands are protected by concrete encasement in the same way as reinforcing steel, it is permitted to apply 2-2/Expression (6.101a) directly, with $A_{s,\text{min}}$ provided by the pretensioning itself. The criterion is then simply that the bending resistance of the cracked section exceeds the cracking moment of the section ignoring the prestress. It seems illogical that the cracking moment, M_{rep}, should here be calculated ignoring the prestress. However, it does mean that the minimum steel area will usually be easily satisfied and since there have been no experience of problems with pretensioned beams in the UK, it is suggested that this easily satisfied check is accepted.

For pretensioned members, 2-2/Expression (6.101a) should then be applied with some modifications. In checking $A_{s,\text{min}}$, two alternatives are possible following 2-2/clause 6.1(110) (ii):

(a) All strands with a concrete cover greater than a certain specified amount may be included. EC2-2 suggests twice the minimum cover given in 2-1-1/Table 4.3, but the applicable cover is a nationally determined parameter. The National Annex is not available at the time of writing, but it is likely that the applicable cover will be chosen so that all strands may be utilized and the criterion will then never govern. f_{yk} should be replaced with $f_{p0.1k}$ in 2-2/Expression (6.101a). The lever arm should be based on the effective strands.
(b) Where strands have stresses lower than $0.6f_{pk}$ after losses in the characteristic combination, they can be considered regardless of cover but limiting their stress increase, $\Delta\sigma_p$, to the lesser of $0.4f_{pk}$ and 500 MPa. 2-2/Expression (6.101a) then becomes $A_{s,\text{min}}f_{yk} + A_p\Delta\sigma_p \geq M_{\text{rep}}/z_s$.

6.2. Shear
This section deals with the design at ultimate limit state of members subject to shear. It is split into the following sub-sections; the last is an additional sub-section that is not provided in EN 1992-2:

- General verification procedure rules   *Section 6.2.1*
- Members not requiring design shear reinforcement   *Section 6.2.2*
- Members requiring design shear reinforcement   *Section 6.2.3*
- Shear between web and flanges of T-sections   *Section 6.2.4*
- Shear at the interface between concrete cast at different times   *Section 6.2.5*
- Shear and transverse bending   *Section 6.2.6*
- Shear in precast concrete and composite construction   *Section 6.2.7*

6.2.1. General verification procedure rules

2-1-1/clause 6.2.1(1)P, (2) and (3)

2-1-1/clause 6.2.1(1)P, (2) and *(3)* define the following general terms for use in shear design calculations:

V_{Ed} is the (ultimate) design shear force in the section considered resulting from external loading and prestressing (either bonded or unbonded)

$V_{Rd,c}$ is the design shear resistance of the member without shear reinforcement

$V_{Rd,s}$ is the design value of the shear force, which can be sustained by the yielding shear reinforcement

V_{Rd} is the design shear resistance of a member with shear reinforcement including the components from inclined compressive and tensile chords

$V_{Rd,max}$ is the design value of the maximum shear force which can be sustained by the member, limited by crushing of the compression struts

V_{ccd} is, in members with inclined compression chords, the design value of the shear component of the force in the compression area

V_{td} is, in members with inclined tensile chords, the design value of the shear component of the force in the tensile reinforcement

The design of reinforced concrete members for shear is usually carried out as a check after the flexural design and therefore basic section sizes and properties should already have been chosen. For flanged beams with thin webs, the maximum shear strength achievable, $V_{Rd,max}$, may, however, need to be considered at the initial sizing stage to ensure that web thicknesses are great enough. It may also be beneficial to increase the section sizes and reduce the shear reinforcement content for economic or buildability reasons.

2-1-1/clause 6.2.1(2)

2-1-1/clause 6.2.1(2) allows the designer to take account of the vertical components of the inclined tension and compression chord forces in the shear design of a member with shear reinforcement. These components, V_{ccd} and V_{td}, are added to the shear resistance based on the links. They should, in theory, be determined from the actual chord forces obtained from the truss model and not the value of M/z from beam theory since the latter would overestimate V_{ccd} and underestimate V_{td}. It is not clear why the clause appears not to allow account to be taken of the inclined chord forces for members without shear reinforcement. It should be noted that these components can also reduce strength, notably in simply supported suspended spans with arched soffits.

The designer should be careful when considering these inclined chord forces in conjunction with inclined prestress, since the shear due to prestress is included within V_{Ed} on the applied actions side of the equation according to *2-1-1/clause 6.2.1(3)* and should not be double-counted. Often, the prestressing force forms the tension chord. The risk of double-counting is compounded by 2-1-1/clause 6.2.1(6), which deducts the inclined chord components from the applied shear when considering the web crushing limit. Care should also be taken with variable depth sections when a varying neutral axis height has been modelled in the global analysis. In this case, the output shear components will then be perpendicular to this inclined axis. The inclined chord forces should relate to a consistent axis.

2-1-1/clause 6.2.1(3)

2-1-1/clause 6.2.1(4)
2-1-1/clause 6.2.1(5)

2-1-1/clause 6.2.1(4) requires only minimum shear reinforcement to be provided where $V_{Ed} \leq V_{Rd,c}$. This is discussed further in section 6.2.2.1 of this guide. Where $V_{Ed} > V_{Rd,c}$, the concrete resistance $V_{Rd,c}$ is assumed to be completely lost and *2-1-1/clause 6.2.1(5)* requires the resistance to be taken as V_{Rd}. This differs from previous UK practice,[9] but is more rational since the cracking needed to mobilize forces in shear reinforcement reduces any contribution to the resistance from the concrete. This does not mean that the shear resistances for a given beam will be less than that found to BS 5400 Part 4,[9] since EN 1992 allows the shear truss angle to be varied, whereas BS 5400 Part 4 restricted the truss angle to 45°. This usually leads to greater resistance for a given link provision in reinforced concrete beams.

2-1-1/clause 6.2.1(6)

2-1-1/clause 6.2.1(6) relates to the maximum shear stress that can be carried by the web. This is limited by crushing of the web compression struts in the truss model. In calculating the shear for comparison against the crushing limit $V_{Rd,max}$, the inclined chord components V_{ccd} and V_{td} are this time included on the applied shear side of the equation, unlike in

2-1-1/clause 6.2.1(2) where they are on the resistance side, as they influence the state of stress in the web.

The use of a truss model for shear design clearly shows that the chord forces differ from those predicted by a flexural analysis, being greater than expected in the tension chord and lower than expected in the compression chord. **2-1-1/clause 6.2.1(7)** therefore requires the longitudinal tension reinforcement to make allowance for this. This is discussed in section 6.2.3.1 of this guide.

2-1-1/clause 6.2.1(7)

Where a member is predominantly subjected to uniform load, **2-1-1/clause 6.2.1(8)** permits the design shear force to not be checked at a distance less than d from the support, provided that the link reinforcement needed d away is continued to the support and web crushing ($V_{\text{Rd,max}}$) is checked at the support. If there are significant point loads applied within a distance d of the support, they should be considered using 2-1-1/clause 6.2.3(8).

2-1-1/clause 6.2.1(8)

Where a load is applied near the bottom of a section, **2-1-1/clause 6.2.1(9)** requires sufficient vertical reinforcement to carry the load to the top of the section, commonly referred to as 'suspension reinforcement', to be provided in addition to any reinforcement required to resist shear. This is similar to the requirement for members with indirect support, discussed further in section 9.2.5.

2-1-1/clause 6.2.1(9)

6.2.2. Members not requiring design shear reinforcement

This section is split into two additional sub-sections dealing with reinforced concrete (section 6.2.2.1) and prestressed concrete (section 6.2.2.2).

6.2.2.1. Reinforced concrete members

The formulae given by EC2 for the design of reinforced concrete members without shear reinforcement are empirical and have been chosen to fit with the extensive test data. The main characteristics governing the shear strength of members without shear reinforcement are concrete strength, amount of longitudinal reinforcement in tension and absolute values of section depths. The longitudinal reinforcement content contributes to the shear resistance in two ways:

(1) Directly by dowel action.
(2) Indirectly by controlling crack widths. These in turn influence the amount of shear that can be transferred across the cracks by aggregate interlock.

The member depths have also been shown to have a significant influence on shear strength and EC2 takes account of this by defining a depth factor, k, given by

$$k = 1 + \sqrt{\frac{200}{d}} \leq 2.0 \qquad \text{(D6.2-1)}$$

where d is the effective depth to longitudinal tension reinforcement (in millimetres).

2-2/clause 6.2.2(101) gives the following expression for calculating the shear strength of sections without shear reinforcement:

2-2/clause 6.2.2(101)

$$V_{\text{Rd,c}} = (C_{\text{Rd,c}} k (100 \rho_1 f_{\text{ck}})^{1/3} + k_1 \sigma_{\text{cp}}) b_w d \qquad \text{2-2/(6.2.a)}$$

where:

f_{ck} is the concrete cylinder strength in MPa

$$\rho_1 = \frac{A_{\text{sl}}}{b_w d} \leq 0.02 \qquad \text{(D6.2-2)}$$

A_{sl} is the area of longitudinal tensile reinforcement which extends a minimum of a design anchorage length and an effective depth beyond the section considered (as indicated in 2-1-1/Fig. 6.3)

b_w is the smallest width of the cross-section in the tensile zone (see section 6.2.2.2). Tests such as those in reference 11 suggest that the use of the minimum width is quite conservative

d is the effective depth to tensile reinforcement

$\sigma_{cp} = N_{Ed}/A_c$ (in MPa)

N_{Ed} is the axial force in the cross-section from the loading or from prestressing
A_c is the gross concrete cross-sectional area

The values of $C_{Rd,c}$ and k_1 may be given in the National Annex. EC2 recommends values of $0.18/\gamma_c$ and 0.15 respectively.

EC2 defines a minimum value of $V_{Rd,c}$ as:

$$V_{Rd,c} = (v_{min} + k_1 \sigma_{cp})b_w d \qquad \text{2-2/(6.2.b)}$$

where v_{min} can also be found in the National Annex and is recommended as:

$$v_{min} = 0.035 k^{3/2} f_{ck}^{1/2} \qquad \text{2-2/(6.3N)}$$

Recent tests (reference 4) have indicated this minimum strength to be unconservative for high strength concretes and those made with limestone aggregate. The UK National Annex is likely to set a limit to f_{ck} of 50 MPa for shear design by way of the nationally determined parameter in 2-2/clause 3.1.2(102)P.

The increase in shear resistance for axial load in the above expressions is greater than the increase in current UK practice. However, because of the general form of 2-2/Expression (6.2.a) and 2-2/Expression (6.2.b), a reduction to resistance is obtained where there is a tensile force. Axial tension can be expected to reduce both dowel action and aggregate interlock. However, providing the member is appropriately designed for the axial tension, tests by Regan[12] showed only a very minor influence of axial tension on shear resistance. The EC2 expressions are therefore conservative. This conservatism was of concern to the drafters of EN 1994-2, because deck slabs of composite bridges often have significant tension and yet there have not been related problems with shear resistance in deck slabs. Clause 6.2.2.5(3) of EN 1994-2 consequently reduces the effect of tension on shear resistance of deck slabs in composite bridges. This is discussed in reference 13. The same problem of conservatism potentially affects concrete box girder bridges and beam and slab bridges.

Only externally applied loads (from prestress or external loads) should be considered when applying 2-2/Expressions (6.2.a) and (6.2.b). The influence of imposed deformations (such as those induced by shrinkage or temperature effects) may be ignored. Indirect actions can be shed at the ultimate limit state and shrinkage effects, at least in part, are also inevitably included in tests themselves.

Worked example 6.2-1: Reinforced concrete deck slab

Find $V_{Rd,c}$ for a 250 mm-thick deck slab assuming 40 mm cover to main reinforcement of 20ϕ bars at 150 mm centres with a concrete of grade C35/45.

Effective depth, $d = 250 - 40 - 20/2 = 200$ mm

Effective width, $b_w = 1000$ mm

$$A_{sl} = \frac{\pi \times (20/2)^2}{0.150} = 2094.4 \text{ mm}^2/\text{m}$$

thus from equation (D6.2-2):

$$\rho_1 = \frac{A_{sl}}{b_w d} = \frac{2094.4}{1000 \times 200} = 0.0105 < 0.02 \text{ limit as required}$$

Assume $\gamma_c = 1.5$ and $C_{Rd,c} = 0.18/\gamma_c$ as recommended, thus $C_{Rd,c} = 0.18/1.5 = 0.12$.

From equation (D6.2-1):

$$k = 1 + \sqrt{200/d} = 1 + \sqrt{200/200} = 2.0 \leq 2.0 \text{ limit as required}$$

Ignoring any axial load, 2-1-1/Expression (6.2.a) becomes:

$$V_{Rd,c} = C_{Rd,c} k (100 \rho_1 f_{ck})^{1/3} b_w d$$

$$= 0.12 \times 2.0 \times (100 \times 0.0105 \times 35)^{1/3} \times 1000 \times 200 \times 10^{-3} = \mathbf{159.9 \text{ kN/m}}$$

CHAPTER 6. ULTIMATE LIMIT STATES

> **Worked example 6.2-2: Reinforced concrete column**
> Find the maximum shear force that can be sustained by the pier in Worked example 6.1-6 without the need for design shear reinforcement. (Note that minimum shear reinforcement would be required for columns in accordance with 2-1-1/clause 9.5.3.)
>
> Effective depth, $d = 540$ mm and effective width, $b_w = 1200$ mm
>
> $f_{ck} = 50$ MPa; therefore, $f_{cd} = 1.0 \times 50/1.5 = 33.3$ MPa (using $\alpha_{cc} = 1.0$ for shear – see discussion in section 3.1.6 of this guide)
>
> $A_{sl} = 11$ no. 32ϕ bars $= 11 \times \pi \times (32/2)^2 = 8846.7$ mm^2
>
> thus from equation (D6.2-2):
>
> $$\rho_l = \frac{A_{sl}}{b_w d} = \frac{8846.7}{1200 \times 540} = 0.01365 < 0.02 \text{ limit as required}$$
>
> Assume $k_1 = 0.15$ and take $\gamma_c = 1.5$ and $C_{Rd,c} = 0.18/\gamma_c$ as recommended to give $C_{Rd,c} = 0.18/1.5 = 0.12$.
>
> From equation (D6.2-1):
>
> $k = 1 + \sqrt{200/d} = 1 + \sqrt{200/540} = 1.609 \leq 2.0$ limit as required
>
> $N_{Ed} = 2600$ kN and $A_c = 600 \times 1200 = 720 \times 10^3$ mm^2
>
> thus
>
> $$\sigma_{cp} = \frac{2600 \times 10^3}{720 \times 10^3} = 3.611 \text{ MPa} < 0.2 f_{cd} \, (= 6.67 \text{ MPa}) \text{ limit therefore okay}$$
>
> Thus from 2-1-1/Expression (6.2.a):
>
> $$V_{Rd,c} = (C_{Rd,c} k (100 \rho_l f_{ck})^{1/3} + k_1 \sigma_{cp}) b_w d$$
>
> $$= (0.12 \times 1.609 \times (100 \times 0.01365 \times 50)^{1/3} + 0.15 \times 3.611) \times 1200 \times 540 \times 10^{-3}$$
>
> $$= (0.789 + 0.542) \times 1200 \times 540 \times 10^{-3}$$
>
> $$= \mathbf{862.3\,kN} \text{ (compared to 511.3 kN without benefit of axial force)}$$

Behaviour close to supports

There has been extensive testing carried out which illustrates that greater shear strengths than those given by 2-1-1/Expression (6.2.a) can be obtained in members where the load is applied close to the support or in short members, such as corbels. The increase in resistance, for sections closer to the support than approximately $2d$, occurs for two reasons:

- a proportion of the load is carried directly to the support by a compression strut;
- the angle of the shear failure plane is steepened from that obtained for a load applied further from the support.

In current UK practice (and earlier drafts of EN 1992-2 and ENV 1992-1-1[14]), this behaviour was accounted for by introducing an enhancement factor of kd/a_v which was applied to the concrete shear resistance. a_v is the distance from the support centreline to the face of the load. Testing suggests that the value of k is not the same for cases with uniformly distributed loads and those with single or multiple concentrated loads. For simply supported beams, $k = 2$ is appropriate for uniformly distributed loads and $k = 2.5$ is appropriate for a single concentrated load. For continuous beams and concentrated loads, $k = 3$ is appropriate. A value for k of 2 is therefore conservative for all loading types.

There are very few tests for situations covering multiple loads. As a result, the final version of EC2 removed the above enhancement factor from the resistance side of the equation and introduced a reduction factor in **2-1-1/clause 6.2.2(6)** of $\beta = a_v/2d$, which is applied to the

2-1-1/clause 6.2.2(6)

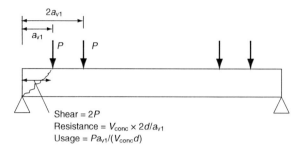

Shear check between support and nearest load to ENV 1992-1-1

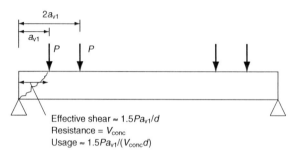

Shear check between support and nearest load to EN 1992-1-1

Fig. 6.2-1. Comparison of treatment of shear enhancement in ENV 1992-1-1 and EN 1992-1-1 for beam with span $\gg a_{v1}$ and d

shear contribution of individual loads to the total shear, V_{Ed}. This is not equivalent to the previous version, other than for cases where there is a single point load. This difference is shown in Fig. 6.2-1 for the case of a beam with span $\gg a_{v1}$ and d, and where the basic concrete resistance in the absence of any enhancement is V_{conc}. The current version of EC2 does not take account of the influence of steepening the shear failure plane, which can, in effect, give enhancement for loads applied further than $2d$ from the support.

For $a_v \leq 0.5d$, a_v should be taken as $0.5d$. The enhancement can only be used where the longitudinal reinforcement is fully anchored at the support. Concrete corbels and half joints are discussed further in Annex J of this guide, which details the special cases of short shear spans covered by EC2-2 in its Annex J.

The approach of 2-1-1/clause 6.2.2(6) does not fit tests for uniformly distributed loads, such as those in reference 15, while the approach of enhancing the resistance side of the equation gives good agreement. 2-1-1/clause 6.2.2(6) is more conservative. It also significantly complicates the practicality of applying support enhancement rules to bridge design with variable load trains as it is necessary to isolate the location of all loads contributing to the shear force. This is usually not convenient for load trains where load cases have been automatically generated by software. For uniformly distributed loads, an integral has to be performed to determine the reduced V_{Ed}. This problem also affects pad foundation design, where successive perimeters or shear lanes need to be considered. Notably, 2-1-1/clause 6.4.4(2) for punching resistance of bases uses the old ENV 1992-1-1 formulation with enhancement to the resistance which avoids the integral but contradicts section 6.2 of EN 1992-2.

2-1-1/clause 6.2.2(6) further limits the applied shear force (without reduction factors) to a maximum of:

$$0.5 b_w d \nu f_{cd} \qquad \qquad 2\text{-}1\text{-}1/(6.5)$$

where:

$$\nu = 0.6\left(1 - \frac{f_{ck}}{250}\right) \text{ is recommended} \qquad \qquad 2\text{-}1\text{-}1/(6.6\text{N})$$

with f_{ck} in MPa. ν is an empirically obtained efficiency factor which accounts for the crushing strength of concrete struts at the ultimate limit state.

CHAPTER 6. ULTIMATE LIMIT STATES

Minimum shear reinforcement
When no shear reinforcement is required on the basis of the design calculations, minimum shear reinforcement should nevertheless be provided in accordance with the detailing requirements of 2-1-1/clause 9.2.2 for beams, 2-1-1/clause 9.3.2 for slabs and 2-1-1/clause 9.5.3 for columns. It would be impractical to put links in flat slabs of bridge decks. 2-1-1/clause 6.2.1(4) allows minimum reinforcement to be omitted from members where transverse redistribution of loads is possible such as slabs (solid, ribbed or voided). For typical voided slabs used in bridge deck applications, it is recommended here that minimum links are always provided in accordance with 2-1-1/clause 9.2.2; they will, in any case, usually be needed for the transverse Vierendeel behaviour of such a deck.

6.2.2.2. Prestressed concrete members
Two types of shear failure can occur in prestressed concrete beams. 'Shear flexure' failures occur in regions of beams that are cracked in flexure and are covered in 2-2/clause 6.2.2(101). 'Shear tension' failures occur in regions of beams that are un-cracked in flexure when the principal tensile stress in the web reaches the concrete tensile strength. This is covered by 2-1-1/clause 6.2.2(2).

Prestressed sections un-cracked in flexure – shear tension
2-1-1/clause 6.2.2(2) defines sections as un-cracked where the maximum flexural tensile stress is smaller than $f_{ctk,0.05}/\gamma_c$. The maximum flexural tensile stress means the maximum tensile fibre stress (i.e. including the axial stress component as well as the bending component). It should also be noted that the definition of f_{ctd} in 2-1-1/Expression (6.4) is taken from 2-1-1/clause 3.1.6(2) and therefore includes the α_{ct} factor. The derivation of un-cracked section resistance should be consistent with this limit and should strictly be $\alpha_{ct} f_{ctk,0.05}/\gamma_c$. If the recommended value of 1.0 is adopted for α_{ct}, there is no inconsistency.

2-1-1/clause 6.2.2(2)

The shear failure criterion for a section with no shear reinforcement assumed in 2-1-1/clause 6.2.2(2) is that the principal tensile stress anywhere in the section exceeds the tensile strength of the concrete, f_{ctd}. (f_{ctd} is a positive number but the sign convention below uses compression as positive, so the appropriate tensile limit is $-f_{ctd}$ with this convention.) Internally self-equilibrating stresses from differential shrinkage and temperature may be ignored by implication as they do not feature in 2-1-1/Expression (6.4).

Equating principal tensile stress to the tensile strength of the concrete gives the following equation:

$$-f_{ctd} = \frac{(\sigma_{cp} + \sigma_{bend})}{2} - \sqrt{\left(\frac{\sigma_{cp} + \sigma_{bend}}{2}\right)^2 + \tau^2}$$

where:

- σ_{cp} is the compressive stress due to axial loading or prestressing (after losses and including appropriate partial safety factors) at the level considered (in MPa, compression taken as positive)
- σ_{bend} is stress due to bending at the level considered (in MPa, compression taken as positive)
- τ is the applied shear stress, where $\tau = V_{Rd,c} A_e \bar{z}/Ib$ (this is only valid where the cross-sectional properties are constant along the beam)
- $V_{Rd,c}$ is the shear resistance determined from the shear force required to cause web cracking
- I is the second moment of area of the section
- b is the web width at the level being checked including allowances for ducts
- $A_e \bar{z}$ is the first moment of area of the concrete above the plane considered about the cross-section centroid level (EC2 refers to this as S when it relates to a plane at the centroidal level)

Substitution for τ in the above leads to the following expression for shear tension resistance:

$$V_{\mathrm{Rd,c}} = \frac{Ib}{A_e \bar{z}}\sqrt{f_{\mathrm{ctd}}^2 + (\sigma_{\mathrm{cp}} + \sigma_{\mathrm{bend}})f_{\mathrm{ctd}}} \qquad (D6.2\text{-}3)$$

Assuming the principal tensile stress occurs at the centroid of the section, $\sigma_{\mathrm{bend}} = 0$, and introducing a factor, α_1, and substituting S for $A_e \bar{y}$, gives the following expression in 2-1-1/clause 6.2.2(2):

$$V_{\mathrm{Rd,c}} = \frac{Ib_{\mathrm{w}}}{S}\sqrt{f_{\mathrm{ctd}}^2 + \alpha_1 \sigma_{\mathrm{cp}} f_{\mathrm{ctd}}} \qquad 2\text{-}1\text{-}1/(6.4)$$

where:

α_1 is a factor introduced to account for the transmission lengths in pretensioned construction

b_{w} is the web width at the centroidal axis including allowances for ducts – see section 6.2.3.2 below. This definition differs from that in 2-2/clause 6.2.2(101). A further slightly different definition is used in 2-1-1/clause 6.2.3

2-1-1/clause 6.2.2(2) applies to 'single-span members', but the reason for this restriction is not clear. An expression similar to 2-1-1/Expression (6.4) was used in BS 5400 Part 4[9] for the shear resistance of sections un-cracked in flexure. It was equally applicable to continuous beams. The limitation is not restrictive close to the internal supports of continuous beams where the section is likely to be cracked in flexure. It is, in any case, unlikely that the shear tension resistance would be adequate on its own to carry the shear at an internal support, whereupon its contribution would be lost and the variable angle truss model would have to be used. There could, however, be economy to be gained in using it near contraflexure regions of continuous beams, but this does not appear to be permitted. 'Single-span members' could include those in fully integral bridges where end hogging moments develop. If the method were to be used in this situation, it would be logical to permit its use in hogging zones of continuous construction. It is also possible that the intention was to restrict its use to 'one-way spanning' members because 2-1-1/Expression (6.4) does not take account of stresses in a transverse direction. It is more likely that the rule was added to remove some conservatism for simply supported beams arising from the use of the other shear rules.

In certain sections, such as I-beams, where the section width varies over height, the maximum principal stress may occur at a level other than at the centroidal axis. In such cases, 2-1-1/clause 6.2.2(2) requires that the minimum value of shear resistance is determined by calculating $V_{\mathrm{Rd,c}}$ at various levels in the cross-section. In such cases the flexural stress term, σ_{bend}, in equation (D6.2-3) above should be included while maintaining the α_1 term applied to the prestress, thus:

$$V_{\mathrm{Rd,c}} = \frac{Ib}{A_e \bar{z}}\sqrt{f_{\mathrm{ctd}}^2 + (\alpha_1 \sigma_{\mathrm{cp}} + \sigma_{\mathrm{bend}})f_{\mathrm{ctd}}} \qquad (D6.2\text{-}4)$$

Other terms are defined above under equation (D6.2-3). For 'I' sections and boxes, it is often sufficient to check the neutral axis and the web–flange junctions using equation (D6.2-4). When inclined tendons are used, 2-1-1/clause 6.2.1(3) allows the vertical component of prestress at a section to be included by adding (if adverse) or subtracting (if beneficial) the component from the applied shear force.

2-1-1/clause 6.2.2(3)

To account for behaviour close to supports, **2-1-1/clause 6.2.2(3)** states that the calculation of shear resistance according to 2-1-1/Expression (6.4) is not required for sections that are nearer to the support than the point which is the intersection of the elastic centroidal axis and a line inclined from the inner edge of the support at an angle of 45°. The reduction to shear force allowed in 2-1-1/clause 6.2.2(6) for beams cracked in flexure should not, however, be applied when using 2-1-1/Expression (6.4).

When, in continuous beams, the effects of imposed deformations are ignored in flexural design at ULS, there is a case for still considering such effects in the shear design as the failure may be non-ductile. This additional complexity should rarely arise in practice as links will usually be required in bridge beams.

Worked example 6.2-3: Post-tensioned concrete box girder, un-cracked in flexure

Consider the following example of a concrete box girder with bonded prestressing tendons.

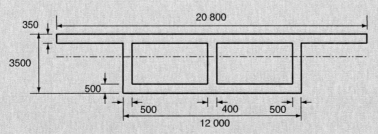

Fig. 6.2-2. Section through concrete box girder

The box girder has the following section properties: $A = 17\,\text{m}^2$, $I = 33.5\,\text{m}^4$, $W_\text{top} = 21.1\,\text{m}^3$, $W_\text{bot} = 17.5\,\text{m}^3$, neutral axis height, $z_\text{na} = 1911.5\,\text{mm}$ and first moment of area of excluded area above and about centroid, $S = A_\text{e}\bar{z} = 11.35\,\text{m}^3$. The prestressing tendons comprise 40 no. tendons each of area $= 2641\,\text{mm}^2$, $f_\text{pk} = 1670\,\text{MPa}$ and $f_\text{p0.1k} = 1468\,\text{MPa}$. Each tendon passes through a steel duct with outer diameter of 102 mm. The tendons at the simply supported end of the bridge are located at the following heights: 25 no. at 250 mm (in the flange), 3 no. at 700 mm, 3 no. at 1000 mm, 3 no. at 1300 mm, 3 no. at 1600 mm and 3 no. at 1900 mm, giving a height to the centroid of the prestressing force of 644 mm. The tendons in the web are stacked vertically so that there is only one tendon across the web width at any given height. $f_\text{ck} = 35\,\text{MPa}$, $f_\text{yk} = 500\,\text{MPa}$, $\gamma_\text{c} = 1.5$ and $\gamma_\text{s} = 1.15$.

The shear resistance of the inner and outer webs is found at the simply supported end assuming the section is un-cracked in flexure in accordance with 2-1-1/clause 6.2.2(2). Assuming initial stressing to 70%, f_pk and total losses of a further 25%, the global prestressing force after losses is $40 \times 2641 \times 1670 \times 0.70 \times 0.75 \times 10^{-3} = 92\,600\,\text{kN}$.

From 2-1-1/Table 3.1, $f_\text{ctk,0.05} = 2.2\,\text{MPa}$ and taking $\alpha_\text{ct} = 1.0$ for shear calculation as recommended, 2-2/Expression (3.16) gives $f_\text{ctd} = \alpha_\text{ct} f_\text{ck}/\gamma_\text{c} = 1.0 \times 2.2/1.5 = 1.467\,\text{MPa}$.

The uncracked shear resistance is calculated from 2-1-1/Expression (6.4):

$$V_\text{Rd,c} = \frac{I b_\text{w}}{S}\sqrt{f_\text{ctd}^2 + \alpha_1 \sigma_\text{cp} f_\text{ctd}}$$

using $\alpha_1 = 1.0$ for post-tensioning, and a partial safety factor, $\gamma_\text{p,fav} = 1.0$ (from 2-1-1/clause 2.4.2.2):

$$\sigma_\text{cp} = N_\text{Ed}/A = 1.0 \times 92\,600 \times 10^3/17 \times 10^6 = 5.45\,\text{MPa}$$

From 2-1-1/Expression (6.16):

$$\sum b_\text{w,nom} = 500 + 400 + 500 - 3 \times 0.5 \times 102 = 1247\,\text{mm}$$

therefore, for the full section:

$$V_\text{Rd,c} = \frac{33.5 \times 10^{12} \times 1247}{11.35 \times 10^9} \times \sqrt{1.467^2 + 1.0 \times 5.45 \times 1.467} = 11\,724\,\text{kN}$$

for all three webs.

As noted in section 6.2.2.2, the above checks the shear resistance at the centroid level. Additionally, the resistance at other levels should also be verified. Now, consider a level just underneath the top slab at a height of 3150 mm. $A_e \bar{z}$ for the concrete above this level about the neutral axis is 10.290 m³. Assuming $M_{Ed} \approx 0$ kNm at the simply supported end, the $(\sigma_{cp} + \sigma_{bend})$ term in equation (D6.2-4), with $\alpha_1 = 1.0$, is given by:

$$(\sigma_{cp} + \sigma_{bend}) = \frac{1.0 \times 92.6 \times 10^3}{17 \times 10^6} + \frac{1.0 \times 92.6 \times 10^6 (644 - 1911.5)(3150 - 1911.5)}{33.5 \times 10^{12}}$$

$$= 5.447 - 4.339 = 1.108 \text{ MPa}$$

From equation (D6.2-4):

$$V_{Rd,c} = \frac{33.5 \times 10^{12} \times 1247}{10.290 \times 10^9} \times \sqrt{1.467^2 + 1.108 \times 1.467} = 7890 \text{ kN} < 11\,724 \text{ kN}$$

for the centroid level resistance and is therefore more critical.

This resistance is shared between the three webs in proportion to their relative widths, thus the resistance for the inner web is $7890 \times (400 - 0.5 \times 102)/1247 = \textbf{2208 kN}$ and the resistance for each outer web is $7890 \times (500 - 0.5 \times 102)/1247 = \textbf{2841 kN}$.

Prestressed sections cracked in flexure

The design of prestressed concrete sections cracked in flexure without designed shear reinforcement is conducted as in section 6.2.2.1 above for reinforced concrete. Bonded prestress may, however, be included in the definition of A_{sl} according to 2-2/clause 6.2.2(101). The effective depth, d, is then based on a 'weighted mean value'. This should be based on the centroid of steel area, irrespective of strength.

Minimum shear reinforcement

Minimum shear reinforcement should always be provided as discussed above for reinforced concrete.

6.2.3. Members requiring design shear reinforcement

This section is split into three additional sub-sections dealing with reinforced concrete (section 6.2.3.1), prestressed concrete (section 6.2.3.2) and segmental construction (section 6.2.3.3).

6.2.3.1. Reinforced concrete members

In regions where $V_{Ed} > V_{Rd,c}$, sufficient shear reinforcement should be provided to ensure $V_{Ed} \leq V_{Rd}$ as stated in 2-1-1/clause 6.2.1(5). 2-2/clause 6.2.3 adopts the truss model illustrated in Fig. 6.2-3 for the calculation of required shear reinforcement.

Consider the portion of a member, of width b_w, illustrated in Fig. 6.2-4 for a general truss model with a compressive strut at an angle θ to the beam axis perpendicular to the shear force

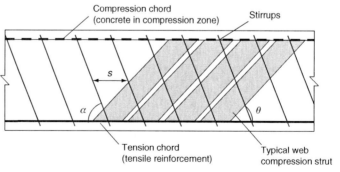

Fig. 6.2-3. Truss model analogy for shear

CHAPTER 6. ULTIMATE LIMIT STATES

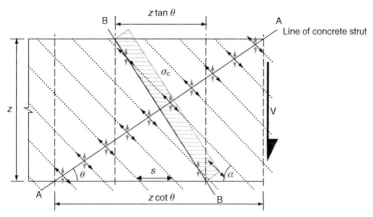

Fig. 6.2-4. Partial smeared truss model for the use of inclined shear reinforcement

and shear reinforcement inclined at an angle of α (to the same axis). 2-1-1/clause 9.2.2(1) limits the angle of the shear reinforcement to be between 45° and 90°. **2-1-1/clause 6.2.3(1)** defines the web width to be used in design calculations for members requiring shear reinforcement as the minimum web width between the tension and compression chords. The shear resistance is obtained by considering vertical equilibrium on section A–A, a plane parallel to the line of the concrete strut force. Since plane A–A is parallel to the concrete force, there is no vertical component of the strut force crossing the plane and therefore only the vertical components of the shear reinforcement legs crossing plane A–A resist the applied shear force, V.

2-1-1/clause 6.2.3(1)

The length of plane A–A is $z/\sin\theta$ and the spacing along plane A–A of the reinforcement legs crossing the plane is $s\sin\alpha/\sin(\theta+\alpha)$; therefore the number of legs crossing the plane is $z\sin(\theta+\alpha)/s\sin\theta\sin\alpha$. Therefore, using the design strength of the shear reinforcement of $f_{ywd}A_{sw}$, the total vertical components of the forces from the reinforcement crossing the plane is given by $f_{ywd}A_{sw}(z\sin(\theta+\alpha)/s\sin\theta\sin\alpha)\sin\alpha$. Thus, vertical equilibrium on section A–A gives:

$$V = \frac{A_{sw}}{s}zf_{ywd}\frac{\sin(\theta+\alpha)}{\sin\theta}$$

$$= \frac{A_{sw}}{s}zf_{ywd}\frac{(\sin\theta\cos\alpha+\cos\theta\sin\alpha)}{\sin\theta}$$

$$= \frac{A_{sw}}{s}zf_{ywd}(\cos\alpha+\cot\theta\sin\alpha)$$

This gives the equation in **2-1-1/clause 6.2.3(4)** (with V replaced with $V_{Rd,s}$):

$$V_{Rd,s} = \frac{A_{sw}}{s}zf_{ywd}(\cot\theta+\cot\alpha)\sin\alpha \qquad \text{2-1-1/(6.13)}$$

2-1-1/clause 6.2.3(4)

For vertical shear reinforcement, $\alpha = 90°$, $\cot\alpha = 0$, $\sin\alpha = 1$ and 2-1-1/Expression (6.13) simplifies to the expression in **2-2/clause 6.2.3(103)**:

$$V_{Rd,s} = \frac{A_{sw}}{s}zf_{ywd}\cot\theta \qquad \text{2-2/(6.8)}$$

2-2/clause 6.2.3(103)

Now consider vertical equilibrium on section B–B, a plane perpendicular to the line of the concrete strut force. The applied shear force, V, is now resisted partly by the vertical component of the concrete strut force and partly by the vertical components of the force from the shear reinforcement crossing plane B–B.

The length of plane B–B is $z/\cos\theta$ and the spacing along plane B–B of the reinforcement legs crossing the plane is $s\sin\alpha/\cos(\theta+\alpha)$; therefore the number of legs crossing the plane is $z\cos(\theta+\alpha)/s\cos\theta\sin\alpha$. Therefore, again using the design strength of the shear reinforcement of $f_{ywd}A_{sw}$, the total vertical components of the forces from the reinforcement crossing the plane is given by $f_{ywd}A_{sw}(z\cos(\theta+\alpha)/s\cos\theta\sin\alpha)\sin\alpha$.

141

The total concrete strut force, F_c, is given by $F_c = \sigma_c b_w z / \cos\theta$ and the vertical component of this force is $F_c \sin\theta = \sigma_c b_w z \tan\theta$.

Thus vertical equilibrium on section B–B gives:

$$V = \frac{A_{sw}}{s} f_{ywd} z \frac{\cos(\theta + \alpha)}{\cos\theta} + \sigma_c b_w z \tan\theta$$

$$= \frac{A_{sw}}{s} f_{ywd} z \frac{(\cos\theta\cos\alpha - \sin\theta\sin\alpha)}{\cos\theta} + \sigma_c b_w z \tan\theta$$

$$= \frac{A_{sw}}{s} f_{ywd} z (\cos\alpha - \tan\theta\sin\alpha) + \sigma_c b_w z \tan\theta$$

$$= \frac{A_{sw}}{s} f_{ywd} z \sin\alpha (\cot\alpha - \tan\theta) + \sigma_c b_w z \tan\theta$$

But from 2-1-1/Expression (6.13):

$$\frac{V}{(\cot\theta + \cot\alpha)} = \frac{A_{sw}}{s} z f_{ywd} \sin\alpha$$

therefore

$$V = \frac{V}{(\cot\theta + \cot\alpha)}(\cot\alpha - \tan\theta) + \sigma_c b_w z \tan\theta$$

so

$$\frac{V}{(\cot\theta + \cot\alpha)}[(\cot\theta + \cot\alpha) - (\cot\alpha - \tan\theta)] = \sigma_c b_w z \tan\theta$$

and

$$\frac{V}{\tan\theta}(\cot\theta + \tan\theta) = \sigma_c b_w z (\cot\theta + \cot\alpha)$$

thus

$$V = \sigma_c b_w z \frac{(\cot\theta + \cot\alpha)}{(1 + \cot^2\theta)} \tag{D6.2-5}$$

For failure of the concrete strut by crushing, $\sigma_c = \alpha_{cw} \nu_1 f_{cd}$ where:

ν_1 is the strength reduction factor for concrete cracked in shear. It is a nationally determined parameter which Note 2 to 2-2/clause 6.2.3(103) recommends be taken as ν given by 2-1-1/Expression (6.6N). If the design stress of the shear reinforcement is limited to a maximum of 80% of the characteristic yield stress, f_{yk}, Note 3 recommends ν_1 be taken as:

0.6 for $f_{ck} \leq 60$ MPa   2-2/(6.10.aN)

or $(0.9 - f_{ck}/200) > 0.5$ for $f_{ck} > 60$ MPa   2-2/(6.10.bN)

This factor is discussed further below.

α_{cw} is a factor used to take account of compression in the shear area and its value may be given in the National Annex. 2-2/clause 6.2.3(103) recommends the following values:

1.0 for non-prestressed structures

$(1 + \sigma_{cp}/f_{cd})$ for $0 < \sigma_{cp} \leq 0.25 f_{cd}$ 2-2/(6.11.aN)

1.25 for $0.25 f_{cd} < \sigma_{cp} \leq 0.5 f_{cd}$ 2-2/(6.11.bN)

$2.5(1 - \sigma_{cp}/f_{cd})$ for $0.5 f_{cd} < \sigma_{cp} \leq 1.0 f_{cd}$ 2-2/(6.11.cN)

where σ_{cp} is the mean compressive stress (measured positive) in the concrete. This factor is discussed further below.

Thus substituting for σ_c in equation (D6.2-5) gives the expression in 2-1-1/clause 6.2.3(4):

$$V_{\text{Rd,max}} = \alpha_{\text{cw}} b_w z \nu_1 f_{\text{cd}} \frac{(\cot\theta + \cot\alpha)}{(1 + \cot^2\theta)} \qquad \text{2-1-1/(6.14)}$$

This expression effectively gives the maximum shear resistance of a section before failure occurs due to crushing of the concrete struts and is therefore designated $V_{\text{Rd,max}}$ in the code.

For practical ranges of axial load in prestressed members, the recommended values for α_{cw} generally result in increasing the maximum shear stress limit by up to 25%. If this is considered together with the recommended benefit from ν_1 where the links are not fully stressed, the maximum shear limit can be around twice that permitted by BS 5400 Part 4[9] and potentially unsafe. If inclined links are used at 45°, $V_{\text{Rd,max}}$ can approach four times the equivalent limit in BS 5400 Part 4! α_{cw}, ν_1 and inclined links are discussed further below.

For beams with vertical links, the authors are unaware of test results which suggest the additional increase for crushing resistance given by α_{cw} is unsafe on its own. It is not, however, supported by the rules for membrane elements in 2-2/clause 6.109(103), where compression reduces the maximum resistance to shear. The physical model behind ν_1 is also hard to understand as it promotes the design of over-reinforced behaviour in shear, which may invalidate the plastic assumptions behind the truss model which relies on rotation of the web compression diagonals. The justification probably relates to limiting the tensile strains acting skew to the struts and thereby enhancing the strut compression limit, as discussed in section 6.5.2 of this guide. In conjunction with the use of the upper values of α_{cw}, the EN 1992-2 results may become unsafe, particularly where the webs are designed to be very slender because of the high permissible stresses. Slender webs (with high height to thickness ratio) may exhibit significant second-order out-of-plane bending effects which would lead to failure at shear forces less than the values based on uniform crushing. There is limited test evidence here and the UK National Annex therefore imposes an upper limit on shear resistance based on web slenderness to safeguard against this. ν_1 is also reduced.

For beams with inclined links, no test results are available to check the high predicted values of $V_{\text{Rd,max}}$. One result is available in reference 12 but this failed prematurely below the load expected from EN 1992-2. It is clear, therefore, that the recommended values in EN 1992-2 should be used with great caution. The UK National Annex therefore reduces both α_{cw} (to 1.0) and ν_1 where inclined links are used.

Where webs carry significant transverse bending in addition to shear, these high shear crushing resistances may not be achievable due to the interaction of the compression fields – section 6.2.6 refers.

Notwithstanding the above, the examples in this guide use the EN 1992-2 recommended values for α_{cw} and ν_1, but with an upper limit of 1.0 imposed on α_{cw}.

For vertical shear reinforcement, $\alpha = 90°$ so $\cot\alpha = 0$ and 2-1-1/Expression (6.14) simplifies to:

$$V_{\text{Rd,max}} = \alpha_{\text{cw}} b_w z \nu_1 f_{\text{cd}} \frac{\cot\theta}{(1 + \cot^2\theta)} = \alpha_{\text{cw}} b_w z \nu_1 f_{\text{cd}} \frac{1}{\left(\tan\theta + \dfrac{\cot^2\theta}{\cot\theta}\right)}$$

giving the expression in 2-2/clause 6.2.3(103):

$$V_{\text{Rd,max}} = \frac{\alpha_{\text{cw}} b_w z \nu_1 f_{\text{cd}}}{(\cot\theta + \tan\theta)} \qquad \text{2-2/(6.9)}$$

Additional longitudinal reinforcement and the shift method
The design of the longitudinal reinforcement in the region cracked in flexure, or where shear reinforcement is provided, is affected by the shear design of members. *2-2/clause 6.2.3(107)* requires additional longitudinal reinforcement to be provided in certain regions over and above that required for the flexural design, as discussed below.

The free body ABD in Fig. 6.2-5 shows a portion of a smeared truss within a member, including both horizontal and vertical forces. As before, the compressive strut is chosen to

2-2/clause 6.2.3(107)

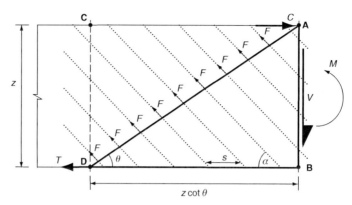

Fig. 6.2-5. Free body diagram for truss analogy

act at an angle θ to the horizontal and the shear reinforcement is inclined at an angle of α. Again, the number of legs of shear reinforcement crossing the plane A–D is $z\sin(\theta + \alpha)/s\sin\theta\sin\alpha$.

First, consider vertical equilibrium: $V = \sum F_{\text{vert}}$ where $F_{\text{vert}} = F\sin\alpha$. Thus

$$V = F\sin\alpha\frac{z\sin(\theta + \alpha)}{s\sin\theta\sin\alpha} = \frac{Fz\sin(\theta + \alpha)}{s\sin\theta} \tag{D6.2-6}$$

Second, consider horizontal equilibrium: $C = T + \sum F_{\text{horiz}}$ where $F_{\text{horiz}} = F\cos\alpha$. Thus

$$C = T + F\cos\alpha\frac{z\sin(\theta + \alpha)}{s\sin\theta\sin\alpha} = T + \frac{Fz\sin(\theta + \alpha)}{s\sin\theta}\cot\alpha \tag{D6.2-7}$$

Finally, consider moment equilibrium about point A:

$$M = Tz + \sum F_{\text{vert}}\frac{z\cot\theta}{2} + \sum F_{\text{horiz}}\frac{z}{2}$$

Therefore, substituting for $\sum F_{\text{vert}}$ from equation (D6.2-6) and $\sum F_{\text{horiz}}$ from equation (D6.2-7) gives:

$$M = Tz + \frac{Fz\sin(\theta + \alpha)}{s\sin\theta}\frac{z\cot\theta}{2} + \frac{Fz\sin(\theta + \alpha)}{s\sin\theta}\cot\alpha\frac{z}{2}$$

which simplifies to:

$$M = Tz + \frac{Fz^2\sin(\theta + \alpha)}{2s\sin\theta}(\cot\theta + \cot\alpha) \tag{D6.2-8}$$

Substituting V from equation (D6.2-6) into equation (D6.2-8) gives:

$$M = Tz + V\frac{z}{2}(\cot\theta + \cot\alpha)$$

and rearranging gives:

$$Tz = M - V\frac{z}{2}\cot\theta - V\frac{z}{2}\cot\alpha = M - Vz\cot\theta + V\frac{z}{2}(\cot\theta - \cot\alpha)$$

But $M - Vz\cot\theta$ is the bending moment at section CD (M_{CD}) therefore:

$$Tz = M_{\text{CD}} + \frac{Vz}{2}(\cot\theta - \cot\alpha)$$

$$T = \frac{M_{\text{CD}}}{z} + \frac{V}{2}(\cot\theta - \cot\alpha) \tag{D6.2-9}$$

hence the longitudinal reinforcement at section CD should be designed for the force from the bending moment at CD, M_{CD}/z, plus $V/2(\cot\theta - \cot\alpha)$. This is equivalent to designing for an effective bending moment of:

$$M = M_{\text{CD}} + \frac{Vz}{2}(\cot\theta - \cot\alpha) = M_{\text{CD}} + Va_1 \tag{D6.2-10}$$

where $a_1 = 0.5z(\cot\theta - \cot\alpha)$.

Equation (D6.2-9) is the basis of the design tensile force given in 2-2/clause 6.2.3(107):

$$\text{Tensile force} = \frac{M_{\text{Ed}}}{z} + \Delta F_{\text{td}} \quad \text{with} \quad \Delta F_{\text{td}} = 0.5 V_{\text{Ed}}(\cot\theta - \cot\alpha) \qquad \text{2-2/(6.18)}$$

For any section, $M_{\text{Ed}}/z + \Delta F_{\text{td}}$ should not be taken as greater than $M_{\text{Ed,max}}/z$, as implied by the shift method below, where $M_{\text{Ed,max}}$ is the maximum moment along the beam (for either the sagging or hogging zones considered).

2-1-1/clause 9.2.1.3(2) allows an alternative method based on equation (D6.2-10). This is to shift the design moment envelope horizontally by a distance $a_l = 0.5z(\cot\theta - \cot\alpha)$ as shown in 2-1-1/Fig. 9.2, which effectively introduces the additional moment Va_l in equation (D6.2-10). The longitudinal reinforcement is designed to this new effective moment envelope. This is equivalent to designing the reinforcement at a section to resist only the real bending moment at that section, but to then continue this reinforcement beyond that section by the distance, a_l. For flanged beams, a further shift may be required as discussed under the comments on 2-1-1/clause 6.2.4(7). For members without shear reinforcement, the value of a_l should be taken as the effective depth, d, at the section considered, as indicated in 2-1-1/Fig. 6.3.

Behaviour for loads applied close to supports
Where a load is applied close to a support, specifically within a distance of $2d$ from the support, **2-1-1/clause 6.2.3(8)** permits the contribution of that load to producing the shear force V_{Ed} to be reduced by a factor, $\beta = a_v/2d$. This is the same factor as for members without shear reinforcement in 2-1-1/clause 6.2.2(6). The shear force calculated in this way must satisfy the following condition:

2-1-1/clause 6.2.3(8)

$$V_{\text{Ed}} \leq A_{\text{sw}} f_{\text{ywd}} \sin\alpha \qquad \text{2-1-1/(6.19)}$$

where $A_{\text{sw}} f_{\text{ywd}}$ is the resistance of the shear reinforcement crossing the inclined shear crack between load and support. (Note that this definition of A_{sw} differs from that elsewhere.) For loads close to supports, there clearly cannot be a free choice of strut angle within the range allowed in 2-1-1/clause 6.2.3(2). Only the reinforcement within the central $0.75a_v$ should be taken into account as shown in Fig. 6.2-6. This limitation is made because tests by Asin[16] indicated that the links adjacent to both load and support do not fully yield. The shear crushing limit according to 2-1-1/clause 6.2.2(6) should be checked without applying the factor $\beta = a_v/2d$. This crushing limit is independent of strut angle.

This concept of considering only the links between the load and the support works only for single loads. Where there are other loads contributing to the shear applied further than $2d$ from the support, the shear design for these loads should be treated in accordance with the remainder of the clauses in section 6.2.3 of EC2. It would be illogical to constrain the location for link provision for these loads according to the location of the load nearest the support. The link requirements from the two systems should then be added. The shear crushing limit should similarly be performed by superposing the results from the two systems. For most bridge applications, unless there is one or more very heavy loads acting within a distance $2d$ from the support which contributes a significant proportion of the total shear, it will usually be sufficient to determine shear reinforcement solely on the basis of the variable strut inclination method in 2-1-1/clause 6.2.3(3).

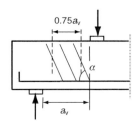

Fig. 6.2-6. Shear reinforcement in short spans

Worked example 6.2-4: Voided reinforced concrete deck slab
The shear reinforcement required for the voided slab in Worked example 6.1-2 is designed at a cross-section with an applied ultimate shear force, V_{Ed}, of 850 kN/m ($= 1190$ kN for 1.4 m width of slab) and a coexistent sagging moment, M_{Ed}, of 1071 kNm/m (i.e. 1500 kNm for a 1.4 m width). The additional longitudinal reinforcement required for shear (assuming $M_{Ed,max}$ is sufficiently greater than M_{Ed}) is also determined. Ignore any Vierendeel action in the transverse direction which could produce transverse moments in the web to be resisted by the links.

From Worked example 6.1-2, $d = 1425$ mm and $z = 1388.1$ mm ($= 0.97d$), $A_s = 4581.5$ mm^2, $f_{ck} = 35$ MPa, $f_{yk} = f_{ywk} = 500$ MPa, $\gamma_c = 1.5$, $\gamma_s = 1.15$ and $b_w = 400$ mm.

Thus $f_{cd} = 1.0 \times 35/1.5 = 23.3$ MPa and $f_{yd} = f_{ywd} = 500/1.15 = 434.8$ MPa. $C_{Rd,c} = 0.18/1.5 = 0.12$ (as recommended).

First, find $V_{Rd,c}$.

From equation (D6.2-2):

$$\rho_1 = \frac{A_{sl}}{b_w d} = \frac{4581.5}{400 \times 1425} = 0.00804 < 0.02 \text{ limit as required}$$

From equation (D6.2-1):

$$k = 1 + \sqrt{200/d} = 1 + \sqrt{200/1425} = 1.375 \leq 2.0 \text{ limit as required}$$

and thus, ignoring any axial load, 2-2/Expression (6.2.a) becomes:

$$V_{Rd,c} = C_{Rd,c} k (100 \rho_1 f_{ck})^{1/3} b_w d$$
$$= 0.12 \times 1.375 \times (100 \times 0.00804 \times 35)^{1/3} \times 400 \times 1425 \times 10^{-3} = 286.0 \text{ kN}$$

$V_{Ed} > V_{Rd,c}$, therefore design shear reinforcement is required.

(a) Try vertical links ($\alpha = 90°$) and a strut angle, θ, of 45°.

Assuming the shear reinforcement to be fully stressed, from Note 2 of 2-2/clause 6.2.3(103): $\nu_1 = \nu = 0.6 \times (1 - f_{ck}/250) = 0.6 \times (1 - 35/250) = 0.516$ (a higher value could be taken if the reinforcement was not designed to be fully stressed in accordance with Note 3 of 2-2/clause 6.2.3(103), although the National Annex may restrict this as discussed in the main text). Thus from 2-2/Expression (6.9):

$$V_{Rd,max} = \frac{\alpha_{cw} b_w z \nu_1 f_{cd}}{(\cot\theta + \tan\theta)} = \frac{1.0 \times 400 \times 1388.1 \times 0.516 \times 23.3}{(\cot 45 + \tan 45)} \times 10^{-3} = 3337 \text{ kN}$$

Rearranging 2-2/Expression (6.8) for reinforcement design gives

$$\frac{A_{sw}}{s} \geq \frac{V_{Rd,s}}{z f_{ywd} \cot\theta}$$

thus to provide sufficient shear reinforcement to just resist V_{Ed}:

$$\frac{A_{sw}}{s} \geq \frac{1190 \times 10^3}{1388.1 \times 434.8 \times \cot 45} = 1.972 \text{ mm}^2/\text{mm}$$

If $s = 200$ mm, $A_{sw} \geq 1.972 \times 200 = 394$ mm^2, thus **adopt 2 legs of 16ϕ links at 200 mm centres** (giving $A_{sw} = 2 \times \pi \times 8^2 = 402$ mm^2).

From 2-2/Expression (6.18): the additional longitudinal tensile force required is $\Delta F_{td} = 0.5 \times 1190 \times (\cot 45 - \cot 90) = 595$ kN. The moment resistance of a 1.4 m wide portion from Worked example 6.1-2 is 2765 kNm. Therefore, the spare force is $(2765 - 1500) \times 10^6/1388.1 = 911$ kN > 595 kN required, so no additional longitudinal reinforcement required.

(b) Alternatively, consider links inclined at an angle of 45°, and a strut angle of 45°.

Thus from 2-1-1/Expression (6.14):

$$V_{Rd,max} = \alpha_{cw} b_w z \nu_1 f_{cd} (\cot\theta + \cot\alpha)/(1+\cot^2\theta)$$

$$= 1.0 \times 400 \times 1388.1 \times 0.516 \times 23.3 \times \frac{(\cot 45 + \cot 45)}{(1 + \cot^2 45)} \times 10^{-3}$$

$$= 6675.6 \text{ kN}$$

This is a very large increase in crushing resistance and the UK National Annex may restrict it as discussed in the main text above.

Rearranging 2-1-1/Expression (6.13) for reinforcement design gives

$$\frac{A_{sw}}{s} \geq \frac{V_{Rd,s}}{z f_{ywd} (\cot\theta + \cot\alpha) \sin\alpha}$$

therefore

$$\frac{A_{sw}}{s} \geq \frac{1190 \times 10^3}{1388.1 \times 434.8 \times (\cot 45 + \cot 45) \times \sin 45} = 1.394 \text{ mm}^2/\text{mm}$$

and using the same two legs of 16ϕ bars as above ($A_{sw} = 402 \text{ mm}^2$) the spacing can be increased from 200 mm to 402/1.394 = 288 mm, say 275 mm, i.e. **adopt 2 legs of 16ϕ links (inclined at 45°) at 275 mm centres**.

From 2-2/Expression (6.18): $\Delta F_{td} = 0.5 \times 1190 \times (\cot 45 - \cot 45) = 0 \text{ kN}$; therefore no additional longitudinal reinforcement is required.

(c) Alternatively, different strut angles, θ, could be chosen and $V_{Rd,s}$ can be maximized, for a given shear reinforcement, by taking the minimum θ limit of 21.8°. However, $V_{Rd,max}$ is reduced (and may then govern) and also the additional longitudinal tensile reinforcement requirement is increased. In practice, the choice of θ is a compromise between the design of longitudinal reinforcement and the design of shear reinforcement.

In this example, if $\theta = 21.8°$ and $\alpha = 90°$, the additional longitudinal force to be resisted increases to $\Delta F_{td} = 0.5 \times 1190 \times (\cot 21.8 - \cot 90) = 1487.5 \text{ kN} > 911 \text{ kN}$ available. The additional longitudinal reinforcement required

$$= \frac{(1487.5 - 911) \times 10^3}{500/1.15} = 1326 \text{ mm}^2 \equiv 947 \text{ mm}^2/\text{m}$$

which would require **an additional 5 no. 16ϕ bars per metre width**.

Shear at points of contraflexure
Near to points of contraflexure, a problem arises in determining the value of shear resistance since the strength is dependent on the lever arm and area of longitudinal tensile reinforcement. It is therefore always necessary to check the maximum shear force coexistent with a sagging moment and the maximum shear force coexistent with a hogging moment when designing shear reinforcement.

6.2.3.2. Prestressed concrete members
Effective web widths
Prestressing ducts can reduce the shear crushing resistance of prestressed members. In calculation, the web width should therefore be reduced to allow for the presence of ducts. **2-1-1/clause 6.2.3(6)** defines values for such nominal web thicknesses as follows:

- For grouted ducts with a diameter, $\phi > b_w/8$:

$$b_{w,nom} = b_w - 0.5 \sum \phi \qquad \text{2-1-1/(6.16)}$$

2-1-1/clause 6.2.3(6)

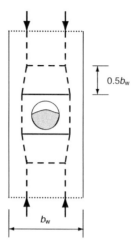

Fig. 6.2-7. Strut-and-tie model for flow of web compression strut around a duct

where ϕ is the outer diameter of the duct and $\sum \phi$ is determined at the most unfavourable level where multiple ducts occur at the same level.
- For grouted metal ducts with a diameter, $\phi \leq b_w/8$, no reduction in web width is required.
- For non-grouted ducts, unbonded tendons and grouted plastic ducts:

$$b_{w,nom} = b_w - 1.2 \sum \phi \qquad \text{2-1-1/(6.17)}$$

The 1.2 factor in the above expression is used to account for the splitting of the concrete struts due to transverse tension. It may be reduced to 1.0 if 'adequate transverse reinforcement' is provided. This may be derived from a strut-and-tie model like that in Fig. 6.2-7. The reinforcement would typically need to be formed as closed links to provide adequate anchorage.

The above expressions penalize the use of plastic ducts compared to steel ducts. Although the use of plastic ducts is often preferable for durability reasons, the nominal web width expressions favour the use of steel ducts as the webs can be made thinner. This is sometimes a significant consideration for long-span post-tensioned bridges where the webs are sized on the shear crushing limit to keep dead weight to a minimum. The reason for the conservative view of the Eurocode is that, even if completely filled with grout, the duct itself must still be stiff enough to transfer the force from the inclined compression struts through its walls rather than around the duct as in Fig. 6.2-7. EC2 assumes that plastic ducts are not stiff enough to be able to achieve this. There is some test evidence to support this[31] but the authors are aware of unpublished tests by a prestressing supplier which indicate virtually no difference between the behaviour of grouted plastic and steel ducts.

Sections cracked in flexure
The definition of cracked in flexure is discussed in section 6.2.2.2. The resistances of prestressed beams that are cracked in flexure are conservatively predicted by 2-2/clause 6.2.2(101), but it will be rare for such beams to not require shear reinforcement. Once shear reinforcement is required, the concrete contribution to resistance is lost and the truss model of section 6.2.3 of EC2 must be used. The design procedure is essentially the same as for reinforced concrete, discussed above, except that crushing resistance $V_{Rd,max}$ is influenced by the axial stress from prestress through the factor α_{cw} in 2-2/clause 6.2.3(103), discussed in section 6.2.3.1 above.

Since prestress is treated in EC2 as an applied force (2-1-1/clause 6.2.1(3) refers), any truss models developed for shear design must include the effects of prestress forces applied at anchorages and arising from tendon curvature. The inclined component of prestress often relieves the shear from other imposed loads. For loads applied after prestressing and

CHAPTER 6. ULTIMATE LIMIT STATES

losses, bonded tendons may be treated in the same way as ordinary reinforcement, contributing to the allowable force in the tensile chords. This is the basis of *2-2/clause 6.2.3(107)* which allows bonded prestressing to be considered in calculation of the additional longitudinal tensile force required for shear. (This means that the force need not necessarily be provided by additional longitudinal reinforcement.) In so doing, the stress increase in bonded tendons should be limited so that the total stress in a tendon does not exceed its design strength. The effects of unbonded tendons should be treated as applied forces acting on the beam (based on the prestressing force after losses), although stress increases from overall structure deflection may be included in calculating these forces where they have been determined. Any additional tensile force provided by unbonded tendons may be considered when applying 2-2/Expression (6.18).

2-2/clause 6.2.3(107)

For members with prestressing tendons at different levels (draped or un-draped) together with un-tensioned reinforcement, the lever arm, z, to use in the shear equations is not obvious. The simplification of $z = 0.9d$ in 2-1-1/clause 6.2.3(1) is restricted to reinforced concrete sections without axial force and might not be conservative in other situations. The simplest approach is to determine the z value from the analysis for the cross-section bending resistance, which usually precedes the design for shear. This is equivalent to basing the lever arm on the centroid of the steel force in the tensile zone. The additional tensile force of 2-2/clause 6.2.3(107) can then be provided by the tensile resistance that is spare after carrying the design bending moment. The additional longitudinal force should be proportioned between the tendons and the reinforcement in the ratio of their respective forces from the bending analysis so as to maintain the value of z used in the flexural design. If there is no spare resistance, then additional longitudinal reinforcement can be provided in the tensile face.

The above is essentially equivalent to the method proposed in the Note to 2-2/clause 6.2.3(107), where the shear strength of the member is calculated by considering the superposition of two different truss models with different geometry and two angles of concrete strut to account for the reinforcement and draped prestressing tendons. This method is subject to variation in the National Annex and is illustrated in 2-2/Fig. 6.102N. Using two angles of concrete strut leads to difficulties in interpreting the rules for $V_{Rd,max}$, both in terms of the appropriate strut angle to use and the value of z. 2-2-1/clause 6.2.3(107) therefore recommends using a weighted mean value for θ. A weighting according to the longitudinal force in each system leads to the same result, as discussed above.

A further simplification, commonly used in prestress design, is to base the bending and shear design on the prestressing tendons alone. The un-tensioned reinforcement can then be taken into account in providing the additional longitudinal reinforcement. Since the centroid of the longitudinal reinforcement is usually at a greater effective depth than that of the prestressing tendons, this is conservative. The approach of considering only the tendons for the flexural design is often used to simplify analysis, so that the reinforcement provided can be used for the torsional design and, in the case of box girders, the local flexural design of deck slabs.

> **Worked example 6.2-5: Post-tensioned concrete box girder without tendon drape**
> At another location in the continuous box girder bridge of Worked example 6.2-3, the tendons are located at the following heights: 25 no. at 3325 mm (in the flange), 3 no. at 2820 mm, 3 no. at 2540 mm, 3 no. at 2260 mm, 3 no. at 1980 mm and 3 no. at 1700 mm, giving a height to the centroid of the prestressing force of 2926 mm. They are at a constant depth in the section (i.e. without drape). The tendons in the web are stacked vertically so that there is only one tendon across the web width at any given height. The tendons and their stressing forces are the same, as in Worked example 6.2-3. The required shear reinforcement to resist an ultimate shear force on the inner web, V_{Ed}, of 5400 kN, is found.

If the calculation of Worked example 6.2-3 is repeated with the revised tendon position, it is found that $V_{Ed} > V_{Rd,c}$, so the section requires links (and the concrete resistance component is lost). From a bending analysis of the section, the lever arm, z, can be shown to be 2630 mm.

(a) Consider vertical links ($\alpha = 90°$) and try a compression strut angle, θ, of 45°.

Rearranging 2-1-1/Expression (6.13) for reinforcement design gives:

$$\frac{A_{sw}}{s} \geq V_{Ed}/z f_{ywd}(\cot\theta + \cot\alpha)\sin\alpha$$

and taking $f_{ywd} = 500/1.15 = 434.8$ MPa gives:

$$\frac{A_{sw}}{s} \geq \frac{5400 \times 10^3}{2630 \times 434.8 \times (\cot 45 + \cot 90)\sin 90} = 4.722\,\text{mm}^2/\text{mm}$$

Using two legs of 20ϕ bars gives $A_{sw} = 2 \times \pi \times 10^2 = 628.3\,\text{mm}^2$ and therefore spacing, $s \leq 628.3/4.722 = 133$ mm, i.e. **adopt 2 legs of 20ϕ vertical links at 125 mm centres**.

(b) Alternatively, using vertical links and trying a strut angle of 41.5° (which will increase the requirement for additional longitudinal tensile force as seen in Worked example 6.2-6) gives:

$$\frac{A_{sw}}{s} \geq \frac{5400 \times 10^3}{2630 \times 434.8 \times (\cot 41.5 + \cot 90)\sin 90} = 4.178\,\text{mm}^2/\text{mm}$$

thus **adopt 2 legs of 20ϕ vertical links at 150 mm centres** (giving $A_{sw}/s = 628.3/150 = 4.18\,\text{mm}^2/\text{mm} > 4.178\,\text{mm}^2/\text{mm}$ as required).

The maximum allowable shear should also be checked. It will be a lower value than would be obtained for (a) above as the maximum value is produced with a strut angle of 45°:

$$V_{Rd,max} = \alpha_{cw} b_w z \nu_1 f_{cd} \frac{(\cot\theta + \cot\alpha)}{(1 + \cot^2\theta)} \quad \text{with} \quad f_{cd} = 1.0 \times 35/1.5 = 23.33\,\text{MPa}$$

(using $\alpha_{cc} = 1.0$). The recommended value of $\alpha_{cw} = 1 + (5.45/23.33) = 1.233$ but will be limited here to a value of 1.0 in the absence of completion of the UK National Annex. ν_1 is also taken as ν, as the reinforcement is fully stressed. Thus

$$V_{Rd,max} = 1.0 \times 349 \times 2630 \times 0.6(1 - 35/250) \times 23.33 \times \frac{(\cot 41.5 + \cot 90)}{(1 + \cot^2 41.5)}$$

$$= 5484\,\text{kN} > 5400\,\text{kN} \text{ so this arrangement is adequate}$$

(c) The greatest shear resistance of the section can be obtained by using links inclined at 45° (although it is not often practical to incline links in real bridges, particularly those built in stages). The strut angle is also then minimized until $V_{Rd,s} \approx V_{Rd,max}$. If 2 legs of 20ϕ links inclined at 45° are provided at 150 mm centres and if f_{ywd} is also limited to 80% of f_{ywk} to allow the term ν_1 to be taken as 0.6 (but see the cautionary note in the main text), using $\theta = 29.5°$ gives (from 2-1-1/Expression (6.13) and 2-1-1/Expression (6.14)):

$$V_{Rd,s} = \frac{A_{sw}}{s} z f_{ywd}(\cot\theta + \cot\alpha)\sin\alpha$$

$$= \frac{628.3}{150} \times 2630 \times 0.8 \times 500 \times (\cot 29.5 + \cot 45)\sin 45 \times 10^{-3} = 8623.4\,\text{kN}$$

$$V_{\text{Rd,max}} = \alpha_{cw} b_w z \nu_1 f_{cd} \frac{(\cot\theta + \cot\alpha)}{(1 + \cot^2\theta)}$$ and with α_{cw} again taken as unity gives:

$$V_{\text{Rd,max}} = 1.0 \times 349 \times 2630 \times 0.6 \times 23.33 \times \frac{(\cot 29.5 + \cot 45)}{(1 + \cot^2 29.5)} \times 10^{-3} = 8623.3\,\text{kN}$$

and V_{Rd} is the minimum value of these, i.e. 8623 kN.

Note that allowing the links to yield with $f_{ywd} = 434.8$ MPa gives $\nu_1 = 0.516$. Minimizing θ to $33.6°$, such that $V_{\text{Rd,s}} \approx V_{\text{Rd,max}}$, gives $V_{\text{Rd}} = V_{\text{Rd,max}} = 8478.2$ kN here which is less than 8623 kN. Thus using the 80% limit on f_{ywd} leads to a greater shear resistance in this instance (but the cautionary note on over-estimation of $V_{\text{Rd,max}}$ made in the main text should be noted).

Worked example 6.2-6: Post-tensioned concrete box girder with draped tendons

Assume the group centroid of the bonded tendons in the continuous box girder of Worked example 6.2-3 now has the elevation profile illustrated in Fig. 6.2-8. It is again assumed that there is only one duct across the width of each web at any given height in the web. (In reality, it would not be practical to provide a drape using a large number of small tendons in this way, but this is not important for the purposes of this example.) The shear resistance of the inner web is checked below.

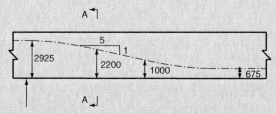

Fig. 6.2-8. Elevation of tendon profile for concrete box girder

The prestressing force after losses is assumed to be 92 600 kN at section A–A. The angle of the tendons to the neutral axis level of the section is $\tan^{-1}(1/5) = 11.3°$; therefore the horizontal component of prestress, $N_{\text{Ed}} = 1.0 \times 92\,600\cos(11.3) = 90\,802$ kN and the vertical component, $V_{\text{p,Ed}} = 1.0 \times 92\,600\sin(11.3) = 18\,160$ kN. Global analysis of the bridge leads to a primary shear from prestress of 18 160 kN (equal to $V_{\text{p,Ed}}$) and a secondary shear from prestress (due to induced support reactions) of 1500 kN. The global analysis shows that these shears due to prestress are approximately equally distributed to the three webs, i.e. 6553 kN per web.

Cross-section bending analysis at section A–A for both hogging and sagging (ignoring any un-tensioned reinforcement provided) gives the following results for the whole cross-section:

Section A–A (hog): $d = 2200$ mm, $z = 1898.4$ mm, $M_{\text{Rd}} = 251\,026$ kNm

Section A–A (sag): $d = 1300$ mm, $z = 1142.3$ mm, $M_{\text{Rd}} = 151\,054$ kNm

The reinforcement (with yield strength of 500 MPa) required to resist an ultimate shear force on the inner web of 10 471 kN (excluding shear from prestress which is of the opposite sign) is now determined. Two cases with global moments of 40 000 kNm hog and 40 000 kNm sag associated with the inner web alone are considered. (The concrete resistance $V_{\text{Rd,c}}$ is exceeded in both cases.)

The external shear force can be reduced to allow for the shear from prestress, so $V_{\text{Ed}} = 10\,471 - 6553 = 3918$ kN. Conservatively ignoring any un-tensioned reinforcement provided, the z values from above lead to the following requirements for vertical link reinforcement based on a truss angle of 45° according to 2-2/Expression (6.8):

Hogging: $\dfrac{A_{sw}}{s} \geq \dfrac{3918 \times 10^3}{1898.4 \times 434.8 \times \cot 45}$

$= 4.75 \, \text{mm}^2/\text{mm}$ (2 × 20ϕ legs at 125 mm centres)

Sagging: $\dfrac{A_{sw}}{s} \geq \dfrac{3918 \times 10^3}{1142.3 \times 434.8 \times \cot 45} = 7.89 \, \text{mm}^2/\text{mm}$

(or $8.58 \, \text{mm}^2/\text{mm}$ if the reinforcement stress is limited to $0.8 f_{yk} = 400 \, \text{MPa}$ in accordance with Note 3 of 2-2/clause 6.2.3(103), but see the cautionary note in the main text).

The limit for maximum effective reinforcement for a 45° truss given by 2-2/Expression (6.12) is:

$$\dfrac{A_{sw,max}}{s} = \dfrac{1}{2}\alpha_{cw}\nu_1 f_{cd}\dfrac{b_w}{f_{ywd}} = \dfrac{1}{2} \times 1.0 \times 0.516 \times 23.333 \times \dfrac{349}{434.8} = 4.833 \, \text{mm}^2/\text{mm}$$

using an upper limit of 1.0 for α_{cw}. (A limit of $6.107 \, \text{mm}^2/\text{mm}$ is obtained if the reinforcement stress is limited to 80% of f_{ywk}, but see the cautionary note in the main text.) The reinforcement needed for the sagging case exceeds this limit, which means that the web crushing limit from 2-2/Expression (6.9) must be exceeded. Checking this:

$$V_{Rd,max} = 1.0 \times 349 \times 1142.3 \times 0.516 \times 23.33 \times 10^{-3}/(\cot 45 + \tan 45) = 2400 \, \text{kN}$$

(or 2791 kN if the reinforcement stress is limited to 80% of f_{ywk}, but see the cautionary note in the main text) which is less than V_{Ed}. The lever arm would therefore need to be increased for the sagging case. This could be done by taking the un-tensioned reinforcement into account in the lever arm calculation.

The additional longitudinal tensile force required for shear in the hogging case is obtained from 2-2/Expression (6.18):

$$\Delta F_{td} = 0.5 \times 3918 \times (\cot 45 - \cot 90) = 1959 \, \text{kN}$$

The force spare in the tendons associated with one web, assuming that each web and attached flange provides one-third of the total moment resistance, is

$$\dfrac{(\frac{1}{3} \times 251\,026 - 40\,000) \times 10^6}{1898.4} = 23\,006 \, \text{kN} \gg 1959 \, \text{kN}$$

required so there is adequate additional tensile force available.

To reduce the required shear reinforcement, the design of the hogging section could be performed by adjusting θ to ensure $V_{Rd,s} \approx V_{Rd,max}$ (noting that longitudinal reinforcement must be increased to suit). In this case, the design reinforcement stress has been limited to 400 MPa to permit ν_1 to be taken as 0.6, but see the cautionary note in the main text. This gives:

$$\theta = 28.9° \Rightarrow \dfrac{A_{sw}}{s} \geq \dfrac{3918 \times 10^3}{1898.4 \times 400.0 \times \cot 28.9} = 2848 \, \text{mm}^2/\text{mm}$$

(2 × 16ϕ legs at 140 mm centres) and from 2-2/Expression (6.9):

$V_{Rd,max} = \alpha_{cw} b_w z \nu_1 f_{cd}/(\cot\theta + \tan\theta)$

$= 1.0 \times 349 \times 1898.4 \times 0.6 \times 23.33 \times 10^{-3}/(\cot 28.9 + \tan 28.9)$

$= 3924 \, \text{kN} > V_{Ed}$ therefore okay

The additional longitudinal tensile force required for shear in the hogging case is obtained from 2-2/Expression (6.18):

$$\Delta F_{td} = 0.5 \times 3918 \times (\cot 28.9 - \cot 90) = 3549 \, \text{kN}$$

CHAPTER 6. ULTIMATE LIMIT STATES

The force spare in the tendons associated with one web is

$$\frac{(\frac{1}{3} \times 251\,026 - 40\,000) \times 10^6}{1898.4} = 23\,006\,\text{kN} \gg 3549\,\text{kN}$$

required so there is adequate additional tensile force available.

Since $V_{\text{Rd,max}}$ reduces with reducing θ, it is not possible to make the sagging case work simply by varying θ.

6.2.3.3. Segmental construction

In segmental construction where there is no bonded prestressing in the tension chord, it is possible for the joints between segments to open once the decompression moment of the section has been reached. The final ultimate moment may not be significantly higher than the decompression moment. This depends primarily on:

- whether the force in the prestressing tendons can increase as a result of the overall structural deflections (see section 5.10.8);
- whether the lever arm between tensile force and concrete compression can increase significantly after decompression. This depends on the section geometry. For solid rectangular cross-sections, the increase in lever arm can be large, approaching a limit of twice that at decompression for low prestress force. For a box girder with very thin webs, the increase in lever arm may be very small as there is little available web compression concrete.

The depth to which the joint opens is governed by the depth of the flexural compression block, h_{red}, and this in turn depends on the above. This is shown in Fig. 6.2-9. *2-2/clause 6.2.3(109)* requires that the prestressing force be assumed to remain constant after decompression, unless a detailed analysis, such as that referred to above, has been done. Clearly, if prestress force increases have been considered in the flexural design, the same must be done here to avoid apparent flexural failure. The opening of the joint introduces a reduced depth through which the web shear compression struts can pass.

Two checks are necessary for a given compression depth, h_{red}. Crushing of the web struts is checked using 2-2/Expression (6.103) which rearranged is:

$$V_{\text{Ed}} < \frac{h_{\text{red}} b_w \nu f_{\text{cd}}}{(\cot\theta + \tan\theta)} \qquad \text{(D6.2-11)}$$

To avoid failure local to the joint, shear reinforcement should be provided in the reduced length, $h_{\text{red}} \cot\theta$, adjacent to the joint as shown in Fig. 6.2-9 according to 2-2/Expression (6.104):

$$\frac{A_{\text{sw}}}{s} = \frac{V_{\text{Ed}}}{h_{\text{red}} f_{\text{ywd}} \cot\theta} \qquad \text{2-2/(6.104)}$$

If either of these checks cannot be satisfied, h_{red} should be increased by increasing the prestressing force and the check repeated.

2-2/clause 6.2.3(109)

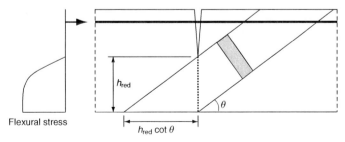

Fig. 6.2-9. Effect of joint opening in precast construction without bonded prestressing

6.2.4. Shear between web and flanges of T-sections
Longitudinal shear

2-1-1/clause 6.2.4(1) allows longitudinal shear in flanges to be checked using a truss model. The check covers the crushing resistance of the concrete struts and the tensile strength of the transverse reinforcement. Clause 6.2.4 applies to planes through the thickness of the flange. It need not be applied to planes through the web at the web–flange junction. If a construction joint has been made between web and flange, however, the similar provisions of 2-1-1/clause 6.2.5 should be checked. Despite the reference to T-sections in the title of the clause, the provisions also apply to other sections, such as boxes, on the basis of the shear per web.

For compatibility with the web shear design and the design of additional longitudinal reinforcement for shear, the flange forces should theoretically be determined considering the same truss model. A typical truss model for the end of an end span is shown in Fig. 6.2-10. This approach would differ from previous UK practice where the longitudinal shear was determined from elastic cross-section analysis using beam theory and the transverse reinforcement placed accordingly, following the envelope of vertical shear force. It can be seen from Fig. 6.2-10 that the transverse reinforcement predicted by a truss model of an 'I' beam does not follow the shear force envelope, but is displaced along the beam from the location of peak shear.

To avoid the need to draw out truss models for every loading situation, the pragmatic simplification is made in *2-2/clause 6.2.4(103)* that the average force increase per metre may be calculated over a length Δx, which should not be taken greater than half the distance between the point of contraflexure and the point of maximum moment in each hog and sag zone. This allows a certain amount of averaging out of the reinforcement from that which would be produced by a detailed truss model. However, where there are point loads, the length Δx should not be taken as greater than the distance between the point loads to avoid significantly underestimating the rate of change of flange force. Since bridges are usually subject to significant point loads from vehicle axles, it appears that this simplification will not usually be appropriate.

2-2/clause 6.2.4(103) offers a further simplified method by which the longitudinal shear is determined directly from the vertical shear per web, V_{Ed}, in the same way as described in section 6.2.5 below. The expression for shear flow given between web and flange, V_{Ed}/z, is only correct where the flexural neutral axis lies at the web–flange junction. In general, the shear flow between web and flange can be taken as $\beta V_{Ed}/z$ as discussed in section 6.2.5 below, where β in this case is the ratio of force in the effective flange to that in the whole

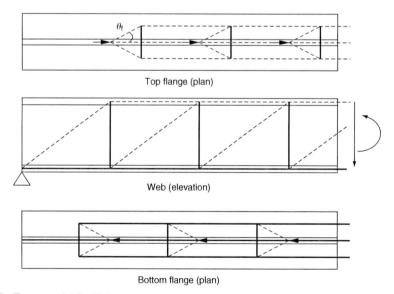

Fig. 6.2-10. Truss model for 'I' beam

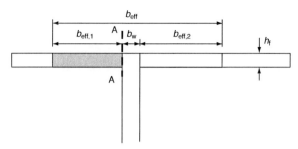

Fig. 6.2-11. Definitions for shear flow calculation

compression or tension zone as appropriate. It is always conservative, however, to take the shear flow as V_{Ed}/z.

Clause 6.2.4 is, however, concerned with the shear flow in the flange itself adjacent to the web. The remainder of 2-2/clause 6.2.4(103) allows a proportion of the flange force to be considered to remain within the width of the web, such that the shear flow on a plane A–A, as in Fig. 6.2-11, is

$$\beta \frac{V_{Ed}}{z} \times \frac{b_{eff,1}}{b_{eff}}$$

and therefore the design shear stress is given by:

$$v_{Ed} = \beta \frac{V_{Ed}}{zh_f} \times \frac{b_{eff,1}}{b_{eff}} \qquad (D6.2\text{-}12)$$

If the full effective flange width, b_{eff}, is not required to resist the bending moment at the section considered, it may be reduced to the actual width required when using equation (D6.2-12) which can lead to a greater proportion of flange force remaining within the web width and a large consequent reduction in transverse reinforcement at ULS. If this is done, a check should be made of cracking at SLS, as wide cracks may open before the spread is limited. For SLS, the stresses in the transverse reinforcement should be determined assuming a spread over the full flange effective width. The elastic spread angle should be used – section 6.9 of this guide is relevant.

It will be noted that the above method of calculating shear stress from the vertical shear force envelope gives no information on where to place the transverse reinforcement along the beam. In clause 6.2.5 for design at construction joints, it is placed according to the shear force envelope. Transverse reinforcement in flanges has been positioned this way in previous UK practice. To achieve greater compatibility with the location of transverse reinforcement indicated by the truss model in Fig. 6.2-10, it would be possible to produce a shifted envelope of transverse reinforcement in the same way as for longitudinal reinforcement as discussed in section 6.2.3.1. For tension flanges, 2-1-1/Fig. 9.2 is appropriate and the calculated transverse reinforcement provision could be extended along the beam by a distance a_l to account for the web truss. It is recommended that the provision is *extended* by a_l rather than translated by a_l as it would be undesirable to have no transverse reinforcement adjacent to supports, as would result if reinforcement was detailed in accordance with Fig. 6.2-10.

The above use of beam theory avoids the need to construct truss models for every load case. It should also be noted that in indeterminate structures, the full truss model cannot be developed without an initial analysis using beam theory to first determine the support reactions.

The transverse reinforcement required per unit length, A_{sf}/s_f, is calculated in accordance with **2-1-1/clause 6.2.4(4)**. It follows simply from a smeared truss model (where the struts and ties are not discrete as in Fig. 6.2-10). Figure 6.2-12 shows a plan of an area ABCD of a concrete flange, assumed to be in longitudinal compression, with shear stress v_{Ed} and transverse reinforcement A_{sf} at spacing s_f. The shear force per transverse bar is:

2-1-1/clause 6.2.4(4)

$$F_v = v_{Ed} h_f s_f \qquad (D6.2\text{-}13)$$

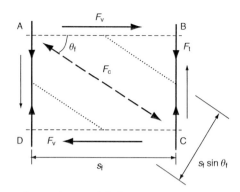

Fig. 6.2-12. Forces acting on flange element (plan)

acting on side AB of the rectangle shown. It is transferred to side CD by a concrete strut AC at angle θ_f to AB, and with edges that pass through the mid-points of AB, etc., as shown, so that the width of the strut is $s_f \sin \theta_f$.

For equilibrium at A, the force in the strut is:

$$F_c = F_v \sec \theta_f \qquad \text{(D6.2-14)}$$

Substituting from equation (D6.2-13) into equation (D6.2-14) gives:

$$F_c = v_{Ed} h_f s_f \sec \theta_f$$

Taking the limiting stress in the strut of width $s_f \sin \theta_f$ to be νf_{cd}, where $\nu = 0.6(1 - f_{ck}/250)$ gives the verification for the strut as:

$$v_{Ed} h_f s_f \sec \theta_f \leq h_f s_f \sin \theta_f \nu f_{cd}$$

which leads to the expression in 2-1-1/clause 6.2.4(4):

$$v_{Ed} \leq \nu f_{cd} \sin \theta_f \cos \theta_f \qquad \text{2-1-1/(6.22)}$$

For equilibrium at C, the force in the transverse bar BC is:

$$F_t = F_c \sin \theta_f = F_v \tan \theta_f$$

Substituting from equation (D6.2-13) gives:

$$F_t = v_{Ed} h_f s_f \tan \theta_f \qquad \text{(D6.2-15)}$$

If the reinforcement is stressed to its design strength f_{yd}, equation (D6.2-15) leads to the expression in 2-1-1/clause 6.2.4(4):

$$A_{sf} f_{yd}/s_f \geq v_{Ed} h_f / \cot \theta_f \qquad \text{2-1-1/(6.21)}$$

2-1-1/clause 6.2.4(4) limits the angle of spread, θ_f, to between 26.5° and 45° for compression flanges and 38.6° to 45° for tension flanges, unless more detailed modelling, such as a non-linear finite element analysis that can consider cracking of concrete, is used.

2-1-1/clause 6.2.4(6)

When the shear stress is less than 40% (a nationally determined parameter) of the design tensile stress, f_{ctd}, **2-1-1/clause 6.2.4(6)** permits the concrete alone to carry the longitudinal shear and no additional transverse reinforcement is required (other than minimum reinforcement). For a concrete with cylinder strength 40 MPa, this gives a limiting stress of 0.67 MPa. For greater shear stress, the concrete's resistance is lost completely as in the main vertical shear design.

It should be noted that interface shear should still be checked according to 2-1-1/clause 6.2.5 where there are construction joints. Different concrete contributions are calculated depending on the degree of roughening of the interface. For a surface prepared by exposing aggregate, classed as 'rough' in 2-1-1/clause 6.2.5(2), the concrete contribution is 45% of the design tensile stress, f_{ctd}, which is a greater proportion than allowed here.

EN 1992-2 does not specify the distribution of the required transverse reinforcement between the upper and lower layers in the slab. It was a requirement of early drafts of

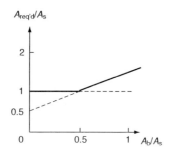

Fig. 6.2-13. Total transverse reinforcement requirements for shear and transverse bending

EN 1992-2 that the transverse reinforcement provided should have the same centre of resistance as the longitudinal force in the slab. This requirement was removed, presumably because it has been common practice to consider the shear resistance to be the sum of the resistances from the two layers. It should be noted that application of Annex MM of EN 1992-2 would necessitate provision of transverse reinforcement with the same centre of resistance as the longitudinal force in the slab.

On the tension flange, the angle of spread of the compression strut into the flange requires the longitudinal tension reinforcement to be required 'earlier' than expected from considerations of the web truss alone, as shown in Fig. 6.2-10. *2-1-1/clause 6.2.4(7)* requires account to be taken of this in determining the curtailment of the longitudinal reinforcement. This may be achieved by introducing a further shift of $e \cdot \cot \theta_f$ in the shift method described in section 6.2.3.1 of this guide. e is the distance of the longitudinal bar considered from the edge of the web plus a distance $b_w/4$.

2-1-1/clause 6.2.4(7)

Flanges with combined longitudinal shear and transverse bending

Flanges forming deck slabs will also usually be subjected to transverse bending from dead and live loads. The flange reinforcement needs to be checked for its ability to carry both in-plane shear and any transverse bending. A simplified rule for combining the reinforcement requirements from shear and bending is given in *2-2/clause 6.2.4(105)*. This requires the amount of transverse reinforcement to be the greater of that required for longitudinal shear alone (case (a)) and half that required for longitudinal shear plus that required for transverse bending (case (b)). This rule is illustrated in Fig. 6.2-13, in which $A_{\text{req'd}}$ is the total reinforcement required, and subscripts s and b refer to the reinforcement required for shear and bending, respectively. The reinforcement provided should not be less than the minimum requirements of 2-1-1/clause 9.2.1.

2-2/clause 6.2.4(105)

Once again, EN 1992-2 does not specify the distribution of the required transverse reinforcement between the upper and lower layers in the slab, particularly with respect to the component due to shear. Clearly the reinforcement must at least be placed so as to resist the transverse bending moment alone. The designer is left to decide where he or she places the centre of gravity of the reinforcement needed for shear. The intention of the drafters was to distribute the reinforcement as shown in Fig. 6.2-14 for the two cases (a) and (b) above. Use of 2-2/Annex MM would suggest that these combination rules are optimistic and indeed they are not allowed for web design. Previous UK practice in BS 5400 Part 4[9] has, however, been to ignore any interaction in flanges.

2-2/clause 6.2.4(105) also requires that the interaction between bending and longitudinal shear is checked in the zone which is in compression under the transverse bending. Compression from strut-and-tie action in the flange resisting shear will add to that from transverse bending and can lead to crushing failure. It has previously been UK practice to ignore this interaction but it can be unsafe to do so. To allow for the interaction, a simplified approach is given, whereby the depth of flange required for compression in transverse bending is deducted from h_f when calculating the crushing resistance. This approach is conservative because the directions of the compressive stresses from bending and from shear do not coincide. If the flange is inadequate when checked according to the above simplified approach, then the sandwich model of 2-2/Annex MM can be used to check the

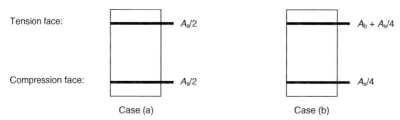

Fig. 6.2-14. Placement of transverse reinforcement for cases (a) and (b)

compressions in the two outer layers as discussed in Annex M of this guide. The reinforcement should also then be derived from 2-2/Annex MM, but this will generally exceed that from the simple rules above.

It was not intended that the *longitudinal* compressive stress from the main beam bending moment be considered when determining the depth to be ignored in shear calculation according to 2-2/clause 6.2.4(105). It would, however, be consistent with the α_{cw} term in 2-2/Expression (6.9) where axial force is high. As a result, for flanges with high compressive axial stress, 2-2/Annex MM may not lead to a more economic check of concrete crushing as the shear stress does interact with compression from both longitudinal and transverse bending – see discussion on 2-2/Annex MM in this guide.

6.2.5. Shear at the interface between concrete cast at different times

Shear stresses across construction joints between concrete elements cast at different times must be checked to ensure that the two concrete components act fully compositely. The bending and shear designs of such members are based on this assumption. 2-1-1/clause 6.2.5 deals specifically with this interface shear requirement, which must be considered in addition to the requirements of 2-2/clause 6.2.4.

2-1-1/clause 6.2.5(1)

2-1-1/clause 6.2.5(1) specifies that the interfaces should be checked to ensure that $v_{Edi} \leq v_{Rdi}$, where v_{Edi} is the design value of shear stress in the interface and v_{Rdi} is the design shear resistance at the interface.

The applied design value of shear stress is given by:

$$v_{Edi} = \beta V_{Ed}/zb_i \qquad \text{2-1-1/(6.24)}$$

where:

β is the ratio of the longitudinal force in the new concrete area and the total longitudinal force either in the compression or the tension zone, both calculated for the section considered

V_{Ed} is the total vertical shear force for the section

z is the lever arm of the composite section

b_i is the width of the interface shear plane (2-1-1/Fig. 6.8 gives examples)

2-1-1/Expression (6.24) is intended to be used assuming all loads are carried on the composite section, which is compatible with the design approach for ultimate flexure. The basic shear stress for design at the interface is related to the maximum longitudinal shear stress at the junction between compression and tension zones given by V_{Ed}/zb_i. This follows from consideration of equilibrium of the forces in either the tension or compression zone. The instantaneous force is given by M_{Ed}/z so the change in force per unit length along the beam (the shear flow), assuming that the lever arm, z, remains constant, is given by:

$$q = \frac{d}{dx}\left(\frac{M_{Ed}}{z}\right) = \frac{V_{Ed}}{z} \qquad \text{(D6.2-16)}$$

The shear stress is then obtained by dividing by the thickness at the interface to give:

$$v_{Ed} = \frac{V_{Ed}}{zb_i} \qquad \text{(D6.2-17)}$$

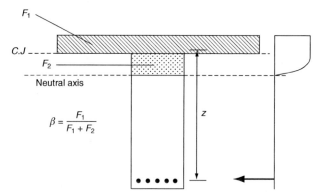

Fig. 6.2-15. Determination of β

If the shear plane checked lies within either the compression or tension zones, the shear stress from equation (D6.2-17) may be reduced by the factor β above. It will always be conservative to take $\beta = 1.0$. For flanged beams, much of the force is contained in the flanges so construction joints at the underside of flange will typically have $\beta \approx 1.0$. In other cases, β can be obtained from the flexural design as shown in Fig. 6.2-15, from the forces F_1 and F_2 (which are shown with the flange in compression).

The question arises as to what to use for the lever arm, z. Strictly, the value of z should reflect the stress block in the beam for the loading considered. In reality, this would be time-consuming to achieve and it will generally be reasonable to use the same value as obtained from the ultimate bending resistance analysis, as shown in Fig. 6.2-15. For cracked sections, the use of the ultimate bending resistance lever arm will slightly overestimate the actual lever arm at lower bending moments. However, for other than very heavily reinforced sections, this difference in lever arm will be small and is compensated for by also basing the value of β on the stress block for ultimate resistance.

The above represents a difference to previous UK practice, where the shear stress distribution was based on the elastic distribution on an uncracked section regardless of flexural stress distribution. If a section remained uncracked in bending and elastic analysis was used to determine the lever arm, then the shear stress determined from 2-1-1/Expression (6.24) would be the same as that from the analysis for the uncracked shear resistance in 2-1-1/clause 6.2.2.

A further point is that if the lever arm used is not taken to be the same as that for the calculations on flexural shear, it would then be possible to find that the maximum shear stress according to this clause was exceeded, while the check against $V_{\text{Rd,max}}$ for flexural shear was satisfied.

The design shear resistance at the interface is based on the CEB Model Code 90[6] provisions and is given in 2-1-1/clause 6.2.5(1) as:

$$v_{\text{Rdi}} = cf_{\text{ctd}} + \mu\sigma_n + \rho f_{\text{yd}}(\mu \sin \alpha + \cos \alpha) \leq 0.5\nu f_{\text{cd}} \qquad \text{2-1-1/(6.25)}$$

where c and μ are factors which depend on the roughness of the interface.

Recommended values in the absence of results from tests are given in **2-1-1/clause 6.2.5(2)**. Other factors are defined in 2-1-1/clause 6.2.5(1). The first term in 2-1-1/Expression (6.25) relates to bond between the surfaces and any mechanical interlock provided by indenting the surfaces, the second relates to friction across the interface under the action of compressive stress, σ_n, and the third term relates to the mechanical resistance of reinforcement crossing the interface. The reinforcement provided for shear in accordance with clauses 6.2.1 to 6.2.4 may be considered in the reinforcement ratio ρ. The reinforcement does not need to be taken as that provided in addition to that needed for ordinary shear.

In order to allow the practical placement of reinforcement across the interface in bands of decreasing longitudinal spacing, **2-1-1/clause 6.2.5(3)** allows a stepped distribution of reinforcement to be used by averaging the shear stress over a given length of the member corresponding to the length of band chosen. This is illustrated in Worked example 6.2-8.

2-1-1/clause 6.2.5(2)

2-1-1/clause 6.2.5(3)

No guidance is given as to by how much the local shear stress may exceed the local calculated resistance. A reasonable approach would be to allow the shear stress to exceed the resistance locally by 10%, provided that the total resistance within the band was equal to or greater than the total longitudinal shear in the same length. This would be consistent with the design of shear connectors in EN 1994-2.

2-2/clause 6.2.5(105)

Where interface shear is checked under dynamic or fatigue loads, *2-2/clause 6.2.5(105)* requires that the roughness coefficient values, c, are taken as zero to account for potential deterioration of the concrete component of resistance across the interface under cyclic loading.

6.2.6. Shear and transverse bending

In webs, particularly those of box girders, transverse bending moments can lead to reductions in the maximum permissible coexistent shear force because the compressive stress fields from shear and from transverse bending have to be combined. The two stress fields do not, however, fully add because they act at different angles. In the UK it has been common to design reinforcement in webs for the combined action of transverse bending and shear, but not to check the concrete itself for the combined effect. The lower crushing limit in shear used in the UK made this a reasonable approximation, but it is potentially unsafe if a less conservative (and more realistic) crushing strength is used.

2-2/clause 6.2.106 formally requires consideration of the above shear–moment interaction, but if the web shear force according to clause 6.2 is less than 20% of $V_{\text{Rd,max}}$ or the transverse moment is less than 10% of $M_{\text{Rd,max}}$, the interaction does not need to be considered. $M_{\text{Rd,max}}$ is defined as the maximum web resistance to transverse bending. The subscript 'max' might suggest that this is the maximum obtainable bending resistance if the web were to be heavily over-reinforced. It was, however, intended to be the actual web bending resistance in the absence of shear, even though the former could be considered more relevant for the crushing check. These criteria are unlikely to be satisfied for box girders, but the allowance for coexisting moment will often be sufficient to negate the need for a check of typical beam and slab bridges. Where the interaction has to be considered, 2-2/Annex MM can be used.

6.2.7. Shear in precast concrete and composite construction (additional sub-section)

EN 1992-2 does not directly cover the vertical shear design of composite construction, such as pretensioned beams made composite with an in situ deck slab. The design of members requiring designed shear reinforcement can be carried out in accordance with 2-2/clause 6.2.3. For members not requiring designed shear reinforcement, which will not be very common, 2-2/clause 6.2.2 is applicable but the provisions of 2-1-1/clause 6.2.2(2) for shear tension resistance need some interpretation.

Precast pretensioned beam made composite with in situ deck slab

As discussed in section 6.2.2.2 of this guide, the un-cracked shear resistance should be determined by limiting the principal tensile stress to f_{ctd}. While the centroid of the composite section may often be critical, the limit applies equally at all levels within the section, which should be checked accordingly. In practice, it is usually sufficient to check the composite centroid and other significant levels of change in section (the web–flange junctions, for example).

The shear force (V_{c1}) due to self-weight and the loads acting on the precast section alone produces a shear stress distribution, as shown in Fig. 6.2-16(b). The shear stress produced by V_{c1} in the precast member at the level considered is τ_s. Figure 6.2-16(c) shows the shear stress distribution in the composite member with the addition of the second stage shear force (V_{c2}) acting on the composite section. The shear stress at the same level under this loading case is τ'_s. Figure 6.2-16(d) shows the total shear stress distribution given by $\tau_s + \tau'_s$.

If the stress (compression taken positive) at the level checked due to the build-up of stress from loads acting on the beam alone (including prestress) and on the composite section (with

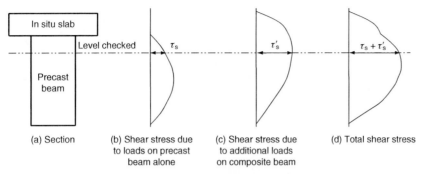

Fig. 6.2-16. Shear stress distribution in a composite beam and slab section

each load component multiplied by appropriate partial safety factors and α_1 considered for the prestress component) is σ_{tot}, then the most tensile principal stress at this level is given by:

$$\frac{\sigma_{tot}}{2} - \sqrt{\left(\frac{\sigma_{tot}}{2}\right)^2 + (\tau_s + \tau'_s)^2}$$

This stress should not exceed the f_{ctd} tensile limit, hence:

$$-f_{ctd} = \frac{\sigma_{tot}}{2} - \sqrt{\left(\frac{\sigma_{tot}}{2}\right)^2 + (\tau_s + \tau'_s)^2}$$

which rearranges to:

$$\tau'_s = \sqrt{f_{ctd}^2 + f_{ctd}\sigma_{tot}} - \tau_s$$

However, for uniform members:

$$\tau'_s = \frac{V_{c2}}{b}\left(\frac{A_e \bar{z}}{I}\right)$$

where $A_e\bar{z}$ is the first moment of area of the excluded area above the plane considered about the cross-section centroid level and I is the second moment of area, both relating to the composite section – see section 6.2.2.2 for definitions. Thus:

$$V_{c2} = \frac{Ib}{A_e\bar{z}}\left(\sqrt{f_{ctd}^2 + f_{ctd}\sigma_{tot}} - \tau_s\right) \tag{D6.2-18}$$

with

$$\tau_s = \frac{V_{c1}}{b}\left(\frac{A_{pc}\bar{z}_{pc}}{I_{pc}}\right)$$

where $A_{pc}\bar{z}_{pc}$ is the first moment of area of the excluded area above the plane considered about the cross-section centroid level and I_{pc} is the second moment of area, both relating to the precast beam alone.

The total shear resistance of the section is thus given by:

$$V_{Rd,c} = V_{c1} + V_{c2} \tag{D6.2-19}$$

V_{c2} could be negative, representing cracking of the precast beam section before being made composite with the deck slab. In this case, the shear tension resistance is inadequate even to carry the shear imposed on the precast beam. The procedure is illustrated in Worked example 6.2-7.

Following the comments made on 2-1-1/clause 6.2.2(2), this approach is valid for sections subjected to sagging moments but it is not clear if it is permitted for hogging moments, even though logically it should be. If it were to be applied to hogging regions, the in situ flange should be considered to be cracked and only the reinforcement considered in the section properties.

The comments made in section 6.2.2.2 above, regarding neglect of imposed deformations, also apply to composite sections. There is a further complication in composite members as the primary effects of differential temperature, differential shrinkage and differential creep cause locked-in self-equilibrating stresses through the section depth. Since 2-1-1/Expression (6.4) effectively ignores such effects for differential temperature, through its definition of σ_{cp}, one could argue that all self-equilibrating stresses in composite construction from differential temperature, shrinkage and creep, could be neglected by analogy.

Worked example 6.2-7: Pretensioned concrete 'M' beam with composite deck slab

The composite 'M' beam from Worked example 6.1-5 forms part of a simply supported deck. The shear resistance is determined under the following internal actions:

$$\left.\begin{array}{l} V_{Ed,precast} = 73\,kN \\ V_{Ed,composite} = 284\,kN \end{array}\right\} V_{Ed} = 357\,kN$$

$$\left.\begin{array}{l} M_{Ed,precast} = 490\,kNm \\ M_{Ed,composite} = 810\,kNm \end{array}\right\} M_{Ed} = 1300\,kNm \text{ (sagging)}$$

Summary of section properties based on the cross-section in Fig. 6.1-10:

	Beam alone	Composite section
Area, A (mm^2)	349×10^3	500×10^3
Height to centroid, z_{na} (mm)	310	474
Second moment of area, I (mm^4)	23.02×10^9	54.5×10^9
First moment of area of concrete above the composite centroid about the relevant centroid (mm^3)	38.5×10^6	76.5×10^6

The strand pattern is as in Fig. 6.1-10 and the centroid of the strands is therefore 100 mm from the bottom of the section.

A factored prestressing force after all losses of 2680 kN is assumed here. For simplicity in this example, this prestressing force after losses is applied solely to the precast beam. In reality, part of the long-term loss of prestressing force would act on the composite section but application of the entire loss of force to the precast beam alone is usually conservative.

For the check of shear tension to 2-1-1/clause 6.2.2(2) to be applicable, it is first necessary to determine if the precast beam is uncracked in flexure. The stress at the extreme bottom fibre is:

$$= \frac{2680 \times 10^3}{349 \times 10^3} + 2680 \times 10^3 \times (310 - 100)$$

$$\times \frac{310}{23.02 \times 10^9} - \frac{490 \times 10^6 \times 310}{23.02 \times 10^9} - \frac{810 \times 10^6 \times 474}{54.50 \times 10^9}$$

$$= 7.68 + 7.58 - 6.60 - 7.04 = 1.61\,MPa$$

As the extreme fibre stress is compressive, the section is clearly uncracked in flexure.

Shear flexure is now checked at various heights in the precast beam:

(a) Check at level of composite centroid

Stress at the composite centroid due to the moment acting on the precast beam alone

$$= 490 \times 10^6 \times \frac{(474 - 310)}{23.02 \times 10^9} = 3.49\,MPa \text{ (compression)}$$

Stress at the composite centroid due to prestress

$$= \frac{2680 \times 10^3}{349 \times 10^3} + 2680 \times 10^3 \times (310 - 100) \times \frac{(310 - 474)}{23.02 \times 10^9} = 3.67\,MPa \text{ (compression)}$$

Therefore total stress at composite centroid from actions on the precast beam alone to be used in equation (D6.2-18) is

$\sigma_{tot} = 3.67 + 3.49 = 7.16\,\text{MPa}$

The design shear force at ULS on the precast beam alone, $V_{c1} = 73\,\text{kN}$, thus the shear stress at the height of the composite centroid is from equation (D6.2-18):

$$\tau_s = \frac{V_{c1}}{b}\left(\frac{A_{pc}\bar{z}_{pc}}{I_{pc}}\right) = \frac{73 \times 10^3}{160} \times \frac{38.5 \times 10^6}{23.02 \times 10^9} = 0.763\,\text{MPa}$$

The additional shear force (V_{c2}) which can be carried by the composite section before the principal tensile stress at the composite centroid reaches the limit of f_{ctd} ($= 1.667\,\text{MPa}$) is given by equation (D6.2-18):

$$V_{c2} = \frac{Ib}{A_e\bar{z}}\left(\sqrt{f_{ctd}^2 + f_{ctd}\sigma_{tot}} - \tau_s\right)$$

thus

$$V_{c2} = \frac{54.5 \times 10^9 \times 160}{76.5 \times 10^6} \times (\sqrt{1.667^2 + 1.667 \times 7.16} - 0.763) \times 10^{-3} = 350\,\text{kN}$$

(b) Check at the top of the precast unit

The check at the composite centroid level above is in this case only 16 mm from the top of the 160 mm-wide web, so no further check will be made at the top of the web.

Stress at top of precast unit due to the moment acting on the beam alone

$$= 490 \times 10^6 \times \frac{(800 - 310)}{23.02 \times 10^9} = 10.43\,\text{MPa (compression)}$$

Stress at top of precast unit due to the moment acting on the composite section

$$= 810 \times 10^6 \times \frac{(800 - 474)}{54.5 \times 10^9} = 4.85\,\text{MPa (compression)}$$

Stress at top of precast unit due to prestress

$$= \frac{2680 \times 10^3}{349 \times 10^3} + 2680 \times 10^3 \times (100 - 310) \times \frac{(800 - 310)}{23.02 \times 10^9} = -4.30\,\text{MPa (tension)}$$

Therefore total stress at the underside of the top slab to be used in equation (D6.2-18) is $\sigma_{tot} = 10.43 + 4.85 - 4.30 = 10.98\,\text{MPa}$.

The shear stress at the top of the precast unit from shear acting on the precast beam alone $= 0\,\text{MPa}$. It is assumed here that the only shear connection between precast unit and in situ slab is across the 300 mm-wide portion at the top of the precast unit.

For the composite section, the first moment of area of the deck slab about the composite section neutral axis $= A_e\bar{z} = 160 \times 1000 \times (930 - 80 - 474) = 60.2\,\text{m}^3$. The additional shear force (V_{c2}) which can be carried by the composite section before the principal tensile stress at the top of the precast unit reaches the limit of f_{ctd} ($= 1.667\,\text{MPa}$) is again given by equation (D6.2-18):

$$V_{c2} = \frac{54.5 \times 10^9 \times 300}{60.2 \times 10^6} \times \left(\sqrt{1.667^2 + 1.667 \times 10.98} - 0\right) = 1247\,\text{kN}$$

This is less critical than the check at the composite centroid.

(c) Check at bottom of 160 mm thick web

A check is made at the bottom of the web as this will be more critical than within the chamfer above the bottom flange.

Stress at bottom of web due to the moment acting on the beam alone

$$= 490 \times 10^6 \times \frac{(290 - 310)}{23.02 \times 10^9} = -0.43\,\text{MPa (tension)}$$

Stress at bottom of web due to the moment acting on the composite section

$$= 810 \times 10^6 \times \frac{(290 - 474)}{54.5 \times 10^9} = -2.73 \text{ MPa (tension)}$$

Stress at bottom of web due to prestress

$$= \frac{2680 \times 10^3}{349 \times 10^3} + 2680 \times 10^3 \times (100 - 310) \times \frac{(290 - 310)}{23.02 \times 10^9} = 8.17 \text{ MPa (compression)}$$

Therefore total stress at the bottom of the web to be used in equation (D6.2-18) is

$$\sigma_{\text{tot}} = 8.17 - 0.43 - 2.73 = 5.01 \text{ MPa}$$

For the precast section, the first moment of area of the bottom flange and chamfer about the precast section neutral axis is:

$$A_{\text{pc}} \bar{z}_{\text{pc}} \approx 185 \times 950 \times (310 - 185/2) + 105 \times 240 \times (310 - 185 - 105/2) = 40.1 \text{ m}^3$$

The design shear force at ULS on the precast beam alone, V_{c1}, is 73 kN, thus the shear stress at the bottom of the web is from equation (D6.2-18):

$$\tau_s = \frac{V_{c1}}{b}\left(\frac{A_{\text{pc}} \bar{z}_{\text{pc}}}{I_{\text{pc}}}\right) = \frac{73 \times 10^3}{160} \times \frac{40.1 \times 10^6}{23.02 \times 10^9} = 0.795 \text{ MPa}$$

For the composite section, the first moment of area of the bottom flange and chamfer about the composite section neutral axis is:

$$A_e \bar{z} \approx 185 \times 950 \times (474 - 185/2) + 105 \times 240 \times (474 - 185 - 105/2) = 73.0 \text{ m}^3$$

The additional shear force (V_{c2}) which can be carried by the composite section before the principal tensile stress at the top of the precast unit reaches the limit of f_{ctd} (=1.667 MPa) is again given by equation (D6.2-18):

$$V_{c2} = \frac{54.5 \times 10^9 \times 160}{73.0 \times 10^6} \times (\sqrt{1.667^2 + 1.667 \times 5.01} - 0.795) = 304 \text{ kN}$$

This is more critical than the check at the composite centroid. It is possible for other heights in the cross-section to be critical but any further reduction in strength will usually be very small.

The above illustrates that the bottom of the web is critical in this case. From equation (D6.2-19):

$$V_{\text{Rd,c}} = V_{c1} + V_{c2} = 73 + 304 = 377 \text{ kN}$$

There is no vertical component of prestress to consider for this section and $V_{\text{Ed}} < V_{\text{Rd,c}}$ (357 kN < 377 kN); therefore no shear reinforcement is required.

Worked example 6.2-8: Pretensioned concrete 'M' beam with composite deck slab

Continuing from Worked example 6.2-7, a constant change of shear force is assumed from 357 kN at section 1 at quarter span to 524 kN at section 2 at the beam end, where the strands deflect up from the values in Worked example 6.2-7. The interface design shear stress for each section is calculated from 2-1-1/Expression (6.24):

Section 1: $V_{\text{Ed}} = 357$ kN and $b_i = 300$ mm (conservative, since the exact seating arrangements are not yet known and therefore the design cannot rely on increasing the interface length beyond just the top surface). From a bending analysis, $z = 710.7$ mm and the compressive force in the top slab can be shown to be 2719.3 kN (of a total compressive force at the ultimate limit state of 3727.8 kN). Thus from Fig. 6.2-15, $\beta = 2719.3/3727.8 = 0.729$; therefore $v_{\text{Edi}} = (0.729 \times 357 \times 10^3)/(710.7 \times 300) = 1.221$ MPa.

Section 2: $V_{Ed} = 524$ kN and $b_i = 300$ mm (conservative, as above).
From a bending analysis, $z = 667.5$ mm and the tensile force in the reinforcement in the top slab can be shown to be 2380.1 kN (of a total tensile force at the ultimate limit state of 3444.7 kN). Thus, from Fig. 6.2-15, $\beta = 2380.1/3444.7 = 0.691$; therefore $v_{Edi} = (0.691 \times 524 \times 10^3)/(667.5 \times 300) = 1.808$ MPa.

Calculate the resistance from 2-1-1/Expression (6.25), assuming rough surface preparation (from 2-1-1/clause 6.2.5(2): $c = 0.45$ and $\mu = 0.70$), ignoring $\mu\sigma_n$ term and assuming vertical links ($\alpha = 90°$). Note that in accordance with 2-2/clause 6.2.5(105), c should be taken as 0 where interface shear is checked under fatigue loads (as in this example).

Beam is class C40/50 concrete and slab is class C30/40; therefore, taking the weakest concrete for the interface shear calculations gives $f_{ck} = 30$ MPa, $f_{cd} = 20$ MPa (using $\gamma_c = 1.5$), $f_{ctk,0.05} = 2.0$ MPa (from 2-1-1/Table 3.1) and $f_{ctd} = 1.333$ MPa.

From 2-1-1/Expression (6.6N): $\nu = 0.6(1 - f_{ck}/250) = 0.6 \times (1 - 30/250) = 0.528$. Limit to 2-1-1/Expression (6.25) is $0.5\nu f_{cd} = 0.5 \times 0.528 \times 20 = 5.28$ MPa $> v_{Edi}$ throughout therefore okay.

Thus resistances are given by $v_{Rdi} = 0.45 \times 1.333 + 0 + \rho f_{yd} \times (0.70 \times \sin 90)$, which can be re-arranged to the following to determine required area of reinforcement crossing the interface plane:

$$A_s \geq \frac{(v_{Edi} - 0.45 \times 1.333)}{0.7 f_{yd}} A_i$$

(which further reduces to $A_s \geq (v_{Edi}/0.7 f_{yd})A_i$ where fatigue loads are present).

Assuming $f_{yk} = 500$ MPa and $\gamma_s = 1.15$ gives $f_{yd} = 434.8$ MPa and the area of the plane per metre length, $A_i = 300 \times 1000 = 300 \times 10^3$ mm^2.

Therefore for section 1:

$$A_s \geq \frac{(1.221 - 0)}{0.7 \times 434.8} \times 300 \times 10^3 = 1203.3 \text{ mm}^2/\text{m}$$

and for section 2:

$$A_s \geq \frac{(1.963 - 0)}{0.7 \times 434.8} \times 300 \times 10^3 = 1782.2 \text{ mm}^2/\text{m}$$

EC2 allows the interface shear reinforcement provision to be averaged over appropriate lengths into suitable bands as discussed in the main text.

Considering a band between section 2 and midway between section 1 and section 2 gives an average required $A_s = 1203.3 + 0.75 \times (1782.2 - 1203.3) = 1637.5$ mm^2/m.

Considering a band between section 1 and midway between section 1 and section 2 gives an average required $A_s = 1203.3 + 0.25 \times (1782.2 - 1203.3) = 1348.0$ mm^2/m.

Adopting 2 legs of 10ϕ links at 95 mm centres (marginally more than that required by shear design for section 2) gives $A_s = 1653$ mm^2/m (and $v_{Rdi} = 1.677$ MPa). Adopting 2 legs of 10ϕ links at 115 mm centres gives $A_s = 1366$ mm^2/m (and $v_{Rdi} = 1.386$ MPa). These values have been included in Fig. 6.2-17, illustrating the acceptability of banding the link provisions in this manner.

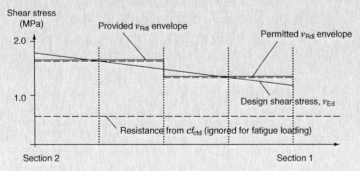

Fig. 6.2-17. Stepped provision of banding shear link reinforcement for interface shear design

6.3. Torsion

This section deals with the design at ultimate limit state of members subject to torsion. It is split into the following sub-sections:

- General   *Section 6.3.1*
- Design procedure   *Section 6.3.2*
- Warping torsion   *Section 6.3.3*
- Torsion in slabs (additional sub-section)   *Section 6.3.4*

6.3.1. General

Torsion does not usually determine the size of bridge members and can usually therefore be checked after the flexural design has been completed. This is a convenient approach as frequently the maximum torsional design moment will not coexist with the maximum flexural moment and shear force, so there will often be excess reinforcement when considering the maximum torsional moment.

In some instances, the equilibrium of the structure depends on torsion. Typical examples include box girders or beams curved in plan. Such torsion is sometimes called 'equilibrium torsion' and it must be considered in the ultimate limit state design. In box girders, it may also be necessary to check webs for the combination of shear and torsion at the serviceability limit state in accordance with 2-2/clause 7.3.1(110).

In other cases, in statically indeterminate structures, torsion may arise from consideration of compatibility with the rest of the structure only. This is sometimes called 'compatibility torsion'. In such cases, it is then possible to neglect torsional stiffness in the analysis at the ultimate limit state and to carry the additional effects caused by neglecting torsion by some other means. Before doing this, it is essential to ensure that torsion plays only a minor part in the behaviour of the bridge to prevent excessive cracking.

2-1-1/clause 6.3.1(1)
2-1-1/clause 6.3.1(2)

This distinction between compatibility torsion and equilibrium torsion is the subject of *2-1-1/clause 6.3.1(1)* and *2-1-1/clause 6.3.1(2)*. Where compatibility torsion is neglected, 2-1-1/clause 6.3.1(2) requires a minimum reinforcement to be provided, in accordance with sections 7.3 and 9.2 of EN 1992-1-1 as modified by EN 1992-2.

In cases where torsional stiffness is considered in analysis, it is important to evaluate it realistically. Decks with discrete beams, such as the M beams in Worked example 6.3-2, have small torsional resistance and would crack under a relatively small torque. According to *Model Code 90*,[6] the cracked stiffness in torsion is typically only about a quarter of the uncracked value and this would significantly reduce any torque attracted to the beams once they had cracked.

When discussing torsion, a further distinction needs to be made between St Venant torsion, circulatory torsion and warping torsion. St Venant torsion arises due to the closed flow of shear stresses around the perimeter of a cross-section. The term 'St Venant torsion' is usually used to cover both a closed flow of shear around a closed hollow section such as a box girder, referred to as circulatory torsion, and also a similar flow around the perimeter of an open section. In EN 1992, a closed flow of shear around a hollow section is referred to as 'circulatory torsion' and a closed flow of shear around an open or solid section is referred to as 'St Venant torsion'. Consideration of circulatory torsion forms the basis of the rules in 2-2/clause 6.3.2. Warping torsion arises from in plane bending of individual walls when there is restraint to longitudinal deformations, such as might occur in an 'I' beam. The total applied torque has to be carried by either or a combination of these two mechanisms. Warping torsion is discussed in section 6.3.3 of this guide, together with guidance on how to apportion torsion between the two mechanisms.

2-1-1/clause 6.3.1(3)

The circulatory torsional resistance covered in 2-2/clause 6.3.2 is calculated on the basis of thin-walled closed sections, in which equilibrium is satisfied by a closed shear flow, even where sections are open – *2-1-1/clause 6.3.1(3)*. The provisions of 2-1-1/clause 6.3.1(3) to (5) regarding torsion calculation for open flanged sections is discussed in section 6.3.2 of this guide, after the basic method has been explained.

6.3.2. Design procedure

2-2/clause 6.3.2 deals with circulatory torsion. The torsional resistance of sections may be calculated on the basis of a thin-walled closed section, even if the section is actually solid. For solid members, the section is idealized as a thin-walled section with a chosen effective wall thickness as discussed below. Figure 6.3-1 shows a generalized section with an idealized thin wall.

The following definitions in *2-1-1/clause 6.3.2(1)* are required to develop the rules for torsion:

2-1-1/clause 6.3.2(1)

- A_k is the area enclosed by the centreline in Fig. 6.3-1
- $\tau_{t,i}$ is the shear stress in wall i
- $t_{ef,i}$ is the effective thickness of wall i discussed below
- A is the area within the outer surface as shown in Fig. 6.3-1
- u is the outer circumference of the cross-section
- z_i is the side length of wall i defined by the distance between the intersection points with the adjacent walls

The effective thickness, $t_{ef,i}$, may be varied to optimize the torsional resistance. It can be made equal to A/u, but not less than twice the distance from edge of section to centre of longitudinal reinforcement. A/u represents a thickness which gives somewhere close to the peak crushing resistance as discussed later in this section. The latter requirement ensures that the centreline of the wall, and hence the line of action of the compression struts, does not lie outside the longitudinal reinforcement. These requirements can lead to difficulties of interpretation with thin solid sections, where the minimum permissible effective thickness may exceed both A/u and the physical half-thickness as seen in Worked example 6.3-2. For hollow sections, $t_{ef,i}$ should obviously not exceed the actual wall thickness. If $t_{ef,i}$ is made as small as possible (i.e. not less than twice the distance from edge of section to centre of longitudinal reinforcement), the resistance based on reinforcement will be maximized and the shear force to be carried in a wall is minimized. However, a small thickness will mean that the limiting torque based on the concrete crushing resistance is reduced. This is illustrated later in Fig. 6.3-3.

Reinforcement requirements

Following the above principles, the following design rules for reinforcement may be derived.

From 2-1-1/Expression (6.26), the torque is obtained as:

$$T_{Ed} = 2A_k(\tau_{t,i} t_{ef,i}) \tag{D6.3-1}$$

where $\tau_{t,i} t_{ef,i}$ is the shear flow around the perimeter which is a constant. The shear force in each wall, $V_{Ed,i}$, is given by 2-1-1/Expression (6.27) as follows:

$$V_{Ed,i} = \tau_{t,i} t_{ef,i} z_i \qquad \text{2-1-1/(6.27)}$$

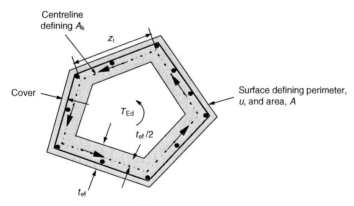

Fig. 6.3-1. General section illustrating definitions

Consequently, eliminating $\tau_{t,i} t_{ef,i}$ using equation (D6.3-1) and 2-1-1/Expression (6.27) gives:

$$V_{Ed,i} = \frac{T_{Ed}}{2A_k} z_i \quad \text{(D6.3-2)}$$

It is noted that *Model Code 90*[6] includes an additional shape-related model factor in equation (D6.3-2), which increases the shear force for a square section by 33% from the value obtained according to equation (D6.3-2). The multiplication factor given for rectangular shapes was $1/(1 - 0.25b/d)$, where d was the greatest section dimension and b the least. Circular sections had a model factor of unity. Some caution is therefore advised with stressing square shapes close to their limit in torsion, although this is seldom likely to arise in practice.

The individual walls of the box are then designed for the shear force in equation (D6.3-2). Equating the above to the shear resistance from 2-1-1/Expression (6.8) for vertical links, but adopting the same definition of z_i as above as a simplification, gives:

$$V_{Ed,i} = \frac{T_{Ed}}{2A_k} z_i = \frac{A_{st}}{s_t} z_i f_{yd} \cot\theta$$

which leads to the following transverse reinforcement requirement for each wall:

$$\frac{A_{st}}{s_t} = \frac{T_{Ed}}{2A_k f_{yd} \cot\theta} \quad \text{(D6.3-3)}$$

where A_{st} is the area of transverse reinforcement in the thickness $t_{ef,i}$, provided at a spacing of s_t. The limits on strut angle, θ, are the same as those for the shear design.

2-2/clause 6.3.2(102)

According to *2-2/clause 6.3.2(102)*, if torsion is combined with shear, the same value of θ should be taken for both checks and the reinforcement requirements added, including any reinforcement for other effects. It is also possible, as may be desirable for box girders, to add the shear force obtained from equation (D6.3-2), with A_k based on the actual wall thickness, directly to that from flexural shear and design each wall for the resulting shear directly. The shear check is then carried out as described in section 6.2 of this guide for the total shear, where $V_{Ed} = V_{shear} + V_{torsion}$.

2-2/clause 6.3.2(103)

2-2/clause 6.3.2(103) gives a similar equation for the area of longitudinal reinforcement required:

$$\frac{\Sigma A_{sl} f_{yd}}{u_k} = \frac{T_{Ed}}{2A_k} \cot\theta \quad \text{2-2/(6.28)}$$

where ΣA_{sl} is the total longitudinal reinforcement required for torsion and u_k is the perimeter of the area A_k. The longitudinal reinforcement generally has to be distributed uniformly along each wall, except for small sections where bars may be concentrated at corners, so an alternative statement of 2-2/Expression (6.28) for each wall is:

$$\frac{A_{sl}}{s_l} = \frac{T_{Ed}}{2A_k f_{yd}} \cot\theta \quad \text{(D6.3-4)}$$

where A_{sl} is the area of longitudinal reinforcement in the thickness $t_{ef,i}$, provided at a spacing of s_l. The need for uniform distribution of reinforcement was demonstrated through tests carried out by Chalioris.[17] These showed that failure to distribute longitudinal reinforcement evenly around the perimeter reduces the torsional resistance.

The above reinforcement requirements are additive to those for bending and shear in tension zones. Bonded prestressing steel may be considered to contribute to this reinforcement, providing the stress increase in the prestressing steel is limited to 500 MPa. In zones of longitudinal compression (as produced in a flange due to flexure, for example), the longitudinal reinforcement for torsion can be reduced in proportion to the available compressive force. No guidance is given on what extent of compression zone should be considered, so it is recommended that a depth equal to twice the cover to the torsion links be considered, as was used in BS 5400 Part 4.[9] This could be taken at each face of a flange in the case of box girders.

2-1-1/clause 6.3.2(5)

2-1-1/clause 6.3.2(5) effectively allows some torsion to be carried without designed torsional reinforcement. Under torsion alone, if the torsional shear stress is less than the design tensile stress f_{ctd}, then only minimum reinforcement is required. The limiting torsion from equation (D6.3-1) is therefore:

$$T_{Rd,c} = 2A_k f_{ctd} t_{ef,min} \quad \text{(D6.3-5)}$$

where $t_{ef,min}$ is the minimum wall thickness.

In the presence of shear, 2-1-1/clause 6.3.2(5) provides an interaction, for approximately rectangular sections, to determine whether designed reinforcement is necessary:

$$T_{Ed}/T_{Rd,c} + V_{Ed}/V_{Rd,c} \leq 1.0 \quad \text{2-1-1/(6.31)}$$

where $V_{Rd,c}$ is the shear resistance without designed reinforcement according to 2-1-1/clause 6.2.2. For box girders, the torsional shear and vertical shear can be summed for each wall as described above and no designed reinforcement need be provided if the total shear is less than $V_{Rd,c}$.

Crushing limit for combined shear and torsion
According to *2-2/clause 6.3.2(104)*, two methods of combining shear and torsion (both essentially the same) are to be used. The first applies to solid sections and is a simple linear addition of the torsion and shear usages, again assuming the same value of compressive strut angle θ for both effects:

$$\frac{T_{Ed}}{T_{Rd,max}} + \frac{V_{Ed}}{V_{Rd,max}} \leq 1.0 \quad \text{2-2/(6.29)}$$

2-2/clause 6.3.2(104)

where $T_{Rd,max} = 2\nu\alpha_{cw} f_{cd} A_k t_{ef,i} \sin\theta \cos\theta$ from 2-2/Expression (6.30) and $V_{Rd,max}$ is the limiting shear for web crushing, as discussed in section 6.2 of this guide. Other symbols are also defined in section 6.2. The combination of torsion and shear within the effective walls of a solid section is illustrated in Fig. 6.3-2. Strictly, 2-2/Expression (6.29) should be checked for each effective wall if their thicknesses differ, as $T_{Rd,max}$ then varies depending on which wall is considered, or if there is a transverse shear force in addition to a vertical shear force. When used in this way, 2-2/Expression (6.29) could be applied to the individual walls of box girders.

The interaction of torsion and compressive axial load on crushing resistance is covered by the factor α_{cw} in 2-2/Expression (6.30) (as discussed in section 6.2 of this guide), which should be based on the average compressive stress in the wall. This would typically be applicable to the bottom flange of continuous beams at supports. The value of α_{cc} applicable to f_{cd} in the definition of α_{cw} is recommended here to be taken as 1.0, following the discussion in section 3.1.6 of this guide. The value finally proposed in the National Annex will, however, need to be observed. Although it appears that the torsional resistance could be seriously reduced in hogging zones due to the bottom flange compression, α_{cw} need not be calculated at locations nearer than $0.5d \cot\theta$ to the edge of the support according to 2-2/clause 6.2.3(103). Consequently, maximum compression from bending will not generally be fully combined with the torsional shear stress, although this benefit will be less for haunched beams.

The second method of combination of shear and torsion, applicable to box girders, is to add the shear force obtained from torsion according to equation (D6.3-2) directly to that

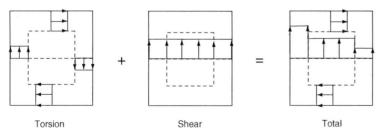

Fig. 6.3-2. Combination of shear stresses within the effective thickness of a solid section

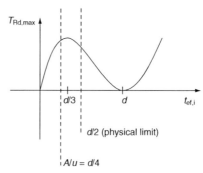

Fig. 6.3-3. Variation of $T_{Rd,max}$ with $t_{ef,i}$ for square section according to 2-2/Expression (6.29)

from shear and check each wall for the resulting shear directly. The shear web crushing check is then carried out, as described in section 6.2 of this guide.

The crushing resistance of solid sections based upon the thin-walled analogy becomes conservative where the effective thickness of the wall is a significant fraction of the available concrete section half-width. This is illustrated by considering a square section of side d. If $t_{ef,i}$ is taken as $A/u = 0.25d$ and $\theta = 45°$, the maximum torque that can be carried acting alone from 2-2/Expression (6.29) is $T_{Rd,max} = \nu\alpha_{cw} f_{cd} \times 0.141 d^3$. If $t_{ef,i}$ is taken as $0.5d$, the maximum physical width available, $T_{Rd,max} = \nu\alpha_{cw} f_{cd} \times 0.125 d^3$. The greatest torque using 2-2/Expression (6.29) occurs if $t_{ef,i}$ is taken as $0.33d$, whereupon $T_{Rd,max} = \nu\alpha_{cw} f_{cd} \times 0.148 d^3$. The torque according to 2-2/Expression (6.29) varies with effective thickness, as shown in Fig. 6.3-3, where $d/2$ represents the maximum physically available wall thickness. It can be seen that utilizing this full available thickness does not give the greatest resistance.

If the theoretical maximum plastic torque is derived by integration of infinitesimally thin-walled sections, as in Fig. 6.3-4, each with shear stress $\tau_{max} = \nu\alpha_{cw} f_{cd}/2$, over the whole cross-section, the maximum torque allowed is $T_{Rd,max} = 2\tau_{max} \int_0^{d/2} (2h)^2 \, dh = \nu\alpha_{cw} f_{cd} \times 0.167 d^3$ and the above anomaly does not arise. For a general truss angle, θ, this resistance becomes $T_{Rd,max} = \nu\alpha_{cw} f_{cd} \sin\theta \cos\theta \times d^3/3$.

Worked example 6.3-2 illustrates this problem, where the minimum value of $t_{ef,i}$ as permitted by 2-1-1/clause 6.3.2(1) actually exceeds the available half width. For crushing resistance, it is suggested here that where $t_{ef,i}$ exceeds the available half width, either:

- the plastic torque resistance above could be used, which more generally for a rectangle of greatest dimension d and least dimension b can be obtained from the sand heap analogy as $T_{Rd,max} = \nu\alpha_{cw} f_{cd} \sin\theta \cos\theta (b^2(d - b/3))/2$, or
- $t_{ef,i}$ could be based on either the actual available half width or A/u.

Open sections

2-1-1/clause 6.3.1(3) allows open flanged sections, such as T-sections, to be divided into a series of component rectangles, each of which is modelled as an equivalent thin-walled section, and the total torsional resistance taken as the sum of the resistances of the individual elements. This sub-division should be done so as to maximize the total torsional stiffness derived for the overall section. The un-cracked torsional stiffness, I_{xx}, of a component

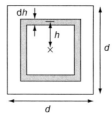

Fig. 6.3-4. Idealization of solid square section into thin-walled sections

Table 6.3-1. Value of k for use in equation (D6.3-6)

b_{max}/b_{min}	k	b_{max}/b_{min}	k	b_{max}/b_{min}	k
1.0	0.141	1.5	0.196	4.0	0.281
1.1	0.153	1.8	0.218	5.0	0.291
1.2	0.165	2.0	0.229	7.5	0.305
1.3	0.177	2.5	0.250	10.0	0.312
1.4	0.187	3.0	0.263	∞	0.333

rectangle may be determined from:

$$I_{xx} = k \cdot b_{max} b_{min}^3 \quad \text{(D6.3-6)}$$

where b_{max} and b_{min} are the length of the longer and shorter sides respectively. k depends on the shape and is determined as follows:

$$k = \frac{1}{3}\left[1 - 0.63 \frac{b_{min}}{b_{max}}\left(1 - \frac{b_{min}^4}{12 \times b_{max}^4}\right)\right] \quad \text{(D6.3-7)}$$

Some particular values are given in Table 6.3-1.

2-1-1/clause 6.3.1(4) requires the share of the total design torsional moment acting on each component rectangle to be based on its percentage contribution to the total uncracked torsional stiffness. Each sub-section may then be designed separately in accordance with *2-1-1/clause 6.3.1(5)*. This procedure is illustrated in Worked example 6.3-2.

Segmental construction

Special care is needed with precast segmental box design where there is no internally bonded prestress or reinforcement crossing the joints. Box girders can carry torsional loading of a type shown in Fig. 6.105 of EN 1992-2 as a combination of pure torsion (shown as 'A' in the figure) and distortion (shown as 'B' in the figure). The pure torsion is carried by a closed flow of shear around the box perimeter (circulatory torsion). If cracks open up at the joints through the thickness of the flanges and the shear keys cannot carry this circulatory torsional shear, the box becomes, in effect, an open section. Open sections are considerably less stiff and strong in torsion than closed sections and both the pure torsional and distortional effects have to be substantially carried by a warping of the box webs, as illustrated in Fig. 6.105 of EN 1992-2. (The circulatory torsional mechanism shown as 'C' in the Fig. is a very weak means of resisting the torsion and hence the warping mechanism prevails.) *2-2/clause 6.3.2/(106)* requires the design for the resulting web effects to be carried out in accordance with Annexes LL and MM of EC2-2. The uses of these Annexes are discussed in the respective sections of this guide.

6.3.3. Warping torsion

The distinction between St Venant and warping torsion is discussed in section 6.3.1 of this guide. The total applied torque has to be carried by either or a combination of these two mechanisms. For closed box beams and solid sections, warping torsion may be ignored at the ultimate limit state according to *2-1-1/clause 6.3.3(1)* since the warping torsion is not itself necessary for equilibrium. It is then, however, necessary for the torsion to be equilibrated entirely by St Venant torsion (referred to as circulatory torsion in EN 1992-2 when applied to hollow sections).

For open sections, warping could again be ignored at the ultimate limit state, but it can be an efficient method of carrying torsion (where the span for transverse bending of individual walls is short due to restraint at intermediate or support diaphragms). In this situation, *2-1-1/clause 6.3.3(2)* essentially recommends that a spaceframe analysis is carried out to determine the distribution of torsion between the two mechanisms. An example idealization for an 'I' beam is shown in Fig. 6.3-5. The design of individual walls would then be carried out for the

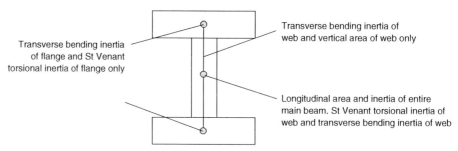

Fig. 6.3-5. Typical idealization of a beam for spaceframe analysis

bending, shear, torsion and axial forces determined therein. Alternatively, texts such as reference 18 could be used to determine the stresses from torsion.

6.3.4. Torsion in slabs (additional sub-section)

The rules on torsion in this section do not apply to slabs. Torsion in slabs is carried by a modification to the moment field caused by the presence of the twisting moments, in addition to the bending moments in two orthogonal directions. In the UK, this has been traditionally achieved in design by using the Wood–Armer equations,[19,20] which combine the twisting moments with flexural moments into 'reinforcement moments' to optimize the amount of reinforcement needed for a particular moment field. This is a special case of more general moment field equations which may be beneficial in assessment, such as those in reference 21. These methods could still be used with EC2 – see the further discussion in section 6.9 of this guide. An alternative method, more directly in compliance with EN 1992, would be to use the rules in 2-2/Annex LL.

Worked example 6.3-1: Box girder bridge

A box girder is idealized as shown in Fig. 6.3-6. The total torque acting on the box is 10 000 kNm. Determine the transverse and longitudinal reinforcement required for torsion in the webs. Reinforcement has $f_{yk} = 500$ MPa and the concrete has a cylinder strength of 40 MPa. Ignore any distortional effects of the applied torque.

The effective thickness of each wall will be taken as the actual box wall thickness, so the enclosed area $A_k = (6000 - 500) \times (3500 - 500/2 - 350/2) = 16.91 \times 10^6$ mm^2.

From equation (D6.3-3) and assuming a truss angle of 45°, the area of reinforcement required from two link legs is:

$$\frac{A_{st}}{s_t} = \frac{T_{Ed}}{2A_k f_{yd} \cot\theta} = \frac{10\,000 \times 10^6}{2 \times 16.91 \times 10^6 \times 500/1.15 \times 1.0} = 0.68 \text{ mm}^2/\text{mm}$$

i.e. 0.34 mm^2/mm for each outer and inner link leg.

Since the truss angle is 45°, an equivalent amount of longitudinal reinforcement will be required according to equation (D6.3-4):

$$\frac{A_{sl}}{s_l} = \frac{T_{Ed}}{2A_k f_{yd}} \cot\theta = \frac{10\,000 \times 10^6}{2 \times 16.91 \times 10^6 \times 500/1.15} \times 1.0 = 0.68 \text{ mm}^2/\text{mm}$$

i.e. 0.34 mm^2/mm for each outer and inner layer.

Crushing resistance must also be checked for the web using 2-2/Expression (6.29):

$$T_{Rd,max} = 2\nu\alpha_{cw}f_{cd}A_k t_{ef,i} \sin\theta \cos\theta$$

$$= 2 \times 0.6[1 - 40/250] \times 1.0 \times 40/1.5 \times 16.91 \times 10^6 \times 500 \times \frac{1}{\sqrt{2}} \times \frac{1}{\sqrt{2}}$$

$$= 113.6 \text{ MNm} \gg 10 \text{ MNm}$$

The concrete section is only stressed to 9% of the crushing load in torsion and may therefore be stressed to 91% of $V_{Rd,max}$ in vertical shear simultaneously if reinforced appropriately.

A similar calculation would be required for the top and bottom flanges.

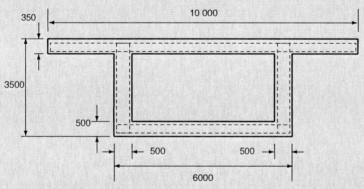

Fig. 6.3-6. Cross-section for Worked example 6.3-1

Worked example 6.3-2: Pretensioned multiple M beam and slab bridge

An M beam is composite with the deck slab as part of a multiple beam and slab bridge. The M beam is idealized as shown in Fig. 6.3-7. The total torque on the beam from a grillage analysis is found to be 40 kNm and the reinforcement required is determined below. Initially, 10 mm bars are assumed for longitudinal and transverse reinforcement for the purposes of determining torsion perimeters. Reinforcement has $f_{yk} = 500$ MPa and the concrete has a cylinder strength of 40 MPa. Cover to reinforcement is 30 mm. The contribution of prestressing to reducing longitudinal reinforcement requirements is ignored.

As a first step, the elastic torsional inertia of each constituent rectangle is evaluated according to equations (D6.3-6) and (D6.3-7) as follows:

Deck slab: $I_{xx} = \frac{1}{2} \times 0.300 \times 1000 \times 160^3 = 6.144 \times 10^8$ mm^4

(Note the initial 'half' term as the deck slab acts in two directions.)

Top flange: $I_{xx} = 0.204 \times 400 \times 250^3 = 1.273 \times 10^9$ mm^4

Web: $I_{xx} = 0.233 \times 335 \times 160^3 = 3.204 \times 10^8$ mm^4

Bottom flange: $I_{xx} = 0.292 \times 950 \times 185^3 = 1.759 \times 10^9$ mm^4

Total torsional inertia is the sum of the above = 3.967×10^9 mm^4. The torque is shared as follows:

Deck slab: $T_{Ed} = 6.144/39.67 \times 40 = 6.2$ kNm

Top flange: $T_{Ed} = 12.73/39.67 \times 40 = 12.8$ kNm

Web: $T_{Ed} = 3.204/39.67 \times 40 = 3.2$ kNm

Bottom flange: $T_{Ed} = 17.59/39.67 \times 40 = 17.7$ kNm

Reinforcement required for torsion
Top slab:
The torque should be combined with the deck slab moments for bending design, as discussed in section 6.3.4 of this guide.

Top flange:
Take the effective thickness as A/u as a first calculation. Consequently:

$$t_{\text{ef},i} = A/u = \frac{400 \times 250}{2 \times (400 + 250)} = 76.9 \text{ mm}$$

However, 2-1-1/clause 6.3.2(1) requires that the effective thickness should not be taken as less than twice the cover to the centre of the longitudinal steel (to avoid the torsion perimeter lying outside the reinforcement), which in this case is 90 mm. This latter figure will therefore be used and A_k calculated accordingly, which is the more correct for reinforcement calculation:

$$A_k = (400 - 90) \times (250 - 90) = 4.96 \times 10^4 \text{ mm}^2$$

From equation (D6.3-3) and assuming a truss angle of 45°, the area of reinforcement required is:

$$\frac{A_{st}}{s_t} = \frac{T_{Ed}}{2 A_k f_{yd} \cot \theta} = \frac{12.8 \times 10^6}{2 \times 4.96 \times 10^4 \times 500/1.15 \times 1.0} = 0.297 \text{ mm}^2/\text{mm}$$

There is no additional effect from shear to add to this reinforcement area, so 8 mm diameter closed links at 100 mm would suffice (0.50 mm²/mm).

Since the truss angle is 45°, an equivalent amount of longitudinal reinforcement will be required according to equation (D6.3-4):

$$\frac{A_{sl}}{s_l} = \frac{T_{Ed}}{2 A_k f_{yd}} \cot \theta = 0.297 \text{ mm}^2/\text{mm}$$

Due to the small size of the section, this could be provided at link corners only, rather than as distributed around the perimeter.

Crushing resistance must also be checked using 2-2/Expression (6.29):

$$T_{Rd,max} = 2 \nu \alpha_c f_{cd} A_k t_{\text{ef},i} \sin \theta \cos \theta$$

$$= 2 \times 0.6(1 - 40/250) \times 1.0 \times 40/1.5 \times 4.96 \times 10^4 \times 90 \times \frac{1}{\sqrt{2}} \times \frac{1}{\sqrt{2}}$$

$$= 60.0 \text{ kNm} > 12.8 \text{ kNm}$$

The concrete section in the top flange is stressed to 21% of the concrete crushing resistance in torsion.

Web:
Try

$$t_{\text{ef},i} = A/u = \frac{335 \times 160}{2 \times (335 + 160)} = 54 \text{ mm}$$

Once again, EC2 requires that the effective thickness should not be taken as less than twice the cover to the centre of the longitudinal steel, which is again 90 mm. Additionally, in this case, this is greater than the actual half-width available of 80 mm.

The reinforcement will be most conservatively and correctly obtained using the greatest $t_{\text{ef},i}$, so 90 mm will be used.

$A_k = (335 - 90) \times (160 - 90) = 1.72 \times 10^4 \text{ mm}^2$ – slightly conservative as transverse reinforcement is outside the 335 mm height so the height used could be increased to the distance between transverse reinforcement.

From equation (D6.3-3) and assuming a truss angle of 45°, the area of reinforcement required is:

$$\frac{A_{st}}{s_t} = \frac{T_{Ed}}{2 A_k f_{yd} \cot \theta} = \frac{3.2 \times 10^6}{2 \times 1.72 \times 10^4 \times 500/1.15 \times 1.0} = 0.215 \text{ mm}^2/\text{mm}$$

This requirement for each vertical leg would have to be combined with the requirement for shear. An equal amount of longitudinal reinforcement will also be required.

Crushing resistance must also be checked, but this time it is not reasonable to take the effective thickness as 90 mm as this exceeds the actual half-width available. Consequently, it is suggested that the actual half-width of 80 mm is used here. As discussed in the text of section 6.3.2, this will still be conservative. The torsional area then becomes $A_k = (335 - 80) \times (160 - 80) = 2.04 \times 10^4 \, \text{mm}^2$ and the resistance is given by:

$$T_{\text{Rd,max}} = 2\nu\alpha_c f_{cd} A_k t_{ef,i} \sin\theta \cos\theta$$

$$= 2 \times 0.6(1 - 40/250) \times 1.0 \times 40/1.5 \times 2.04 \times 10^4 \times 80 \times \frac{1}{\sqrt{2}} \times \frac{1}{\sqrt{2}}$$

$$= 21.9 \, \text{kNm} > 3.2 \, \text{kNm}$$

The concrete section in the web is stressed to 15% of the concrete crushing stress in torsion, so the web may be stressed to 85% of its crushing strength in shear at the same time. If calculated, the maximum torques for $t_{ef,i}$ of 54 mm (A/u) and 90 mm are 21.6 kNm and 20.8 kNm respectively. The reason for the latter figure actually being lower than the one used for a thinner wall is that the calculation method presented in EC2 for thin-walled sections becomes conservative for thick walls where a significant part of the section is used, as discussed in section 6.3.2 of this guide.

Similar calculations would be performed for the bottom flange.

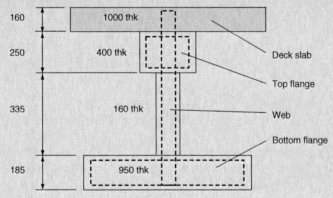

Fig. 6.3-7. Beam for Worked example 6.3-2

6.4. Punching

This section deals with punching shear in slabs including foundations. It is split into the following sub-sections:

- General — Section 6.4.1
- Load distribution and basic control perimeter — Section 6.4.2
- Punching shear calculation — Section 6.4.3
- Punching shear resistance of slabs and bases without shear reinforcement — Section 6.4.4
- Punching shear resistance of slabs and bases with shear reinforcement — Section 6.4.5
- Pile caps (additional sub-section) — Section 6.4.6

There are no modifications to the general rules of EN 1992-1-1 given in EN 1992-2.

6.4.1. General

Punching shear is a local shear failure around concentrated loads on slabs. *2-1-1/clause 6.4.1(1)P* states that the punching shear rules essentially only cover solid slabs and complement the flexural shear rules in 2-2/clause 6.2. *2-1-1/clause 6.4.1(2)P* calls the area on which a concentrated load acts, 'the loaded area', A_{load}. The most common situations where the punching shear rules are relevant in bridge applications are in the design of pile caps, pad footings and deck slabs subjected to local wheel loads.

2-1-1/clause 6.4.1(1)P

2-1-1/clause 6.4.1(2)P

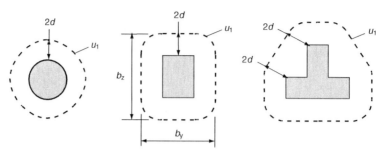

Fig. 6.4-1. Typical basic control perimeters

The remainder of 2-1-1/clause 6.4.1 defines general rules including identifying appropriate verification models and the perimeters on which to check punching shear. These general rules are covered in greater detail in the following sections of this guide.

6.4.2. Load distribution and basic control perimeter

Punching shear failures occur on perimeters within the slab surrounding the loaded area and have the form of truncated cones through the depth of the concrete slab. It should be noted that the specified perimeters in the code are not intended to coincide with actual failure surfaces which occur in tests. The 'basic control perimeter' was chosen so that the allowable design shear stress values could be taken to be the same as for the flexural shear design.

2-1-1/clause 6.4.2(1) defines the basic control perimeter, u_1, which may 'normally' be taken to be at a distance of $2d$ from the loaded area and is constructed to minimize its length. The word 'normally' is used because of the exception identified in *2-1-1/clause 6.4.2(2)*, where the concentrated force is opposed by one or more reactions within the control perimeter. This leads to the necessity to check perimeters inside $2d$, as in 2-1-1/clause 6.4.4(2). Typical basic perimeters are illustrated in Fig. 6.4-1. It is often the case that at the top of a column, in a pile cap or in the vicinity of concentrated wheel loads, the slab reinforcement is different in the two reinforcement directions. The effective depths to the reinforcement will also be different in each direction. A problem then arises as to what area of reinforcement and what effective depth to use in the calculation of the punching shear resistance. The effective depth is also used to determine the critical perimeter geometry. As a simplification, EC2 allows the designer to use the rooted average of the percentage reinforcement areas (see section 6.4.4) and the average of the effective depths in calculation – 2-1-1/Expression (6.32). This could be inappropriate for highly elongated loaded areas where the reinforcement in one direction dominates.

2-1-1/clause 6.4.2(3) defines reductions to the length of the control perimeter for loaded areas close to openings in the slab. More common for bridge applications are situations where loaded areas are close to the edges of a slab or a corner. For such situations, *2-1-1/clause 6.4.2(4)* requires the control perimeters illustrated in Fig. 6.4-2 to be used if the lengths are smaller than those obtained from the basic perimeters defined above. Where the load is within d of a free edge, *2-1-1/clause 6.4.2(5)* requires the edge of the slab to be properly closed off by reinforcement (with U-bars for example).

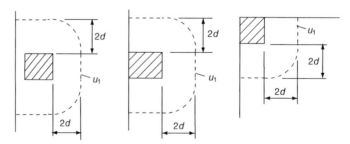

Fig. 6.4-2. Control perimeters for loads close to slab edges or corners

CHAPTER 6. ULTIMATE LIMIT STATES

Where the depth of a slab varies uniformly (as in 2-1-1/Fig. 6.16), the effective depth is determined by considering the depth of concrete through which the shear plane passes. ***2-1-1/clause 6.4.2(6)*** therefore allows the effective depth to be based on that at the face of the loaded area in 2-1-1/Fig. 6.16. Where it is necessary to check perimeters other than the basic control perimeter (for example, because of soil pressure acting on the reverse side of the slab, as noted in the comments to 2-1-1/clause 6.2.4(2)), ***2-1-1/clause 6.4.2(7)*** requires the perimeters to have the same shape as the basic one.

2-1-1/clause 6.4.2(6)

2-1-1/clause 6.4.2(7)

If the section depth varies in steps (as in 2-1-1/Fig. 6.18) rather than uniformly, the assumption for effective depth above may be invalid as failure may occur either in the deeper section or outside it in the shallower section. ***2-1-1/clauses 6.4.2(8) to (11)*** are relevant to this case. They cover the special case of slabs with circular or rectangular column heads, where there is a local deepening of the slab (or local widening of the column) at the top of the column. In such cases, two control perimeters may need to be checked, as indicated in 2-1-1/Fig. 6.18 – one within the local slab thickening (with punching through an increased slab thickness) and one outside it (with punching through the smaller slab thickness).

2-1-1/clauses 6.4.2(8) to (11)

6.4.3. Punching shear calculation

2-1-1/clause 6.4.3(1)P defines the following general shear stresses for use in punching shear design calculations:

2-1-1/clause 6.4.3(1)P

v_{Ed} is the design shear stress for the perimeter considered
$v_{Rd,c}$ is the design value (along the control section considered) of the punching shear resistance of a slab without punching shear reinforcement
$v_{Rd,cs}$ is the design value (along the control section considered) of the punching shear resistance of a slab with punching shear reinforcement
$v_{Rd,max}$ is the design value of the maximum punching shear resistance along the control section considered

2-1-1/clause 6.4.3(2) requires checks of punching to be carried out at the perimeter of the loaded area (against $v_{Rd,max}$ defined in 2-1-1/clause 6.4.5(3)) and on other perimeters, as discussed in sections 6.4.4 and 6.4.5. Tests have shown that punching shear resistance can be significantly reduced in the presence of a coexisting bending moment, M_{Ed}, transmitted to the slab. To allow for the adverse effect of moment, which gives rise to a non-uniform distribution of shear around the control perimeter, ***2-1-1/clause 6.4.3(3)*** gives the design shear stress to be used in punching shear calculations as:

2-1-1/clause 6.4.3(2)

2-1-1/clause 6.4.3(3)

$$v_{Ed} = \beta \frac{V_{Ed}}{u_i d} \qquad \text{2-1-1/(6.38)}$$

where:

V_{Ed} is the ultimate design punching shear force resulting from external loading
u_i is the length of the control perimeter under consideration
d is the average effective depth of the slab which, as discussed above, may be taken as $(d_y + d_z)/2$ (where d_y and d_z are the effective depths of the reinforcement in the y- and z-directions of the control section respectively)
β is a factor used to include for the effects of eccentric loads and bending moments

Essentially, the design shear stress in 2-1-1/Expression (6.38) is:

$$v_{Ed} = \beta \frac{V_{Ed}}{u_i d} = \frac{V_{Ed}}{u_i d} + \Delta v_{M,Ed}$$

where $\Delta v_{M,Ed}$ is the additional shear stress from the bending moment M_{Ed}. This leads to the result that:

$$\beta = 1 + \frac{\Delta v_{M,Ed}}{V_{Ed}} u_i d \qquad \text{(D6.4-1)}$$

If a plastic stress distribution, such as that in 2-1-1/Fig. 6.19, is used to calculate $\Delta v_{M,Ed}$, it is found that:

$$\Delta v_{M,Ed} = \frac{M_{Ed}}{d \int_0^{u_i} |e|\, dl} = \frac{M_{Ed}}{d \cdot W_i}$$

where $W_i = \int_0^{u_i} |e|\, dl$ is defined in 2-1-1/Expression (6.40) as $\int_0^{u_1} |e|\, dl$ for the u_1 perimeter and is a summation of the moments of the control perimeter lengths dl about the axis of bending, with all offsets, e, taken as positive. Substitution for $\Delta v_{M,Ed}$ in equation (D6.4-1) produces:

$$\beta = 1 + \frac{M_{Ed}}{V_{Ed}} \frac{u_i}{W_i} \tag{D6.4-2}$$

The expression for β in EC2 is given in 2-1-1/Expression (6.39). It differs from equation (D6.4-2) in that it is written for perimeter u_1 (although, for column bases, it applies to other perimeters when substitution is made for the relevant u_i and W_i in place of u_1 and W_1 – see Worked example 6.4.1) and an additional factor k has been introduced:

$$\beta = 1 + k \frac{M_{Ed}}{V_{Ed}} \frac{u_1}{W_1} \qquad\qquad 2\text{-}1\text{-}1/(6.39)$$

where:

u_1 is the length of the basic control perimeter

k is a coefficient dependent on the ratio between the column dimensions (c_1 and c_2); its value is a function of the proportions of the unbalanced moment transmitted by uneven shear and by bending and torsion. Suitable values are tabulated in 2-1-1/Table 6.1

M_{Ed} is the ultimate design moment transmitted to the slab and coexistent with the punching shear force

W_1 $= \int_0^{u_1} |e|\, dl$ and corresponds to a plastic distribution of shear stress produced in resisting a moment, as shown in 2-1-1/Fig. 6.19 and discussed above. Its solution is presented for rectangular columns and indirectly for circular columns in 2-1-1/Expressions (6.41) and (6.42) respectively.

The factor β is relevant to all cases where the punching force is transferred to the slab through a monolithic connection transmitting moments. In bridges this would include pier to pilecap connections and integral column to deck slab connections. It would not be relevant to wheel loads on slabs, although the shear from such a wheel load can be unevenly distributed around the control perimeter when the wheel load is adjacent to a support. In this case, the uneven shear distribution should be considered through an additional check of flexural shear in a similar way to that discussed in section 6.4.6 for pile caps.

2-1-1/clauses 6.4.3(4) to (6) provide some simplified alternatives to full calculation of β from 2-1-1/Expression (6.39). They will rarely be applicable to bridge design.

Where a concentrated load is applied close to a support (such as a wheel load adjacent to a diaphragm or a pile load adjacent to a column), the concept of shear enhancement is incompatible with a check of punching shear resistance based on the entire control perimeter. In this case, shear is unevenly distributed around the perimeter and it would be unsafe to allow for a further enhancement by reducing the shear force in accordance with 2-1-1/clause 6.2.2(6). In such circumstances, the control section should be checked without enhancement in accordance with *2-1-1/clause 6.4.3(7)*. No guidance is given on the definition of this control perimeter. It is possible to reduce the control perimeter to avoid the support. Although this will result in a lower punching shear resistance than that based on the basic control perimeter, the reduction may be appropriate since the shear will be concentrated on the part of the perimeter adjacent to the support. An additional check of flexural shear should always be performed in such cases, but this time taking account of shear enhancement for support proximity where appropriate.

This grey area between 'punching' shear and 'flexural' shear leads, in particular, to problems in pile cap design. Some suggestions are made in section 6.4.6 of this guide.

Where the applied punching shear force is in a foundation slab, *2-1-1/clause 6.4.3(8)* allows the force to be reduced by the favourable action of the soil but the provisions of 2-1-1/clause 6.4.2(2), regarding the check of several perimeters, then apply. Similarly, *2-1-1/clause 6.4.3(9)* allows the applied shear force to be reduced by the favourable action of any vertical components of inclined prestressing tendons crossing the control section where relevant.

6.4.4. Punching shear resistance of slabs and bases without shear reinforcement

The expression for punching shear resistance for slabs without shear reinforcement in *2-1-1/clause 6.4.4(1)* is similar in form to that for the flexural shear design of beams in 2-1-1/clause 6.2.2:

$$v_{Rd,c} = C_{Rd,c}k(100\rho_l f_{ck})^{1/3} + k_1\sigma_{cp} \geq (v_{min} + 0.10\sigma_{cp}) \qquad \text{2-1-1/(6.47)}$$

where:

$k = 1 + \sqrt{(200/d)}$ (with d in millimetres)

$\rho_l = \sqrt{\rho_{ly}\rho_{lz}} \leq 0.02$

ρ_{ly}, ρ_{lz} relate to the bonded tension steel in the y- and z-directions respectively. The values should be calculated as mean values taking into account a slab width equal to the loaded area or column width plus $3d$ each side

d should be taken as the average effective depth obtained in each orthogonal direction from 2-1-1/Expression (6.32)

$\sigma_{cp} = (\sigma_{cy} + \sigma_{cz})/2$ (in MPa, positive for compression)

σ_{cy}, σ_{cz} are the normal concrete stresses from longitudinal forces in the critical section in the y- and z-directions respectively (in MPa, compression taken as positive)

Other terms have the same definitions as in 2-1-1/clause 6.2.2.

The values of $C_{Rd,c}$, v_{min} and k_1 may be given in the National Annex. EC2 recommends values of $0.18/\gamma_c$ and 0.10 (compared to 0.15 for flexural shear) for these respectively for punching shear. The value of v_{min} is recommended to be the same as for flexural shear.

In general, the punching shear resistance for a slab should be assessed for the basic control section discussed in section 6.4.2 using 2-1-1/Expression (6.47) and at the perimeter of the loaded area against $v_{Rd,max}$ defined in 2-1-1/clause 6.4.5(3). *2-1-1/clause 6.4.4(2)*, however, requires that the punching shear resistance of column bases be verified at control perimeters within the $2d$ periphery of the column and the lowest value taken. This is because the angle of the punching cone may be steeper in this situation due to the favourable reaction from the soil. A check of punching on the basic perimeter ignoring the relieving force from the soil would be conservative. On any given perimeter, and in the absence of transmitted moment, v_{Ed} is as follows:

$$v_{Ed} = V_{Ed,red}/ud \qquad \text{2-1-1/(6.49)}$$

with $V_{Ed,red} = V_{Ed} - \Delta V_{Ed}$ \qquad 2-1-1/(6.48)

where V_{Ed} is the column load and ΔV_{Ed} is the net upward force within the control perimeter considered (i.e. upward pressure from the soil calculated excluding the self-weight of the base).

For steeper shear planes, an enhancement of the basic resistance to 2-1-1/Expression (6.47) is applicable, whereupon the resistance becomes:

$$v_{Rd} = C_{Rd,c}k(100\rho_l f_{ck})^{1/3} \times \frac{2d}{a} \geq v_{min} \times \frac{2d}{a} \qquad \text{2-1-1/(6.50)}$$

where a is the distance from the periphery of the column to the control perimeter considered. This formulation for shear enhancement, where the *resistance* is enhanced, is at odds with that for flexural shear, where the shear itself is reduced. This is discussed in section 6.2.2.1 of this guide.

For most bases, column axial load will be accompanied by some bending moment (due to moment fixity at deck level or horizontal forces applied to the column top through bearings). In such cases, an increase in the design shear stress to account for uneven shear distribution around the control perimeter is necessary. 2-1-1/Expression (6.51) is provided to do this and it is equivalent to 2-1-1/Expression (6.38) for cases where there is no reacting soil pressure:

$$v_{Ed} = \frac{V_{Ed,red}}{ud}\left[1 + k\frac{M_{Ed}}{V_{Ed,red}}\frac{u}{W}\right] \qquad \text{2-1-1/(6.51)}$$

2-1-1/Expression (6.51) contains an equivalent term to the term

$$\beta = 1 + k\frac{M_{Ed}}{V_{Ed}}\frac{u_1}{W_1}$$

in 2-1-1/Expression (6.39), but with V_{Ed} replaced by $V_{Ed,red}$. It would be logical to allow M_{Ed} to similarly be replaced with a reduced value allowing for the soil pressure, but this is conservatively (and probably unintentionally) not done in 2-1-1/Expression (6.51). u_1 and W_1 are also replaced by u and W which relate to the actual perimeter being checked.

Worked example 6.4-1: Reinforced concrete pad footing
Consider the design of the pad foundation for a bridge pier illustrated in Fig. 6.4-3.

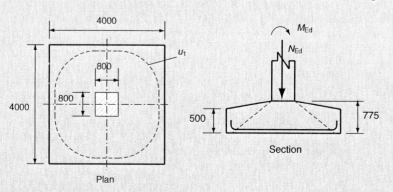

Fig. 6.4-3. Reinforced concrete pad footing

The punching shear resistance is checked assuming an ultimate applied axial force, $N_{Ed} = 3000$ kN and a bending moment M_{Ed} of 1200 kNm. $f_{ck} = 35$ MPa, $f_{yk} = 500$ MPa, $\gamma_c = 1.5$ and $\gamma_s = 1.15$.

The average base pressure from the axial force excluding pad self-weight is $3000/(4 \times 4) = 187.5$ kN/m^2. The peak of the linear varying pressure from the moment is $1200/(1/6 \times 4^3) = 112.5$ kN/m^2. The maximum total pressure at the edge of the pad is 300 kN/m^2.

The maximum bending moment in the base is $187.5 \times 2.0 \times 1.0 + 112.5/2 \times 2.0 \times 1.333 = 525$ kNm/m width (conservatively calculated at the centre of the column without rounding).

Assuming 50 mm cover to the outer reinforcement, a bending analysis of the section gives the following results using 25ϕ bars at 200 mm centres in both directions:

Layer 1: $d_y = 775 - 50 - 25/2 = 712.5$ mm, $A_{sy} = 2454$ mm^2/m, $M_{y,Rd} = 730.8$ kNm/m

Layer 2: $d_z = 712.5 - 25 = 687.5$ mm, $A_{sz} = 2454$ mm^2/m, $M_{z,Rd} = 704.1$ kNm/m

The effective depth for punching calculations is $d = (712.5 + 687.5)/2 = 700$ mm.

The basic control perimeter at $2d$ has perimeter length, $u_1 = 4\pi \times 700 + 2 \times (800 + 800) = 11\,996$ mm and the area within u_1 is $(4\pi \times 700^2 + 4 \times 700 \times 2 \times 800 + 800^2) \times 10^{-6} = 11.28$ m^2. (Note that d is taken as the depth at the face of the column despite the reducing depth, since the shear plane must pass through the full depth.)

Net applied force is given by 2-1-1/Expression (6.48): $V_{Ed,red} = V_{Ed} - \Delta V_{Ed}$

Relieving pressure from soil for basic perimeter: $\Delta V_{Ed} = 187.5 \times 11.28 = 2115\,\text{kN}$

Thus $V_{Ed,red} = 3000 - 2115 = 885\,\text{kN}$. For the 4000 mm wide footing:

$$A_{sy} = A_{sz} = \pi \times 12.5^2 \times 4000/200 = 9817.5\,\text{mm}^2$$

thus

$$\rho_{ly} = 9817.5/(4000 \times 712.5) = 0.34\%$$

and

$$\rho_{lz} = 9817.5/(4000 \times 687.5) = 0.36\%$$

and in applying 2-1-1/Expression (6.47):

$$\rho_l = \sqrt{0.34 \times 0.36} = 0.35\% \quad k = 1 + \sqrt{(200/700)} = 1.534$$

$$v_{min} = 0.035 \times 1.534^{3/2} \times 35^{1/2} = 0.393\,\text{MPa}$$

Assuming $C_{Rd,c} = 0.18/\gamma_c = 0.18/1.5 = 0.12$ as recommended, 2-1-1/Expression (6.47) becomes:

$$v_{Rd,c} = 0.12 \times 1.534 \times (100 \times 0.0035 \times 35)^{1/3} = 0.424\,\text{MPa}$$

($>v_{min}$ so is used as the resistance).

The effect of the column moment on the shear distribution around the control perimeter now needs to be considered.

From 2-1-1/Expression (6.41) for a square column:

$$W_1 = \frac{c_1^2}{2} + c_1 c_2 + 4 c_2 d + 16 d^2 + 2\pi d c_1$$

$$= \frac{800^2}{2} + 800^2 + 4 \times 800 \times 700 + 16 \times 700^2 + 2\pi \times 700 \times 800 = 14.56 \times 10^6\,\text{mm}^2$$

From 2-1-1/Expression (6.51) with $k = 0.6$ from 2-1-1/Table 6.1:

$$v_{Ed} = \frac{V_{Ed,red}}{u_1 d}\left[1 + k\frac{M_{Ed}}{V_{Ed,red}}\frac{u_1}{W_1}\right] = \frac{885 \times 10^3}{11\,996 \times 700} \times \left[1 + 0.6\frac{1200 \times 10^6}{885 \times 10^3}\frac{11\,996}{14.56 \times 10^6}\right]$$

$$= 0.105 \times 1.67 = \mathbf{0.175\,MPa} < v_{Rd,c}$$

therefore okay. It is noted that the use of the full moment M_{Ed} in 2-1-1/Expression (6.51) is conservative, as discussed in the main text.

In accordance with 2-1-1/clause 6.4.4(2), the base should also be checked at perimeters within $2d$. The above check is therefore repeated for a perimeter at d. As above, $d = 700\,\text{mm}$; therefore the perimeter length, $u_2 = 2\pi \times 700 + 2 \times (800 + 800) = 7598\,\text{mm}$, and the area within this perimeter is $(\pi \times 700^2 + 4 \times 700 \times 800 + 800^2) \times 10^{-6} = 4.42\,\text{m}^2$. (The subscript '2' is used to denote this perimeter.)

The net applied force is given by 2-1-1/Expression (6.48): $V_{Ed,red} = V_{Ed} - \Delta V_{Ed}$

Relieving pressure from soil for basic perimeter: $\Delta V_{Ed} = 187.5 \times 4.42 = 829\,\text{kN}$

thus $V_{Ed,red} = 3000 - 829 = 2171\,\text{kN}$

The concrete resistance within the $2d$ perimeter is given by 2-1-1/Expression (6.50):

$$v_{Rd,c} = C_{Rd,c}k(100\rho_l f_{ck})^{1/3} \times 2d/a$$

which is equivalent to the basic resistance calculated above multiplied by $2d/a$. For this case, $a = d$, therefore the concrete resistance becomes: $v_{Rd,c} = 0.424 \times 2 = 0.848\,\text{MPa}$.

The effect of the column moment on the shear distribution around the control perimeter is first considered. 2-1-1/Expression (6.41) cannot be used for the perimeter at d unless d is replaced by $d/2$ in the expression as done here:

$$W_2 = \frac{800^2}{2} + 800^2 + 4 \times 800 \times \frac{700}{2} + 16 \times \left(\frac{700}{2}\right)^2 + 2\pi \times \frac{700}{2} \times 800$$

$$= 5.799 \times 10^6 \, \text{mm}^2$$

From 2-1-1/Expression (6.51) with $k = 0.6$ from 2-1-1/Table 6.1:

$$v_{Ed} = \frac{V_{Ed,red}}{u_2 d}\left[1 + k\frac{M_{Ed}}{V_{Ed,red}}\frac{u_2}{W_2}\right] = \frac{2171 \times 10^3}{7598 \times 700} \times \left[1 + 0.6\frac{1200 \times 10^6}{2171 \times 10^3}\frac{7598}{5.799 \times 10^6}\right]$$

$$= 0.408 \times 1.43 = \mathbf{0.586 \, MPa} < 0.848 \, \text{MPa so okay}$$

Finally for punching checks, punching shear at the face of the column is checked against $v_{Rd,max}$ from 2-1-1/Expression (6.53):

$$v_{Rd,max} = 0.5\nu f_{cd} = 0.5 \times \left(0.6 \times \left(1 - \frac{35}{250}\right)\right) \times \frac{35}{1.5} = 6.02 \, \text{MPa}$$

$$u_0 = 4 \times 800 = 3200 \, \text{mm}$$

$$W_0 = \frac{800^2}{2} + 800^2 = 9.60 \times 10^5 \, \text{mm}^2 \text{ (setting } d = 0 \text{ in 2-1-1/Expression (6.41))}$$

From 2-1-1/Expressions (6.38) and (6.39) with $k = 0.6$ from 2-1-1/Table 6.1:

$$v_{Ed} = \frac{V_{Ed}}{u_0 d}\left[1 + k\frac{M_{Ed}}{V_{Ed}}\frac{u_0}{W_0}\right] = \frac{3000 \times 10^3}{3200 \times 700} \times \left[1 + 0.6\frac{1200 \times 10^6}{3000 \times 10^3}\frac{3200}{9.60 \times 10^5}\right]$$

$$= \mathbf{2.41 \, MPa} < v_{Rd,max} \text{ therefore okay}$$

The foundation should also be checked for flexural shear. From 2-1-1/clause 6.2.1(8), planes nearer than d need not be checked where there is approximately uniform load, so the resistance will be checked at a section at a distance d from the column face.

Base pressure at d from column face $= 187.5 + (400 + 687.5)/2000 \times 112.5 = 249 \, \text{kN}/\text{m}^2$. It is assumed that the moment acts in a direction so as to put the bottom reinforcement with smallest d value into tension.

Conservatively ignoring any support enhancement:

Applied shear force at (minimum) d away from support

$$= \frac{(300 + 249)}{2} \times 4000 \times (4000/2 - 800/2 - 687.5) \times 10^{-6} = 1002 \, \text{kN}$$

Taking $d = 687.5 \, \text{mm}$ for the plane under consideration and $A_{sz} = 9817.5 \, \text{mm}^2$ as before gives $\rho_{lz} = 9817.5/(4000 \times 687.5) = 0.357\%$.

From equation (D6.2-1): $k = 1 + \sqrt{(200/d)} = 1 + \sqrt{(200/687.5)} = 1.539$

From 2-2/Expression (6.2.a): $V_{Rd,c} = C_{Rd,c}k(100\rho_l f_{ck})^{1/3}b_w d$

so

$$V_{Rd,c} = 0.12 \times 1.539 \times (100 \times 0.00357 \times 35)^{1/3} \times 4000 \times 687.5 \times 10^{-3}$$

$$= \mathbf{1179 \, kN} > 1002 \, \text{kN applied therefore okay}$$

This is conservative as enhancement was not considered to avoid the need to check planes further away than d. Thus **no shear reinforcement is required**.

6.4.5. Punching shear resistance of slabs and bases with shear reinforcement

Where v_{Ed} exceeds the value of $v_{Rd,c}$ for the perimeter considered, usually the basic perimeter u_1, punching shear reinforcement is required in accordance with 2-1-1/clause 6.4.5 and three zones are required to be checked:

- the zone immediately adjacent to the loaded area (against the shear crushing limit);
- the zone in which the shear reinforcement is placed;
- the zone outside the shear reinforcement. 2-1-1/clause 6.4.5(4) includes a definition for an outer perimeter where the concrete resistance alone is sufficient for the punching shear and shear reinforcement has to be provided within this zone as discussed below.

Where shear reinforcement is required, the following equation is defined in **2-1-1/clause 6.4.5(1)** to calculate the punching shear resistance of slabs or column bases:

$$v_{Rd,cs} = 0.75 v_{Rd,c} + 1.5\left(\frac{d}{s_r}\right) A_{sw} f_{ywd,ef} \left(\frac{1}{u_1 d}\right) \sin\alpha \qquad \text{2-1-1/(6.52)}$$

2-1-1/clause 6.4.5(1)

where:

A_{sw} is the area of one perimeter of shear reinforcement around the column or loaded area according to 2-1-1/Fig. 6.22

s_r is the radial spacing of perimeters of shear reinforcement

$f_{ywd,ef}$ is the effective design strength of the punching shear reinforcement allowing for anchorage efficiency $= 250 + 0.25d \leq f_{ywd}$ (in MPa) with d taken, as before, as the average effective depth (in millimetres)

α is the angle between the shear reinforcement and the plane of the slab

2-1-1/Expression (6.52) differs from the formulae for flexural shear in that a concrete term is added to a shear reinforcement term. However, 2-1-1/Expression (6.52) does not fully combine the concrete resistance and the link resistance in the $2d$ perimeter. The reasons for this are entirely test-based. The 0.75 factor on the concrete term represents a reduced concrete contribution as one might expect when reinforcement is yielding with the associated implied concrete cracking and deformation. The use of $1.5d$ rather than $2d$ is also needed for adequate calibration with test results and reflects observations that shear reinforcement at the ends of shear planes is less effective. It doesn't imply a steeper failure plane. The reduced shear reinforcement strength, $f_{ywd,ef}$, is a further anchorage efficiency factor affecting shallower slabs.

The above expression has been presented assuming a constant area of shear reinforcement on each perimeter moving away from the loaded area – Fig. 6.4-4 refers. In bridges, reinforcement is not usually placed like this, but rather on an orthogonal rectangular grid, coinciding with horizontal reinforcement arrangements. This is a necessity where moving loads are to be catered for. Such arrangements inevitably lead to the area of reinforcement increasing on successive perimeters away from the loaded area. One solution is to apply 2-1-1/Expression (6.52) by considering only the reinforcement bars that are located as in Fig. 6.4-4(b) and ignoring other reinforcement provided between the arms of the cruciform shape (which would increase the reinforcement area on successive perimeters moving away from the loaded area). There would be no need to reduce the effective concrete perimeter as indicated in Fig. 6.4-4(b) if additional reinforcement were so placed to reduce the circumferential spacing. An alternative, less conservative, approach is given later.

The control perimeter at which shear reinforcement is not required (u_{out} or $u_{out,ef}$ in Fig. 6.4-4) is defined in **2-1-1/clause 6.4.5(4)** as the perimeter where the concrete resistance alone is sufficient to resist the applied shear stress:

2-1-1/clause 6.4.5(4)

$$u_{out,ef} = \beta \frac{V_{Ed}}{v_{Rd,c} d} \qquad \text{2-1-1/(6.54)}$$

with $v_{Rd,c}$ from 2-1-1/Expression (6.47).

The outermost perimeter of shear reinforcement should be placed at a distance not greater than $kd = 1.5d$ (which may be varied in the National Annex) within this outer perimeter (as illustrated in Fig. 6.4-4) to ensure an inclined failure plane cannot develop within this

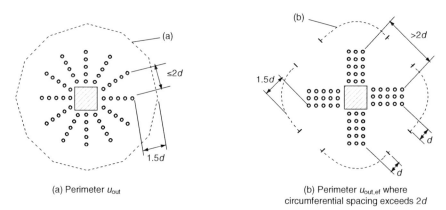

(a) Perimeter u_{out}

(b) Perimeter $u_{out,ef}$ where circumferential spacing exceeds $2d$

Fig. 6.4-4. Control perimeters adjacent to loaded area with shear reinforcement

perimeter over a radial distance of $2d$ without passing through a set of shear reinforcement legs. If the perimeter u_{out} or $u_{out,ef}$ is less than $3.0d$ from the face of the loaded area, shear reinforcement should, however, still be provided out to at least a perimeter $1.5d$ from the face of the loaded area such that the required resistance according to 2-1-1/Expression (6.52) can be achieved. According to 2-1-1/Fig. 9.10, the innermost perimeter of shear reinforcement should not be placed nearer than $0.3d$ to the face of the loaded area. This is similar to the reason for putting reinforcement in the middle $0.75a_v$ of a shear span, as discussed under clause 6.2.3(8). The radial spacing of reinforcement should also not exceed $0.75d$ in accordance with 2-1-1/clause 9.4.3(1).

Because of the difficulties with matching available reinforcement to perimeters for real reinforcement layouts and control perimeter shapes, an alternative approach is proposed here which allows successive control perimeters to be checked if necessary. In general, shear failure is deemed to occur over a radial distance of $2d$. Consequently, to enhance resistance, shear reinforcement of area $\sum A_{sw}$ should be placed within an area enclosed between the control perimeter chosen and one $2d$ inside it. To correspond to the $1.5d$ in 2-1-1/Expression (6.52), it is desirable to consider only the reinforcement within a radial band of $1.5d$. To comply with the need to consider only reinforcement further than $0.3d$ from the loaded perimeter, only reinforcement further than $0.3d$ from the inner perimeter should be considered. Consequently, only reinforcement further than $0.2d$ inside the control perimeter should be included. These two limits are consistent with the fact that reinforcement at each end of a failure plane is unlikely to be fully effective. This reinforcement zone is shown in Fig. 6.4-5.

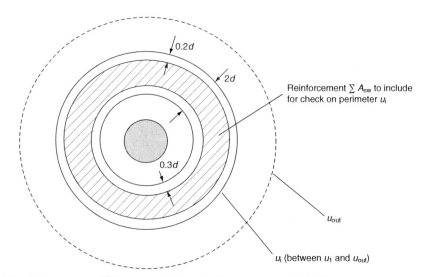

Fig. 6.4-5. Reinforcement $\sum A_{sw}$ to include in check to equation (D6.4-3)

If the above method is followed, successive perimeters, u_i, between the basic control perimeter at $2d$ and the perimeter u_{out} are checked to ensure that the reinforcement in each $2d$ zone above satisfies:

$$\sum A_{sw} = \frac{(v_{Ed} - 0.75v_{Rd,c})u_i d}{f_{ywd,ef} \sin \alpha} \qquad \text{(D6.4-3)}$$

It will be noted that if the above is applied to the control perimeter at $2d$, the same total reinforcement requirement as in 2-1-1/Expression (6.52) is produced. If it is applied at the perimeter u_{out}, some reinforcement requirement will still be predicted because of the 0.75 factor on $v_{Rd,c}$ in equation (D6.4-3). This is unfortunate, but as long as reinforcement is detailed so that it is stopped no further than $1.5d$ inside the perimeter u_{out} as required by 2-1-1/clause 6.4.5(4), some reinforcement will be available for this check.

Maximum punching shear stress
2-1-1/clause 6.4.5(3) requires the shear stress at any section to be less than $v_{Rd,max}$. This check is equally applicable to sections with or without shear reinforcement, but is only likely to be critical in slabs with shear reinforcement. Clearly, the most critical section to check is the column perimeter or the perimeter of the loaded area (and 2-1-1/clause 6.4.5(3) includes specific u_0 values for the special cases close to slab edges or corners). $v_{Rd,max}$ may be given in the National Annex but the recommended value in 2-1-1/clause 6.4.5(3) is $v_{Rd,max} = 0.5\nu f_{cd}$; the same as for flexural shear design in 2-1-1/clause 6.2.2(6).

2-1-1/clause 6.4.5(3)

6.4.6. Pile caps (additional sub-section)
EN 1992-2 provides no specific guidance for checking punching shear in pile caps. The general rules can be applied to pile caps where the edges of the piles are located further than $2d$ from the pier face, but this situation is rare in practice. In other cases, a lot of interpretation is required. The revised thinking on shear enhancement in the final drafting of EC2, discussed in section 6.2.2.1, has made matters more complicated. Some suggestions are made below and in Worked example 6.4-2.

Where pile edges are closer to the piers than $2d$, some of the shear force will be transmitted directly into the support by way of a strutting action. The basic punching perimeter cannot be constructed without encompassing a part of the support – perimeter (a) in Fig. 6.4-6. 2-1-1/clause 6.4.2(2) requires reduced perimeters to be checked, such that the support reaction is excluded, suggesting a perimeter like that of type (b) in Fig. 6.4-6. 2-1-1/clause 6.4.3(7) does not allow any enhancement to be taken on such perimeters. On the one hand, it would not be reasonable to enhance resistance for support proximity on the whole of such a punching perimeter without making reductions to the effectiveness of other parts of the perimeter, as the shear will be unevenly distributed around that perimeter. On the other hand, some degree of enhancement must take place because the failure surface is, at least locally, steepened by the presence of the support and some load can strut directly into the

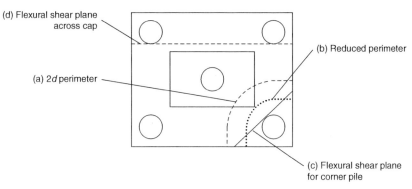

Fig. 6.4-6. Corner pile within $2d$ of a column base

support. In the limit, if the pile is very close to the pier face, the load will transfer straight between support and pile in compression and the very short perimeter of type (b) may underestimate resistance if enhancement is not permitted.

The above problems stem from the subjective distinction between 'punching' and 'flexural' shear. It is arguable that the above situation is more of a flexural shear problem. The following procedure is therefore proposed:

- First, a failure plane of type (d) in Fig. 6.4-6, extending the full width of the pile cap, should be checked for flexural shear. A method considering shear enhancement of the concrete resistance over the sections where reinforcement crosses the pile head as in Worked example 6.4-2 is recommended.
- Second, axial loads from corner piles can be checked for punching at the pile face to check the maximum shear stress and against the minimum resistance from:
    (i) punching at a $2d$ perimeter (without support enhancement) ignoring the presence of the support, and;
    (ii) a diagonal flexural shear plane at the edge of the pile of type (c) in Fig. 6.4-6, which was the approach used in BS 5400 Part 4.[9] A method considering shear enhancement of the concrete resistance over the pile head as in Worked example 6.4-2 is recommended.

Check (i) will generally be less critical than check (ii).

For all of the checks above, where support proximity is included to enhance part of the shear resistance, it is suggested here that, in keeping with current UK practice, a_v is taken to be the distance between the face of a column or wall and the nearer edge of the piles plus 20% of the pile diameter.

This approach is pursued in Worked example 6.4-2.

Worked example 6.4-2: Reinforced concrete pile cap
Consider the design of the pile cap for a bridge pier illustrated in Fig. 6.4-7. For the purpose of this example, the pile cap depth has been deliberately kept shallow to ensure shear reinforcement is required.

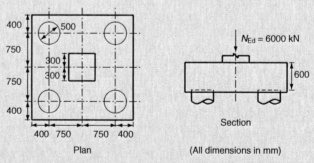

Fig. 6.4-7. Reinforced concrete pile cap

Check the punching shear design assuming an ultimate applied load, $N_{Ed} = 6000\,kN$, $f_{ck} = 35\,MPa$, $f_{yk} = 500\,MPa$, $\gamma_c = 1.5$ and $\gamma_s = 1.15$. For simplicity in this example, no column moment or lateral force is considered. The effect of moment on reducing punching shear resistance is illustrated in Worked example 6.4-1.

Assuming 50 mm cover to the outer reinforcement from the top of the piles, a bending analysis of the section gives the following results using 25ϕ bars at 150 mm centres in both directions ($A_{sy} = A_{sz} = 3272\,mm^2/m$):

Layer 1: $d_y = 600 - 50 - 25/2 = 537.5\,mm$, $z_y = 500.6\,mm$, $M_{y,RD} = 712.3\,kNm/m$

Layer 2: $d_z = 537.5 - 25 = 512.5\,mm$, $z_z = 475.6\,mm$, $M_{z,Rd} = 676.7\,kNm/m$

This is sufficient to resist the applied bending moment at the face of the pier, taken as the maximum of:

(i) $2 \times \dfrac{6000}{4} \times \dfrac{(0.75 - 0.30)}{2.30} = 587\,\text{kNm/m}$

and

(ii) $0.65 \times 0.75 \times 2 \times \dfrac{6000}{4} \times \dfrac{1}{2.30} = 636\,\text{kNm/m}$ (from 2-1-1/clause 5.3.2.2(3))

i.e. **636 kNm/m**

The following four checks are made for shear:

(1) The flexural shear plane across the face of the support is checked (as illustrated in Fig. 6.4-8 below):

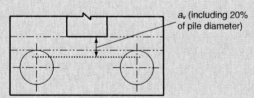

Fig. 6.4-8. Flexural shear plane at face of support

Load per pile $= 6000/4 = 1500\,\text{kN}$

Pile cap width $= (0.4 + 0.75) \times 2 = 2.3\,\text{m}$

Conservatively, consider only the plane with the smallest effective depth to the reinforcement:

$d_z = 512.5\,\text{mm}$, $A_{sz} = 3272\,\text{mm}^2/\text{m}$

therefore

$$\rho_{lz} = \dfrac{A_{sz}}{bd_z} = \dfrac{3272}{1000 \times 512.5} = 0.00638$$

From equation (D6.2-1), $k = 1 + \sqrt{200/d} = 1 + \sqrt{200/512.5} = 1.625$. $\gamma_c = 1.5$ and taking $C_{Rd,c} = 0.18/\gamma_c$ gives $C_{Rd,c} = 0.18/1.5 = 0.12$. From 2-2/Expression (6.2.a):

$$V_{Rd,c} = C_{Rd,c} k (100 \rho_l f_{ck})^{1/3} bd$$

thus $V_{Rd,c}$ per metre width of slab $= 0.12 \times 1.625 \times (100 \times 0.00638 \times 35)^{1/3} \times 1000 \times 512.5 \times 10^{-3} = 281.4\,\text{kN/m}$.

Shear span, $a_v = 750 - 250 - 300 + 0.20 \times 500 = 300\,\text{mm}$ (including 20% of pile diameter); therefore enhancement factor, $1/\beta = 2d/a_v = 2 \times 512.5/300 = 3.417$.

Previous UK practice has been to enhance the shear resistance of the concrete, but only over the width of the plane where the reinforcement passes over the pile heads. If this approach is followed, then the enhanced resistance, $V^*_{Rd,c}$, is given by:

$$V^*_{Rd,c} = 281.4 \times (3.417 \times 2 \times 0.5 + 1.0 \times (2.3 - 2 \times 0.5)) = 961.4 + 365.8 = 1327\,\text{kN}$$

$V^*_{Rd,c} < V_{Ed}\ (= 3000\,\text{kN})$ therefore shear reinforcement is required

The current wording of 2-1-1/clause 6.2.2(6), however, requires the shear force to be reduced by $\beta = 1/3.417 = 0.293$, rather than an enhancement to be made to the concrete resistance. Consequently, $V_{Ed} = 0.293 \times 3000 = 878\,\text{kN}$ compared to a resistance of $V_{Rd,c} = 281.4 \times 2.3 = 647\,\text{kN}$. $V_{Rd,c} < V_{Ed}\ (= 878\,\text{kN})$, therefore shear reinforcement is still required.

Both checks lead to the requirement for shear reinforcement, but the latter is less conservative as it effectively implies enhancement of the concrete resistance across the full width of the shear plane.

As the shear span is less than '$2d$', the use of 2-2/Expression (6.8) is not appropriate to design the reinforcement. The current wording of 2-1-1/clause 6.2.3(8) requires the shear force to be reduced by $\beta = 1/3.417 = 0.293$, rather than an enhancement to be made to the concrete resistance. This method, however, effectively enhances the concrete resistance across the whole shear plane. Consequently, $V_{Ed} = 0.293 \times 3000 = 878$ kN as above. From 2-1-1/Expression (6.19):

$$A_{sw} \geq \frac{V_{Ed}}{f_{ywd} \times \sin \alpha} = \frac{878 \times 10^3}{434.8 \times \sin 90} = 2019 \text{ mm}^2$$

This reinforcement must be provided within the central $0.75a_v$ of the shear span.

Previous UK practice for $a_v \leq d$, and the requirement of earlier drafts of EC2, was to add the shear resistances of the concrete and shear reinforcement together. In doing this, only the shear reinforcement within the central $0.75a_v$ should again be included and it is recommended here that the concrete resistance should only be enhanced over the width of the plane where the reinforcement passes over the pile heads. The enhanced shear resistance, V_{Rd}^*, can then be calculated from:

$$V_{Rd}^* = V_{Rd,c}'^* + A_{sw} f_{ywd} \sin \alpha \tag{D6.4-4}$$

Re-arranging this for reinforcement design gives:

$$A_{sw} \geq \frac{V_{Ed} - V_{Rd,c}'^*}{f_{ywd} \sin \alpha}$$

where A_{sw} must be placed within the middle $0.75a_v$ (i.e. $0.75 \times 300 = 225$ mm).

$V_{Rd,c}'^*$ is here taken as the contribution from the enhanced parts of the section only for consistency when using this formula, as it was only permissible to add the concrete resistance to the reinforcement resistance in enhanced zones in earlier drafts of EC2.

Thus

$$A_{sw} \geq \frac{3000 \times 10^3 - 961.4 \times 10^3}{434.8 \times \sin 90} = 4689 \text{ mm}^2$$

Using link legs in 2 rows at 150 mm centres requires an area per leg of $4689/(15 \times 2) = 156$ mm², i.e. 16ϕ links. The results of this method are taken forward as it is more similar to practice to BS 5400 Part 4.[9] This is a considerably greater amount of shear reinforcement than required by the above method using 2-1-1/clause 6.2.3(8). If the whole of the concrete shear plane had been enhanced, however, very similar reinforcement requirements, by coincidence, would have been produced to the method of 2-1-1/clause 6.2.3(8) (slightly less).

The maximum shear force also needs to be checked. This is independent of the way in which shear enhancement is catered for:

From 2-2/Expression (6.9):

$$V_{Rd,max} = \frac{\alpha_{cw} b_w z \nu_1 f_{cd}}{(\cot \theta + \tan \theta)}$$

$$V_{Rd,max} = \frac{1.0 \times 2300 \times 475.6 \times 0.6 \times (1 - 35/250) \times 1.0 \times 35/1.5}{(\cot 45 + \tan 45)} = 6585 \text{ kN}$$

> 3000 kN as required therefore okay

Thus **adopting 16ϕ bars for links on a 2(min) by 16 grid** (150 mm centres in each direction is adequate).

(2) Face shear at corner pile and column is checked (using u_0 perimeters):
From 2-1-1/Expression (6.53), taking $\beta = 1.0$ since the pile head moment is here assumed to be small compared to the axial load:

$$v_{Ed} = \frac{V_{Ed}}{u_0 d} \quad \text{where} \quad d = \frac{537.5 + 512.5}{2} = 525\,\text{mm}$$

$u_0 = \pi \times 500/4 + 2 \times 400 = 1193\,\text{mm}$ based on a perimeter passing around one-quarter of the pile perimeter and then extending out to the free edges of the pile cap.

Thus

$$v_{Ed} = \frac{1500 \times 10^3}{525 \times 1193} = 2.39\,\text{MPa}$$

$$v_{Rd,max} = 0.5 \nu f_{cd} \quad \text{where} \quad f_{cd} = \frac{\alpha_{cc} f_{ck}}{\gamma_c} = \frac{1.0 \times 35}{1.5} = 23.3\,\text{MPa}$$

and

$$\nu = 0.6\left(1 - \frac{f_{ck}}{250}\right) = 0.6 \times \left(1 - \frac{35}{250}\right) = 0.516$$

Thus $v_{Rd,max} = 0.5 \times 0.516 \times 23.3 = 6.02\,\text{MPa} > 2.39\,\text{MPa}$ as required.

In this example, the size of the column is relatively small and thus punching at the face of the column should also be checked against the maximum shear stress limit:

$d = 525\,\text{mm}$ as above and $u_0 = 4 \times 600 = 2400\,\text{mm}$

thus

$$v_{Ed} = \frac{6000 \times 10^3}{525 \times 2400} = 4.76\,\text{MPa} < v_{Rd,max} \text{ as required}$$

(3) Flexural shear across the corner pile is checked (as illustrated in Fig. 6.4-9 below):

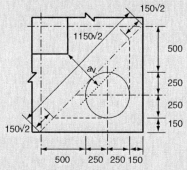

Fig. 6.4-9. Flexural shear plane across corner pile

Again

$$d = \frac{537.5 + 512.5}{2} = 525\,\text{mm}$$

The reinforcement is equal in the two orthogonal directions; therefore $A_s = 3272\,\text{mm}^2/\text{m}$ and

$$\rho_l = \frac{A_s}{bd} = \frac{3272}{1000 \times 525} = 0.00623$$

$\gamma_c = 1.5$ and again taking $C_{Rd,c} = 0.18/\gamma_c$ gives $C_{Rd,c} = 0.18/1.5 = 0.12$.

From equation (D6.2-1): $k = 1 + \sqrt{200/d} = 1 + \sqrt{200/525} = 1.617$

From 2-2/Expression (6.2.a): $V_{Rd,c} = C_{Rd,c} k (100 \rho_l f_{ck})^{1/3} bd$ (as there is no axial load) thus for a 1 m width of slab:

$$V_{Rd,c} = 0.12 \times 1.617 \times (100 \times 0.00623 \times 35)^{1/3} \times 1000 \times 525 \times 10^{-3} = 284.7\,\text{kN/m}$$

Shear span, $a_v = 1150\sqrt{2} - 300\sqrt{2} - 400\sqrt{2} - 250 + 0.20 \times 500 = 486.4\,\text{mm}$ (including 20% of pile diameter). Following on from the flexural shear check in (1) above, two methods of considering shear enhancement are possible. Only the more conservative method of enhancing the resistance side of the equation is carried out here to determine whether or not links are required:

The enhancement factor $1/\beta = 2d/a_v = 2 \times 525/486.4 = 2.16$

and the enhanced shear resistance $= 2.16 \times 284.7 = 614.6\,\text{kN/m}$

For the shear plane illustrated in Fig. 6.4-9, it is reasonable to enhance the resistance where the longitudinal reinforcement crosses the pile head, thus the length of the enhanced section is $1150\sqrt{2} - 2 \times 150\sqrt{2} = 1202.1\,\text{mm}$ and the length of the un-enhanced section is $2 \times 150\sqrt{2} = 424.3\,\text{mm}$.

Thus the shear resistance of the plane is given as

$$V^*_{Rd,c} = 284.7 \times 0.4243 + 614.6 \times 1.2021$$
$$= 120.8 + 734.8 = 859.6\,\text{kN} < V_{Ed} = 1500\,\text{kN}$$

so shear reinforcement is required.

Using the same approach as in (1) above based on enhancing the concrete resistance:

$$A_{sw} \geq \frac{V_{Ed} - V'^*_{Rd,c}}{f_{ywd}\sin\alpha}$$

where $V'^*_{Rd,c}$ is here taken as the contribution from the enhanced parts of the section only, as discussed in (1) above. Thus

$$A_{sw} \geq \frac{1500 \times 10^3 - 734.8 \times 10^3}{434.8 \times \sin 90} = 1760\,\text{mm}^2$$

which must be placed within the middle $0.75a_v$.

(If 2-1-1/clause 6.2.3(8) is applied, the shear force has to be reduced by $\beta = 0.463$, rather than an enhancement being made to the concrete resistance. Consequently, $V_{Ed} = 0.463 \times 1500 = 695\,\text{kN}$. From 2-1-1/Expression (6.19):

$$A_{sw} \geq \frac{V_{Ed}}{f_{ywd} \times \sin\alpha} = \frac{695 \times 10^3}{434.8 \times \sin 90} = 1597\,\text{mm}^2$$

which is a slightly lower amount of shear reinforcement than the above.)

The reinforcement provision of $1760\,\text{mm}^2$ from the first approach is carried forward here again as it is more representative of previous UK practice. Reinforcement is placed in the enhanced zone indicated in Fig. 6.4-10, although one could also include links in the zones marked 'X'.

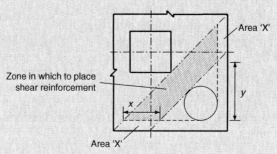

Fig. 6.4-10. Zones for shear reinforcement

An orthogonal grid of links (say 150 mm by 150 mm centres) is adopted. The horizontal dimension x is given by:

$$x = (1150\sqrt{2} - 300\sqrt{2})\sqrt{2} - (400\sqrt{2} + 250)\sqrt{2} = 546 \text{ mm}$$

and the number of legs at 150 mm centres in this length is $546/150 = 3.64$.

In the orthogonal direction, the length covered by the enhanced zone of width x is given by dimension y, where $y = 1150 - 150 - 150 = 850$ mm (and the number of legs at 150 mm centres in this length is $850/150 = 5.67$).

Thus the total number of legs in the zone indicated is $3.64 \times 5.67 + 0.5 \times 3.64^2 = 27$.

Required area per leg is therefore $1760/27 = 65 \text{ mm}^2$ (i.e. **use 10ϕ legs**)

(4) Punching shear on a $2d$ perimeter, as illustrated in Fig. 6.4-11, is checked ignoring enhancement for proximity to the column as required by 2-1-1/clause 6.4.3(7). (This is the only punching shear plane explicitly covered in EC2.) The perimeter is not, however, reduced to exclude the column load as required in 2-1-1/clause 6.4.2(2), as this is effectively checked in (3) above by way of a flexural shear plane.

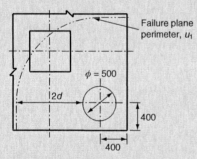

Fig. 6.4-11. $2d$ punching shear plane for corner pile

Again

$$d = \frac{537.5 + 512.5}{2} = 525 \text{ mm} \quad \text{therefore} \quad 2d = 2 \times 525 = 1050 \text{ mm}$$

(Adjacent perimeters overlap, therefore a combined perimeter should also be checked. This is covered by (1) and (3), however.)

$$u_1 = \frac{1}{4} \times 2\pi \left(2d + \frac{\phi}{2}\right) + 2 \times 400 = \frac{\pi}{2}(1050 + 250) + 800 = 2842 \text{ mm}$$

From 2-1-1/Expression (6.38) taking $\beta = 1.0$:

$$v_{Ed} = \frac{V_{Ed}}{u_1 d} = \frac{1500 \times 10^3}{2842 \times 525} = 1.01 \text{ MPa}$$

$$A_{sy} = A_{sz} = 3272 \text{ mm}^2$$

therefore

$$\rho_{ly} = \frac{A_{sy}}{bd_y} = \frac{3272}{1000 \times 537.5} = 0.00609$$

and

$$\rho_{lz} = \frac{A_{sz}}{bd_z} = \frac{3272}{1000 \times 512.5} = 0.00638$$

$\gamma_c = 1.5$ and again taking $C_{Rd,c} = 0.18/\gamma_c$ gives $C_{Rd,c} = 0.18/1.5 = 0.12$.

From 2-1-1/Expression (6.47):

$$v_{Rd,c} = C_{Rd,c}k(100\rho_l f_{ck})^{1/3} \text{ (no axial load)}$$

$$\rho_l = \sqrt{0.00609 \times 0.00638} = 0.00623$$

$$k = 1 + \sqrt{200/d} = 1 + \sqrt{200/525} = 1.617$$

thus

$$v_{Rd,c} = 0.12 \times 1.617 \times (100 \times 0.00623 \times 35)^{1/3} = 0.542 \text{ MPa}$$

$v_{Ed} > v_{Rd,c}$, therefore punching shear reinforcement is required.

Calculate u_{out} – the perimeter where the concrete resistance alone is sufficient:

$$v_{Rd,c} = \frac{V_{Ed}}{u_{out}d} \Rightarrow u_{out} = \frac{V_{Ed}}{v_{Rd,c}d} = \frac{1500 \times 10^3}{0.542 \times 525} = 5269 \text{ mm}$$

This represents a perimeter of $(5269 - 2 \times 400) \times 4/2\pi - 500/2 = 2595$ mm, i.e. $4.94d$ instead of $2d$. Links should extend to $4.94d - 1.5d = 3.44d$. Clearly, this is beyond the edge of the pile cap and is therefore not meaningful; the links provided will therefore be extended throughout the width of the pile cap.

2-1-1/Expression (6.52) is used to find the required punching shear reinforcement:

$$v_{Rd,cs} = 0.75 v_{Rd,c} + 1.5 \left(\frac{d}{s_r}\right) A_{sw} f_{ywd,ef} \left(\frac{1}{u_1 d}\right) \sin \alpha$$

Re-arranging for reinforcement design gives:

$$\frac{A_{sw}}{s_r} \geq (v_{Ed} - 0.75 v_{Rd,c}) \frac{u_1}{1.5 f_{ywd,ef} \sin \alpha}$$

with $f_{ywd,ef} = 250 + 0.25 \times 525 = 381.25$ MPa < 434.8 MPa as required.

Therefore

$$\frac{A_{sw}}{s_r} \geq (1.01 - 0.75 \times 0.542) \times \frac{2842}{1.5 \times 381.25 \times \sin 90} = 3.00 \text{ mm}^2/\text{mm}$$

For a radial spacing of 150 mm, $A_{sw} \geq 3.00 \times 150 = 449.9 \text{ mm}^2$.

Assuming 6 legs per perimeter (i.e. 3 no. bars associated with each perpendicular direction), required area per leg $= 449.9/6 = 75.0 \text{ mm}^2$ (i.e. 10ϕ links as for the flexural shear check above).

Check final resistance from 2-1-1/Expression (6.52):

$$v_{Rd,cs} = 0.75 \times 0.542 + 1.5 \times \frac{525}{150} \times 6 \times \pi \times 5^2 \times \frac{381.25}{2842 \times 525} \times \sin 90 = 1.039 \text{ MPa}$$

which is greater than v_{Ed} ($= 1.01$ MPa) as required.

Thus adopt 6 legs of 10ϕ links at 150 mm radial centres.

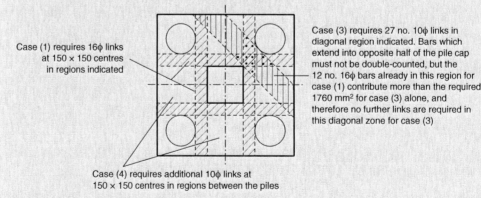

Fig. 6.4-12. Summary of punching shear reinforcement requirements

This reinforcement is arguably unnecessary because of the other checks performed under cases (1) and (3). The reinforcement calculated here, however, is treated as secondary reinforcement to be provided in the areas not reinforced for cases (1) and (3).

The final arrangement for punching shear reinforcement in the pile cap is therefore as illustrated in Figs 6.4-12 and 6.4-13:

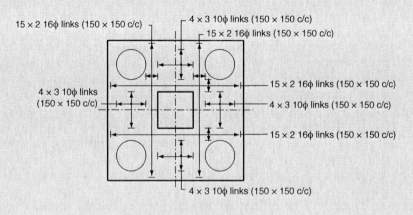

6.5. Design with strut-and-tie models
6.5.1. General
Section 5.6.4 of this guide provides general background to the use of strut-and-tie rules and the circumstances when they should be used, such as in the case of non-linear strain distribution identified by *2-1-1/clause 6.5.1(1)P*. EN 1992-2 makes no amendment to the strut-and-tie rules in EN 1992, other than possibly indirectly through the National Annex definition of f_{cd}.

2-1-1/clause 6.5.1(1)P

6.5.2. Struts
The allowable compressive stress that a concrete strut can carry is strongly affected by its multiaxial state of stress. Transverse compression is beneficial (particularly if acting in both transverse directions, as in the partially loaded area effect discussed in section 6.7), while transverse tension reduces the concrete's compressive resistance. The reduction in limiting compressive stress is worse where the tension is not perpendicular to the compression strut, as this leads to cracks that are not parallel to the direction of the compression so that the compressive force has to transfer across the cracks in shear. Quantifying the effects of transverse tension is difficult. Section 5.6 of the *AASHTO LRFD Bridge Design Specifications*[22] relates the compressive strength to the principal tensile strain and its direction relative to the direction of compression. However, this tensile strain is not always readily available. EC2 therefore gives two simplified and conservative limits for allowable compressive stress in *2-1-1/clause 6.5.2(1)* and *2-1-1/clause 6.5.2(2)* as follows.

2-1-1/clause 6.5.2(1)
2-1-1/clause 6.5.2(2)

(i) Transverse stress is zero or compressive

$$\sigma_{Rd,max} = f_{cd} \qquad \text{2-1-1/(6.55)}$$

This value should give the same limiting stress as that for a compression chord of a beam so the value of f_{cd} should be taken as that for compression in 2-2/clause 3.1.6, i.e. $f_{cd} = 0.85 f_{ck}/\gamma_c$ with $\alpha_{cc} = 0.85$ recommended. It will not often be possible to use this limit as transverse tension can occur simply by the bulging of a compression strut between nodes, as shown in Figs 6.5-1 and 6.5-2. Higher limiting stresses could, however, be taken in areas of triaxial compression, as discussed in 2-1-1/clause 3.1.9.

(ii) Transverse stress is tensile and concrete is cracked

$$\sigma_{Rd,max} = 0.6\nu' f_{cd} \qquad \text{2-1-1/(6.56)}$$

where ν' is a nationally determined parameter whose value is recommended to be $\nu' = 1 - f_{ck}/250$. This limit has two main significances. First, it corresponds approximately to the minimum stress at which vertical cracking is expected to occur in an un-reinforced strut with the geometry of that in Fig. 6.5-1. The actual cracking stress depends on the geometry of the load and the supporting member, as discussed in section 6.7 of this guide, and can be higher than the value in 2-1-1/Expression (6.56), as illustrated in Fig. 6.7-3. Second, the limit corresponds to the same limiting stress as that for a compression strut in a web under shear away from the supports, where link reinforcement carries tensile forces across the compression band. It is therefore important that the factor α_{cc} (which is intended to apply to beam–column bending and compression checks only) is given the same value in 2-1-1/Expression (6.56) as for shear design. It is recommended here, therefore, that $f_{cd} = f_{ck}/\gamma_c$ is used in 2-1-1/Expression (6.56), as it is envisaged that α_{cc} will be made equal to unity for shear design, i.e. $\sigma_{Rd,max} = 0.6(1 - f_{ck}/250)f_{ck}/\gamma_c$.

The limit in 2-1-1/Expression (6.56) covers the more detrimental skew cracking, so it will be conservative for cases where cracks are actually parallel to the compression. Schlaich et al.[8] recommend a higher limit of $f_{cd} = 0.68 f_{ck}/\gamma_c$ (which implicitly includes a factor of 0.85 for sustained loading similar to α_{cc}), where the tensile forces are perpendicular to the compression struts. The same limit is recommended therein for nodes with one member in tension (CCT node), as shown in Fig. 6.5-4(b). The equivalent node limit in EC2 is given by 2-1-1/Expression (6.61) which gives a limiting stress, this time incorporating $\alpha_{cc} = 0.85$, of:

$$\sigma_{Rd,max} = 0.85\nu' f_{cd} = 0.85(1 - f_{ck}/250) \times 0.85 f_{ck}/\gamma_c = 0.72(1 - f_{ck}/250)f_{ck}/\gamma_c$$

In summary, the limit in 2-1-1/Expression (6.56) does not distinguish between perpendicular cracking and skew cracking or between applied transverse tensile forces that are carried by reinforcement, and those which arise purely from an elastic bulging (spreading) of the struts between nodes as in Figs 6.5-1 and 6.5-2. In the latter case, the compressive stress should be checked in the neck region.

The limit in 2-1-1/Expression (6.56) also does not account for the actual magnitude of tensile strain, which is also relevant. Essentially, it relates to a safe lower-bound stress that can be assumed for all compression struts, whether reinforced or un-reinforced transversely, providing the strut-and-tie idealization does not depart significantly from elastic stress trajectories. In reality, different limits would apply in different situations and the following are suggested here, providing that the reinforcement yield strength does not exceed 500 MPa:

(a) Applied transverse tension (cracked) and reinforced transversely, both perpendicular to the strut:

$$\sigma_{Rd,max} = 0.72(1 - f_{ck}/250)f_{ck}/\gamma_c \text{ as for a CCT node}$$

(b) Applied transverse tension (cracked) and reinforced transversely, either or both skew to the strut:

$$\sigma_{Rd,max} = 0.60(1 - f_{ck}/250)f_{ck}/\gamma_c$$

(c) Transverse tension resulting from spread of load only but unreinforced transversely or reinforced by skew reinforcement:

$$\sigma_{Rd,max} = 0.60(1 - f_{ck}/250)f_{ck}/\gamma_c$$

or higher value derived allowing for the concrete tensile strength in accordance with Fig. 6.7-3 up to a limiting value of

$$\sigma_{Rd,max} = 0.72(1 - f_{ck}/250)f_{ck}/\gamma_c$$

Benefit will only arise for certain strut geometries and the lowest limit with varying geometry (based on transverse cracking and not usually final failure) is close to

$$\sigma_{Rd,max} = 0.60(1 - f_{ck}/250)f_{ck}/\gamma_c$$

The tensile strength of concrete should only be used to derive a higher strength if the concrete will not be cracked under other actions during the life of the structure.

(d) Transverse tension resulting from spread of load only and reinforced transversely perpendicular to the strut:

$$\sigma_{Rd,max} = 0.60(1 - f_{ck}/250)f_{ck}/\gamma_c$$

or higher value governed by the resistance of the transverse ties (see section 6.5.3) up to a limiting value of

$$\sigma_{Rd,max} = 0.72(1 - f_{ck}/250)f_{ck}/\gamma_c$$

A higher limit could be obtained using the partially loaded area method discussed in section 6.7, where the node at the end of the strut is triaxially constrained.

It can be seen from the above that there are difficulties in applying the strut rules consistently. However, the compressive limit of $\sigma_{Rd,max} = 0.6\nu' f_{cd}$ can be used conservatively in all the above cases with transverse tension unless specific rules elsewhere can be used to allow higher limits. These include the rules for partially loaded areas and for maximum shear stress in members loaded close to supports. This recognition is the basis of *2-1-1/clause 6.5.2(3)*, which allows reference to be made to 2-1-1/clause 6.2.2 and 6.2.3 for the design of members with short shear span. Another good illustration is the compression limit for a flanged beam in bending. Strictly, as transverse tension develops in a compression flange due to the spread of load across the flange, the lower limit of 2-1-1/Expression (6.56) should be used for the flange. However, experience shows that the limit in 2-1-1/Expression (6.55) is the correct one. When the limit in 2-1-1/Expression (6.56) is used for the design of struts, it is unlikely that the nodes will ever govern the design.

Finally, the compression limits in 2-1-1/clause 6.5.2 assume that the strut-and-tie model approximately follows the flow of force from an un-cracked elastic analysis. Tests have shown in some cases (reference 8) that lower limits are applicable if the angle of concrete struts departs significantly from their un-cracked elastic directions. This reduction is taken into account in the compression limit in the membrane rules of 2-2/clause 6.109. However, it is re-emphasized that testing has also shown in other cases that quite large departures from the elastic stress trajectories can be tolerated without reduction to the crushing resistance – the shear rules of 2-1-1/clause 6.2.3 provide one such instance.

2-1-1/clause 6.5.2(3)

6.5.3. Ties

Reinforcement ties may be used up to their design yield strength, f_{yd}, at the ultimate limit state and must be adequately anchored at nodes. A reduced stress limit is applicable for the serviceability limit state to control cracks in accordance with 2-1-1/clause 7.3.1(8). Prestressing steel may be similarly utilized. Ties may be discrete (as in the case of bottom reinforcement in a pilecap) or smeared (as in the case of transverse tension in the bursting zone adjacent to a concentrated load). Where ties are smeared, they should be distributed over the length of tension zone arising from the curved compression stress trajectories, as illustrated in Fig. 6.5-1.

Formulae for tie forces are given in *2-1-1/clause 6.5.3(3)* for the two simple cases of a 'partial discontinuity' and a 'full discontinuity'. These are sometimes referred to as 'bottle' distributions due to the bulging of the compression field from a neck region. The partial discontinuity is discussed in greater detail in section 6.7 of this guide. The strut-and-tie model shown in Fig. 6.5-1 can easily be shown to produce the tension force given in 2-1-1/Expression (6.58), thus:

2-1-1/clause 6.5.3(3)

$$T = \frac{1}{4}\frac{b-a}{b}F \qquad \qquad 2\text{-}1\text{-}1/(6.58)$$

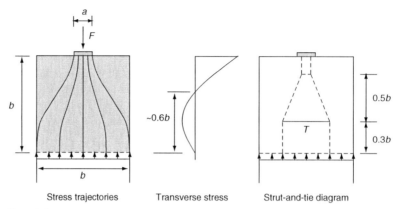

Fig. 6.5-1. Partial discontinuity

For the full discontinuity as in Fig. 6.5-2, a similar tension is produced according to 2-1-1/Expression (6.59), thus:

$$T = \frac{1}{4}\left[1 - 0.7\frac{a}{(H/2)}\right]F \qquad \text{2-1-1/(6.59)}$$

It is generally not safe to ignore this transverse tension and design the strut based on the 'a' dimension to a stress of f_{cd}, ignoring the spread of load, as discussed in section 6.7. This is because the transverse tension and associated cracking can lead to premature compression failure unless it is resisted by either reinforcement or the tensile strength of the concrete. A method is given for calculating the strut resistance without transverse reinforcement in section 6.7 of this guide. It is based on the tensile resistance of the concrete and a partial discontinuity, but it could easily be modified for the full discontinuity case above.

6.5.4. Nodes

A node is a volume of concrete containing the intersections of the struts and ties. Its dimensions are determined from the geometry of the struts, ties and external forces. Nodes may be smeared or concentrated in the same way as ties above. 2-1-1/clause 6.5.4 gives limiting stresses, $\sigma_{Rd,max}$, based on the greatest compressive stress framing into the node for three different types of concentrated nodes discussed below. The compressive resistance of nodes is also affected by bands of tension passing through them as discussed for struts. Smeared nodes generally require no check of concrete stress, but anchorage of

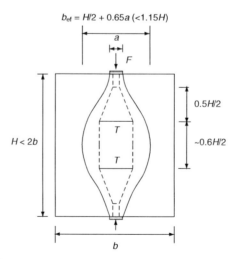

Fig. 6.5-2. Full discontinuity

CHAPTER 6. ULTIMATE LIMIT STATES

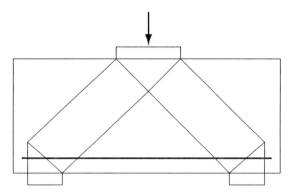

Fig. 6.5-3. Nodes under concentrated loads

bars should still be checked. *2-1-1/clause 6.5.4(2)P* requires that nodes must always be detailed so that they are in equilibrium. To use the stress limits in 2-1-1/clause 6.5.4, nodes must always be detailed without eccentricities.

According to *2-1-1/clause 6.5.4(1)P*, the rules for nodes should also be applied to the verification of bearing stresses at concentrated loads in members where the strut-and-tie rules have not been used elsewhere in the member design. This can, however, potentially lead to incompatibility. A typical example occurs in members where the shear design has been performed using the rules of section 6.2. If the strut-and-tie node limits were applied to check web crushing for loads applied near to supports as in Fig. 6.5-3, they would often give a lower maximum shear resistance than the shear rules. (Where there are shear links, Fig. 6.5-3 does not allow for some additional steeper compression struts which would develop, allowing some load to be suspended by links to the top of the section again and then carried back to the supports on further steep compression struts.) In this instance, the node rules could be applied to the bearing surfaces of the nodes only, on the basis that the shear design itself has been validated by testing.

The following cases are covered by *2-1-1/clause 6.5.4(4)*:

(a) All members at a node are in compression (CCC node)

$$\sigma_{Rd,max} = k_1 \nu' f_{cd} \qquad \text{2-1-1/(6.60)}$$

where k_1 is a nationally determined parameter whose recommended value is 1.0 and ν' is as defined in section 6.5.2 above. The node region is assumed to be limited by a polygon with sides typically, although not necessarily, at right angles to the strut directions. This type of node can occur, for example, at beam internal supports and at the compression faces of frame corners with closing moment (Fig. 5.6-6(a)). For nodes comprising three compression struts (as in Fig. 6.5-4(a)), a useful guide to sizing the node is to assume that the node boundaries are perpendicular to the struts and that hydrostatic pressure exists. This leads to dimensions with $F_{cd,1}/a_1 = F_{cd,2}/a_2 = F_{cd,3}/a_3$, as suggested in *2-1-1/clause 6.5.4(8)*. It is not necessary to achieve this hydrostatic state and generally a ratio of stresses on adjacent faces of a node of only 0.5 will still be satisfactory. Dimensioning of the node can therefore be modified to suit. Greater departures from hydrostatic conditions will, however, reduce the allowable stress limit from that given above.

To construct nodes at internal supports, as in Fig. 6.5-4(a), the two short vertical struts going down to the bearing plate must obviously have a centre of gravity at the position of the actual bearing reaction. It will usually only be necessary to check the bearing pressure on each node face. However, if there are additional struts passing through the node horizontally (as there would be at internal supports), the stress on a vertical section through the nodes should also be checked. Some enhancement may additionally be made for the effects of triaxial compression, as discussed below, whether produced by confining reinforcement or applied stress. (The partially loaded area case discussed in section 6.7 of this guide is one such example.)

(b) One member at a node in tension, others in compression (CCT node)

$$\sigma_{Rd,max} = k_2 \nu' f_{cd} \qquad \text{2-1-1/(6.61)}$$

where k_2 is a nationally determined parameter whose recommended value is 0.85. This type of node can occur at end supports or in deep beams as in Fig. 5.6-6(c). The reinforcement must be fully anchored from the start of the node, as shown (i.e. where the compressive stress trajectories of a strut first intersect the anchored bar). **2-1-1/clause 6.5.4(7)** requires the bar to extend over the entire node length, but the reinforcement may also be anchored behind the node if the anchorage length exceeds the node length. 2-1-1/Expression (6.61) makes allowance for development of bond stresses and hence cracking in the node. A higher stress could therefore be justified where the bar is anchored by an end plate behind the node. The height of the node is fixed as shown. If s_0 is not fixed by proximity to the edge of the member, a reasonable dimension would be 6 bar diameters as permitted in reference 22. The reinforcement should preferably be distributed over the height of the node, which has the benefit of maximizing the width, a_2, of the incoming strut. 2-1-1/clause 6.5.4(5) also allows an increase in the allowable concrete stress of 10% to be taken where the reinforcement is distributed in multiple layers over the node height.

(c) Two members at a node in tension formed by bent bar, others in compression (CTT node)

$$\sigma_{Rd,max} = k_3 \nu' f_{cd} \qquad \text{2-1-1/(6.62)}$$

where k_3 is a nationally determined parameter whose recommended value is 0.75. This type of node is illustrated in Fig. 6.5-4(c). If the compression strut does not bisect the bend equally, some force is also transmitted to the reinforcement through concrete bond. In addition to the above check, the bearing stress on the concrete inside the bend should also be checked in accordance with 2-1-1/clause 8.3.

It is recommended here that α_{cc} is taken as 0.85 in (a) to (c) above for compatibility with the strut limits above and therefore $f_{cd} = 0.85 f_{ck}/\gamma_c$; a National Annex may, however, direct otherwise. The increasing reduction to allowable stress at nodes with increasing tension reinforcement in (a) to (c) is again due to the detrimental effect of

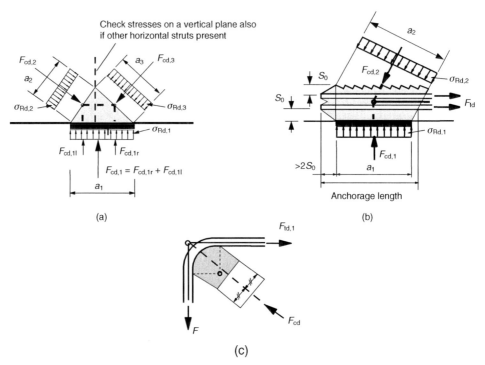

Fig. 6.5-4. Different types of nodes

transverse tensile strain on limiting compressive stress. If prestressing steel was used to form ties, such that decompression did not occur, no such reduction would be required.

In accordance with **2-1-1/clause 6.5.4(5)**, the above design compressive stresses can be increased by 10% where any of the following apply:

2-1-1/clause 6.5.4(5)

- triaxial compression is assured;
- all angles between struts and ties are $\geq 55°$;
- the stresses applied at supports or at point loads are uniform, and the node is confined by stirrups, although no criteria are given for their design. The rules for confined concrete in 2-1-1/clause 3.1.9 could be used with sufficient reinforcement provided to get a 10% increase in concrete resistance. Care would be required in detailing the confining reinforcement so that it was effective;
- the reinforcement is arranged in multiple layers;
- the node is reliably confined by means of bearing arrangement or friction, although no criteria are given for the degree of restraint required.

If the node resistances are enhanced as above, the struts themselves will usually govern so there will often be little benefit in invoking these recommendations.

Additionally, **2-1-1/clause 6.5.4(6)** allows triaxially compressed nodes to be checked using a limiting stress based on the confined strength, $f_{ck,c}$, when checking each direction, subject to $\sigma_{Rd,max} \leq k_4 \nu' f_{cd}$, where k_4 is a nationally determined parameter whose recommended value is 3.0. This is equivalent to using the rules for partially loaded areas, where the triaxial compressed state arises from ring tension in the surrounding concrete, as discussed in section 6.7. It would therefore seem reasonable to use the rules for partially loaded areas to determine maximum allowable pressures at supports where there is no applied transverse tension (Fig. 6.5-4(a)) and where the compression spreads in both transverse directions, provided that the bulging compression struts are reinforced according to 2-1-1/Expression (6.58) or 2-1-1/Expression (6.59).

2-1-1/clause 6.5.4(6)

Worked example 6.5-1: Diaphragm design
A bridge diaphragm is loaded as shown in Fig. 6.5-5. The presence of a large access hole, together with deep beam behaviour, leads to the use of strut-and-tie analysis as indicated on the figure. The concrete has cylinder strength 40 MPa and the reinforcement has a yield strength of 500 MPa. Struts A and B, ties C and D, and the pressure at the bearings, are now verified at ULS.

The loads in elements A to D under the ULS load case shown are as follows:

Strut A = 21.7 MN (compression)
Strut B = 15.9 MN (compression)
Tie C = 13.7 MN (tension)
Tie D = 1.2 MN (actually compression under this particular load case so treated as a strut)

Part of a strut-and-tie diagram is drawn for the relevant area in Fig. 6.5-6. The node region has been detailed with the vertical bearing reaction split into two components, reflecting the contributions induced by strut A and B. This has been done largely for consistency with the presentation in EN 1992-1-1 Fig. 6.26; it slightly modifies the strut angles from those in Fig. 6.5-5, although insignificantly. It would, however, be acceptable to detail the node in a more traditional manner directly from the intersections of the members in Fig. 6.5-5.

Bearing pressure
The node at the bearings is type (a) from Fig. 6.5-4 for this load case. This gives a limiting stress of $\sigma_{Rd,max} = k_1 \nu' f_{cd} = 1.0 \times (1 - 40/250) \times 0.85 \times 40/1.5 = 19.04$ MPa. The applied stress $= 29/(1.8 \times 1.8) = 9.0$ MPa < 19.4 MPa, so bearing pressure is

adequate. (In the real bridge, the bearing stress under the maximum torsion load case was considerably higher and was further increased by the forces from deviated prestressing, not included here for simplicity.)

Since the bearing dimensions are both smaller than those of the diaphragm and there is no tension through the node, the beneficial partially loaded area effect of 2-1-1/Expression (6.63) could be used to enhance the allowable pressure. The effect will, however, be small as the bearing dimension of 1800 mm in the span direction is almost the same as the diaphragm width of 2000 mm, so the allowable pressure could be increased to $2000/1800 f_{cd} = 1.1 f_{cd}$. (It should be noted that the load configuration here is not the same as in Fig. 6.5-1 for which the partially loaded area rules were derived, although the derivation of triaxial restraint from the surrounding concrete would still hold.) If benefit is taken from the partially loaded area effect at the node, each radiating strut must still be checked for the tensions developed from the spread of the compression and this will often effectively limit the bearing pressure as illustrated in the strut checks below.

Strut A

The limiting stress at the node itself will be $\sigma_{Rd,max} = k_1 \nu' f_{cd} = 19.04$ MPa. However, the strut bulges away from the bearing node, as shown in Fig. 6.5-6, generating transverse tension which can be quantified from 2-1-1/Expression (6.59) and Fig. 6.5-2. (Additionally, the suspension force in the outer links will generate some skew tension across the strut.) Since no reinforcement is provided specifically for this purpose (other than minimum surface reinforcement), the limiting compression in the strut is conservatively taken here as $\sigma_{Rd,max} = 0.6 \nu' f_{cd}$ from 2-1-1/Expression (6.56). Hence $\sigma_{Rd,max} = 0.6(1 - 40/250)40/1.5 = 13.4$ MPa (noting the 0.85 factor has been omitted from f_{cd} as discussed in the main text above). Actual strut stress $= 21.7/(1.8 \times 1.4) = 8.61$ MPa < 13.4 MPa, so the strut is adequate. (The strut dimension in the through-thickness direction is taken as 1.8 m equal to the bearing width.) It would be unwise in this situation to invoke the concrete's tensile strength according to Fig. 6.7-3 (which is based on a partial discontinuity and is approximate for a full discontinuity) to increase the allowable compressive limit, due to the adverse transverse tension from the adjacent web zone. In any case, the geometry in the transverse direction would not lead to a higher limit in this instance as the governing ratio b/a in the plane of the diaphragm is about 2.0 here, which gives a minimum resistance according to Fig. 6.7-3. The complex node region at the top of strut A was not critical by inspection due to the low compressive limit within the strut itself.

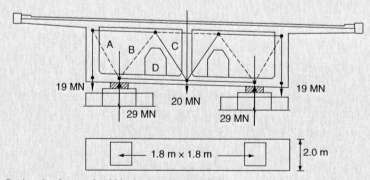

Fig. 6.5-5. Bridge diaphragm for Worked example 6.5-1

Strut B

The limiting stress at the node will again not govern as transverse tension develops from bulging of the strut, so the lower limit of $\sigma_{Rd,max} = 0.6 \nu' f_{cd}$ again applies. This time, however, the strut actually tapers down to pass the access hole. For compatibility with the limiting stress for the bottle case discussed in sections 6.5.3 and 6.7, the stress should be calculated at the narrower neck section 750 mm wide. Actual strut stress $= 15.9/(0.75 \times 2) = 10.6$ MPa < 13.4 MPa, so is adequate. A higher concrete

CHAPTER 6. ULTIMATE LIMIT STATES

compression limit using Fig. 6.7-3 could be justified here. The node at the top of strut B, type (c) to 2-1-1/clause 6.5.4, will not govern because a higher concrete stress limit is permitted at the node itself. Bearing pressure inside the bend of each bar must also be checked in general, but this was not critical here by inspection due to the large bend radius specified.

Tie C
Tensile area required $= 13.7 \times 10^6/(500/1.15) = 31\,510\,\text{mm}^2$, which is equivalent to 26 no. 40ϕ bars in two layers as indicated.

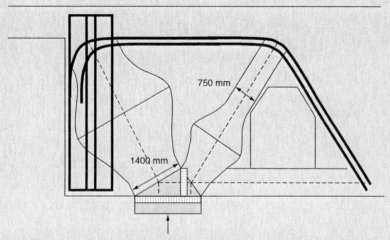

Fig. 6.5-6. Detail of node region and design strut widths for strut-and-tie model for Worked example 6.5-1. (Note: the bulging of the struts is intended only to show the actual dispersal of stress, not the local width to use in the check of the struts)

6.6. Anchorage and laps
The design of laps and anchorage length are discussed in detail in Chapter 9 of this guide.

6.7. Partially loaded areas
The rules in this section apply typically to bearing zones on both superstructures and substructures, but their derivation is based on a single bearing on a column where the axial stress simply spreads out over the column area. The rules for nodes in section 6.5.4 are also relevant to bearing areas. The compression resistance is governed by the lesser of the local crushing strength of the concrete and the strength of reinforcement resisting the transverse tensile (bursting) forces generated by transverse load dispersal. *2-1-1/clause 6.7(1)P* requires both mechanisms to be considered. The regions where these two failure mechanisms can occur are shown in Fig. 6.7-1 for a typical solid column. Subsequent discussions relate to this case.

2-1-1/clause 6.7(1)P

Crushing and spalling
For a uniformly distributed load F_{Rdu} acting on an area A_{c0} as shown in Fig. 6.7-2, the resistance is determined from *2-1-1/clause 6.7(2)*:

2-1-1/clause 6.7(2)

$$F_{Rdu} = A_{c0} f_{cd} \sqrt{A_{c1}/A_{c0}} \leq 3.0 f_{cd} A_{c0} \qquad 2\text{-}1\text{-}1/(6.63)$$

where:

A_{c0} is the loaded area

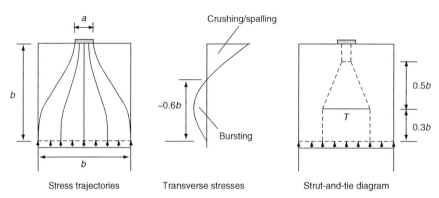

Fig. 6.7-1. Stress field under concentrated load

A_{c1} is the design distribution area which must be centred on the area A_{c0} and be similar in shape. This area must remain within the physical concrete section, which may limit the area for a load near an edge as shown in Fig. 6.7-2.

Compliance with the above limiting force prevents spalling near the loaded face caused by transverse expansion of the compressed concrete core. The form of 2-1-1/Expression (6.63) is derived from considerations of the confinement provided to the core by the surrounding concrete, whose perimeter is defined by b_2 and d_2 in Fig. 6.7-2. The surrounding area resists transverse expansion of the core by acting in 'ring' tension at its tensile resistance just prior to spalling. This ring tension causes a triaxial stress state and the enhanced compressive strength for confined concrete in 2-1-1/clause 3.1.9 becomes applicable. There is no benefit from this effect if the loaded area has a smaller dimension than the loaded element in one direction only, as the ring tension cannot then develop. It also follows that pressures higher than $f_{cd}\sqrt{A_{c1}/A_{c0}}$ could be permitted if sufficient confining reinforcement or ring prestressing were placed near the loaded face.

In addition to restraint from ring tension or confining reinforcement, some further restraint occurs from shear stress resisting splitting on the surface A_{c1}. Model Code 90[6] gives an approximate derivation of 2-1-1/Expression (6.63) based on the principles above and making qualitative allowance for this shear stress.

The distribution of load should be such that adjacent areas, A_{c1}, do not overlap. The distribution should also not exceed 1H:2V. This leads to the requirement that the available height, h, over which the distribution occurs must be greater than both $(b_2 - b_1)$ and $(d_2 - d_1)$. The upper limits given on b_2 and d_2 only apply for this crushing check and are there to produce the limiting strength of $F_{Rdu} \leq 3.0 f_{cd} A_{c0}$.

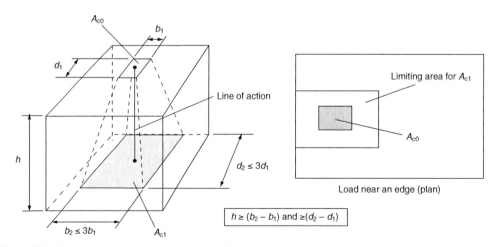

Fig. 6.7-2. Distribution to check crushing/spalling only

The value of F_{Rdu} should be reduced in accordance with *2-1-1/clause 6.7(3)* if the load is not uniformly distributed on the area A_{c0} or if high shear forces exist. If the load is not uniform, the bearing pressure check could be based on the peak pressure. No guidance is given on the effects of shear, but shear force could reasonably be ignored if it is less than 10% of the vertical force, which is consistent with 2-1-1/clause 10.9.4.3 which deals with precast elements. For higher shear, the vector resultant, F_r, of the shear force, F_h, and the vertical force, F_v, could be used in the bearing check, as recommended in *Model Code 90*,[6] according to:

$$F_r = F_v \sqrt{1 + \left(\frac{F_h}{F_v}\right)^2} \tag{D6.7-1}$$

2-1-1/clause 6.7(3)

This shear force would have to be tied into the surrounding structure by tie reinforcement at the loaded face.

It may be tempting to assume no distribution of load and set A_{c0} equal to A_{c1}, whereupon the bearing resistance becomes $F_{Rdu} = A_{c0} f_{cd}$. Many UK designers have in the past effectively taken this as the limiting pressure where bursting reinforcement has not been provided. However, a check of reinforcement is strictly still required for bursting as discussed below since the transverse spread of load in un-reinforced concrete leads to cracking when the concrete's tensile strength is reached, and this can give rise to premature failure at stresses less than f_{cd}. No guidance is given in EC2 on bearing pressure in the absence of any suitably placed reinforcement. The bearing pressure could safely be limited to $\sigma_{Rd,max} = 0.6(1 - f_{ck}/250) f_{ck}/\gamma_c$ as discussed in section 6.5 of this guide, or the tensile resistance of the concrete could be considered to increase resistance as discussed below. For piers with geometry such that the load has to spread in one direction only, it is likely that the minimum perimeter reinforcement would give a reasonable bursting resistance and hence a limiting bearing pressure in excess of $0.6(1 - f_{ck}/250) f_{ck}/\gamma_c$.

The rules for nodes, as discussed in section 6.5.4 of this guide, may also apply in cases other than this simple column case and is illustrated in both Worked examples 6.5-1 and 6.7-1.

Bursting
The tensile forces generated by the transverse spread of load can be resisted as shown in the strut-and-tie model in Fig. 6.7-1. The depth over which stresses become uniform can be taken equal to the dimension b, which in this case is the width of the section, or twice the distance from centre of load to a free edge in the direction considered for eccentric loads.

The strut-and-tie model shown produces a tension force as follows:

$$T = \frac{1}{4} \frac{b-a}{b} F \qquad \text{2-1-1/(6.58)}$$

where F is the applied vertical force. This tension needs to be calculated for both transverse directions and reinforcement detailed accordingly. Where the load spreads out from an applied load but tapers back into another node without spreading to the full cross-section in between, the tension in 2-1-1/Expression (6.58) should be replaced by the slightly modified expression for a 'full discontinuity' as given in 2-1-1/Expression (6.59) in section 6.5 of this guide. Worked example 6.7-1 illustrates the use of this expression.

EC2 gives no guidance where there is no (or insufficient) reinforcement to resist this tie force. *Model Code 90*[6] permits the force to be resisted by the concrete tensile resistance. For the case shown in Fig. 6.7-1, this would lead to a concrete tensile resistance of:

$$T_{max} = 0.6b \cdot L \cdot f_{ctd} \tag{D6.7-2}$$

and a limiting bearing reaction of:

$$F_{max} = \frac{2.4b^2 \cdot L \cdot f_{ctd}}{b - a} \tag{D6.7-3}$$

where L is the length of the loaded area perpendicular to the side a, $0.6b$ is the height of the tensile zone in Fig. 6.7-1 and f_{ctd} is the design tensile strength of the concrete. The limiting

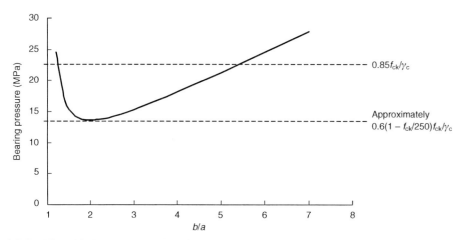

Fig. 6.7-3. Allowable bearing stress for 40 MPa unreinforced concrete allowing for tensile strength

bearing pressure then becomes:

$$f_{max} = \frac{2.4(b/a)^2 f_{ctd}}{(b/a - 1)} \tag{D6.7-4}$$

Equation (D6.7-4) needs to be applied in both perpendicular directions and the lowest resistance taken. Figure 6.7-3 shows equation (D6.7-4) plotted for a concrete with 40 MPa cylinder strength which has $f_{ctd} = 1.42$ MPa (including $\alpha_{ct} = 0.85$, as this is a case of sustained compression). The minimum strength, based on transverse cracking, occurs at $a/b = 2.0$ where the predicted allowable stress is very close to $\sigma_{Rd,max} = 0.6(1 - f_{ck}/250)f_{ck}/\gamma_c$ for a strut with transverse tension in accordance with 2-1-1/clause 6.5.2. The real failure load observed in tests on this configuration is usually greater than the derived value based on cracking.

Equation (D6.7-4) can also be applied to individual bulging compression struts between nodes, as discussed in section 6.5.2 of this guide, where limits on the allowable compression are suggested. It would not be appropriate to use this method of allowing for the tensile strength of the concrete where the concrete is expected to be cracked from other effects, such as flexure. In this case, the bearing stress should be limited to $\sigma_{Rd,max} = 0.6(1 - f_{ck}/250)f_{ck}/\gamma_c$, as discussed above, for cases where there is no appropriate reinforcement.

Where the load is eccentric to the supporting area, further strut-and-tie idealization would be necessary to distribute the stresses to their values remote from the loaded end, using the methods discussed in section 6.5 of this guide. Similarly, alternative strut-and-tie solutions will have to be developed where the section remote from the applied load is not the same as at the loaded end. This might, for example, occur in hollow piers made solid at the top only as in Fig. 6.7-4. In this case, the load has to spread out to the pier walls for equilibrium so reinforcement must be provided at the tie location shown. The bearing resistance may

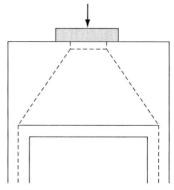

Fig. 6.7-4. Strut-and-tie system for hollow pier with solid top

CHAPTER 6. ULTIMATE LIMIT STATES

then effectively be governed by the compression limit for individual bulging compression struts, as discussed in section 6.5, if they are not themselves reinforced transversely, or by the nodes themselves. More complex geometries, such as those in Worked examples 6.5-1 and 6.7-1, will generally require a check of the struts and nodes.

2-2/clause 6.7(105) makes reference to EN 1992 Annex J for further guidance on bridge bearing areas. 2-2/clause J.104 confirms that 2-1-1/clauses 6.5 and 6.7 are relevant to the design of bearing areas and adds some requirements on edge distances and high strength concrete. These are discussed in Annex J of this guide.

2-2/clause 6.7(105)

Worked example 6.7-1: Bearing loads on a leaf pier

The bearing pressure is checked and the transverse reinforcement designed for the tall concrete pier illustrated in Fig. 6.7-5 under the concentrated vertical loads from the bearings. The load in each bearing is 11.5 MN. It is assumed that the concrete has a cylinder strength of 30 MPa and the reinforcement has a yield strength of 500 MPa.

The overall strut-and-tie idealization is shown in Fig. 6.7-5.

Bearing pressure at node 1
Node 1 is a CCT node according to 2-1-1/clause 6.5.4(4)(b) with limiting stress of $\sigma_{Rd,max} = k_2 \nu' f_{cd} = 0.85 \times (1 - 30/250) \times 0.85 \times 30/1.5 = 12.72$ MPa. This could be increased by 10% to 14 MPa in accordance with 2-1-1/clause 6.5.4(5) as the angle between strut-and-tie is greater than 55°. The partially loaded area rules cannot be directly applied here due to the presence of Tie 1 passing through the node, generating tensile stress.

The applied stress at the bearing surface $= 11.5 \times 10^6/(1200 \times 800) = 11.98$ MPa < 12.72 MPa so OK. The stress at the node edge meeting strut A must also be checked. The applied stress

$$= \frac{11.5 \times 10^6 / \cos 11.3°}{1216 \times 800} = 12.06 \text{ MPa} < 12.72 \text{ MPa so OK}$$

This check could be made less critical by distributing the reinforcement for Tie 1 vertically and thereby increasing the size of the node and width of strut A.

Reinforcement for Tie 1
Taking the lever arm $= 0.5b = 0.5 \times 8000 = 4000$ mm, the top face reinforcement must carry a force of $11.5 \times (2000 - 1200)/4000 = 2.3$ MN.

This can be resisted by:

$$A_s = F/(f_{yk}/\gamma_s) = 2.3 \times 10^6/(500/1.15) = 5290 \text{ mm}^2$$

(11 no. 25ϕ bars) at the top face. The reinforcement is provided as shown in Fig. 6.7-5.

Strut A
Strut A bulges towards node 2 so transverse tension is generated. This must be resisted by either reinforcement or the tensile strength of the concrete. From 2-1-1/clause 6.5.2(2), the compression limit for a strut with transverse tension is:

$$\sigma_{Rd,max} = 0.6\nu' f_{cd} = 0.6 \times (1 - 30/250) \times 30/1.5 = 10.56 \text{ MPa}$$

This will not be satisfied for strut A. However, as discussed in section 6.5.2 of this guide, it would be reasonable to limit the strut stress to the value at the node provided that either perpendicular reinforcement is provided or the tensile resistance of the concrete is not exceeded. Two directions need to be considered.

x direction
In the x direction, the ratio $b/a = 1.8/0.8 = 2.25$. From Fig. 6.7-3, it can be seen that the tensile strength of the concrete will not permit a compressive stress of much in excess of

$\sigma_{\text{Rd,max}} = 0.6\nu' f_{\text{cd}}$, so reinforcement will need to be designed. Reinforcement to resist the tensile force from partial discontinuity can be determined from 2-1-1/Expression (6.58) according to:

$$T = \frac{1}{4}\frac{(b-a)}{b}F$$

Thus in the x direction:

$$T = \frac{1}{4} \times \frac{(1.8 - 0.8)}{1.8} \times 11.5 = 1.60\,\text{MN}$$

(spreading out over the full width of the pier). This force can be resisted by a reinforcement area of:

$$A_s = F/(f_{yk}/\gamma_s) = 1.60 \times 10^6/(500/1.15) = 3680\,\text{mm}^2$$

EC2 gives no guidance about where this reinforcement must be placed. From Fig. 6.5-6, it is recommended to place it within a depth of $0.6b$, i.e. between $0.4b$ and b from the loaded surface. The reinforcement can be provided as shown in Fig. 6.7-5.

y direction

In the y direction, the ratio b/a will be approximately equal to 3 for the bulging strut at the point where tensile transverse stresses begin to develop. Figure 6.7-3 indicates there may be some benefit to the compression limit. From equation (D6.7-4):

$$f_{\max} = \frac{2.4(b/a)^2 f_{\text{ctd}}}{(b/a - 1)} \quad \text{where} \quad f_{\text{ctd}} = \alpha_{\text{ct}} f_{\text{ctk},0.05}/\gamma_C = \frac{0.85 \times 2.0}{1.5} = 1.13$$

(allowing for $\alpha_{\text{ct}} = 0.85$ for sustained compression). Therefore:

$$f_{\max} = \frac{2.4 \times 3^2 \times 1.13}{(3-1)} = 12.2\,\text{MPa}$$

which is just greater than the applied stress of 12.06 MPa, so it would be acceptable to provide only nominal transverse reinforcement. If the concrete tensile strength had been inadequate or the concrete was expected to be cracked under other loadings, reinforcement would need to be provided perpendicular to the strut, again using 2-1-1/Expression (6.58). The 'b' dimension could be taken as 4.0 m. Engineering judgement might be exercised in this case and the reinforcement placed horizontally due to the very small skew. This would allow the perimeter bars to be used. As discussed in section 6.5.2 of this guide, the use of reinforcement skew to the strut will weaken the compressive strength but small skews will only make a small reduction. In section 5.6 of the *AASHTO LRFD Bridge Design Specifications*,[22] the allowable stress reduces with angle α between tie and strut and with steel strain ε_s according to the factor:

$$\frac{1}{0.85\{0.8 + 170[\varepsilon_s + (\varepsilon_s + 0.002)\cot^2\alpha]\}} \leq 1.0$$

For

$$\varepsilon_s = \frac{500/1.15}{200 \times 10^3} = 0.0022$$

and $\alpha = 88.7°$, this reduction factor is only 0.98.

Edge sliding reinforcement to 2-2/Annex J

An additional check of edge sliding is required to 2-2/clause J.104.1(105). In the y direction, the area required

$$= \frac{11.5 \times 10^6/2}{500/1.15} = 13\,225\,\text{mm}^2$$

which has to be distributed over a length of $2.0/\tan 30° = 3.46\,\text{m}$. This is equivalent to 23×2 no. 20ϕ horizontal bars at 150 mm centres down the pier over the first 3.46 m. This could be reduced to allow for the steel already provided in Tie 1. This amount of steel is likely to be necessary in any case as perimeter reinforcement to control early thermal cracking in the pier. A similar amount of steel is required in the x direction, but it must be distributed over a shorter distance of $1.3/\tan 30° = 2.25\,\text{m}$. Assuming that the reinforcement for the y direction is provided around the whole perimeter, 15×4 legs of 16ϕ links at 150 mm centres down the pier at each bearing would provide this reinforcement. This could be reduced to allow for the steel calculated above for bursting in the x direction. These reinforcement requirements are not shown in Fig. 6.7-5.

The reinforcement to prevent edge sliding is not additive to reinforcement required for other effects, e.g. the strut-and-tie reinforcement above, shear, early thermal cracking. It can, however, be a significant amount of reinforcement and it is likely that the UK National Annex will relax the criterion.

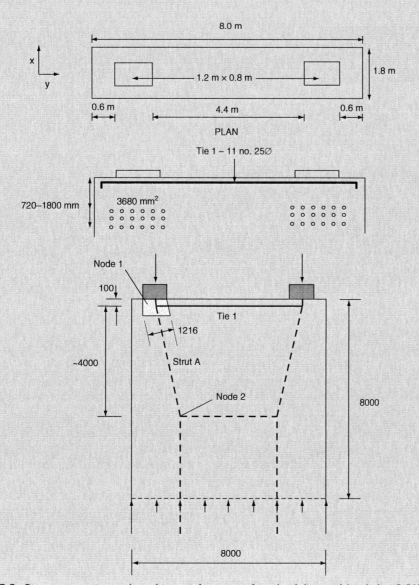

Fig. 6.7-5. Pier cross-section and resulting reinforcement from load dispersal (excluding 2-2/Annex J requirements) for Worked example 6.7-1

6.8. Fatigue

This section covers the rules given in section 6.8 of EC2-2. Guidance on damage equivalent stress calculation is given in the commentary on 2-2/Annex NN.

6.8.1. Verification conditions

Throughout the life of a bridge, constant road or rail traffic loading will produce large numbers of repetitive loading cycles in bridge elements. Both steel (reinforcing and prestressing) and concrete components which are subjected to large numbers of repetitive loading cycles can become susceptible to fatigue damage. As a consequence, *2-2/clause 6.8.1(102)* requires that fatigue assessment is undertaken for structures and structural components which are subjected to regular load cycles. Some exceptions, where fatigue verification is generally unnecessary, are provided in the note to that clause as follows:

- Footbridges, except those components very sensitive to wind action. The most common cause of wind-induced fatigue is vortex shedding. EN 1991-1-4 covers wind-induced fatigue.
- Buried arch and frame structures with a minimum earth cover of 1.0 m (road bridges) or 1.5 m (railway bridges). This assumes a certain amount of arching of the soil, which suggests that span should also be relevant.
- Foundations.
- Piers and columns not rigidly connected to bridge superstructures. 'Rigid' in this context is intended to refer to moment connection as pinned connections will not usually lead to cycles of significant live, load stress range.
- Retaining walls of embankments for roads and railways.
- Abutments which are not rigidly connected to bridge superstructures (with the exception of the slabs of hollow abutments).
- Prestressing and reinforcing steel in regions where, under the frequent combination of actions and P_k (presumably $P_{k,\text{inf}}$), only compressive stresses occur at the extreme concrete fibres. This is because the strain and hence stress range in the steel is typically small while the concrete remains in compression.

The National Annex may give other rules.

6.8.2. Internal forces and stresses for fatigue verification

2-1-1/clause 6.8.2(1)P requires stresses to be calculated assuming cracked concrete sections, neglecting the tensile strength of the concrete. Shear lag should be taken into account where relevant (2-1-1/clause 5.3.2.1 refers). *2-1-1/clause 6.8.2(2)P* additionally requires the effect of different bond behaviour of prestressing and reinforcing steel to be taken into account in the calculation of reinforcement stress. This results in an increase in stress in the reinforcing steel from that calculated using a cracked elastic cross-section analysis by a factor, η, given in 2-1-1/Expression (6.64).

2-1-1/clause 6.8.2(3) requires fatigue verification to be undertaken for the design of shear reinforcement, which is a new check for UK practice. Steel forces are calculated from the truss analogy using a compressive strut angle of θ_{fat}. For fatigue calculation, it is important to use a realistic estimate of the stress range. It is therefore appropriate that this angle is taken greater than that assumed for the ultimate limit state design (within the angular limits of 2-1-1/clause 6.2.3(2)), since the latter is the angle at the ultimate limit state after a certain amount of plastic redistribution has taken place to reduce the stress in the links and to use them optimally. As a result, θ_{fat} may be taken as:

$$\tan \theta_{\text{fat}} = \sqrt{\tan \theta} \leq 1.0 \qquad \text{2-1-1/(6.65)}$$

where θ is the angle of concrete compression struts to the beam axis assumed in the ultimate limit state shear design. For shear reinforcement inclined at an angle α to the horizontal, the

CHAPTER 6. ULTIMATE LIMIT STATES

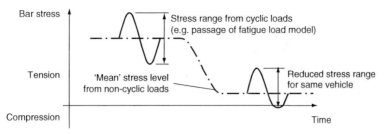

Fig. 6.8-1. Stress ranges for reinforcement fatigue verification caused by same cyclic action at different mean stress levels

steel force can be determined by rearranging 2-1-1/Expression (6.13), thus:

$$\Delta\sigma_s = \frac{\Delta V \cdot s}{A_{sw} z (\cot\theta_{fat} + \cot\alpha)\sin\alpha} \tag{D6.8-1}$$

where ΔV is the shear force range.

6.8.3. Combination of actions
The calculation of the stress ranges for fatigue verification to EC2 requires the applied load to be divided into non-cyclic and fatigue-inducing cyclic action effects. The basic combination of the non-cyclic load is defined by Expressions (6.66) and (6.67) of EN 1992-1-1 and is equivalent to the definition of the frequent combination for the serviceability limit state. The cyclic action is then combined with the unfavourable non-cyclic action to determine the stress ranges – 2-1-1/Expressions (6.68) and (6.69) refer.

The non-cyclic action gives a mean stress level upon which the cyclic part of the action effect is superimposed, as illustrated in Fig. 6.8-1 for reinforcement. Mean stress is important as it determines whether the sign of the stress in an element reverses in the course of a cycle of loading. In Fig. 6.8-1, the reinforcement stress range for a given cyclic action is less for the smaller tensile mean stress as part of the cyclic loading then causes compression in the concrete, which reduces the stress in the reinforcement for that part of the cycle.

6.8.4. Verification procedure for reinforcing and prestressing steel
The number of cycles to fatigue failure of a steel component is a function of the stress that each loading cycle induces in the component and the type of component. Since the relationship of stress range ($\Delta\sigma$) to the number of cycles to failure (N) is exponential, the relationship is normally plotted graphically in the form of a $\log\Delta\sigma$–$\log N$ curve. These types of curve are commonly referred to as S–N curves.

2-1-1/clause 6.8.4(1) allows the damage produced by cycles of a single stress range of amplitude $\Delta\sigma$ to be determined by using the corresponding S–N curves for reinforcing and prestressing steel. The form of these curves is illustrated in Fig. 6.8-2 for reinforcement; the diagram for prestressing steel is similar, using 0.1% proof stress in place of yield stress.

2-1-1/clause 6.8.4(1)

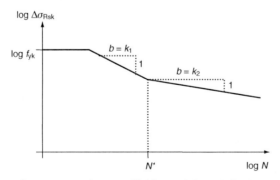

Fig. 6.8-2. Characteristic fatigue strength curve (S–N curve) for reinforcing steel

Recommended values defining the appropriate S–N curve geometry for the steel component under consideration are given in 2-1-1/Tables 6.3N and 6.4N for reinforcement and prestressing steel respectively. The recommended parameters therein may be modified in the National Annex.

2-1-1/clause 6.8.4(1) and also 2-1-1/clause 2.4.2.3(1) require a partial factor, $\gamma_{F,fat}$, to be applied to all fatigue loads when calculating the stress range. The value for $\gamma_{F,fat}$ is defined in the National Annex and is recommended by EC2 to be taken as 1.0. The resisting stress range at N^* cycles, $\Delta\sigma_{Rsk}$, given in 2-1-1/Tables 6.3N and 6.4N, also has to be divided by the material partial safety factor $\gamma_{s,fat}$. The recommended value for $\gamma_{s,fat}$ from 2-1-1/clause 2.4.2.4(1) is 1.15.

In real fatigue assessment situations for concrete bridge design, there will be more than one stress range acting on the steel element throughout its design life. **2-1-1/clause 6.8.4(2)** allows multiple amplitudes to be treated by using a linear cumulative damage calculation, known as the Palmgren–Miner summation:

$$D_{Ed} = \sum_i \frac{n(\Delta\sigma_i)}{N(\Delta\sigma_i)} < 1.0 \qquad \text{2-1-1/(6.70)}$$

where:

$n(\Delta\sigma_i)$ is the applied number of cycles for a stress range of $\Delta\sigma_i$
$N(\Delta\sigma_i)$ is the resisting number of cycles for a stress range of $\Delta\sigma_i$, i.e. the number of loading cycles to fatigue failure

For most bridges, the above is a complex calculation because the stress in each component usually varies due to the random passage of vehicles from a spectrum. Details on a road or rail bridge could be assessed using the above procedure if the loading regime is known at design. This includes the weight and number of every type of vehicle that will use each lane or track of the bridge throughout its design life, and the correlation between loading in each lane or track. In the majority of cases this would require lengthy calculations.

As an alternative to the use of 2-1-1/Expression (6.70), 2-1-1/clause 6.8.5 allows the use of simplified fatigue Load Models 3 and 71, from EN 1991-2, for road and rail bridges respectively, in order to reduce the complexity of the fatigue assessment calculation. It is assumed that the fictitious vehicle/train alone causes the fatigue damage. The calculated stress from the vehicle is then adjusted by factors to give a single stress range which, for N^* cycles, causes the same damage as the actual traffic during the bridge's lifetime. This is called the 'damage equivalent stress' and is discussed in section 6.8.5.

2-1-1/clause 6.8.4(3) requires that, where prestressing or reinforcing steel is exposed to fatigue loads, the calculated stresses shall not exceed the design yield strength of the steel as EC2 does not cover cyclic plasticity.

2-1-1/clause 6.8.4(5) relates to assessment of existing structures, which is strictly outside the scope of EC2, so its inclusion is curious. Its reference to corrosion is not explicit about either the degree of corrosion or its nature (e.g. general or pitting), so a single value of stress exponent to cover all situations is dubious. Nevertheless, it was not intended that any such allowance for corrosion be made in new design.

2-2/clause 6.8.4(107) permits no fatigue check to be conducted for external and unbonded tendons lying within the depth of the concrete section. This is because the strain, and hence stress, variation under service loads is small in such tendons. Consideration should be given to fatigue in external tendons which are outside the depth of the structure (such as in extradosed bridges) as the fluctuation in stress might be more significant here. This situation is covered by EN 1993-1-11.

6.8.5. Verification using damage equivalent stress range

In the damage equivalent stress range method, described by **2-1-1/clause 6.8.5(1)** and **2-1-1/clause 6.8.5(2)**, the real operational loading is represented by N^* cycles of an equivalent single amplitude stress range, $\Delta\sigma_{s,equ}(N^*)$, which causes the same damage as the actual

CHAPTER 6. ULTIMATE LIMIT STATES

traffic during the bridge's lifetime. This stress range may be calculated for reinforcing or prestressing steel using 2-2/Annex NN, as illustrated in Worked example 6.8-1. **2-1-1/ clause 6.8.5(3)** contains a verification formula for reinforcing steel, prestressing steel and splicing devices:

$$\gamma_{F,fat}\Delta\sigma_{s,equ}(N^*) \leq \frac{\Delta\sigma_{Rsk}(N^*)}{\gamma_{s,fat}}$$ 2-1-1/(6.71)

2-1-1/clause 6.8.5(3)

where:

$\Delta\sigma_{s,equ}(N^*)$ is the appropriate damage equivalent stress range (converted to N^* cycles) from 2-2/Annex NN

$\Delta\sigma_{Rsk}(N^*)$ is the resisting stress range limit at N^* cycles from the appropriate S–N curves given in 2-1-1/Tables 6.3N or 6.4N

2-1-1/Expression (6.71) does not cover concrete fatigue verification. 2-2/Annex NN3.2 provides a damage equivalent verification for concrete in railway bridges, but there is no similar verification for highway bridges. For highway bridges, concrete can be verified using the methods in 2-2/clause 6.8.7, as illustrated in Worked example 6.8-2.

> **Worked example 6.8-1: Damage equivalent stress range for a concrete road bridge**
> Use the damage equivalent stress method to verify the fatigue performance of straight longitudinal reinforcement in the sagging and hogging zones of a two-span concrete road bridge with continuous spans of 30 m, carrying 3 lanes of long-distance motorway traffic with a high flow rate of lorries. The stress ranges caused by the passage of fatigue Load Model 3 across the bridge are 75 MPa (sagging zone) and 40 MPa (hogging zone at intermediate support). Assume 100 years design life and good surface conditions.
>
> (1) First consider the sagging section.
> For in-span sections, the axle loads of Load Model 3 should be multiplied by 1.4 according to 2-2/Annex NN.2.1(101). This produces a stress range of $\Delta\sigma_{s,Ec} = 1.40 \times 75 = 105$ MPa. In this case, using the load combination in 2-1-1/ clause 6.8.3, this increased stress range still did not cause a reversal of the sign of stress (i.e. it did not induce a net compression during part of the cycle) so it is acceptable to factor the stress range in this way. If the sign of the stress had reversed, it would be necessary to consider this non-linear response in determining the stress range.
> From 2-2/Fig. NN.2 for straight reinforcing steel bars and an influence line length of 30 m (curve 3a), $\lambda_{s,1} = 1.19$.
> From 2-2/Annex (NN.103): $\lambda_{s,2} = \bar{Q} \times \sqrt[k_2]{N_{obs}/2.0}$ and from Table 4.5 of EN 1991-2, $N_{obs} = 2.0 \times 10^6$.
> From 2-1-1/Table 6.3N, $k_2 = 9$ for straight bars. Motorway traffic will generally be 'long-distance' traffic, so from 2-2/Table NN.1, $\bar{Q} = 1.0$.
> Thus
> $$\lambda_{s,2} = 1.0 \times \sqrt[9]{2.0/2.0} = 1.0$$
> From 2-2/Annex (NN.104):
> $$\lambda_{s,3} = \sqrt[k_2]{N_{Years}/100} = \sqrt[9]{100/100} = 1.0$$
> From 2-2/Annex (NN.105):
> $$\lambda_{s,4} = \sqrt[k_2]{\frac{\sum N_{obs,i}}{N_{obs,1}}}$$
> and from Table 4.5 of EN 1991-2, $N_{obs,1} = 2.0 \times 10^6$. Note 1 to clause 4.6.1(3) in EN 1991-2 indicates that for each fast lane, 10% of N_{obs} from Table 4.5 of EN 1991-2 for the slow lane may be taken into account. Thus $N_{obs,2} = N_{obs,3} =$

0.2×10^6 and therefore

$$\lambda_{s,4} = \sqrt[9]{\frac{2.0 + 0.2 + 0.2}{2.0}} = 1.02$$

(The National Annex for EN 1991-2 may modify the number of vehicles in the fast lane.)

For surfaces of good roughness (i.e. a maintained surface like a motorway), $\phi_{fat} = 1.2$.

From 2-2/Annex (NN.102):

$$\lambda_s = \phi_{fat}\lambda_{s,1}\lambda_{s,2}\lambda_{s,3}\lambda_{s,4} = 1.2 \times 1.19 \times 1.0 \times 1.0 \times 1.02 = 1.457$$

and therefore from 2-2/Annex (NN.101), $\Delta\sigma_{s,equ} = \Delta\sigma_{s,Ec}\lambda_s = 105 \times 1.457 =$ **153 MPa**.

From 2-1-1/Table 6.3N for straight bars, $N^* = 10^6$ and $\Delta\sigma_{Rsk}(10^6) = 162.5$ MPa.
From 2-1-1/Expression (6.71):

$$\gamma_{F,fat}\Delta\sigma_{s,equ}(N^*) = 1.0 \times 153 = 153 \text{ MPa}$$

$$\frac{\Delta\sigma_{Rsk}(N^*)}{\gamma_{s,fat}} = 162.5/1.15 = 141.3 \text{ MPa} < 153 \text{ MPa so sagging section has inadequate fatigue resistance.}$$

(2) Second, consider the hogging section over an intermediate support.

For support sections, the axle loads and hence stress range from the fatigue load model should be multiplied by 1.75, hence $\Delta\sigma_{s,Ec} = 1.75 \times 40 = 70$ MPa. (The note in (1) above regarding non-linearity where the sign of stress changes applies here again.)

The 'critical length of the influence line' for hogging bending is here taken equal to the span length of 30 m, as discussed in Annex NN of this guide. From 2-2/Fig. NN.1 for straight, reinforcing steel bars and 30 m critical influence line length (curve 3), $\lambda_{s,1} = 0.98$:

$\lambda_{s,2} = 1.0$ as before,

$\lambda_{s,3} = 1.0$ as before,

$\lambda_{s,4} = 1.02$ as before,

and $\phi_{fat} = 1.2$ as before.
From 2-2/Annex (NN.102):

$$\lambda_s = \phi_{fat}\lambda_{s,1}\lambda_{s,2}\lambda_{s,3}\lambda_{s,4} = 1.2 \times 0.98 \times 1.0 \times 1.0 \times 1.02 = 1.20$$

and therefore from 2-2/Annex (NN.101), $\Delta\sigma_{s,equ} = \Delta\sigma_{s,Ec}\lambda_s = 70 \times 1.20 =$ **84 MPa**.
Again, from 2-1-1/Table 6.3N for straight bars, $N^* = 10^6$ and $\Delta\sigma_{Rsk}(10^6) = 162.5$ MPa.

From 2-1-1/Expression (6.71):

$$\gamma_{F,fat}\Delta\sigma_{s,equ}(N^*) = 1.0 \times 84 = 84 \text{ MPa}$$

$$\frac{\Delta\sigma_{Rsk}(N^*)}{\gamma_{s,fat}} = 162.5/1.15 = 141.3 \text{ MPa} > 84 \text{ MPa so hogging section at intermediate support has adequate fatigue resistance.}$$

6.8.6. Other verification methods

2-1-1/clause 6.8.6(1) and (2) give alternative rules for fatigue verifications of reinforcing and prestressing steel components. These methods are intended as an alternative to checking fatigue resistance using 2-1-1/clauses 6.8.4 or 6.8.5.

2-1-1/clause 6.8.6(1) allows the fatigue performance of reinforcement or prestressing steel to be deemed satisfactory if the stress range under the frequent cyclic load combined with the

2-1-1/clause
6.8.6(1)

basic combination is less than k_1 for unwelded reinforcement or k_2 for welded reinforcement. The values of k_1 and k_2 may be given in the National Annex and EC2 recommends taking values of 70 MPa and 35 MPa respectively. The meaning of 'frequent cyclic loading' is not explained but it implies a calculation based on the fatigue load models in EN 1991-2. Assuming this to be the case, it will usually be preferable to perform a damage equivalent stress calculation using 2-2/Annex NN as this also uses the fatigue load models of EN 1991-2 and will lead to a more economic answer.

2-1-1/clause 6.8.6(2) allows the stress range alternatively to be calculated directly from the frequent load combination to avoid the need to calculate stress ranges from fatigue load models or directly from traffic data. However, the recommended allowable stress ranges above would mean that elements would rarely pass such a check.

Where welded joints or splicing devices are used in prestressed concrete construction, *2-1-1/clause 6.8.6(3)* requires that no tension exists in the concrete section within 200 mm of the prestressing tendons or reinforcing steel under the frequent load combination when a reduction factor of k_3 is applied to the mean value of the prestressing force. The value of k_3 is defined in the National Annex. EC2 recommends taking a value of 0.9. This value is increased to 1.0 in the UK's National Annex (in the same manner as r_{sup} and r_{inf} in 2-1-1/clause 5.10.9) to limit the number of load cases to be considered for SLS and fatigue design. The criterion ensures that the stress range for such details is kept small for the majority of cycles since the concrete will generally remain in compression.

6.8.7. Verification of concrete under compression or shear

The general fatigue verification procedure for concrete given in *2-2/clause 6.8.7(101)* requires a cumulative damage summation, like that in 2-1-1/clause 6.8.4, to be carried out using traffic data. For rail bridges, this lengthy calculation can be avoided by using the simplified damage equivalent stress verification of 2-2/Annex NN.3.2. Neither the Annex nor 2-2/clause 6.8.7(101) itself, however, gives appropriate data for road bridges.

As a simpler alternative, *2-1-1/clause 6.8.7(2)* gives a conservative verification based on the non-cyclic loading used for the static design:

$$\frac{\sigma_{c,max}}{f_{cd,fat}} \leq 0.5 + 0.45 \frac{\sigma_{c,min}}{f_{cd,fat}} \qquad \text{2-1-1/(6.77)}$$

but limited to 0.9 for $f_{ck} \leq 50$ MPa or 0.8 for $f_{ck} > 50$ MPa, where:

$\sigma_{c,max}$ is the maximum compressive stress at a fibre under the frequent load combination (compression measured as positive)

$\sigma_{c,min}$ is the minimum compressive stress under the frequent load combination at the same fibre where $\sigma_{c,max}$ occurs. $\sigma_{c,min}$ should be taken as 0 if negative (in tension)

$f_{cd,fat}$ is the concrete design fatigue compressive strength defined in the code as:

$$f_{cd,fat} = k_1 \beta_{cc}(t_0) f_{cd}\left(1 - \frac{f_{ck}}{250}\right) \qquad \text{2-2/(6.76)}$$

where:

k_1 is a coefficient defined in the National Annex and is recommended by EC2 to be taken as 0.85

$\beta_{cc}(t_0)$ is the coefficient for concrete strength at first cyclic loading from 2-1-1/clause 3.1.2(6)

t_0 is the age of the concrete in days upon first cyclic loading, i.e. the age at which live load is first applied

f_{cd} is the design compressive strength of concrete. A value for α_{cc} of 1.0 is intended to be used here in conjunction with $k_1 = 0.85$, as k_1 performs a similar function of accounting for sustained loading

For concrete road bridges, this alternative concrete fatigue verification is unlikely to govern design, other than possibly for very short spans where the majority of the concrete stress is produced by live load. It will therefore generally be appropriate to use this simplified check. No guidance is given on the calculation of the concrete stresses; ignoring concrete in tension will be a conservative assumption.

2-1-1/clause 6.8.7(3) permits the above simplified verification of concrete to be applied to the compression struts of members subjected to shear and requiring shear reinforcement. Since the compression struts have transverse tension passing through them (see discussions in section 6.5 of this guide), $f_{cd,fat}$ has to be reduced by the factor ν, defined in 2-1-1/clause 6.2.2(6), and the verification becomes:

$$\frac{\sigma_{c,max}}{\nu f_{cd,fat}} \leq 0.5 + 0.45 \frac{\sigma_{c,min}}{\nu f_{cd,fat}} \tag{D6.8-2}$$

The stresses $\sigma_{c,max}$ and $\sigma_{c,min}$ can be calculated for reinforced concrete beams, with shear reinforcement inclined at an angle α to the horizontal, from the following expression obtained by re-arranging 2-1-1/Expression (6.14):

$$\sigma_c = \frac{V_{Ed}}{b_w z}\left(\frac{1 + \cot^2\theta}{\cot\theta + \cot\alpha}\right) \tag{D6.8-3}$$

V_{Ed} is the relevant shear force under the frequent load combination and the other symbols are defined in 2-1-1/clause 6.2.3. The concrete stress increases with reducing strut angle θ so, in this case, it is conservative to base θ on its ULS value in the above calculation rather than the larger angle θ_{fat} from 2-1-1/Expression (6.65).

For members subjected to shear but not requiring shear reinforcement, *2-1-1/clause 6.8.7(4)* provides the following expressions for assuming satisfactory fatigue resistance in shear:

$$\text{For } \frac{V_{Ed,min}}{V_{Ed,max}} \geq 0: \quad \frac{|V_{Ed,max}|}{|V_{Rd,c}|} \leq 0.5 + 0.45 \frac{|V_{Ed,min}|}{|V_{Rd,c}|} \qquad \text{2-1-1/(6.78)}$$

but limited to 0.9 for $f_{ck} \leq 50$ MPa or 0.8 for $f_{ck} > 50$ MPa:

$$\text{or, for } \frac{V_{Ed,min}}{V_{Ed,max}} < 0: \quad \frac{|V_{Ed,max}|}{|V_{Rd,c}|} \leq 0.5 - \frac{|V_{Ed,min}|}{|V_{Rd,c}|} \qquad \text{2-1-1/(6.79)}$$

where:

$V_{Ed,max}$ is the design value of the maximum applied shear force under the frequent load combination

$V_{Ed,min}$ is the design value of the minimum applied shear force under the frequent load combination in the cross-section, where $V_{Ed,max}$ occurs

$V_{Rd,c}$ is the design shear resistance from 2-2/Expression (6.2.a)

Worked example 6.8-2: Concrete fatigue verification for a concrete road bridge

Check the fatigue resistance of the concrete under compression in Worked example 6.8-1 for both the hogging and sagging sections. Assume, from a cracked elastic section analysis, the values of the maximum and minimum stresses in the top fibre of the sagging zone and the bottom fibre of the hogging zone under the frequent load combination have been calculated as follows:

	Maximum stress, $\sigma_{c,max}$ (MPa)	Minimum stress, $\sigma_{c,min}$ (MPa)
(1) Sagging section (top fibre stress)	4.0	1.0
(2) Hogging section (bottom fibre stress)	4.5	0.7

Assume that the bridge deck is constructed using rapid hardening, normal strength cement. Concrete is C35/45.

Since the maximum and minimum fibre stresses under the frequent load combination have been calculated, 2-1-1/Expression (6.77) must be satisfied:

$$\frac{\sigma_{c,max}}{f_{cd,fat}} \leq \begin{cases} 0.5 + 0.45\dfrac{\sigma_{c,min}}{f_{cd,fat}} \\ 0.9 \text{ for } f_{ck} \leq 50\,\text{MPa} \end{cases}$$

where:

from 2-2/Expression (6.76): $f_{cd,fat} = k_1\beta_{cc}(t_0)f_{cd}(1 - f_{ck}/250)$ with $k_1 = 0.85$ (recommended value).

It is conservatively assumed here that construction traffic uses the bridge from an age of 7 days so the age at first cyclic loading is $t_0 = 7$ days. Generally, load would not be applied this early.

From 2-1-1/clause 3.1.2(6): $\beta_{cc}(t) = \exp(s(1 - \sqrt{(28/t)}))$

$s = 0.25$ for rapid hardening, normal strength cement and from 2-1-1/clause 3.1.6(1):

$f_{cd} = \alpha_{cc}f_{ck}/\gamma_c = 1.0 \times 35/1.5 = 23.33\,\text{MPa}$

(taking $\alpha_{cc} = 1.0$ for fatigue as discussed in the main text).
Therefore

$\beta_{cc}(7) = \exp(0.25 \times (1 - \sqrt{(28/7)})) = 0.7788$

and

$f_{cd,fat} = 0.85 \times 0.7788 \times 23.33 \times (1 - 35/250) = 13.3\,\text{MPa}$

The final checks are therefore as follows:

(1) Sagging section:

$\sigma_{c,max}/f_{cd,fat} = 4.0/13.3 = 0.30$

This is ≤ 0.9 and $\leq 0.5 + 0.45\sigma_{c,min}/f_{cd,fat} = 0.5 + 0.45 \times 1.0/13.3 = 0.53$ as required. Therefore, the sagging section at mid-span has adequate fatigue resistance.

(2) Hogging section:

$\sigma_{c,max}/f_{cd,fat} = 4.5/13.3 = 0.34$

This is ≤ 0.9 and $\leq 0.5 + 0.45\sigma_{c,min}/f_{cd,fat} = 0.5 + 0.45 \times 0.7/13.3 = 0.52$ as required. Therefore, the hogging section at an intermediate support has adequate fatigue resistance.

6.9. Membrane elements

A problem encountered when using linear elastic finite element techniques to analyse concrete bridges is that the results produced are usually in the form of stresses, while the code resistance rules are presented in terms of stress resultants, such as shear force and bending moment. This applies to the rules for bending and shear given in sections 6.1 and 6.2 respectively of EC2. The rules for membrane elements presented in **2-2/clause 6.109(101)** provide a way of designing directly from the stresses produced by a two-dimensional linear elastic finite element model. The sign convention for stresses in 2-2/clause 6.109 is shown in Fig. 6.9-1. The rules are also intended for use with elements under out-of-plane bending and torsion in conjunction with the sandwich model of Annex LL. Annex MM gives specific recommendations for the design of box girder webs in shear and transverse bending.

It should be noted that the use of these membrane rules where other member resistance formulae could be used (such as the shear model in section 6.2) will generally lead to a

2-2/clause 6.109(101)

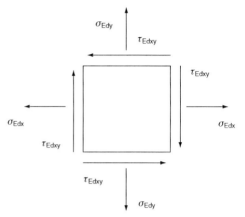

Fig. 6.9-1. Sign convention used in membrane rules

lower calculated resistance. This is because the membrane rules do not consider plastic redistribution within the cross-section or allow for the beneficial results of physical testing used specifically to derive the other member resistance rules.

To design the reinforcement and check the concrete compressive stresses, **2-2/clause 6.109(102)** requires a lower-bound solution, based on the lower-bound theorem of plasticity, to be used. 2-2/Annex F gives equations for designing the reinforcement. Unfortunately, the sign convention for direct stress in Annex F differs from that in 2-2/clause 6.109 and Fig. 6.9-1 (as compression is taken as positive in Annex F but negative in 2-2/clause 6.109). The general equations (F.8) to (F.10) of Annex F are therefore reproduced below with a modification to make them compatible with Fig. 6.9-1:

$$\rho_x \sigma_{sx} = |\tau_{Edxy}| \cot\theta + \sigma_{Edx} \leq \rho_x f_{yd,x} \qquad \text{(D6.9-1)}$$

$$\rho_y \sigma_{sy} = |\tau_{Edxy}| \tan\theta + \sigma_{Edy} \leq \rho_y f_{yd,y} \qquad \text{(D6.9-2)}$$

$$\sigma_{cd} = -|\tau_{Edxy}|(\tan\theta + \cot\theta) \leq \sigma_{cd,max} \qquad \text{(D6.9-3)}$$

ρ, σ_s and f_{yd} are the reinforcement ratio, reinforcement stress and reinforcement design yield stress in each direction respectively. θ is the angle of the assumed plastic compression field to the x-axis. $\sigma_{cd,max}$ is the maximum compressive stress in the concrete stress field. Significant limitations of Annex F and the above expressions are that the reinforcement must be aligned with the X and Y directions in the analysis model (although it is possible to rotate the output stress field to align with the reinforcement using Mohr's circle) and the reinforcement must not be skew. In the latter case, either the expressions need to be modified or analogous lower-bound methods used. A modified set of equations for skew reinforcement is presented beneath Worked example 6.9-1. Where it is required to design reinforcement for moment fields in slabs with or without skew reinforcement and without a net in plane axial force or shear force, methods such as those in references 19, 20 and 21 could be used, provided a check on the concrete compression field discussed below can be included as necessary.

2-2/Annex F gives versions of the above, by way of its expressions (F.2) to (F.7), which are optimized to minimize the reinforcement provision (taking $\tan\theta = 1$ if the greatest compressive stress is less in magnitude than the shear stress or $\tan\theta = |\tau_{Edxy}/\sigma_{Edx}|$ otherwise). Unfortunately, these optimized equations are often not valid according to the rules in 2-2/clause 6.109, as the optimized angle may lie outside the allowable limits below.

Since concrete has limited ductility, it is not possible to indiscriminately apply the lower-bound theorem of plasticity and therefore limits are set in **2-2/clause 6.109(103)** on the deviation of the assumed plastic compression field direction at angle θ, from the angle θ_{el} of the un-cracked elastic principal compressive stress direction. (This is similar to the qualitative requirements for strut-and-tie idealization discussed in section 6.5 of this guide.) Both angles are measured from the x-axis. If both principal stresses are tensile, the least tensile of these principal stresses is chosen to define θ_{el}. θ_{el} can be calculated from textbook formulae or from Mohr's circle, as illustrated in Worked example 6.9-1. At the ultimate limit state, $|\theta - \theta_{el}|$ is limited to $15°$ by 2-2/clause 6.109(103). In addition, the allowable concrete stress reduces with departure of the plastic compression field direction from the elastic principal compressive stress direction. This is again, in principle, consistent with much of the various sources of advice available on strut-and-tie modelling, although the magnitude of the reduction in strength in EC2-2 is very severe.

The limitations on angle of compression fields does, however, sometimes lead to conflict with other resistance models in EC2-2, such as the shear truss model in section 6.2, where the allowable limit of $\cot\theta = 2.5$ corresponds to $|\theta - \theta_{el}| = 23.2°$ at the neutral axis, where only shear stress is present. In this case, shear tests show no variation in the concrete crushing strength with angle. The limitations on angle and the use of $\theta = \theta_{el}$ lead to other anomalies, as illustrated in Worked example 6.9-1. In general, established resistance models should be used in preference to the membrane rules where both options are possible.

The calculated compressive stress according to equation (D6.9-3) should not exceed a limit which depends on the calculated elastic principal stresses and the assumed direction of the compressive stress field. EC2-2 identifies three limiting situations. In all cases, $\alpha_{cc} = 1.0$ in the definition of f_{cd} for compatibility with the case of direct sustained compression given in 2-2/clause 3.1.6 and the shear crushing limit of 2-2/clause 6.2:

(a) For uniaxial or biaxial compression and shear, if the maximum and minimum principal stresses from elastic analysis σ_1 and σ_2, respectively, are both compressive, the maximum allowable compression in the concrete stress field is:

$$\sigma_{cd\,max} = 0.85 f_{cd} \frac{1 + 3.80\alpha}{(1+\alpha)^2} \qquad \text{2-2/(6.110)}$$

where $\alpha = \sigma_2/\sigma_1$. This will apply if both σ_{Edx} and σ_{Edy} are compressive and $\sigma_{Edx}\sigma_{Edy} \geq \tau_{Edxy}^2$. No reinforcement is then needed, as identified by 2-1-1/Annex F.1(3). The design compressive stress in this case is the maximum principal compressive stress, as equation (D6.9-3) will not be applicable.

(b) If at least one principal stress is in tension and the stresses in the reinforcement determined from equations (D6.9-1) and (D6.9-2) **with $\theta = \theta_{el}$** are both less than or equal to yield, the maximum allowable compression in the concrete stress field varies with the reinforcement stress as follows:

$$\sigma_{cd\,max} = f_{cd}\left[0.85 - \frac{\sigma_s}{f_{yd}}(0.85 - \nu)\right] \qquad \text{2-2/(6.111)}$$

where σ_s is the maximum tensile stress in the reinforcement and $\nu = 0.6(1 - f_{ck}/250)$. The allowable compressive stress therefore varies from $0.85 f_{cd}$, where there is no tension across the compression band (which is consistent with the allowable stress of $0.85 f_{ck}/1.5$ in the design of members in bending and axial load) to νf_{cd} for members where there is yield tension in the reinforcement crossing the compression band (which is consistent with the allowable compressive stress of νf_{cd} in the design of members in shear).

(c) If the reinforcement is designed using equations (D6.9-1) and (D6.9-2) **with $\theta = \theta_{el}$**, the maximum allowable compression in the concrete stress field is:

$$\sigma_{cd\,max} = \nu f_{cd}(1 - 0.032|\theta - \theta_{el}|) \qquad \text{2-2/(6.112)}$$

This reduction is very severe and will give more conservative results than the shear truss model of 2-2/clause 6.2, where the allowable stress can effectively always be taken as νf_{cd} regardless of strut angle. When departing from the elastic compression angle, no provision is made for increasing the compression limit in 2-2/Expression (6.112) when the reinforcement is not fully utilized.

In summary, if reinforcement is designed using $\theta = \theta_{el}$ and is fully stressed, the compression limit is νf_{cd}. If it is necessary to depart from $\theta = \theta_{el}$ to produce the desired distribution of reinforcement then the compression limit will be less than νf_{cd}. It is always therefore necessary to calculate θ_{el} to determine which compression limit applies. The use and limitations of these membrane rules is illustrated in Worked example 6.9-1. Their use in the design of webs in shear and transverse bending is covered in Annex M of this guide.

> **Worked example 6.9-1: Use of membrane rules**
> The stress cases below use the sign convention of Fig. 6.9-1 (i.e. compressive stress negative). The concrete element has concrete grade C40/50 and is 400 mm thick. Reinforcement has $f_{yd} = 435$ MPa. The required reinforcement in the x and y directions is calculated for a number of stress fields and the adequacy of the concrete in compression is checked.
>
> **(1) Biaxial compression and shear**
> Assume the following stresses are obtained:
>
> $\sigma_{Edx} = -20$ MPa
>
> $\sigma_{Edy} = -4$ MPa
>
> $\tau_{Edxy} = 7$ MPa
>
> From Fig. 6.9-2(a), the maximum and minimum principal stresses are -22.6 MPa and -1.4 MPa so the concrete is wholly in compression. (This could also be predicted as $\sigma_{Edx}\sigma_{Edy} \geq \tau_{Edxy}^2 \rightarrow 20 \times 4 = 80 \geq 7^2 = 49$.)
>
> The ratio of principal stresses is given by $\alpha = \sigma_2/\sigma_1 = 1.4/22.6 = 0.06$. From 2-2/Expression (6.110):
>
> $$\sigma_{cd,max} = 0.85 f_{cd} \frac{1 + 3.80\alpha}{(1+\alpha)^2}$$
>
> $$= 0.85 \times 1.0 \times \frac{40}{1.5} \times \frac{1 + 3.80 \times 0.06}{(1+0.06)^2} = 24.8 \text{ MPa} > 22.6 \text{ MPa}$$
>
> (Note $\sigma_{cd,max} = 0.93 f_{cd}$ in this case and $\alpha_{cc} = 1.0$ as discussed in the main text.)
> The concrete is adequate and no reinforcement is needed.
>
> **(2) Uniaxial compression and shear (web–flange longitudinal shear check)**
> Assume the following stresses are obtained:
>
> $\sigma_{Edx} = -20$ MPa
>
> $\sigma_{Edy} = 0$ MPa
>
> $\tau_{Edxy} = 6.72$ MPa
>
> The shear stress above is the maximum allowable according to the rules for web-flange shear in 2-1-1/clause 6.2.4, with strut angle $\theta_f = 45°$ used to obtain maximum crushing resistance as follows.
> From 2-1-1/Expression (6.21):
>
> $$\frac{A_{sf} f_{yd}}{s_f h_f} = \rho_y f_{yd} = \tau_{Edxy} = 6.72 \text{ MPa}$$
>
> so the reinforcement ratio $\rho_y = 6.72/435 = 0.0154$. The required transverse reinforcement is therefore:
>
> $0.0154 \times 400 = 6.16 \text{ mm}^2/\text{mm}$
>
> From 2-1-1/Expression (6.22), the concrete resistance to shear stress with $\theta_f = 45°$ is given by:
>
> $$\nu f_{cd}/2 = \frac{0.6(1 - 40/250) \times 40/1.5}{2} = 6.72 \text{ MPa}$$

which is equal to the applied shear stress so the concrete stress is just adequate. ($\alpha_{cc} = 1.0$ for shear in the above as discussed in section 3.1.6 of this guide.) Variations to the angle will reduce the transverse shear reinforcement, but would also reduce the crushing resistance so the concrete would be overstressed.

The check is now repeated using the membrane rules and setting the angle of the plastic compression field to its elastic value.

From Fig. 6.9-2(b), the maximum and minimum principal stresses are -22.0 MPa and 2.0 MPa:

$$2\theta_{el} = \tan^{-1}\left(\frac{6.72}{\frac{1}{2}(20-0)}\right) = 33.90° \text{ so } \theta_{el} = 16.95°$$

$$\rho_x \sigma_{sx} = |\tau_{Edxy}|\cot\theta + \sigma_{Edx} = 6.72 \times \cot 16.95° - 20 = 2.05 \text{ MPa}$$

$$\rho_y \sigma_{sy} = |\tau_{Edxy}|\tan\theta + \sigma_{Edy} = 6.72 \times \tan 16.95° + 0 = 2.05 \text{ MPa}$$

Assuming the reinforcement to be fully stressed, the reinforcement ratio in each direction is therefore $\rho = 2.05/435 = 0.00475$ which is equivalent to reinforcement of $0.00475 \times 400 = 1.90 \text{ mm}^2/\text{mm}$. Less transverse reinforcement is therefore required than from the rules of 2-1-1/clause 6.2.4, but some longitudinal reinforcement is also required:

$$\sigma_{cd} = -|\tau_{Edxy}|(\tan\theta + \cot\theta) = -6.72 \times (\tan 16.95° + \cot 16.95°) = -24.1 \text{ MPa}$$

Assuming the reinforcement is to be designed to be fully stressed by providing the reinforcement ratio above, then the allowable concrete stress from 2-2/Expression (6.111) is:

$$\sigma_{cd\,max} = \nu f_{cd} = 0.6(1 - 40/250) \times \frac{40}{1.5} = 13.44 \text{ MPa} \ll 24.1 \text{ MPa}$$

so the concrete compression resistance is not adequate. There are three possibilities to improve the concrete verification:

(a) If the reinforcement quantity is increased so that its stress reduces, the concrete resistance from 2-2/Expression (6.111) increases but it cannot increase beyond:

$$\sigma_{cd\,max} = 0.85 f_{cd} = 0.85 \times \frac{40}{1.5} = 22.67 \text{ MPa} < 24.1 \text{ MPa}$$

so the solution using a compression field angle equal to that of the elastic principal stress angle results in concrete failure.

(b) If the compression angle is taken as $45°$, as in the design to 2-2/clause 6.2.4, the check would result in the same reinforcement and applied concrete compressive stress as to 2-2/clause 6.2.4 and therefore an apparent adequate verification. However, this solution is not allowed by 2-2/clause 6.109(103) as $|\theta - \theta_{el}| = 45 - 16.95 = 28.05 > 15°$, which is the maximum angular departure allowed.

(c) If the compression angle is increased from the elastic angle by the maximum $15°$ allowed, so that $\theta = 15 + 16.95 = 31.95°$, then

$$\sigma_{cd} = -|\tau_{Edxy}|(\tan\theta + \cot\theta)$$

$$= -6.72 \times (\tan 31.95° + \cot 31.95°) = -14.97 \text{ MPa}$$

which is a large reduction. However, the allowable compression is also reduced according to 2-2/Expression (6.112) so that:

$$\sigma_{cd\,max} = \nu f_{cd}(1 - 0.032|\theta - \theta_{el}|)$$

$$= 0.6(1 - 40/250) \times 40/1.5 \times (1 - 0.032 \times 15)$$

$$= 6.99 \text{ MPa} \ll 14.97 \text{ MPa}$$

Consequently, adequacy of the concrete cannot be demonstrated using the membrane rules, even though the concrete was adequate using the rules of 2-2/clause 6.4.2. This illustrates that it is usually preferable to use code resistance formulae, where available.

(3) **Uniaxial compression**
Assume the following stresses are obtained:

$\sigma_{Edx} = -20\,\text{MPa}$

$\sigma_{Edy} = 0\,\text{MPa}$

$\tau_{Edxy} = 0\,\text{MPa}$

The angle of the compression field is set to the elastic angle and the expressions for reinforcement are first derived for small shear stress: From Fig. 6.9-2(c),

$$\tan 2\theta_{el} = \frac{2\tau_{Edxy}}{-\sigma_{Edx}} \quad \text{so as } \tau_{Edxy} \to 0\,\text{MPa}, \quad \tan \theta_{el} = \frac{\tau_{Edxy}}{-\sigma_{Edx}}$$

(Note that care is needed with signs to keep $\tan\theta$ positive.) The reinforcement is then determined from equations (D6.9-1) and (D6.9-2) as follows:

$$\rho_x \sigma_{sx} = |\tau_{Edxy}|\cot\theta + \sigma_{Edx} = |\tau_{Edxy}|\frac{-\sigma_{Edx}}{\tau_{Edxy}} + \sigma_{Edx} = 0$$

$$\rho_y \sigma_{sy} = |\tau_{Edxy}|\tan\theta + \sigma_{Edy} = |\tau_{Edxy}|\frac{\tau_{Edxy}}{-\sigma_{Edx}} + \sigma_{Edy} = \sigma_{Edy} = 0 \text{ for } \tau_{Edxy} = 0\,\text{MPa}$$

The compressive stress is then:

$$\sigma_{cd} = -|\tau_{Edxy}|(\tan\theta + \cot\theta) = -|\tau_{Edxy}|\left(\frac{\tau_{Edxy}}{-\sigma_{Edx}} + \frac{-\sigma_{Edx}}{\tau_{Edxy}}\right) = \sigma_{Edx} = -20\,\text{MPa}$$

Since no reinforcement is required, the compression limit can be checked using 2-2/Expression (6.111):

$$\sigma_{cd\,max} = f_{cd}\left[0.85 - \frac{\sigma_s}{f_{yd}}(0.85 - \nu)\right]$$

$$= 0.85 f_{cd} = 0.85 \times 40/1.5 = 22.67\,\text{MPa} > 20\,\text{MPa}$$

The concrete is therefore adequate as expected.

(4) **Uniaxial tension**
Assume the following stresses are obtained:

$\sigma_{Edx} = 5\,\text{MPa}$

$\sigma_{Edy} = 0\,\text{MPa}$

$\tau_{Edxy} = 0\,\text{MPa}$

The angle of the compression field is set to the elastic angle and the expressions for reinforcement are first derived for small shear stress to illustrate the problem of indiscriminately selecting $\theta = \theta_{el}$. From Fig. 6.9-2(d):

$$\tan 2\theta_{el} = \tan\left(\pi - \frac{2\tau_{Edxy}}{\sigma_{Edx}}\right) \text{ as } \tau_{Edxy} \to 0\,\text{MPa}$$

so $\tan\theta_{el} = \tan\left(\pi/2 - \frac{\tau_{Edxy}}{\sigma_{Edx}}\right) = \frac{\sigma_{Edx}}{\tau_{Edxy}}$ from trigonometric rules as $\tau_{Edxy} \to 0\,\text{MPa}$

$$\rho_x \sigma_{sx} = |\tau_{Edxy}|\cot\theta + \sigma_{Edx} = |\tau_{Edxy}|\frac{\tau_{Edxy}}{\sigma_{Edx}} + \sigma_{Edx} = \sigma_{Edx} = 5\,\text{MPa}$$

for $\tau_{Edxy} = 0$ MPa. Similarly:

$$\rho_y \sigma_{sy} = |\tau_{Edxy}| \tan\theta + \sigma_{Edy} = |\tau_{Edxy}|\frac{\sigma_{Edx}}{\tau_{Edxy}} + \sigma_{Edy} = \sigma_{Edx} + \sigma_{Edy} = 5\,\text{MPa}$$

The reinforcement required is therefore $5.0/435 \times 400 = 4.60\,\text{mm}^2/\text{mm}$ in *both* directions. This is clearly not sensible and results from the selection of $\theta = \theta_{el}$. If the compression field is assumed to be at 45°, as Annex F recommends for optimized reinforcement, the correct result of reinforcement being required only in the direction of tension is obtained. The compression is as follows:

$$\sigma_{cd} = -|\tau_{Edxy}|(\tan\theta + \cot\theta) = -|\tau_{Edxy}|\left(\frac{\sigma_{Edx}}{\tau_{Edxy}} + \frac{\tau_{Edxy}}{\sigma_{Edx}}\right) = -\sigma_{Edx} = -5\,\text{MPa}$$

This is adequate, although physically has no meaning. The use of $\theta = \theta_{el}$ is further unsatisfactory here as **2-2/clause F.1(104)** requires the reinforcement derived to be neither more than twice nor less than half that from the case with the compression field at 45°, which gives reinforcement in the applied tension direction only. The 45° case is not, however, allowed by 2-2/clause 6.109(103)!

2-2/clause F.1(104)

In this particular instance, the problem above can be overcome by varying the angle from the elastic within the 15° limit. If $\theta = 85°$, then:

$$\rho_x \sigma_{sx} = |\tau_{Edxy}| \cot\theta + \sigma_{Edx} = 0 + 5 = 5\,\text{MPa}$$

$$\rho_y \sigma_{sy} = |\tau_{Edxy}| \tan\theta + \sigma_{Edy} = 0 + 0 = 0\,\text{MPa}$$

$$\sigma_{cd} = -|\tau_{Edxy}|(\tan\theta + \cot\theta) = 0\,\text{MPa}$$

The correct solution is then yielded. This example is only intended to highlight the problems that can occur by sticking rigidly to the elastic compression angle, which might occur where spreadsheets are used to automate calculations for example. The example of zero shear stress may seem artificial, but a similar procedure of departing from the elastic angle is required if the above example is repeated with a small value of shear stress.

(5) Uniaxial tension (repeated with tension in the Y direction)
Assume the following stresses are obtained:

$\sigma_{Edx} = 0$ MPa

$\sigma_{Edy} = 5$ MPa

$\tau_{Edxy} = 0$ MPa

This time the angle of the compression field to the X axis is equal to 0:

$$\tan 2\theta_{el} = \frac{2\tau_{Edxy}}{\sigma_{Edy}} \text{ as } \tau_{Edxy} \to 0\,\text{MPa} \quad \text{so } \tan\theta_{el} = \frac{\tau_{Edxy}}{\sigma_{Edy}}$$

If $\theta = \theta_{el}$, this yields the same unsatisfactory results for reinforcement and compression as case (4) above:

$$\rho_x \sigma_{sx} = |\tau_{Edxy}| \cot\theta + \sigma_{Edx} = |\tau_{Edxy}|\frac{\sigma_{Edy}}{\tau_{Edxy}} + \sigma_{Edx} = \sigma_{Edy} + \sigma_{Edx} = 5\,\text{MPa}$$

$$\rho_y \sigma_{sy} = |\tau_{Edxy}| \tan\theta + \sigma_{Edy} = |\tau_{Edxy}|\frac{\tau_{Edxy}}{\sigma_{Edy}} + \sigma_{Edy} = \sigma_{Edy} = 5\,\text{MPa}$$

$$\sigma_{cd} = -|\tau_{Edxy}|(\tan\theta + \cot\theta) = -|\tau_{Edxy}|\left(\frac{\tau_{Edxy}}{\sigma_{Edy}} + \frac{\sigma_{Edy}}{\tau_{Edxy}}\right) = -\sigma_{Edy} = -5\,\text{MPa}$$

The same remedy as in (4) above can be applied.

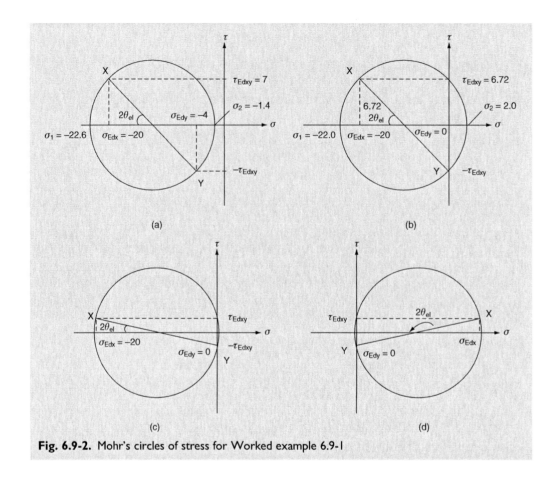

Fig. 6.9-2. Mohr's circles of stress for Worked example 6.9-1

Skew reinforcement

It is possible to derive similar expressions to equations (D6.9-1) to (D6.9-3) for cases with skew reinforcement. Such expressions have been presented by Bertagnoli, G., Carbone, V.I., Giordano, L. and Mancini, G. Unfortunately, their expressions as presented at the C.I. Premier Congress on 1–2 July 2003 in Milan entitled '2nd International Speciality Conference on the Conceptual Approach to Structural Design' contained a typographical mistake and cannot therefore be referenced. They are reproduced below with modifications to correct the error and to make the notation compatible with EN 1992. The sign convention is shown in Fig. 6.9-3.

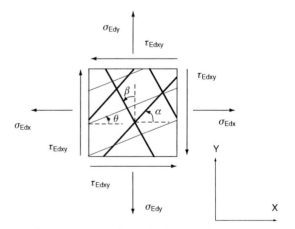

Fig. 6.9-3. Sign convention for membrane rules with skew reinforcement

$$\rho_\alpha \sigma_{s\alpha} = \frac{\sigma_{Edx} \sin\theta \cos\beta - \sigma_{Edy} \cos\theta \sin\beta + \tau_{Edxy} \cos(\theta + \beta)}{\sin(\theta - \alpha) \cos(\alpha - \beta)} \tag{D6.9-4}$$

$$\rho_\beta \sigma_{s\beta} = \frac{\sigma_{Edx} \sin\theta \sin\alpha + \sigma_{Edy} \cos\theta \cos\alpha + \tau_{Edxy} \sin(\theta + \alpha)}{\cos(\theta - \beta) \cos(\alpha - \beta)} \tag{D6.9-5}$$

$$\sigma_{cd} = \sigma_{Edx} - \tau_{Edxy} \tan\theta - \rho_\alpha \sigma_{s\alpha} \cos(\theta - \alpha) \frac{\cos\alpha}{\cos\theta} + \rho_\beta \sigma_{s\beta} \sin(\theta - \beta) \frac{\sin\beta}{\cos\theta} \tag{D6.9-6}$$

In using the above equations, it is vital that the sign convention for the angles and stresses in Fig. 6.9-3 (anti-clockwise positive) is observed and that the direction of the plastic compression field, θ, is taken to be in the same X, Y quadrant as the angle of the principal compressive stress in the un-cracked state, θ_{el}. The former is not required in equations (D6.9-1) to (D6.9-3), where θ is always taken as positive because of the mod sign introduced on the shear stress terms.

CHAPTER 7

Serviceability limit states

This chapter deals with the design at service limit states of members as covered in section 7 of EN 1992-2 in the following clauses:

- General   Clause 7.1
- Stress limitation   Clause 7.2
- Crack control   Clause 7.3
- Deflection control   Clause 7.4

An additional section 7.5 is included to discuss early thermal cracking.

7.1. General

EN 1992-2 section 7 covers only the three serviceability limit states relating to clause 7.2 to 7.4 above. *2-1-1/clause 7.1(1)P* notes that other serviceability limit states 'may be of importance'. EN 1990/A2.4 is relevant in this respect. It covers partial factors, serviceability criteria, design situations, comfort criteria, deformations of railway bridges and criteria for the safety of rail traffic. Most of its provisions are qualitative but some recommended values are given in various Notes, as guidance for National Annexes.

EN 1990 is of general relevance. From clause 6.5.3 of EN 1990, the relevant combination of actions for serviceability limit states is 'normally' either the characteristic, frequent, or quasi-permanent combination. These are all used in EC2-2 and the general forms of these combinations, together with examples of use, are given in Table 7.1, but reference to section 2 and Annex A2 of EN 1990 is recommended for a detailed explanation of the expressions and terms. Specific rules for the combinations of actions (e.g. actions that need not be considered together), recommended combination factors and partial safety factors for bridge design are also specified in Annex A2 of EN 1990. Section 2 of this guide gives further commentary on the basis of design and the use of partial factors and combinations of actions.

The general expressions in Table 7.1 have been simplified assuming that partial factors of 1.0 are used throughout for all actions at the serviceability limit state, as recommended in Annex A2 of EN 1990, but they may be varied in the National Annex.

Appropriate methods of global analysis for determining design action effects are discussed in detail in section 5. For serviceability limit state verification, the global analysis may be either elastic without redistribution (clause 5.4) or non-linear (clause 5.7). Elastic global analysis is most commonly used and it is not normally then necessary to consider the effects of cracking within it – section 5.4 refers.

2-1-1/clause 7.1(2) permits an un-cracked concrete cross-section to be assumed for stress and deflection calculation provided that the flexural tensile stress under the relevant combination of actions considered does not exceed $f_{ct,eff}$. $f_{ct,eff}$ may be taken as either f_{ctm}

2-1-1/clause 7.1(1)P

2-1-1/clause 7.1(2)

Table 7.1. Combinations of actions for serviceability limit states

Combination	Notes	General expression
Characteristic	Combination of actions with a fixed (small) probability of being exceeded during normal operation within the structure's design life, e.g. combination appropriate to checks on stress in reinforcement as it is undesirable for inelastic deformation of reinforcement to occur at any time during the service life.	$\sum G_{k,j} + P + Q_{k,1} + \sum \psi_{0,i} Q_{k,i}$
Frequent	Combination of actions with a fixed probability of being exceeded during a reference period of a few weeks, e.g. combination used for checks of cracking and decompression in prestressed bridges with bonded tendons.	$\sum G_{k,j} + P + \psi_{1,1} Q_{k,1} + \sum \psi_{1,i} Q_{k,i}$
Quasi-permanent	Combination of actions expected to be exceeded approximately 50% of the time, i.e. a time-based mean. For example, combination appropriate to crack width checks in reinforced concrete members on the basis that durability is influenced by average crack widths, not the worst crack width ever experienced.	$\sum G_{k,j} + P + \psi_{2,1} Q_{k,1} + \sum \psi_{2,i} Q_{k,i}$

or $f_{\text{ctm,fl}}$ but should be consistent with the value used in the calculation of minimum tension reinforcement (see section 7.3). For the purpose of calculating crack widths and tension stiffening effects, f_{ctm} should be used.

7.2. Stress limitation

Stresses in bridges are limited to ensure that under normal conditions of use, assumptions made in design models (e.g. linear-elastic behaviour) remain valid, and to avoid deterioration such as the spalling of concrete or excessive cracking leading to a reduction of durability. For persistent design situations, it is usual to check stresses soon after the opening of the bridge to traffic, when little creep has occurred, and also at a later time when creep and shrinkage are substantially complete. This affects the loss of prestress in prestressed structures and the modular ratio for stress and crack width calculation in reinforced concrete structures. It may be necessary to include part of the long-term shrinkage effects in the first check, because up to half of the long-term shrinkage can occur in the first 3 months after the end of curing of the concrete. Calculation of an effective concrete modulus allowing for creep is discussed below.

2-1-1/clause 7.2(1)P
2-2/clause 7.2(102)

2-1-1/clause 7.2(1)P requires compressive stresses in the concrete to be limited to avoid longitudinal cracking, micro-cracking or excessive creep. The first two can lead to a reduction of durability. *2-2/clause 7.2(102)* addresses longitudinal cracking by requiring the stress level under the characteristic combination of actions to not exceed a limiting value of $k_1 f_{ck}$ (for areas with exposure classes of XD, XF or XS), where k_1 is a nationally determined parameter with recommended value of 0.6. The clause identifies that the limit can be increased where specific measures are taken, such as increasing the cover to reinforcement (from the minimum values discussed in section 4) or by providing confinement by transverse reinforcement. The improvement from confining reinforcement is quantified as an increase in allowable stress of 10%, but this may be varied in the National Annex. The design of this reinforcement is not covered by EC2-2, but the strut-and-tie rules in 2-2/clause 6.5 and discussions in section 6.5 of this guide are relevant. Such reinforcement would need to operate at low stresses to have any significant effect in limiting the width of compression-induced cracks.

Micro-cracking typically begins to develop in concrete where the compressive stress exceeds approximately 70% of the compressive strength. Given the limits above to control longitudinal cracking, no further criteria are given to control micro-cracking.

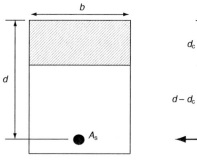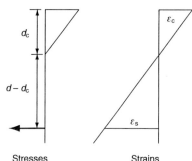

Fig. 7.1. Notation for a rectangular beam

2-1-1/clause 7.2(3) addresses non-linear creep as covered by 2-2/clause 3.1.4(4). It requires non-linear creep to be considered where the stress under the quasi-permanent combination of actions exceeds $k_2 f_{ck}$, where k_2 is a nationally determined parameter with recommended value of 0.45. 2-2/clause 3.1.4(4) gives the same limiting stress, but it is not subject to national variation in that clause so must be deemed to take precedence.

2-1-1/clause 7.2(4)P requires stresses in reinforcement and prestressing steel to be limited to ensure inelastic deformations of the steel are avoided under serviceability loads, which could result in excessive concrete crack widths and invalidate the assumptions on which the calculations within EC2 for cracking and deflections are based. *2-1-1/clause 7.2(5)* requires that the tensile stress in reinforcement under the characteristic combination of actions does not exceed $k_3 f_{yk}$. Where the stress is caused by imposed deformations, the tensile stress should not exceed $k_4 f_{yk}$, although it will be rare for tensile stress to exist solely from imposed deformations. The mean value of stress in prestressing tendons should not exceed $k_5 f_{pk}$. The values of k_3, k_4 and k_5 are nationally determined parameters and are recommended to be taken as 0.8, 1.0 and 0.75 respectively. The higher stress limit for reinforcement tension under indirect actions reflects the ability for stresses to be shed upon concrete cracking.

The following method can be used to determine stresses in cracked reinforced concrete beams and slabs. The concrete modulus to use for section analysis depends on the ratio of permanent (long-term) actions to variable (short-term) actions. The short-term modulus is E_{cm} and the long-term modulus is $E_{cm}/(1+\phi)$. The effective concrete modulus for a combination of long-term and short-term actions can be taken as:

$$E_{c,eff} = \frac{(M_{qp} + M_{st})E_{cm}}{M_{st} + (1+\phi)M_{qp}} \qquad (D7\text{-}1)$$

where M_{st} is the moment due to short-term actions and M_{qp} is the moment from quasi-permanent actions. The neutral axis depth and steel strain can be derived from a cracked elastic analysis, assuming plane sections remain plane. For a rectangular beam, from Fig. 7.1:

Strains

$$\varepsilon_s = \frac{d - d_c}{d_c}\varepsilon_c \qquad (D7\text{-}2)$$

Forces

$$F_s = F_c$$

so

$$A_s E_s \varepsilon_s = 0.5 b d_c \varepsilon_c E_{c,eff} \qquad (D7\text{-}3)$$

Putting equation (D7-2) into equation (D7-3) gives:

$$d_c = \frac{-A_s E_s + \sqrt{(A_s E_s)^2 + 2 b A_s E_s E_{c,eff} d}}{b E_{c,eff}} \qquad (D7\text{-}4)$$

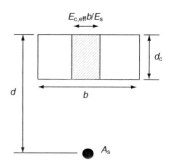

Fig. 7.2. Cracked section transformed to steel units

The second moment of area of the cracked section, in steel units, is derived from the cross-section in Fig. 7.2:

$$I = A_s(d - d_c)^2 + \frac{1}{3}\frac{E_{c,eff}}{E_s}bd_c^3 \qquad \text{(D7-5)}$$

The elastic section moduli are:

Concrete: $z_c = I/d_c$ (D7-6)

Steel: $z_s = I/(d - d_c)$ (D7-7)

For a given moment, M_{Ed}, the stresses are therefore:

Concrete: $\sigma_c = \dfrac{M_{Ed}}{z_c}\dfrac{E_{c,eff}}{E_s}$ (D7-8)

Steel: $\sigma_s = \dfrac{M_{Ed}}{z_s}$ (D7-9)

The strains are:

Concrete: $\varepsilon_c = \dfrac{M_{Ed}}{z_c}\dfrac{1}{E_s}$ (D7-10)

Steel: $\varepsilon_s = \dfrac{M_{Ed}}{z_s}\dfrac{1}{E_s}$ (D7-11)

The above may also be applied to flanged beams where either the neutral axis remains in the compression flange (when b is the flange width) or remains in the web when the flange is wholly in tension (whereupon b is the web width).

The procedure for checking stresses is illustrated in Worked example 7.1. In particular, this illustrates the treatment of creep on modular ratio.

Worked example 7.1: Reinforced concrete deck slab

A reinforced concrete deck slab, 250 mm thick and with class C35/45 concrete, is subjected to a transverse sagging moment of 85 kNm/m under the characteristic combination of actions at SLS. This moment comprises 15% permanent actions from self-weight and superimposed dead load and 85% transient actions from traffic. The ultimate design requires a reinforcement area of 2010 mm^2/m (16 mm diameter bars at 100 mm centres) at an effective depth of 192 mm. The serviceability stresses are checked against the relevant limits.

The section is first checked to see if it is un-cracked in accordance with 2-1-1/clause 7.1(2):

Depth to neutral axis $h/2 = 250/2 = 125$ mm

Second moment of area, $I = bh^3/12 = 1000 \times 250^3/12 = 1.302 \times 10^9$ mm^4/m.

Therefore, the un-cracked section gives the following compressive and tensile stress at the top and bottom of the section respectively:

$\sigma_{top} = \sigma_{bot} = My/I = 85 \times 10^6 \times 125/1.302 \times 10^9 = 8.16$ MPa

From 2-1-1/Table 3.1 for class C35/45 concrete, $f_{ctm} = 3.2$ MPa $< \sigma_{bot}$; therefore the section is cracked. Stresses therefore will be calculated ignoring concrete in tension. The relevant modular ratio, α_{eff}, depends on the proportion of long-term and short-term loading.

(a) First, a check is performed at an age when the bridge first opens, assuming minimal creep has occurred, and therefore the short-term modulus is used for all loading:

$$E_s = 200 \text{ GPa, and from 2-1-1/Table 3.1}, E_{cm} = 22\left(\left(\frac{f_{ck}+8}{10}\right)^{0.3}\right)$$

thus

$$E_{c,eff} = E_{cm} = 22\left(\left(\frac{35+8}{10}\right)^{0.3}\right) = 34.08 \text{ GPa}$$

The depth of concrete in compression from equation (D7-4) is:

$$d_c = \frac{-A_s E_s + \sqrt{(A_s E_s)^2 + 2bA_s E_s E_{c,eff} d}}{bE_{c,eff}}$$

$$= \frac{-2010 \times 200 \times 10^9 + \sqrt{(2010 \times 200 \times 10^9)^2 + 2 \times 1000 \times 2010 \times 200 \times 10^9 \times 34.08 \times 10^9 \times 192}}{1000 \times 34.08 \times 10^9}$$

$$= 56.53 \text{ mm}$$

The cracked second moment of area in steel units from equation (D7-5) is:

$$I = A_s(d - d_c)^2 + \frac{1}{3}\frac{E_{c,eff}}{E_s}bd_c^3$$

$$= 2010 \times (192 - 56.53)^2 + \frac{1}{3} \times \frac{34.08}{200} \times 1000 \times 56.53^3 = 47.15 \times 10^6 \text{ mm}^4$$

The concrete stress at the top of the section from equation (D7-8) is:

$$\sigma_c = \frac{M_{Ed}}{z_c}\frac{E_{c,eff}}{E_s} = \frac{85 \times 10^6}{47.15 \times 10^6/56.54} \times \frac{34.08}{200} = 17.37 \text{ MPa}$$

From 2-2/clause 7.2(102), the compression limit $= k_1 f_{ck} = 0.6 \times 35 = 21$ MPa >17.37 MPa as required.

The reinforcement stress from equation (D7-9) is:

$$\sigma_s = \frac{M_{Ed}}{z_s} = 85 \times 10^6 \times (192 - 56.54)/(47.15 \times 10^6) = 244.2 \text{ MPa}$$

From 2-1-1/clause 7.2(5), the tensile limit $= k_3 f_{yk} = 0.8 \times 500 = 400$ MPa >244.2 MPa as required.

(b) Second, a check is performed after all the creep has taken place. The creep factor is determined for the long-term loading using 2-1-1/clause 3.1.4 and is found to be $\phi = 2.2$. This is used to calculate an effective modulus of elasticity for the concrete under the specific proportion of long-term and short-term actions defined using equation (D7-1):

$$E_{c,eff} = \frac{(M_{qp} + M_{st})E_{cm}}{M_{st} + (1+\phi)M_{qp}} = \frac{(0.15 + 0.85) \times 34.08}{0.85 + (1+2.2) \times 0.15} = 25.62 \text{ GPa}$$

Repeating the calculation process in (a) above, the depth of concrete in compression from equation (D7-4) is 63.50 mm and the cracked second moment of area in steel units from equation (D7-5) is 44.12×10^6 mm^4.

> The concrete stress at the top of the section from equation (D7-8) is:
>
> $$\sigma_c = \frac{M_{Ed}}{z_c}\frac{E_{c,eff}}{E_s} = \frac{85 \times 10^6}{44.12 \times 10^6/63.50} \times \frac{25.62}{200} = 15.67\,\text{MPa} < 21\,\text{MPa as required}$$
>
> The reinforcement stress from equation (D7-9) is:
>
> $$\sigma_s = \frac{M_{Ed}}{z_s} = 85 \times 10^6 \times (192 - 63.5)/(44.12 \times 10^6)$$
>
> $$= 247.6\,\text{MPa} < 400\,\text{MPa as required}$$
>
> The effect of creep here is to reduce the concrete stress and slightly increase the reinforcement stress.

Treatment of differential temperature in stress calculation

For bridge beams, a calculation of the stresses induced by non-uniform temperature distributions needs to be considered. Strictly, these stresses (which include primary self-equilibriating stresses and secondary stresses due to restraint of deflection) need to be included in the stress checks discussed above. For cracked sections, the analysis to determine self-equilibriating stresses is complicated and highly iterative. However, since cracking results in a reduction in stiffness of the section, cracking of a section will lead to a substantial relaxation of the stresses induced by temperature. It is therefore generally satisfactory to ignore temperature-induced self-equilibriating stresses in cracked sections and to consider only the secondary effects.

The neglect of temperature-induced self-equilibriating stresses applies also to crack width calculation for reinforced concrete sections in 2-2/clause 7.3, where temperature is also included in the quasi-permanent combination for bridge members. For prestressed members where decompression is being checked (and therefore sections are expected to be substantially uncracked), both primary and secondary effects should, however, be included in the calculation of stresses. This approach is in accordance with previous UK practice and the UK's National Annex.

7.3. Crack control

The consistency in use of symbols in this section of EN 1992-2 and EN 1992-1-1 is poor, with the same symbols sometimes changing definition from sub-clause to sub-clause. Care is therefore needed to use the correct definition in the relevant clause.

7.3.1. General considerations

2-1-1/clause 7.3.1(1)P states that cracking shall be limited to the extent that it should not impair the proper functioning or durability of the structure, or cause its appearance to be unacceptable. *2-1-1/clause 7.3.1(2)* is, however, a reminder that cracking is inevitable in reinforced concrete bridges subjected to bending, shear, torsion or tension. Cracking may arise from the result of either direct loading, from traffic actions, for example, or restraint of imposed deformations, such as shrinkage or temperature movements. The rules in EC2 cover the control of cracking from these causes and are discussed in detail in this section. Section 7.5 below discusses early thermal cracking. Additionally, cracks may arise from other causes such as plastic shrinkage, corrosion of reinforcement or expansive chemical reactions (such as alkali–silica reaction). *2-1-1/clause 7.3.1(3)* notes that the control of such cracks is beyond the scope of EC2, even though they could be very large if they occur.

2-1-1/clause 7.3.1(4) and *2-2/clause 7.3.1(105)* both essentially require the design crack width to be chosen such that cracking does not impair the functioning of the structure. Cracking normally 'impairs' the function of the structure by either helping to initiate reinforcement corrosion or by spoiling its appearance. The relationship between cracking

2-1-1/clause 7.3.1(1)P
2-1-1/clause 7.3.1(2)

2-1-1/clause 7.3.1(3)
2-1-1/clause 7.3.1(4)
2-2/clause 7.3.1(105)

CHAPTER 7. SERVICEABILITY LIMIT STATES

and corrosion in reinforced concrete has been extensively researched. The alkalinity of fresh concrete protects reinforcement from corrosion. This protection can be destroyed, however, by carbonation or ingress of chlorides. Cracks can lead to an acceleration of both of these processes by providing a path for carbon dioxide and chloride ions to the reinforcement. The size of the cracks also has an influence on the time to initiation of reinforcement corrosion. Noticeable cracking in structures causes concern to the public and it is therefore prudent to limit crack widths to a size that is not readily noticeable.

The above considerations have led to the crack width limitations specified in 2-2/Table 7.101N, which is subject to variation in a National Annex. 2-2/clause 7.3.1(105) notes that although compliance with the crack width calculation methods and adherence to these limiting crack widths should guarantee adequate performance, the calculated crack widths themselves should not be considered as real values. For reinforced concrete, the crack width check is recommended to be performed under the quasi-permanent load combination. This effectively excludes traffic for highway bridges when the recommended value of $\psi_2 = 0$ from Annex A2 of EN 1990 is used. The quasi-permanent combination does, however, include temperature. In checking crack widths in reinforced concrete members, only the secondary effects of temperature difference need to be considered as discussed in section 7.2 above. For bonded prestressed members, however, the self-equilibriating stresses should also be included in decompression checks.

Prestressing steels are much more sensitive to damage from corrosion than normal reinforcement, mostly due to their smaller diameter and higher level of stress under which they normally operate. It is therefore widely accepted that it is necessary to have more onerous rules for protection of prestressed concrete members against corrosion. This is reflected in stricter crack control criteria for prestressed members with bonded tendons in 2-2/Table 7.101N. It also specifies requirements for decompression checks for prestressed members with bonded tendons and defines under which relevant combination of actions the decompression check is required. For XC2, XC3 and XC4 environments, it is the quasi-permanent combination while for XD and XS classes, it is the frequent combination. Members with only unbonded tendons are treated in the same way as reinforced concrete members – *2-2/clause 7.3.1(6)*.

2-2/clause 7.3.1(6)

In order to safeguard bonded tendons from corrosion, it would be logical for two-way spanning elements with prestressing in one direction only, such as a deck slab in a prestressed concrete box girder, to also have stricter crack criteria in the direction transverse to the prestressing. This is not, however, explicitly required by 2-2/Table 7.101N and was not required in previous UK codes.

The decompression limit check requires that no tensile stresses occur in any concrete within a certain distance, recommended to be 100 mm, of the tendon or its duct. This ensures that there is no direct crack path to the tendon for contaminants. The 100 mm requirement is not a cover requirement. It simply means that if the cover is less than 100 mm, it must all be in compression. Lesser covers may be acceptable, providing the minimum requirements of 2-2/clause 4 are met. Conversely, tensile stresses are permitted in the cover as long as the concrete within 100 mm (or amended value in the National Annex) of the tendons or ducts is in compression. If, in checking decompression, the extreme fibre is found to be cracked, the check of decompression at the specified distance from the tendons becomes iterative. Additionally, although not stated in 2-2/Table 7.101N, if decompression is not checked at the surface for XD and XS environmental classes, a crack width check should also be performed if untensioned reinforcement is present. It is therefore simpler and conservative to check decompression at the surface of the member. If a crack width check is performed, the criterion for reinforced concrete in Table 7.101N can be adopted. Stress checks in a pre-tensioned beam are illustrated in Worked example 5.10-3 in section 5.10 of this guide.

In deep beams and elements with geometrical discontinuities, where strut-and-tie analysis is required, it is still necessary to check crack widths. *2-1-1/clause 7.3.1(8)* allows the bar forces thus determined to be used to calculate reinforcement stresses to verify crack widths in accordance with the remainder of 2-1-1/clause 7.3. *2-1-1/clause 7.3.1(9)* in

2-1-1/clause 7.3.1(8)
2-1-1/clause 7.3.1(9)

general permits either a direct calculation of crack widths using 2-1-1/clause 7.3.4 or a check of allowable reinforcement stress for a given crack width in accordance with 2-1-1/clause 7.3.3. The latter is simpler as many of the parameters needed in 2-1-1/clause 7.3.4 relate to beam geometry. Where strut-and-tie modelling is used to verify crack widths in this way, the results will only be representative if the strut-and-tie model is based on the elastic stress trajectories in the uncracked state. This is discussed in section 6.5.1 and is noted in 2-1-1/clause 7.3.1(8).

2-2/clause 7.3.1(110) suggests that 'in some cases it may be necessary to check and control shear cracking in webs'. These 'cases' are not defined and 2-2/Annex QQ, which is referenced for further information, is equally vague other than to imply that a check is most relevant for prestressed members, perhaps partly because the longitudinal web compression reduces the tensile strength of the concrete in the direction of maximum principal tensile stress. Previous UK design standards have not required a verification of cracking due to shear in webs, but the shear design for reinforced concrete members at ULS differs in EC2 in two ways. First, higher crushing resistances are possible which means greater forces need to be carried by the links if the web concrete is fully stressed. Second, shear design was previously based on a truss model with web compression struts fixed at 45°. Since the Eurocode permits the compression struts to rotate to flatter angles, fewer links might be provided using EC2 in some cases to mobilize a given shear force, thus creating greater link stresses at the serviceability limit state.

7.3.2. Minimum areas of reinforcement

In deriving the expressions for the calculation of crack widths and spacings in section 7.3.3, a fundamental assumption is that the reinforcement remains elastic. If the reinforcement yields, deformation will become concentrated at the crack where yielding is occurring, and this will inevitably invalidate the formulae.

For a section subjected to uniform tension, the force necessary for the member to crack is $N_{cr} = A_c f_{ctm}$, where N_{cr} is the cracking load, A_c is the area of concrete in tension and f_{ctm} is the mean tensile strength of the concrete. The strength of the reinforcement is $A_s f_{yk}$. To ensure that distributed cracking develops, the steel must not yield when the first crack forms hence:

$$A_s f_{yk} > A_c f_{ctm} \tag{D7-12}$$

Equation (D7-12) needs to be modified for stress distributions other than uniform tension. *2-2/clause 7.3.2(102)* introduces a variable, k_c, to account for different types of stress distribution which has the effect of reducing the reinforcement requirement when the tensile stress reduces through the section depth. A further factor, k, is included to allow for the influence of internal self-equilibrating stresses which arise where the strain varies non-linearly through the member depth. Common sources of non-linear strain variation are shrinkage (where the outer concrete shrinks more rapidly than the interior concrete) and temperature difference (where the outer concrete heats up or cools more rapidly than the interior concrete). The self-equilibrating stresses that are produced can increase the tension at the outer fibre, thus leading to cracking occurring at a lower load than expected. This in turn means that less reinforcement is necessary to carry the force at cracking and thus to ensure distributed cracking occurs. The factor k therefore reduces the reinforcement necessary where self-equilibrating stresses can occur. These stresses are more pronounced for deeper members and thus k is smaller for deeper members.

The minimum required reinforcement area is thus given as:

$$A_{s,\min}\sigma_s = k_c k f_{ct,eff} A_{ct} \qquad \text{2-2/(7.1)}$$

where:

$A_{s,\min}$ is the required minimum area of reinforcing steel within the tensile zone

A_{ct} is the area of concrete within the tensile zone. The tensile zone should be taken as that part of the concrete section which is calculated to be in tension just before the formation of the first crack

σ_s is the absolute value of the maximum stress permitted in the reinforcement immediately after formation of the first crack. This will generally need to be taken as the value to satisfy the maximum bar size or bar spacing requirements of 2-1-1/Tables 7.2N or 7.3N respectively

$f_{ct,eff}$ is the mean value of tensile strength of the concrete effective at the time when the cracks are expected to first occur, i.e. f_{ctm} or lower ($f_{ctm}(t)$) if cracking is expected earlier than 28 days. In many cases, where the dominant imposed deformations result from dissipation of the heat of hydration, cracking may occur within 3 to 5 days from casting. However, **2-2/clause 7.3.2(105)** requires that $f_{ctm}(t)$ is not taken as less than 2.9 MPa, which corresponds to the mean 28-day tensile strength of grade C30/37 concrete. The use of mean tensile strength (rather than the more conservative upper characteristic value which was used in *Model Code 90*[6]) has, in part, been used to produce similar minimum reinforcement provisions to those obtained with previous European practice

2-2/clause 7.3.2(105)

k is a coefficient allowing for the effect of non-uniform self-equilibriating stresses and should be taken as:
1.0 for webs with $h \leq 300$ mm or flanges with widths less than 300 mm
0.65 for webs with $h \geq 800$ mm or flanges with widths greater than 800 mm
Intermediate values should be interpolated. A value of k of 1.0 can always conservatively be used

k_c is a coefficient allowing for the nature of the stress distribution within the section immediately prior to cracking and the change of the lever arm, calculated from 2-2/Expression (7.2) or 2-2/Expression (7.3) for webs and flanges of flanged beams respectively. It depends on the mean direct stress (whether tensile or compressive) acting on the part of the cross-section being checked. It is equal to 0.4 for rectangular beams without axial force. Prestressing has the effect of reducing k_c for webs by way of 2-2/Expression (7.2), while direct tension increases its value. A value of 1.0 is always conservative. It should be noted that 2-2/Expression (7.2) contains a term, k_1, which differs in definition and value from that in 2-1-1/Expression (7.2) and another in 2-1-1/clause 7.3.4

For flanged beams, such as T-beams or box girders, 2-2/clause 7.3.2(102) requires that the minimum reinforcement provision is determined for each individual part of the section (webs and flanges, for example), and 2-2/Fig. 7.101 identifies how the section should be sub-divided for this purpose; the web height, h, is taken to extend over the full height of the member. The sub-division used clearly affects the calculation of some of the terms in 2-2/Expressions (7.2) and (7.3), particularly concrete area and mean concrete stress.

Despite the apparent similarity between 2-2/(7.1) and the minimum reinforcement reqirements in 2-1-1/9.2.1.1, the former is associated with limiting crack widths while the latter is associated with preventing steel yield upon cracking of the cross-section. Both checks must therefore be performed.

2-1-1/clause 7.3.2(3) allows any bonded tendons located within the effective tension area to contribute to the area of minimum reinforcement required to control cracking, provided they are within 150 mm of the surface to be checked. 2-2/Expression (7.1) then becomes:

2-1-1/clause 7.3.2(3)

$$A_{s,min}\sigma_s + \xi_1 A_p \Delta\sigma_p = k_c k f_{ct,eff} A_{ct} \tag{D7-13}$$

A_p is the area of bonded pre- or post-tensioned tendons within the effective tensile area, $A_{c,eff}$ (discussed under clause 7.3.4), and $\Delta\sigma_p$ is the stress increase in the tendons from the state of zero strain of the concrete at the same level (i.e. the increase in stress in the tendons after decompression of the concrete at the level of the tendons). ξ_1 is the adjusted ratio of bond strength taking into account the different diameters of prestressing and reinforcing steel, $\xi_1 = \sqrt{\xi \phi_s/\phi_p}$, where:

ξ is the ratio of bond strength of prestressing and reinforcing steel, given in 2-1-1/clause 6.8.2

ϕ_s is the largest bar diameter of reinforcing steel

ϕ_p is the equivalent diameter of the tendon in accordance with 2-1-1/clause 6.8.2. If only prestressing steel is used to control cracking, $\xi_1 = \sqrt{\xi}$

2-1-1/clause 7.3.2(4) allows minimum reinforcement to be omitted where, in prestressed concrete members, the stress at the most tensile fibre is limited to a nationally determined value, recommended to be $f_{ct,eff}$, under the characteristic combination of actions and the characteristic value of prestress. This does not remove the need to consider the provision of reinforcement to control early thermal cracking prior to application of the prestressing.

7.3.3. Control of cracking without direct calculation

The basis of the crack width calculation method in EN 1992 is presented in section 7.3.4. *2-2/clause 7.3.3(101)*, however, allows 'simplified methods' to be used for the control of cracking without direct calculation and, undesirably for pan-European consistency, allows the National Annex to specify a method. The recommended method is that given in 2-1-1/clause 7.3.3. In this method, *2-1-1/clause 7.3.3(2)* requires the reinforcement stress to be determined from a cracked section analysis (see Worked example 7.1) under the relevant combination of actions (see section 7.1 of this guide). The relevant effective concrete modulus for long-term and short-term loading should also be used. It is assumed that minimum reinforcement according to 2-1-1/clause 7.3.2 will be provided. An advantage of this simplified approach is that many of the difficulties of interpretation of parameter definition involved in direct calculations to 2-2/clause 7.3.4 for non-rectangular cross-sections (such as for circular sections, discussed below) can be avoided.

For cracks caused mainly by direct actions (i.e. imposed forces and moments), cracks may be controlled by limiting reinforcement stresses to the values in either 2-1-1/Table 7.2N or 2-1-1/Table 7.3N. It is not necessary to satisfy both. The former sets limits on reinforcement stress based on bar diameter and the latter based on bar spacing. For cracks caused mainly by restraint (for example, due to shrinkage or temperature), only Table 7.2N can be used; cracks have to be controlled by limiting the bar size to match the calculated reinforcement stress immediately after cracking.

Tables 7.2N and 7.3N of EN 1992-1-1 were produced from parametric studies carried out using the crack width calculation formulae in 2-1-1/clause 7.3.4, discussed below. They were based on reinforced concrete rectangular sections ($h_{cr} = 0.5h$) in pure bending ($k_2 = 0.5$, $k_c = 0.4$) with high bond bars ($k_1 = 0.8$) and C30/37 concrete ($f_{ct,eff} = 2.9$ MPa). The cover to the centroid of the main reinforcement was assumed to be $0.1h$ ($h - d = 0.1h$). The values in brackets above refer to the assumptions given in Note 1 of 2-1-1/Table 7.2N. (h_{cr} and h are defined in 2-1-1/clause 7.3.3(2), k_1 and k_2 are defined in 2-1-1/clause 7.3.4(3) and k_c is defined in 2-2/clause 7.3.2(102).) Correction for other geometries can be made, as discussed below.

The use of these tables for bridges was criticized by some countries because they have been derived for members with covers more typical of those found in buildings (specifically 25 mm), whereas bridge covers are typically much greater. Cover is a significant contributor to crack spacing and hence crack width, as can be seen in section 7.3.4 and Worked example 7.3. This potentially leads to greater calculated crack widths for bridges. This criticism was one reason for the allowance of national choice in the calculation method to be employed. Detailed arguments over the parameters to use in crack width calculation, however, tend to attribute a greater implied accuracy to the crack width calculation than is really justified. Of greater significance is the load combination used to calculate the crack widths, as discussed in section 7.3.1 above. There is a strong argument that adequate durability is achieved by specifying adequate cover and by limiting reinforcement stresses to sensible values below yield. The former is achieved through compliance with 2-2/clause 4 and the latter by following the reinforcement stress limits in 2-1-1/clause 7.3.3(2) and 2-1-1/clause 7.2(5).

For members with geometry, loading or concrete strength other than as in the assumptions above, the maximum bar diameters in 2-1-1/Table 7.2N strictly need to be modified. The following two equations are given:

$$\phi_s = \phi_s^*(f_{ct,eff}/2.9)\frac{k_c h_{cr}}{2(h-d)} \quad \text{for sections at least partly in compression} \qquad 2\text{-}1\text{-}1/(7.6\text{N})$$

$$\phi_s = \phi_s^*(f_{ct,eff}/2.9)\frac{h_{cr}}{8(h-d)} \quad \text{for sections completely in tension} \qquad 2\text{-}1\text{-}1/(7.7\text{N})$$

where:

- ϕ_s is the adjusted bar diameter
- ϕ_s^* is the maximum bar size given in 2-1-1/Table 7.2N
- h is the overall depth of the section
- h_{cr} is the depth of the tensile zone immediately prior to cracking, considering the characteristic values of prestress and axial forces under the quasi-permanent combination of actions
- d is here defined as the effective depth to the centroid of the *outer layer* of reinforcement. This is not intended to be a general definition of d, but rather a clarification that if the whole section depth is in tension, the effective depth should be measured to the centroid of the steel in one face and not to the centroid of the steel in both faces. The latter could lead to a centroid at mid-depth of the member. Generally, d is the depth to the centroid of the area of the reinforcement in tension, as clarified in 2-1-1/Fig. 7.1a), where there are two layers of bars

While 2-1-1/clause 7.3.3(2) states that this adjustment of bar diameter 'should' be made, earlier drafts of EN 1992-1-1 and *Model Code 90*,[6] stated that this adjustment 'may' be made. It should be noted that the adjustment can be either beneficial or detrimental to limiting stress, depending on circumstance, so the distinction between 'may' and 'should' is important. The former implies that benefit can be taken from the adjustment, ignoring it in other cases, while the latter implies that it should be considered also where it is more onerous to do so. Whether or not this level of sophistication is merited is a matter for debate, as mentioned above. It is noted that no such adjustment to bar diameter is made in crack checks in EN 1994-2, where the same two tables of limiting bar stress appear. In general, the whole issue of adjusting bar diameters can be avoided by using Table 7.3N to determine limiting stresses. For a 0.3 mm crack width and bar spacing up to 200 mm, it will be advantageous to use 2-1-1/Table 7.3N for all bar diameters from 16 mm upwards.

The adjustment to bar diameter is not practical when strut-and-tie modelling is used to determine reinforcement stresses as the terms all relate to beam behaviour. Direct use of EN 1992-1-1 Table 7.2N or 7.3N would probably be reasonable in such cases.

Differences in cover cannot be accommodated by the adjustments of 2-1-1/Expression (7.6N) and 2-1-1/Expression (7.7N) which, ironically, is the most significant factor in determining crack widths when using the direct calculation method of 2-1-1/clause 7.3.4. Calculations on reinforced concrete slabs and rectangular beams indicate that the simple method of calculation based on the use of EN 1992-1-1 Table 7.2N and 7.3N remain conservative relative to the more accurate calculation method of 2-1-1/clause 7.3.4 for covers up to about 35 mm. For greater covers, the simple method becomes increasingly un-conservative by comparison. However, considering the limited accuracy of both crack width calculation methods and the benefits to durability associated with the provision of greater covers as discussed above, the simplified method probably remains acceptable for greater covers. There is certainly no stated limit to its use in EN 1992 based on a maximum cover.

Where there is a mixture of prestressing steel and un-tensioned reinforcement, the prestress can conservatively be treated as an external force applied to the cross-section (ignoring the stress increase in the tendons after cracking) and the stress determined in the reinforcement, ignoring concrete in tension as usual. The reinforcement stress derived can then be compared against the tabulated limits. For pre-tensioned beams with relatively little untensioned reinforcement, where crack control is to be provided mainly by the bonded tendons themselves, the Note to 2-1-1/clause 7.3.3(2) permits Tables 7.2N and 7.3N to be used with the steel stress taken as the total stress in the tendons after cracking, minus the initial prestress after losses. This is approximately equal to the stress increase in the tendons after decompression at the level of the tendons.

For beams with depth greater than 1000 mm and main reinforcement concentrated in only a small proportion of the depth, *2-1-1/clause 7.3.3(3)* requires additional minimum reinforcement to be evenly distributed over the side faces of the beams in the tension zone to control cracking. It is normal in bridge design to distribute steel around the perimeter of sections to control early thermal cracking and reinforcement for this purpose should generally be sufficient to meet the requirements of this clause; section 7.5 refers.

2-1-1/clause 7.3.3(4) is a reminder that there is a particular risk of large cracks occurring in sections where there are sudden changes of stress such as at changes of section, near concentrated loads, where bars are curtailed or at areas of high bond stresses such as at the end of laps. While sudden changes of section should normally be avoided (by introducing tapers), compliance with the reinforcement detailing rules given in clauses 8 and 9, together with the crack control rules of clause 7, should normally give satisfactory performance.

2-1-1/clause 7.3.3(3)

2-1-1/clause 7.3.3(4)

Worked example 7.2: Crack checks without direct calculation and minimum reinforcement in reinforced concrete deck slab

The 250 mm thick reinforced concrete deck slab analysed in Worked example 7.1 is again considered, assuming the same reinforcement (16 mm diameter bars at 100 mm centres with 50 mm cover) and concrete class (C35/45). The exposure class is XC3. The method of 2-2/clause 7.3.3 (without direct calculation) is used to check crack control and minimum reinforcement is checked in accordance with 2-2/clause 7.3.2.

2-2/Table 7.101N requires crack widths to be limited to 0.3 mm under the quasi-permanent load combination for an exposure class of XC3. According to EN 1990 Table A.2.1, traffic actions have a quasi-permanent combination factor of $\psi_2 = 0$ and so are not considered in crack checks. Thermal actions have $\psi_2 = 0.5$ and so should be considered. Only the secondary effects of temperature difference, however, need to be considered; the primary self-equilibriating stresses may be ignored.

For convenience here, the same moments as in Worked example 7.1 of 85 kNm/m, comprising 15% permanent actions and 85% transient actions will be taken. The make-up would, however, be very different as discussed above; it is unlikely that temperature difference would produce effects anywhere near as severe as those from characteristic traffic actions.

From the above example, the serviceability stress in the reinforcement has already been calculated as 247.6 MPa. To comply with 2-2/clause 7.3.3, either:

(1) the maximum bar size must be limited to 12 mm (from 2-1-1/Table 7.2N); or
(2) the maximum bar spacing must be limited to 190 mm (interpolating within 2-1-1/Table 7.3N).

The provision of 16 mm diameter bars at 100 mm centres complies with the limit on bar spacing in (2) above (which permits a reinforcement stress of 320 MPa for bars at 100 mm centres) and the design is therefore acceptable. It does not matter that it does not comply with the limit in (1) as well.

Additionally, it is necessary to check that the reinforcement complies with the minimum reinforcement area required in accordance with 2-2/clause 7.3.2. This will rarely govern at peak moment positions, but may do so near points of contraflexure if reinforcement is curtailed:

From 2-2/Expression (7.1): $A_{s,min}\sigma_s = k_c k f_{ct,eff} A_{ct}$

A_{ct} is the area of concrete within the tensile zone just before the first crack forms. The section behaves elastically until the tensile fibre stress reaches f_{ctm}, therefore, for a rectangular section, the area in tension is half the slab depth, thus:

$A_{ct} = 250/2 \times 1000 = 125 \times 10^3 \text{ mm}^2$

$f_{ct,eff} = f_{ctm}$ but not less than 2.9 MPa – 2-2/clause 7.3.2(105). From 2-1-1/Table 3.1 for C35/45 concrete, $f_{ctm} = 3.2$ MPa so $f_{ct,eff} = 3.2$ MPa.

For rectangular sections of less than 300 mm depth, k should be taken as 1.0 and can in general be taken as 1.0 conservatively. For sections with no axial load, i.e. $\sigma_c = 0$ MPa, 2-1-1/Expression (7.2) reduces to $k_c = 0.4 \times (1 - 0) = 0.4$.

σ_s may in general be based on the maximum allowable value from either Table 7.2N (240 MPa for 16 mm diameter bars) or Table 7.3N (320 MPa for 100 mm bar centres). However, for minimum reinforcement calculation, it is possible that cracking may arise mainly from restraint, rather than load and, therefore, the value from Table 7.2N is used here in accordance with the Note to 2-1-1/clause 7.3.3(2). Therefore $\sigma_s = 240$ MPa assuming 16 mm bars and so:

$A_{s,min} = 0.4 \times 1.0 \times 3.2 \times 125 \times 10^3 / 240 = 667 \text{ mm}^2/\text{m}$

The 16 mm bars at 100 mm centres provide $A_s = 2010 \text{ mm}^2/\text{m}$, which exceeds this minimum value, so the design is adequate. From minimum reinforcement considerations alone, the bar centres could be increased or the bar diameter reduced in zones of low moment, but further crack control and ultimate limit state checks would then be required at these curtailment locations.

7.3.4. Control of crack widths by direct calculation

The basis of the crack width calculation to EN 1992 is presented here, first considering the simplified case of a reinforced concrete prism in tension as in Fig. 7.3. The member will first crack when the tensile strength of the weakest section is reached. Cracking leads to a local redistribution of stresses adjacent to the crack as indicated in Fig. 7.3 by the strain distributions. At the crack, the entire tensile force is carried by the reinforcement. Moving away from the crack, tensile stress is transferred from the reinforcement by bond to the surrounding concrete and, therefore, at some distance L_e from the crack, the distribution of stress is unaltered from that before the crack formed. At this location, the strain in concrete and reinforcement is equal and the stress in the concrete is just below its tensile strength. The redistribution of stress local to the crack results in an extension of the member which is taken up in the crack, causing it to open. This also leads to a reduction in the member stiffness.

With increasing tension, a second crack will form at the next weakest section. This will not be within a distance L_e of the first crack due to the reduction in concrete stresses in that region associated with the first crack. With further increase in tension, more cracks will develop until the maximum crack spacing anywhere is $2L_e$. No further cracks will then

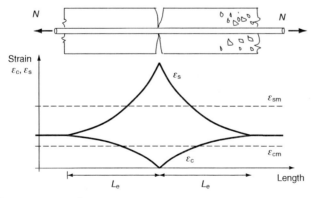

Fig. 7.3. Strains adjacent to a crack

form but further loading will cause the existing cracks to widen. This is called 'stabilized' cracking. The member stiffness will continue to reduce, tending towards that of the fully cracked section, considering reinforcement alone in the tension zone.

The crack width formulae in EC2 are based on the discussions above. The crack width stems from the difference in extension of the concrete and steel over a length equal to the crack spacing. The crack width, w, is thus given in **2-1-1/clause 7.3.4(1)** as:

2-1-1/clause 7.3.4(1)

$$w_k = s_{r,max}(\varepsilon_{sm} - \varepsilon_{cm}) \qquad \text{2-1-1/(7.8)}$$

where:

- w_k is the characteristic crack width
- $s_{r,max}$ is the maximum crack spacing
- ε_{sm} is the mean strain of the reinforcement in the length $s_{r,max}$ under the relevant combination of loads, including the effect of imposed deformations and taking into account the effects of tension stiffening. Only the additional tensile strain beyond zero strain in the concrete is considered
- ε_{cm} is the mean strain in the concrete in the length $s_{r,max}$ between cracks

These terms are discussed in detail below.

Crack spacing

From the above discussion, the minimum crack spacing is L_e and the maximum crack spacing is $2L_e$. The average crack spacing, s_{rm}, is therefore somewhere between these two values. Figure 7.3 illustrates that L_e and thus s_{rm} depend on the rate at which stress can be transferred from the reinforcement to the concrete. Assuming a constant bond stress, τ, along the length L_e, the stress at a distance L_e from a crack will reach the tensile strength of the concrete when:

$$\tau \pi \phi L_e = A_c f_{ct} \qquad \text{(D7-14)}$$

where A_c is the concrete area and f_{ct} is the tensile strength of the concrete. Introducing the reinforcement ratio $\rho = \pi\phi^2/4A_c$ into equation (D7-14) gives:

$$L_e = \frac{f_{ct}}{\tau} \frac{\phi}{4\rho} \qquad \text{(D7-15)}$$

The mean crack spacing can then be expressed as:

$$s_{rm} = 0.25 k_1 \phi / \rho \qquad \text{(D7-16)}$$

where k_1 is a constant which takes account of the bond properties of the reinforcement in the concrete f_{ct}/τ, and the difference between minimum and mean crack spacing. Equation (D7-16) does not fit test data well, so an additional term for reinforcement cover, c, needs to be included in the expression for crack spacing:

$$s_{rm} = kc + 0.25 k_1 \phi / \rho \qquad \text{(D7-17)}$$

The need for the reinforcement cover term probably arises because, although equation (D7-14) assumes that the tensile stress is constant in the area A_c, the concrete stress will actually be greatest adjacent to the bar and will reduce with distance from it. This reduces the cracking load of the area A_c.

Equation (D7-17) applies to concrete sections in pure tension and a further factor is therefore necessary to allow for other cases where the stress distribution varies through the depth of the member. It is also necessary to define an effective reinforcement ratio, $\rho_{p,eff}$ since the appropriate concrete area is not that of the whole member but rather must be related to the actual tension zone. This leads to:

$$s_{rm} = kc + 0.25 k_1 k_2 \phi / \rho_{p,eff}$$

2-1-1/clause 7.3.4(3)

Finally, a further factor α is required to obtain the upper characteristic crack width rather than an average one. The formula presented in **2-1-1/clause 7.3.4(3)** for the maximum final crack spacing is therefore:

$$s_{r,max} = \alpha(kc + 0.25k_1k_2\phi/\rho_{p,eff}) = k_3c + k_1k_2k_4\phi/\rho_{p,eff} \qquad \text{2-1-1/(7.11)}$$

where:

c is the cover to reinforcement; the relevant definition of cover to use is not stated in EN 1992, but clause 4.3(2) of EN 1990 states that 'the dimensions specified in the design may be taken as characteristic values'. This implies the cover to use is c_{nom}. This may produce a conservative design where cover in excess of that required by 2-2/clause 4 is provided

ϕ is the bar diameter. Where a mixture of bar diameters is used in a section, an equivalent diameter, ϕ_{eq}, should be used as defined in 2-1-1/Expression (7.12)

$$\rho_{p,eff} = \frac{A_s + \xi_1^2 A_p}{A_{c,eff}} \qquad \text{2-1-1/(7.10)}$$

A_s is the area of reinforcing steel within the effective area $A_{c,eff}$

A_p is the area of pre- or post-tensioned tendons within the effective area $A_{c,eff}$

$A_{c,eff}$ is the effective tension area, i.e. the area of concrete surrounding the tension reinforcement of depth $h_{c,ef}$ (2-1-1/Fig. 7.1 refers), where $h_{c,ef}$ is the lesser of:
- one-third of the tension zone depth of the cracked section, $(h - x)/3$, with x negative when the whole section is in tension
- half the section depth, $h/2$, or
- $2.5(h - d)$

d is the depth to the centroid of the area of the reinforcement in tension. This is clarified in 2-1-1/Fig. 7.1(a). The definition of d in the Note to 2-1-1/clause 7.3.3(2), which refers to the depth to the outer layer of reinforcement, is not intended to be a general definition but rather a clarification of what to do when the whole section depth is in tension

ξ_1 is the adjusted ratio of bond strength taking into account the different diameters of prestressing and reinforcing steel. It is defined under the comments on 2-1-1/clause 7.3.2(3) above

k_1, k_2, k_3, k_4 are coefficients whose purpose is discussed above. Values of k_1 and k_2 are specified in 2-1-1/clause 7.3.4(3), while k_3 and k_4 are nationally determined parameters

The above definitions are readily applicable to rectangular sections, but for general sections, for example a circular column, considerable interpretation is required in choosing appropriate values to use. This makes the use of the simplified method in section 7.3.3 appealing, as there is no such difficulty of interpretation; the reinforcement stress is simply compared to the allowable limit in 2-1-1/Table 7.3N. The use of the tables is not prohibited for circular sections, although Note 1 beneath 2-1-1/Table 7.2N gives the assumptions on which they are based, implying they were derived for selected rectangular sections, as discussed in section 7.3.3 of this guide.

For a circular section, the maximum crack widths will occur between the most highly stressed bars. A possible calculation method would therefore be to consider a thin slice in the plane of bending through the diameter, thus producing a narrow rectangular beam with width equal to the bar spacing. $h_{c,ef}$, $A_{c,ef}$ and $\rho_{p,eff}$ can then be calculated for a rectangular beam of these proportions, with d referring to the centroid of the reinforcement in this slice. Alternatively, $h_{c,ef}$ could be calculated with d taken equal to the effective depth of the reinforcement in the tension zone of the complete cross-section. $A_{c,ef}$ and $\rho_{p,eff}$ would then be determined for the concrete and reinforcement within this height $h_{c,ef}$. The first method tends to lead to the most conservative prediction of crack width and both often lead to greater crack widths than the simplified method of 2-1-1/clause 7.3.3 for typical bridge covers.

The above analysis for $s_{r,max}$ does not hold for members without bonded reinforcement in the tension zone (as the reinforcement ratio is needed). In such cases, the concrete stresses are modified by the presence of the crack over a length approximately equal to the crack height each side of the crack. Using the same arguments as above, the average crack spacing will then be equal to between one and two times the crack height. In this case, 2-1-1/clause 7.3.4(3) presents this as a maximum crack spacing, thus:

$$s_{r,max} = 1.3(h - x) \qquad \text{2-1-1/(7.14)}$$

2-1-1/Expression (7.14) also applies where the reinforcement spacing is wide (defined as greater than $5(c + \phi/2)$ in EN 1992-1-1) because the strain in the concrete then varies appreciably between bars. A similar rule is presented in *2-1-1/clause 7.3.4(5)*.

When the angle between the axes of principal stress and the direction of the reinforcement in a member reinforced in two orthogonal directions exceeds 15°, such as may occur in skew deck slabs for example, the following expression in *2-1-1/clause 7.3.4(4)* can be used to determine the maximum crack spacing:

$$s_{r,max} = \left(\frac{\cos\theta}{s_{r,max,y}} + \frac{\sin\theta}{s_{r,max,z}} \right)^{-1} \qquad \text{(D7-18)}$$

where θ is the angle between the reinforcement in the y-direction and the direction of principal tensile stress and $s_{r,max,y}$ and $s_{r,max,z}$ are the crack spacings calculated in the y- and z-directions according to 2-1-1/Expression (7.11).

Average strains for crack prediction

To calculate the crack width, the average strains in the steel and concrete between cracks must be considered. Figure 7.4 illustrates the situation for a section subjected to pure tension. The total axial tension, N, is given by the force in the reinforcement at the crack so that $N = E_s \varepsilon_{s2} A_s$. The average steel and concrete forces are given by $N_s = E_s \varepsilon_{sm} A_s$, where ε_{sm} is the average strain in the steel, and $N_c = E_c \varepsilon_{cm} A_c$ where ε_{cm} is the average strain in the concrete between the cracks. The total axial tension can therefore be expressed as $N = N_s + N_c$ so that:

$$E_s \varepsilon_{s2} A_s = E_s \varepsilon_{sm} A_s + E_c \varepsilon_{cm} A_c$$

or

$$\varepsilon_{s2} = \varepsilon_{sm} + \frac{E_c}{E_s} \frac{A_c}{A_s} \varepsilon_{cm}$$

Substituting for $\rho = A_s/A_c$ and $\alpha_e = E_s/E_c$ and re-arranging gives $\varepsilon_{sm} = \varepsilon_{s2} - \varepsilon_{cm}/\alpha_e \rho$, thus:

$$\varepsilon_{sm} - \varepsilon_{cm} = \varepsilon_{s2} - \frac{\varepsilon_{cm}}{\alpha_e \rho}(1 + \alpha_e \rho) \qquad \text{(D7-19)}$$

At section 1, the axial force, N, is shared between the reinforcement and the concrete such that the strains are equal. For the concrete to be just un-cracked, this occurs at a concrete stress of f_{ctm}; therefore $\varepsilon_{s1} = \varepsilon_{c1} = f_{ctm}/E_c$. The average concrete strain between the cracks, ε_{cm}, is some proportion of this value, thus:

$$\varepsilon_{cm} = k_t f_{ctm}/E_c \qquad \text{(D7-20)}$$

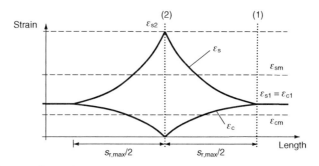

Fig. 7.4. Strains for a single crack

CHAPTER 7. SERVICEABILITY LIMIT STATES

k_t is an empirically determined factor which is dependent on the duration of the applied load. 2-1-1/clause 7.3.4(2) assigns it values of 0.6 for short-term loading and 0.4 for long-term loading. It would be possible to interpolate between these values where the total applied load comprises a proportion of both short- and long-term loads.

Substituting equation (D7-19) into equation (D7-20) gives:

$$\varepsilon_{sm} - \varepsilon_{cm} = \varepsilon_{s2} - \frac{k_t f_{ctm}}{E_c \alpha_e \rho}(1 + \alpha_e \rho)$$

$$= \varepsilon_{s2} - \frac{k_t f_{ctm}}{E_s \rho}(1 + \alpha_e \rho)$$

and since $\varepsilon_{s2} = \sigma_{s2}/E_s$:

$$\varepsilon_{sm} - \varepsilon_{cm} = \frac{\sigma_{s2} - \frac{k_t f_{ctm}}{\rho}(1 + \alpha_e \rho)}{E_s} \qquad (D7\text{-}21)$$

This equation appears in **2-1-1/clause 7.3.4(2)** as:

$$\varepsilon_{sm} - \varepsilon_{cm} = \frac{\sigma_s - k_t \dfrac{f_{ct,eff}}{\rho_{p,eff}}(1 + \alpha_e \rho_{p,eff})}{E_s} \geq 0.6 \frac{\sigma_s}{E_s} \qquad 2\text{-}1\text{-}1/(7.9)$$

2-1-1/clause 7.3.4(2)

where:

- σ_s is the stress in the tension reinforcement assuming a cracked section. For pre-tensioned members, σ_s may be replaced by $\Delta\sigma_p$ which is the increase in stress in the tendons after decompression at the level of the tendons
- $f_{ct,eff}$ is the mean value of tensile strength of the concrete discussed in section 7.3.3 below. It should be noted that the definition of $f_{ct,eff}$ in EC2 is given in 2-2/clause 7.3.2(102) regarding minimum area of reinforcement where maximizing $f_{ct,eff}$ is conservative. Here, however, $f_{ct,eff}$ is contributing to the tension stiffening effect of the concrete and is therefore beneficial for crack width predictions, so a smaller value is conservative. An argument for using the lower characteristic value could therefore be made, but, for similar reasons to those given under clause 7.3.2(102), a value of f_{ctm} has been adopted to produce values of crack width that do not differ significantly from current practice
- α_e is the modular ratio E_s/E_c
- $\rho_{p,eff}$ is the effective reinforcement ratio from 2-1-1/Expression (7.10) reproduced above
- k_t is the factor dependent on the duration of the load, as discussed above
- ξ_1 is the bond strength adjustment factor discussed under the comments on 2-1-1/clause 7.3.2(3) above

The limit of $0.6\sigma_s/E_s$ in 2-1-1/Expression (7.9) is made to place a limit on the benefit of the concrete tension stiffening.

Worked example 7.3: Crack checks in reinforced concrete deck slab by direct calculation

Worked example 7.2 is repeated using direct calculation of the crack width:

From 2-1-1/Expression (7.8): $w_k = s_{r,max}(\varepsilon_{sm} - \varepsilon_{cm})$

From 2-1-1/Expression (7.11): $s_{r,max} = k_3 c + k_1 k_2 k_4 \phi/\rho_{p,eff} = 3.4c + 0.425 k_1 k_2 \phi/\rho_{p,eff}$

$c = 50$ mm and $\phi = 16$ mm; therefore $d = 250 - 50 - 16/2 = 192$ mm and the depth to the neutral axis, $x = 63.5$ mm (from Worked example 7.1). From 2-1-1/clause 7.3.4(3), this equation is valid providing the actual bar spacing is less than $5(c + \phi/2) = 5 \times (50 + 16/2) = 290$ mm, which it clearly is here.

From 2-1-1/Expression (7.10): $\rho_{p,eff} = \dfrac{A_s + \xi_1^2 A_p}{A_{c,eff}}$

where A_s = area of reinforcement = $\pi \times 8^2/0.10 = 2010\,\text{mm}^2/\text{m}$. $A_p = 0$ since no prestress. $A_{c,eff}$ = effective tension area = $bh_{c,ef}$ with $h_{c,ef}$ taken as the lesser of:

$2.5(h-d) = 2.5 \times (250-192) = 145\,\text{mm}$

or

$(h-x)/3 = (250-63.5)/3 = 62.2\,\text{mm}$

or

$h/2 = 250/2 = 125\,\text{mm}$

Thus $h_{c,ef} = 62.2\,\text{mm}$ and $A_{c,eff} = 1000 \times 62.2 = 62.2 \times 10^3\,\text{mm}^2/\text{m}$

Therefore $\rho_{p,eff} = \dfrac{2010}{62.2 \times 10^3} = 0.0323$

$k_1 = 0.8$ for high bond bars and $k_2 = 0.5$ for bending, therefore:

$s_{r,max} = 3.4 \times 50 + 0.425 \times 0.8 \times 0.5 \times 16/0.0323 = 254.2\,\text{mm}$

(It should be noted that the concrete cover term, $3.4c$, contributes 170 mm of the total 254 mm crack spacing here, so is very significant. The cover assumed in 2-1-1/Table 7.2N is only 25 mm; its use would reduce the crack spacing to 169 mm, assuming that the change to the depth of the compression zone, x, is minimal.)

From 2-1-1/Expression (7.9):

$$\varepsilon_{sm} - \varepsilon_{cm} = \dfrac{\sigma_s - k_t \dfrac{f_{ct,eff}}{\rho_{p,eff}}(1 + \alpha_e \rho_{p,eff})}{E_s} \geq 0.6 \dfrac{\sigma_s}{E_s}$$

From Worked example 7.2, the reinforcement stress assuming a fully cracked section is 247.6 MPa, so the minimum value of 2-1-1/Expression (7.9) is $0.6 \times 247.6/(200 \times 10^3) = 0.7428 \times 10^{-3}$.

$k_t = 0.6$ for short-term loading or 0.4 for long-term loading, thus, interpolating for 85% transient loading, $k_t = 0.57$. From 2-1-1/Table 3.1 for C35/45 concrete, $f_{ct,eff} = f_{ctm} = 3.2\,\text{MPa}$. $\alpha_e = 200/34.08 = 5.869$.

Therefore:

$$\varepsilon_{sm} - \varepsilon_{cm} = \dfrac{247.6 - 0.57 \times \dfrac{3.2}{0.0323}(1 + 5.869 \times 0.0323)}{200 \times 10^3} = \dfrac{247.6 - 67.2}{200 \times 10^3} = 0.902 \times 10^{-3}$$

which is greater than the minimum value of 0.7428×10^{-3}.

Therefore, maximum crack width, $w_k = 254.2 \times 0.902 \times 10^{-3} = \mathbf{0.23\,mm} < 0.3\,\text{mm}$ limit from 2-2/Table 7.101N, as required.

Repeating the above calculation, the reinforcement stress could be increased to 303 MPa until a crack width of 0.3 mm is reached, assuming the same ratio of short-term and long-term moments as used in this example. This compares with an allowable stress of 320 MPa from 2-1-1/Table 7.3N for bars at 100 mm centres as used in Worked example 7.2. The method without direct calculation therefore gives a more economic answer here, which is largely because the cover it assumes is smaller than the real cover in this example, as noted above. The difference is, however, not too great, which, given the low accuracy of crack width calculations, generally reinforces the recommendation in EN 1992-2 to permit either method. If Worked examples 7.2 and 7.3 are repeated for the same slab, but with 25 mm cover to reinforcement (as assumed in 2-1-1/Table 7.3N), it is found that the direct calculation of crack width becomes more economic.

CHAPTER 7. SERVICEABILITY LIMIT STATES

7.4. Deflection control

2-1-1/clause 7.4.1(1)(P) requires that the deformation of a member or structure shall not be such that it adversely affects its proper function or appearance. Excessive deformations under permanent actions can give a visual impression of inadequate strength and disrupt the intended drainage path. Excessive deformations under live load can damage surfacing and waterproofing systems, and can also lead to dynamic problems including both motion discomfort and structural damage.

2-1-1/clause 7.4.1(2) requires that appropriate limiting values of deflections should be established for each individual structure considering the above. Excessive sagging deflections under permanent actions can generally be overcome by precambering and dynamic considerations are addressed in EN 1990 and EN 1991-1-4 for live load and wind-induced oscillations respectively. Resonance of bridges can become an ultimate limit state if sufficient fatigue damage occurs or if divergent-amplitude wind-induced motion, such as galloping and flutter, occur. No guidance is given in EC2 on appropriate limits for other effects (such as disruption to drainage) as these will need to be checked on a structure-by-structure basis.

2-1-1/clause 7.4.2 (identifying situations where no check of deflection is required) does not apply to bridge design and deflections should be calculated following the rules in 2-1-1/clause 7.4.3. In practice, it is usually necessary to calculate deformations of concrete bridges in order, for example, to calculate rotations and translations in the design of bearings. *2-1-1/clause 7.4.3(1)* and *2-1-1/clause 7.4.3(2)* effectively require the actions and structural behaviour to be assessed realistically and to an accuracy merited by the type of check being carried out. The remainder of 2-1-1/clause 7.4.3 gives recommendations for achieving a realistic calculation. These include:

2-1-1/clause 7.4.1(1)(P)

2-1-1/clause 7.4.1(2)

2-1-1/clause 7.4.3(1)
2-1-1/clause 7.4.3(2)

- adopting un-cracked section properties where the tensile strength of the concrete will not be exceeded anywhere in the member;
- using member behaviour intermediate between the un-cracked and fully cracked conditions for members which are expected to crack. This behaviour is dependent on duration of loading and stress level. An allowance for the tension stiffening effect of a cracked section is also given;
- using an effective modulus of elasticity of the concrete to allow for creep;
- consideration of shrinkage curvature.

7.5. Early thermal cracking (additional sub-section)

Early thermal cracking is the cracking that occurs in concrete members due to restraint as the heat of hydration dissipates and while the concrete is still young. Restraint can occur from *internal* or *external* restraint:

- **Internal restraint** – one part of a concrete pour expands or contracts relative to another. This is most likely in thick sections where thermal gradients develop. Excessive reinforcement quantities can also restrain contraction and give rise to very fine cracking aligned with the reinforcement. This will generally be avoided if the maximum percentages of reinforcement specified in EN 1992-2 section 9 are observed.
- **External restraint** – member being cast is restrained by the member or external element to which it is being cast against.

EC2-2 and EC2-1-1 do not provide guidance on early thermal cracking, but EC2-3 for liquid retaining and containing structures does. The UK National Annex for EC2-2 is likely to refer to this. The early thermal and shrinkage strains calculated can be used in conjunction with 2-1-1/clause 7.3.4 to check crack widths.

CHAPTER 8

Detailing of reinforcement and prestressing steel

This chapter deals with the detailing of reinforcement and prestressing steel as covered in section 8 of EN 1992-2 in the following clauses:

• General	*Clause 8.1*
• Spacing of bars	*Clause 8.2*
• Permissible mandrel diameters for bent bars	*Clause 8.3*
• Anchorage of longitudinal reinforcement	*Clause 8.4*
• Anchorage of links and shear reinforcement	*Clause 8.5*
• Anchorage by welded bars	*Clause 8.6*
• Laps and mechanical couplers	*Clause 8.7*
• Additional rules for large diameter bars	*Clause 8.8*
• Bundled bars	*Clause 8.9*
• Prestressing tendons	*Clause 8.10*

8.1. General

Section 8 of EN 1992-2 gives general rules for the detailing of reinforced and prestressed concrete members. The detailing rules in EC2 apply to ribbed reinforcement, mesh and prestressing tendons only – plain round bars are not covered.

Many of the detailing requirements of EC2 relate to bond and mechanical anchorage properties which are influenced mainly by:

- the concrete type and strength;
- the surface characteristics of the bars;
- the shape of the bars and the inclusion of bends or hooks;
- the amount of concrete cover in relation to the size of the bars;
- the presence of welded transverse bars;
- additional confinement offered by links or other non-welded transverse bars;
- additional confinement offered by transverse pressure.

2-1-1/clause 8.1(1)P indicates that the rules in section 8 may not be adequate for coated reinforcement bars. This is because the coating may affect the bond characteristics of the reinforcement. The rules may also be insufficient for bridges with seismic loading. Additional confining links are often provided in this latter case, particularly around compression members, to give greater ductility.

2-1-1/clause 8.1(2)P is a reminder that the detailing rules given in EC2 assume the minimum cover requirements have been complied with in accordance with 2-1-1/clause

2-1-1/clause 8.1(1)P

2-1-1/clause 8.1(2)P

4.4.1.2. This is particularly important for anchorage and lap lengths whose strength could be reduced by substandard cover (due to an increased tendency for concrete splitting and the greater potential for reinforcement corrosion).

2-1-1/clause 8.1(3) and *2-1-1/clause 8.1(4)* refer to additional rules for the use of lightweight aggregate concrete and for members subjected to fatigue loading.

Many of the requirements in 2-2/clause 8 are self-explanatory and are not discussed in their entirety here. 2-2/clause 9 also deals with detailing, but this is concerned with detailing of entire members or zones of members rather than individual bars.

2-1-1/clause 8.1(3)
2-1-1/clause 8.1(4)

8.2. Spacing of bars

2-1-1/clause 8.2 defines practical rules for the minimum spacing of reinforcement. These rules are designed to ensure that concrete can be placed and satisfactorily compacted, enabling adequate bond strength to be developed along the full length of the bar. The minimum spacings in *2-1-1/clause 8.2(2)* should certainly not be detailed throughout as it would lead to inadequate access for vibrators as required by *2-1-1/clause 8.2(3)*. They should only be resorted to in localized regions of congestion, e.g. short regions where links overlap.

2-1-1/clause 8.2(2)
2-1-1/clause 8.2(3)

8.3. Permissible mandrel diameters for bent bars

2-1-1/clause 8.3(2) gives recommended minimum mandrel diameters to which reinforcement can be bent. These permissible mandrel diameters have been chosen to avoid bending cracks being formed in the bar. EC2 recommends minimum values of 4ϕ for bar diameters less than or equal to 16 mm and 7ϕ for bar diameters greater than 16 mm, but these may be amended in the National Annex. These are the same values as presented in BS 8666: 2000.

2-1-1/clause 8.3(3) does not require an explicit check of concrete crushing in the bend if the following are satisfied:

2-1-1/clause 8.3(2)

2-1-1/clause 8.3(3)

- the anchorage of the bar does not require a length more than 5ϕ beyond the end of the bend (i.e. the bar is not fully stressed within the bend);
- the plane of the bend of the bar is not positioned 'close', presumably adjacent, to a concrete face (so that spalling of the cover cannot occur) and there is a bar of at least the same diameter positioned inside the bend and running along the axis of the bend (which locally increases the concrete compressive resistance);
- the mandrel diameter is at least equal to the recommended values in 2-1-1/clause 8.3(2).

Where the above conditions cannot be met, the following expression for mandrel diameter is given:

$$\phi_{m,min} \geq \frac{F_{bt}}{f_{cd}}\left(\frac{1}{a_b} + \frac{1}{2\phi}\right) \qquad \text{2-1-1/(8.1)}$$

where:

ϕ is the bar diameter

F_{bt} is the tensile force from ultimate loads in the bar or group of bars in contact at the start of the bend

a_b for any given bar (or group of bars) is half of the centre-to-centre distance between bars (or groups of bars) perpendicular to the plane of the bend. For a bar or group of bars adjacent to the face of the member, a_b should be taken as the cover plus $\phi/2$

f_{cd} is the concrete design compressive strength, but should not be taken greater than that for concrete of class C55/67

The concrete stress inside the bend of a bar is $F_{bt}/(\phi \phi_{m,min}/2)$ and thus the requirement of 2-1-1/Expression (8.1) corresponds to a limiting concrete stress of $4f_{cd}/(1 + (2\phi/a_b))$. The allowable stress is reduced for proximity to another adjacent bend or to a free surface by the $1/a_b$ term. The applicability of this high limiting bearing stress depends on the concrete's ability to resist splitting stresses by way of its tensile strength. As tensile strength does not increase in proportion with compressive strength (it increases rather more slowly), the limit on f_{cd} is necessary to avoid overestimation of bearing strength. The use of $\alpha_{cc} = 0.85$ is most appropriate here as the relevant concrete stress is a sustained compression.

8.4. Anchorage of longitudinal reinforcement
8.4.1. General
2-1-1/clause 8.4.1(1)P requires reinforcement to be adequately anchored to ensure that the bond stresses are safely transmitted to the concrete without longitudinal cracking or spalling. The transfer of these stresses is mainly by mechanical locking of the reinforcement ribs with the surrounding concrete, and is dependent on several factors including the shape of the bar, concrete cover, presence of transverse reinforcement and any confinement from transverse pressure. The rules in the remainder of this section ensure compliance with 2-1-1/clause 8.4.1(1)P.

2-1-1/clause 8.4.1(1)P

Figure 8.4-1 illustrates the permissible methods of anchorage covered in EC2, other than by a straight length of bar. Other mechanical devices are permitted, but must be designed and tested in accordance with the relevant product standards and are beyond the scope of this guide.

Anchorage failure can take one of three forms, illustrated in Fig. 8.4-2. The mechanical locking of the reinforcement ribs with the surrounding concrete leads to resultant forces inclined away from the bar acting on the concrete. These forces have a radial component which can be considered analogous to a pressure acting inside a thick-walled cylinder. The cylinder has an inner diameter equal to the reinforcing bar diameter and an outer diameter equal to the smaller of the bottom cover or half the clear spacing between adjacent bars.

Where the clear bottom cover is greater than half the clear spacing to the adjacent bar, a horizontal split develops at the level of the bars. This failure is known as a side split failure. Where the bottom cover is less than half the clear spacing to the adjacent bar, face-and-side split failure occurs, with longitudinal cracking through the bottom cover followed by splitting along the plane of the bars. If the bottom cover is significantly less than half the clear spacing to the adjacent bar, V-notch failure occurs, with longitudinal cracking through the bottom cover followed by inclined cracking.

The strength of the transverse reinforcement and its spacing along the anchorage length are significant parameters influencing the anchorage strength of the main bar. The greater the transverse restraint, the greater is the anchorage strength up to a certain limit beyond which the transverse reinforcement contributes no further.

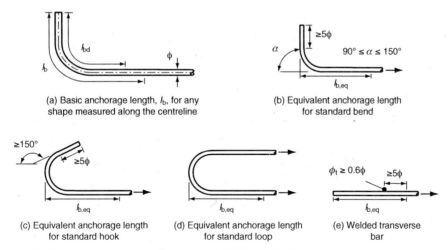

Fig. 8.4-1. Methods of anchorage covered in EN 1992-1-1 Fig. 8.1

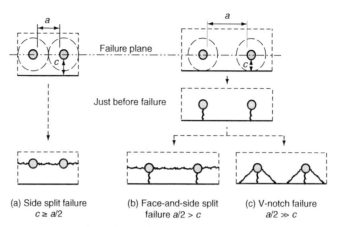

Fig. 8.4-2. Failure patterns of single anchored bars

EC2 introduces factors adjusting the basic anchorage length to include for the above effects. The anchorage rules are empirical, however, and have been chosen to give a reasonable match to test data.

Due to the lack of relevant test data, *2-1-1/clause 8.4.1(3)* does not allow bends or hooks to contribute to compression anchorages. Since no allowance is given in the code for partial transmission of the bar force by end bearing, the anchorage requirements for bars in compression are always at least as onerous as for bars in tension. A further conservatism is that the additional clamping from lateral expansion of the bar under compression is ignored. Where bars with bends are used to anchor tensile forces, *2-1-1/clause 8.4.1(4)* requires that concrete failure inside bends is prevented by complying with the rules discussed in section 8.3 above.

8.4.2. Ultimate bond stress

Bond stress is a function of the tensile strength of the concrete, bar size and position of bar during concreting. The design value of the ultimate bond stress, f_{bd}, is given in *2-1-1/clause 8.4.2(2)* for ribbed bars as:

$$f_{bd} = 2.25\eta_1\eta_2 f_{ctd} \qquad 2\text{-}1\text{-}1/(8.2)$$

where:

- η_1 is a coefficient related to the quality of the bond condition and the position of the bar during concreting (see 2-1-1/Fig. 8.2)
  $\eta_1 = 1.0$ for 'good' bond conditions (e.g. bottom bars and vertical bars)
  $\eta_1 = 0.7$ for all other cases (e.g. top bars, where the concrete is less well compacted), including for bars in structural elements built with slip-forms, unless it can be shown that 'good' conditions exist
- η_2 is related to the bar size and should be taken as $(132 - \phi)/100$ for bar diameters greater than 32 mm or 1.0 otherwise
- f_{ctd} is the design value of concrete tensile strength with $a_{ct} = 1.0$, as discussed in section 3.1.6. Due to the increasing brittleness of high-strength concrete, EC2 recommends that the value of f_{ctd} taken here is limited to the value obtained for class C60/75 concrete

8.4.3. Basic anchorage length

The basic anchorage length of a bar, $l_{b,rqd}$, is obtained by assuming an average bond stress, equal to the ultimate bond stress, which acts over the full perimeter of the bar and uniformly along its length. This leads to the following expression given in *2-1-1/clause 8.4.3(2)*:

$$l_{b,rqd} = (\phi/4)(\sigma_{sd}/f_{bd}) \qquad 2\text{-}1\text{-}1/(8.3)$$

where ϕ is the bar diameter, f_{bd} is the design value of the ultimate bond stress and σ_{sd} is the design stress of the bar at the ultimate limit state at the position from where the anchorage is measured.

It is desirable that, generally, anchorage lengths should be based on the design strength of the bar considered (i.e. $\sigma_{sd} = f_{yd}$), although the code allows a lower stress in the bar to be used where reinforcement is not intended to be fully stressed. The reason for this is that redistribution of moments at the ultimate limit state may increase the stress in the bar beyond the envisaged design value. This recommendation would avoid the potential for sudden, brittle failure where the distribution of moments is not as anticipated in the design.

The basic anchorage length applies to straight bars with cover equal to the bar diameter (the minimum permitted by 2-1-1/Table 4.2 regardless of environmental requirements) and no transverse reinforcement between the anchored bar and the concrete surface. The basic anchorage length can therefore always be conservatively taken as the design anchorage length – see Table 8.4-1. 2-1-1/clause 8.4.4, however, may allow a reduction in this length as discussed below.

Where bent bars are used, **2-1-1/clause 8.4.3(3)** requires that the basic anchorage length and design anchorage length should be measured along the centreline of the bar as in Fig. 8.4-1(a).

2-1-1/clause 8.4.3(3)

8.4.4. Design anchorage length

The design anchorage length, l_{bd}, is obtained by reducing the basic anchorage length to allow for the beneficial effects of additional cover, transverse reinforcement, transverse clamping pressure and bar shape for bent bars. The reductions are made by the introduction of coefficients to take account of these effects. The design anchorage length is given in **2-1-1/clause 8.4.4(1)** as:

2-1-1/clause 8.4.4(1)

$$l_{bd} = \alpha_1 \alpha_2 \alpha_3 \alpha_4 \alpha_5 l_{b,rqd} \geq l_{b,min} \qquad \text{2-1-1/(8.4)}$$

where:

- α_1 accounts for the shape of the bars (equal to 1.0 for a straight bar in tension)
- α_2 accounts for the effect of concrete cover (equal to 1.0 for a straight bar in tension with cover equal to the bar diameter and 0.7 with cover equal to three times the bar diameter). The relevant definition of cover to use is not stated in EN 1992 but clause 4.3(2) of EN 1990 states that 'the dimensions specified in the design may be taken as characteristic values'. This implies the cover to use is c_{nom}. The use of c_{min} will produce a more conservative anchorage length
- α_3 accounts for the effect of confinement by transverse reinforcement (equal to 1.0) for bars in tension with minimum transverse reinforcement, $\sum A_{st,min}$, in accordance with 2-1-1/Table 8.2. For greater amounts of transverse reinforcement placed between the anchoring bar and the concrete surface, α_3 reduces to a minimum of 0.7. Transverse reinforcement placed inside the anchoring bar is, however, ineffective as it does not cross the relevant planes in Fig. 8.4-2(b) and (c) and $\alpha_3 = 1.0$. This is forced by the value of $K = 0$ from 2-1-1/Fig. 8.4 (reproduced here as Fig. 8.4-3) to be used in the expression $\alpha_3 = 1 - K\lambda$
- α_4 accounts for the influence of one or more welded transverse bars (of diameter, $\phi_t > 0.6\phi$) along the design anchorage length (equal to 1.0 in the absence of welded transverse reinforcement and 0.7 otherwise)
- α_5 accounts for the effect of any pressure along the design anchorage length, transverse to the plane of splitting as indicated in Fig. 8.4-2 (equal to 1.0 for bars in tension in the absence of transverse pressure)
- $l_{b,min}$ is the minimum anchorage length if no other limitation is applied:
  for anchorages in tension: $\quad l_{b,min} \geq \max\{0.3 l_{b,rqd};\ 10\phi;\ 100\,\text{mm}\}$
  for anchorages in compression: $l_{b,min} \geq \max\{0.6 l_{b,rqd};\ 10\phi;\ 100\,\text{mm}\}$

The full range of values for α_1, α_2, α_3, α_4 and α_5 are given in 2-1-1/Table 8.2. 2-1-1/clause 8.4.4(1) requires that the product $\alpha_2 \alpha_3 \alpha_5$ is not taken less than 0.7. For compression

Table 8.4-1. Basic anchorage lengths $l_{b,rqd}$ for B500 straight bars ($\alpha_{ct} = 1.0$)

Concrete cylinder strength (MPa)	Bar diameter (mm)	Good bond conditions $l_{b,rqd}$ (mm)	Poor bond conditions $l_{b,rqd}$ (mm)
25	10	404	577
	12	484	692
	16	646	922
	20	807	1153
	25	1009	1441
	32	1291	1845
	40	1755	2507
30	10	357	511
	12	429	613
	16	572	817
	20	715	1021
	25	893	1276
	32	1144	1634
	40	1554	2220
35	10	322	461
	12	387	553
	16	516	737
	20	645	921
	25	806	1152
	32	1032	1474
	40	1402	2003
40	10	295	421
	12	354	506
	16	472	674
	20	590	843
	25	738	1054
	32	944	1349
	40	1283	1832
45	10	273	390
	12	327	468
	16	436	623
	20	545	779
	25	682	974
	32	873	1247
	40	1186	1694
50	10	254	363
	12	305	436
	16	407	581
	20	508	726
	25	636	908
	32	814	1162
	40	1105	1579

anchorages, only the welding of transverse bars to the bar being anchored reduces the anchorage length from $l_{b,rqd}$. Often, it will not be practical to take advantage of the above reduction factors in design since, if this is done, the anchorage and lap lengths become dependent on the amount of transverse reinforcement and cover. For a given bar diameter, these parameters may vary throughout the bridge and a change to the size of transverse bars during the design process could lead to re-detailing all the laps and anchorage lengths for the main bars. Table 8.4-1 therefore gives basic anchorage lengths $l_{b,rqd}$ for straight B500

CHAPTER 8. DETAILING OF REINFORCEMENT AND PRESTRESSING STEEL

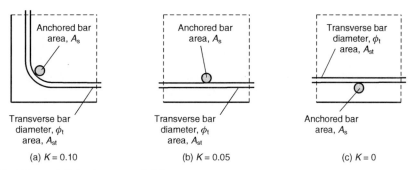

Fig. 8.4-3. Values of K for beams and slabs

reinforcement bars assuming all the above coefficients are equal to 1.0, taking $\alpha_{ct} = 1.0$ (for the reasons discussed in section 3.1.6 of this guide) and assuming the bars are fully stressed to $500/1.15 = 435$ MPa.

For bent bars, the design anchorage length calculated above is generally measured around the bend, as shown in Fig. 8.4-1(a). As a simplified conservative alternative to the above rules, EC2 allows the tension anchorage of certain shapes of bar shown in 2-1-1/Fig. 8.1 to be provided as an equivalent anchorage length, $l_{b,eq}$. The length $l_{b,eq}$ is defined in Fig. 8.1 and may be taken as the following from **2-1-1/clause 8.4.4(2)**:

$\alpha_1 l_{b,rqd}$ for shapes shown in 2-1-1/Figs 8.1(b) to 8.1(d)
$\alpha_4 l_{b,rqd}$ for shapes shown in 2-1-1/Fig. 8.1(e)

where α_1, α_4 and $l_{b,rqd}$ are as defined above.

2-1-1/clause 8.4.4(2)

8.5. Anchorage of links and shear reinforcement

All links and shear reinforcement need to be adequately anchored and this can be achieved by means of bends and hooks or by welded reinforcement. In order to avoid the failure of the concrete on the inside of the bend, or at least to avoid an explicit check of concrete pressure in the bend, **2-1-1/clause 8.5(1)** requires a longitudinal bar to be placed in the corners of hooks and bends. The bar size is not stated, but, for compatibility with the requirements of 2-1-1/clause 8.3, it should have a diameter at least equal to that of the link. **2-1-1/clause 8.5(2)** refers to 2-1-1/Fig. 8.5 for allowable details. Those for bent and hooked bars are shown in Fig. 8.4-4.

2-1-1/clause 8.5(1)
2-1-1/clause 8.5(2)

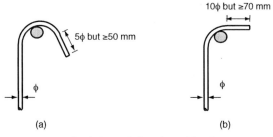

Fig. 8.4-4. Anchorage requirements for links with bends and hooks

8.6. Anchorage by welded bars

EC2 allows additional anchorage to the basic resistances in 2-1-1/clauses 8.4 and 8.5 to be obtained by utilizing transverse bars welded to the main reinforcement. Such details effectively allow some load to be taken in bearing by the transverse bar which reduces the required bond length for a given force. 2-1-1/clause 8.6 gives rules for determining the force taken by the welded transverse bar and hence the reduction in required anchorage length.

8.7. Laps and mechanical couplers

8.7.1. General

2-1-1/clause 8.7.1(1)P allows the transfer of force from one bar to another to be achieved by:

- lapping the bars, with or without bends, hooks or loops (see 2-1-1/Fig. 8.1 for definitions and section 8.4.4 for equivalent anchorage lengths);
- welding;
- mechanical devices such as couplers.

Commentary on the rules for each of these methods is given in the following sections.

8.7.2. Laps

The considerations of anchorage failure discussed in section 8.4 above can also be applied to lapped bars. Where two bars are spliced side by side, the two concrete cylinders to be considered for each bar interact to form an oval ring instead of two circular rings in section. The three potential failure patterns of side splitting, face-and-side splitting and V-notch failure are similar to those indicated in Fig. 8.4-2 for single bars.

For spliced bars lapped side by side, the above consideration is complicated by the uneven distribution of bond stress along the length of the two bars and the uncertainty in inclination of the resultant forces from the mechanical interlocking of the bar ribs within the surrounding concrete. The treatment of the design of laps is therefore again based on matching test data. This shows that for the same bar diameter, bar centres, cover and concrete strength, the same lap length is required for well staggered laps as for anchorage lengths. However, where laps are not sufficiently staggered, the percentage of bars lapped at a section does become influential on required lap lengths. The factors used in EC2 to adjust the basic anchorage length to obtain the design anchorage lengths are therefore augmented by a further factor to account for the percentage of bars lapped at a section.

The design of laps and requirements for lap lengths are covered in 2-1-1/clauses 8.7.2 and 8.7.3. Provided that the design of laps is in accordance with these rules, the detailing is deemed to be sufficient to adequately transmit the forces from one bar to the next without spalling the concrete in the vicinity of the joint and without generating unacceptably large cracks, as required by *2-1-1/clause 8.7.2(1)P*.

2-1-1/clause 8.7.2(2) generally requires that laps between should not 'normally' be located in areas of high stress and that, wherever possible, laps are staggered symmetrically at a section. While this has been adopted as good practice in the UK, there are inevitably some situations in bridge design where these requirements cannot be fulfilled. In in-situ balanced cantilever design, for example, bars must be lapped at segment joints, regardless of stress level. Although it can be attempted to stagger the length of bars projecting from a previous segment, where larger diameter bars are required, such detailing often becomes impossible within the constraints of the segment lengths. These problems lead to the use of the word 'normally' above.

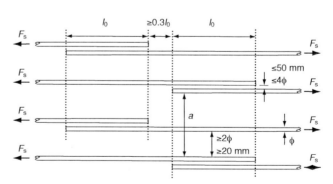

Fig. 8.7-1. Requirements for adjacent laps from 2-1-1/Fig. 8.7

Bearing the practicalities of staggering laps in mind, the arrangement of lapped bars should be in accordance with 2-1-1/Fig. 8.7 (reproduced as Fig. 8.7-1), as specified in **2-1-1/clause 8.7.2(3)** where:

- the clear transverse distance between two lapped bars should not be greater than 4ϕ or 50 mm, otherwise the lap length should be increased by a length equal to the clear space between the bars. This increase is required in order to take account of the inclination of the compression struts by which force is transferred from one bar to the other;
- the clear distance between adjacent bars in adjacent laps should not be less than 2ϕ or 20 mm;
- the longitudinal distance between two adjacent laps should not be less than $0.3l_0$. This is effectively the definition of a staggered lap.

2-1-1/clause 8.7.2(3)

Where the provisions of 2-1-1/clause 8.7.2(3) are met, **2-1-1/clause 8.7.2(4)** allows 100% of the bars in tension to be lapped where the bars are all in one layer. Examples of 100% lapping in this way are shown in Fig. 8.7-3 of Worked example 8.7-1. It is unfortunate that 2-1-1/Table 8.3 and 2-1-1/Fig. 8.8 introduce another definition of the percentage of laps. This latter definition is used for determining α_6 and transverse reinforcement requirements in 2-1-1/clause 8.7.4. If the bars are in several layers, only 50% lapping is permitted by clause 8.7.2(4). In this situation, it is not clear what distance must be left between laps in adjacent layers or indeed over what layer separation the restriction applies. For longitudinal separation, it was simply the intention not to have identical lapping arrangements overlying each other in each layer, i.e. the staggering requirement of maintaining $0.3l_0$ between adjacent laps should be maintained between layers as well as within layers.

2-1-1/clause 8.7.2(4)

Layer separation is clearly relevant to the requirement to stagger laps in different layers. The requirement need not apply to bars in opposite faces of a beam. Where a slab is in flexure alone, the top and bottom layers of reinforcement in one direction will not be in tension together, so 100% of the bars in each layer could be lapped together without consideration of the interaction between layers. However, for deck slabs with local and global effects, such that both top and bottom layers could be in tension together, strictly the laps should also be staggered between layers so that only 50% of the bars are lapped.

2-1-1/clause 8.7.2(4) also allows all bars in compression and secondary (distribution) reinforcement to be lapped in one section. However, for two-way spanning slabs typical in bridge decks, it is recommended that, wherever possible, laps are staggered in both directions since there is not usually one direction that can clearly be considered as 'secondary'.

8.7.3. Lap length

The following equation is given in **2-1-1/clause 8.7.3(1)** for the design lap length, l_0:

$$l_0 = \alpha_1 \alpha_2 \alpha_3 \alpha_5 \alpha_6 l_{b,rqd} \geq l_{0,min} \qquad \text{2-1-1/(8.10)}$$

2-1-1/clause 8.7.3(1)

where:

$$l_{0,min} > \max\{0.3\alpha_6 l_{b,rqd}; 15\phi; 200 \text{ mm}\} \qquad \text{2-1-1/(8.11)}$$

and $\alpha_6 = \sqrt{(\rho_1/25)} \leq 1.5$ and ≥ 1.0, where ρ_1 is the percentage of reinforcement in a section lapped within $0.65l_0$ from the centre of the lap length considered. Considering the laps illustrated in Fig. 8.7-1 as an example, the centres of the right-hand set of laps are further away from the centres of the left-hand set than $0.65l_0$ and therefore only 50% of the reinforcement is lapped at any one section; thus α_6 can be taken as 1.41.

$l_{b,rqd}$ is calculated from 2-1-1/Expression (8.3) and values for α_1, α_2, α_3 and α_5 are taken from 2-1-1/Table 8.2, as discussed in section 8.4.4 above. In the calculation for α_3, however, 2-1-1/clause 8.7.3(1) specifies that $\Sigma A_{st,min}$ should be taken as $1.0A_s(\sigma_{sd}/f_{yd})$, with A_s taken as the area of one lapped bar. This requirement is discussed further below.

8.7.4. Transverse reinforcement in the lap zone

The effect of the presence of transverse reinforcement in preventing splitting at the ends of lapped bars is taken into account by the use of the α_3 factor in calculating the design lap length in 2-1-1/Expression (8.10). As discussed in section 8.4.4, it will generally not be practical to take benefit in reducing lap lengths for excess transverse reinforcement.

Where the diameter of the lapped bars is less than 20 mm or less than 25% of the reinforcement is lapped in one section (presumably less than or equal to 25% was intended so that quadruple staggering, as in Worked example 8.7-1, arrangement (b) complies), ***2-1-1/clause 8.7.4.1(2)*** deems the minimum transverse reinforcement already provided as links or distribution reinforcement to be sufficient to cater for the transverse tensile forces which occur at laps without any further justification.

2-1-1/clause 8.7.4.1(2)

Where the diameter of the lapped bars is greater than or equal to 20 mm, ***2-1-1/clause 8.7.4.1(3)*** requires the transverse reinforcement to have a total area, ΣA_{st}, of not less than the area, A_s, of one lapped bar assuming that the lapped bar is fully stressed. If $A_s = \Sigma A_{st}$, $\alpha_3 = 1.0$ due to its modified definition in 2-1-1/clause 8.7.3(1). The transverse reinforcement is required to be placed perpendicular to the direction of the lapped reinforcement at no more than 150 mm centres. For skew reinforcement, the quantity of skew transverse reinforcement required could be increased so as to provide either the equivalent axial resistance or stiffness perpendicular to the lapped bar as if the bars were placed perpendicularly. If the reinforcement ratio for a set of skew bars is ρ_x at an angle ϕ from the perpendicular to the lapping bars, the former possibility leads to an effective reinforcement ratio transverse to the lapping bars of $\rho_x \cos^2 \phi$, while the latter gives $\rho_x \cos^4 \phi$. The latter option is the more conservative and is recommended here.

2-1-1/clause 8.7.4.1(3)

Transverse reinforcement is required by 2-1-1/clause 8.7.4.1(3) to be placed between the lapped bars and the concrete surface. Clearly, this is only really practical in beams and one-way spanning slabs; for two-way spanning slabs, thin deck slabs where transverse reinforcement needs to be on the outer layer to maximize bending resistance and in the webs of box girders where the links are also providing bending resistance, it is inevitable that laps will sometimes need to be made in the main reinforcement in an outer layer. In such circumstances, where it is impractical or impossible to comply with this requirement, the bars could be quadruply staggered as in Fig. 8.7-3(b) so as to comply with the criterion in 2-1-1/clause 8.7.4.1(2) for needing no special consideration of transverse reinforcement.

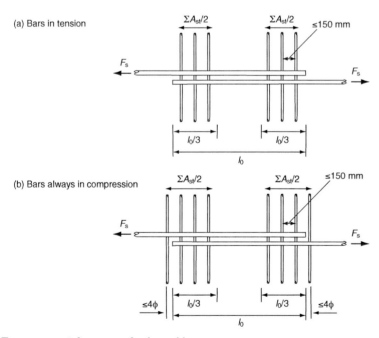

Fig. 8-7.2. Transverse reinforcement for lapped bars

CHAPTER 8. DETAILING OF REINFORCEMENT AND PRESTRESSING STEEL

Alternatively, a value of α_3 of 1.0 could reasonably be used, thus ignoring any benefits from transverse reinforcement in line with the similar treatment of anchorages. Strictly, the latter would not comply with 2-1-1/clause 8.7.4.1(3).

Where more than 50% of the reinforcement is lapped at any one section, **and** the distance between adjacent laps (defined as dimension a in Fig. 8.7-1) is less than or equal to 10ϕ, 2-1-1/clause 8.7.4.1(3) requires that the transverse bars should be formed by links or U-bars anchored into the body of the section. This has not previously been required by UK practice so the need for this provision is questionable. It is not practical to provide links in thin slabs. It is also not clear whether every bar should be enclosed by a link leg or perhaps should be within 150 mm of a bar enclosed by a link, as assumed for restraint of compression bars in section 9.5.3 of this guide. Staggering laps in accordance with Fig. 8.7-1 ensures that there is not more than 50% of reinforcement lapped at a section and the requirement for transverse reinforcement to be provided as links does not then arise.

2-1-1/clause 8.7.4.1(4) requires that the transverse reinforcement calculated above should be positioned at the ends of the laps, as shown in Fig. 8.7-2.

2-1-1/clause 8.7.4.1(4)

Worked example 8.7-1: Lap lengths for reinforced concrete deck slab
A 300 mm thick deck slab of class C35/45 concrete requires 25 mm diameter B500 transverse reinforcement at 150 mm centres on the top and bottom faces to be fully utilized in hogging and sagging zones. The bars are in an outer layer with 45 mm cover. Find suitable arrangements for laps.

First, the bond strength and the basic anchorage length are determined:

From 2-1-1/Table 3.1 for $f_{ck} = 35$ MPa, $f_{ctk,0.05} = 0.7 \times 0.3 \times 35^{(2/3)} = 2.25$ MPa

From 2-1-1/Expression (3.16): $f_{ctd} = \alpha_{ct} f_{ctk,0.05}/\gamma_c = 1.0 \times 2.25/1.5 = 1.50$ MPa, taking $\alpha_{ct} = 1.0$

Design value for ultimate bond stress is given by 2-1-1/clause 8.2: $f_{bd} = 2.25\eta_1\eta_2 f_{ctd}$

From 2-1-1/Fig. 8.2:

Top bars have 'poor' bond conditions therefore $\eta_1 = 0.7$

Bottom bars have 'good' bond conditions therefore $\eta_1 = 1.0$

$\eta_2 = 1.0$ for all bars not greater than 32 mm

Top bars: $f_{bd} = 2.25 \times 0.7 \times 1.0 \times 1.50 = 2.36$ MPa

Bottom bars: $f_{bd} = 2.25 \times 1.0 \times 1.0 \times 1.50 = 3.37$ MPa

Conservatively, assuming the bars are lapped where they are fully stressed:

$\sigma_{sd} = f_{yd} = f_{yk}/\gamma_s = 500/1.15 = 434.8$ MPa

The basic anchorage length is given by 2-1-1/clause 8.3: $l_{b,rqd} = (\phi/4)(\sigma_{sd}/f_{bd})$

Top bars: $l_{b,rqd} = (25/4)(434.8/2.36) = \mathbf{1151.5\,mm}$

Bottom bars: $l_{b,rqd} = (25/4)(434.8/3.37) = \mathbf{806.5\,mm}$

From 2-1-1/Table 8.2, for straight bars, $\alpha_1 = 1.0$.
For straight bars in a slab with clear spacing, $a = 150 - 25 = 125$ mm and cover, $c = 45$ mm (assuming edge cover $c_1 > c$), from 2-1-1/Fig. 8.3, $c_d = 45$ mm therefore from 2-1-1/Table 8.2:

$\alpha_2 = 1 - 0.15(c_d - \phi)/\phi = 1 - 0.15(45 - 25)/25 = 0.88$ (>0.7 and <1.0 as required)

From Fig. 8.4-3(c), for no transverse bars on the outside (i.e. transverse bars are placed on the inside), $K = 0$ and therefore $\alpha_3 = 1.0$. Alternatively, assuming that the minimum transverse reinforcement according to 2-1-1/clause 8.7.4.1(3) is provided on the outside,

α_3 is still 1.0 due to its modified definition in 2-1-1/clause 8.7.3(1). Assuming no transverse pressure, $p = 0$, and therefore $\alpha_5 = 1.0$.

For each layer of reinforcement, consider the two alternative staggered lap arrangements shown in Fig. 8.7-3.

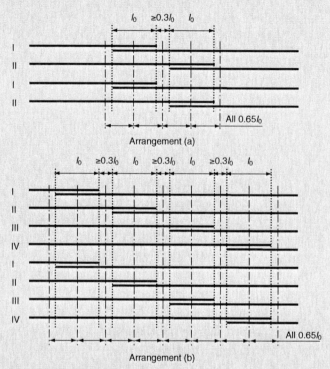

Fig. 8.7-3. Alternative lap arrangements for deck slab example

For arrangement (a), row II is lapped outside $0.65l_0$ of bars lapped in row I; therefore 50% of bars are lapped at a single section and $\alpha_6 = 1.4$. For arrangement (b), no other bars are lapped within $0.65l_0$ of bars lapped in row I (and similarly for rows II, III and IV); therefore only 25% of bars are lapped at a single section and $\alpha_6 = 1.0$.

The minimum value of lap length limit is calculated from 2-1-1/Expression (8.11):

$$l_{0,\min} > \max\{0.3\alpha_6 l_{b,\text{rqd}}; 15\phi; 200 \text{ mm}\}$$

Thus for worst case (Top bars, arrangement (a)):

$$l_{0,\min} > \max\{0.3 \times 1.4 \times 1151.5; 15 \times 25; 200\} = 484 \text{ mm}$$

The required lap lengths are given by 2-1-1/Expression (8.10):

$$l_0 = \alpha_1 \alpha_2 \alpha_3 \alpha_5 \alpha_6 l_{b,\text{rqd}} \geq l_{0,\min}$$

Top bars:

Arrangement (a) $l_0 \geq 1.0 \times 0.88 \times 1.0 \times 1.0 \times 1.4 \times 1151.5 = \mathbf{1419\,mm}$

Arrangement (b) $l_0 \geq 1.0 \times 0.88 \times 1.0 \times 1.0 \times 1.0 \times 1151.5 = \mathbf{1014\,mm}$

Bottom bars:

Arrangement (a) $l_0 \geq 1.0 \times 0.88 \times 1.0 \times 1.0 \times 1.4 \times 806.5 = \mathbf{994\,mm}$

Arrangement (b) $l_0 \geq 1.0 \times 0.88 \times 1.0 \times 1.0 \times 1.0 \times 806.5 = \mathbf{710\,mm}$

2-1-1/clause 8.7.4.1(2) would not require any specific calculation of transverse reinforcement for arrangement (b), since only 25% of bars are staggered at a section and therefore the requirement of 2-1-1/clause 8.7.4.1(3) is not invoked. 2-1-1/clause 8.7.4.1(3) requires transverse reinforcement to be provided outside of the lapping bars. Since the lapped bars are themselves in an outer layer, it is not possible to comply (without adding a third layer) with this placement requirement in the case of arrangement (a). The calculations above have therefore assumed $\alpha_3 = 1.0$, thus ignoring the benefit of transverse reinforcement. Strictly, arrangement (a) does not comply with 2-1-1/clause 8.7.4.1(3). The actual area of transverse reinforcement required would need to be greater than or equal to A_s over two-thirds of the lap length, which for the shortest lap of 994 mm above for arrangement (a) would be $\pi \times 12.5^2 = 491 \text{ mm}^2$ in 663 mm (741 mm^2/m). This can be provided by 12 mm diameter bars at 150 mm centres (754 mm^2/m).

8.7.5. Laps of welded mesh fabrics made of ribbed wires
The use of welded mesh fabric reinforcement is permitted by EC2 but is not common in bridge design due to its reduced fatigue performance. Of the two options for lapping given in 2-1-1/clause 8.7.5.1, intermeshing or layering, intermeshing will be most common in bridge design as it has to be used where fatigue loads are expected. The requirements for lapping intermeshed bars are the same as for bars, except that the transverse reinforcement on the mesh cannot be counted in the calculation of α_3, but no further transverse reinforcement need be provided.

8.7.6. Welding (additional sub-section)
Welding requirements are not mentioned in section 8 of EC2-1-1. The requirements for welding two bars together are given by EC2 in 2-1-1/clause 3.2.5.

8.8. Additional rules for large diameter bars
For cases where large diameter bars are used, tests have shown that the splitting forces and dowel action are significantly greater than for smaller diameter bars. ***2-1-1/clause 8.8(1)*** defines large diameter bars as those with a diameter greater than ϕ_{large}. The value of ϕ_{large} is a nationally determined parameter and EC2-1-1 recommends a value of 32 mm. This will be increased to 40 mm in the UK National Annex to be consistent with current practice, where bars of 40 mm diameter are often used in abutments, cellular decks, piles and pile caps for example. This will mean that the provisions of 2-1-1/clause 8.8 should rarely apply.

2-1-1/clause 8.8(1)

The rest of 2-1-1/clause 8.8 defines additional detailing rules that must be satisfied when using bars of a large diameter. The rules have been developed to account for the increased splitting forces and dowel action and have been based on the recommendations given in *Model Code 90*.[6] The additional detailing rules cover the following:

- crack control and surface reinforcement requirements;
- additional transverse reinforcement at straight anchorages;
- laps.

Where large diameter bars are used, ***2-1-1/clause 8.8(2)*** requires that crack control is achieved by either direct calculation or by incorporating additional surface reinforcement (see Annex J of this guide and 2-1-1/clause 8.8(8)). Direct calculation will normally be chosen as it will often be impractical to provide additional surface reinforcement (e.g. in piles).

2-1-1/clause 8.8(2)

2-1-1/clause 8.8(4) requires that large diameter bars are generally not lapped unless sections are at least 1 m thick or where the stress in the bars is not greater than 80% of the design ultimate strength.

2-1-1/clause 8.8(4)

Where straight anchorages are used for large diameter bars, ***2-1-1/clause 8.8(5)*** requires links, in addition to those required for shear, to be provided as confining reinforcement in

2-1-1/clause 8.8(5)

zones where transverse compression is not present. This additional reinforcement is defined in **2-1-1/clause 8.8(6)** as being not less than the following:

2-1-1/clause 8.8(6)

$$A_{sh} = 0.25 A_s n_1 \text{ in the direction parallel to the tension face} \quad \text{2-1-1/(8.12)}$$

$$A_{sv} = 0.25 A_s n_2 \text{ in the direction perpendicular to the tension face} \quad \text{2-1-1/(8.13)}$$

where A_s is the cross-sectional area of a single anchored bar, n_1 is the number of layers anchored at the same point in the section considered and n_2 is the number of bars anchored in each layer. This additional transverse reinforcement is required to be uniformly distributed along the anchorage zone with bars at centres not exceeding five times the size of the longitudinal reinforcement. EC2 does not indicate how many legs of links are required per longitudinal bar, but *Model Code 90*[6] suggests that 2 legs of a stirrup can surround 3 bars per layer at most.

8.9. Bundled bars

Generally, the rules for individual bars also apply for bundled bars as long as the additional EC2 detailing requirements for bundled bars are met. **2-2/clause 8.9.1(101)** requires that bundled bars are of the same type and grade, but may be of different sizes providing the ratio of diameters does not exceed 1.7.

2-2/clause 8.9.1(101)

For design purposes, the bundle of bars should be treated as a single bar having the same area and centre of gravity as the bundle, but the spacing and cover requirements should be applied to the outer edge of the bundle. For same size bars, the equivalent diameter, ϕ_n, is given in **2-1-1/clause 8.9.1(2)** as $\phi_n = \phi\sqrt{n_b} \leq 55\,\text{mm}$, where n_b is the number of bars in the bundle. The number of bars in a bundle should be limited to no more than 4 for vertical bars in compression (where bond conditions have been shown to be best) or at laps, or to 3 for all other cases. This is to ensure the bond and anchorage characteristics of bundled bars do not stray too far from test cases against which the empirical formulae have been calibrated. Where two touching bars are positioned above one another in regions classified as having good bond conditions, the bars need not be treated as bundled.

2-1-1/clause 8.9.1(2)

2-1-1/clauses 8.9.2 and 8.9.3 cover anchorage and laps respectively. These give requirements for staggering bars within the bundle themselves in some situations.

8.10. Prestressing tendons

8.10.1. Tendon layouts

EC2 gives rules for the spacing of post-tensioning ducts and pre-tensioned strands. These spacings are intended to ensure that placing and compacting of the concrete can be carried out satisfactorily and that sufficient bond is available between the concrete and the tendons, as required by **2-1-1/clause 8.10.1.1(1)P**.

2-1-1/clause 8.10.1.1(1)P

Pre-tensioned tendons

Minimum clear horizontal and vertical spacings of individual pre-tensioned strands are shown in Fig. 8.10-1, where d_g is the maximum aggregate size. Other layouts, including bundling, can be used, but only if test results demonstrate satisfactory ultimate behaviour and that placement and compaction of the concrete is possible – **2-1-1/clause 8.10.1.2(1)**.

2-1-1/clause 8.10.1.2(1)

Post-tensioning ducts

2-1-1/clause 8.10.1.3(1)P requires ducts to be positioned such that concrete can be placed without damaging the ducts and that the concrete can resist the forces imposed by tendons in curved ducts. Minimum clear spacing between ducts, based on the former requirement, are shown in Fig. 8.10-1. These duct spacings are, however, unlikely to be adequate to comply with the latter requirement where the tendon profiles are tightly curved and there is a tendency for the concrete between ducts to split under the bursting stresses generated by the radial force, as shown in Fig. 8.10-2(a). In this situation, either the spacings should be

2-1-1/clause 8.10.1.3(1)P

CHAPTER 8. DETAILING OF REINFORCEMENT AND PRESTRESSING STEEL

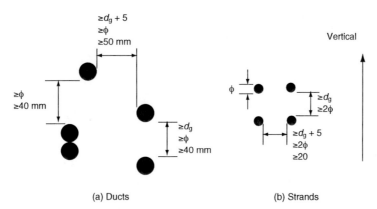

Fig. 8.10-1. Minimum clear spacing between ducts and between pre-tensioned strands

increased or the concrete should be reinforced between the ducts. Curved ducts may also require an increase in cover perpendicular to the plane of the bend. Recommended duct spacings and covers for different duct diameters and bend radii were given in BS 5400 Part 4,[9] and these are reproduced as Tables 8.10-1 and 8.10-2. Tendons will sometimes also need tying back into the main body of the section if the bend is such that the duct pulls against the cover. This is common where tendons are placed in webs curved in plan or flanges curved in elevation and a suitable reinforcement detail is shown in Fig. 8.10-2(b).

2-1-1/clause 8.10.1.3(2) states that ducts for post-tensioned members should not normally be bundled except in the case of a pair of ducts placed vertically one above the other. Particular care should be taken if this is done in a thin deck slab, such as may occur in balanced cantilever construction, as the transverse shear resistance of the slab can be significantly reduced at the duct positions, particularly while ducts are un-grouted.

2-1-1/clause 8.10.1.3(2)

8.10.2. Anchorage of pre-tensioned tendons
8.10.2.1. General
The bond strength applicable to the design of anchoring of pre-tensioned tendons depends on the type of loading. The highest values are applicable to the initial transmission length, l_{pt}, since the tendons thicken against the concrete in the transmission zone at transfer as the stress in them reduces. Lower values are applicable for calculation of the anchorage length, l_{bd}, at ultimate limit states where the force in the tendon increases and the tendon diameter reduces, thus shrinking away from the concrete. These different bond values are reflected in the calculation of initial transmission length and ultimate limit state anchorage length, as shown in Fig. 8.10-4.

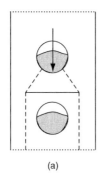 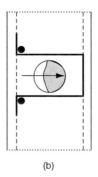

(a)   (b)

Fig. 8.10-2. (a) Splitting of concrete between ducts due to curvature forces and (b) reinforcement resisting pull-out

259

Table 8.10-1. Minimum distance between centrelines of ducts in plane of curvature, in mm

	Duct internal diameter (mm)															
	19	30	40	50	60	70	80	90	100	110	120	130	140	150	160	170
Radius of curvature of duct (m)	Tendon force (kN)															
	296	387	960	1337	1920	2640	3360	4320	5183	6019	7200	8640	9424	10 336	11 248	13 200
2	110	140	350	485	700	960								Radii not normally used		
4	55	70	175	245	350	480	610	785	940							
6	38	60	120	165	235	320	410	525	630	730	870	1045				
8			90	125	175	240	305	395	470	545	655	785	855	940		
10			80	100	140	195	245	315	375	440	525	630	685	750	815	
12						160	205	265	315	365	435	525	570	625	680	800
14						140	175	225	270	315	375	450	490	535	585	785
16							160	195	235	275	330	395	430	470	510	600
18								180	210	245	290	350	380	420	455	535
20									200	220	265	315	345	375	410	480
22											240	285	310	340	370	435
24												265	285	315	340	400
26												260	280	300	320	370
28																345
30																340
32																
34																
36																
38																
40	38	60	80	100	120	140	160	180	200	220	240	260	280	300	320	340

Note 1. The tendon force shown is the maximum normally available for the given size of duct. (Taken as 80% of the characteristic strength of the tendon.)
Note 2. Values less than 2 × duct internal diameter are not included.

8.10.2.2. Transfer of prestress

2-1-1/clause 8.10.2.2(1)

The transmission length at transfer is determined from **2-1-1/clause 8.10.2.2(1)** assuming a constant bond stress f_{bpt}, where:

$$f_{bpt} = \eta_{p1} \eta_1 f_{ctd}(t) \qquad \text{2-1-1/(8.15)}$$

with:

η_{p1} = 2.7 for indented wires or 3.2 for 3- and 7-wire strands
η_1 = 1.0 for good bond conditions (as defined in EC2-1-1 Fig. 8.2) or 0.7 otherwise
$f_{ctd}(t)$ is the concrete design tensile strength at time of release

2-1-1/clause 8.10.2.2(2)

The basic value of transmission length is given in **2-1-1/clause 8.10.2.2(2)** by:

$$l_{pt} = \alpha_1 \alpha_2 \phi \sigma_{pm0} / f_{bpt} \qquad \text{2-1-1/(8.16)}$$

where:

α_1 = 1.0 for gradual release or 1.25 for sudden release
α_2 = 0.25 for circular tendons or 0.19 for 3- and 7-wire strands
ϕ is the nominal diameter of tendon
σ_{pm0} is the tendon stress just after release

2-1-1/clause 8.10.2.2(3)

The design value of the transmission length should be taken from **2-1-1/clause 8.10.2.2(3)** as either $l_{pt1} = 0.8 l_{pt}$ or $l_{pt2} = 1.2 l_{pt}$, whichever is most adverse for the check being carried out. The shorter length, l_{pt1}, will usually be used for checking stresses at transfer at beam ends, as the increasing dead load sagging moments away from the supports help to

CHAPTER 8. DETAILING OF REINFORCEMENT AND PRESTRESSING STEEL

Table 8.10-2. Minimum cover to ducts perpendicular to plane of curvature, in mm

Radius of curvature of duct (m)	Duct internal diameter (mm)															
	19	30	40	50	60	70	80	90	100	110	120	130	140	150	160	170
	Tendon force (kN)															
	296	387	960	1337	1920	2640	3360	4320	5183	6019	7200	8640	9424	10336	11248	13200
2	50	55	155	220	320	445								Radii not normally used		
4		50	70	100	145	205	265	350	420							
6			50	65	90	125	165	220	265	310	375	460				
8				55	75	95	115	150	185	220	270	330	360	395		
10				50	65	85	100	120	140	165	205	250	275	300	330	
12					60	75	90	110	125	145	165	200	215	240	260	315
14					55	70	85	100	115	130	150	170	185	200	215	260
16					55	65	80	95	110	125	140	160	175	190	205	225
18					50	65	75	90	105	115	135	150	165	180	190	215
20						60	70	85	100	110	125	145	155	170	180	205
22						55	70	80	95	105	120	140	150	160	175	195
24						55	65	80	90	100	115	130	145	155	165	185
26						50	65	75	85	100	110	125	135	150	160	180
28							60	75	85	95	105	120	130	145	155	170
30							60	70	80	90	105	120	130	140	150	165
32							55	70	80	90	100	115	125	135	145	160
34							55	65	75	85	100	110	120	130	140	155
36							55	65	75	85	95	110	115	125	140	150
38							50	60	70	80	90	105	115	125	135	150
40	50	50	50	50	50	50	50	60	70	80	90	100	110	120	130	145

Note: The tendon force shown is the maximum normally available for the given size of duct. (Taken as 80% of the characteristic strength of the tendon.)

prevent overstress in tension at the beam top and in compression at the bottom under the effects of prestress.

2-1-1/clause 8.10.2.2(4) allows the concrete stresses to be assumed to have a linear distribution outside the dispersion length, l_{disp}, as shown in Fig. 8.10-3:

2-1-1/clause 8.10.2.2(4)

$$l_{\text{disp}} = \sqrt{l_{\text{pt}}^2 + d^2} \qquad \text{2-1-1/(8.19)}$$

8.10.2.3. Anchorage of tensile force for the ultimate limit state

At the ultimate limit state, the force in the strands increases due to bending and shear and the anchorage of the strands needs to be checked. A check is only, however, necessary according to *2-1-1/clause 8.10.2.3(1)*, where the concrete tensile stress exceeds $f_{\text{ctk},0.05}$ so this check will generally only affect beams with strands which are de-bonded at the beam ends, such that these strands are anchored at higher moment positions.

2-1-1/clause 8.10.2.3(1)

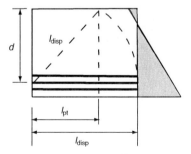

Fig. 8.10-3. Distance at which concrete stresses are uniform

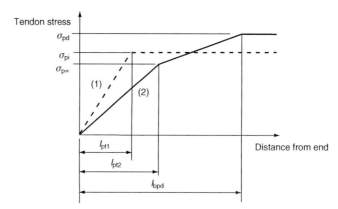

Fig. 8.10-4. Stresses in anchorage zone of pre-tensioned tendons at release (1) and at ultimate limit state (2)

2-1-1/clause 8.10.2.3(2)

The bond strength for anchorage in the ultimate limit state is obtained from *2-1-1/clause 8.10.2.3(2)*:

$$f_{bpd} = \eta_{p2}\eta_1 f_{ctd} \qquad \text{2-1-1/(8.20)}$$

where:

η_{p2} = 1.4 for indented wires or 1.2 for 7-wire strands
η_1 is as above
f_{ctd} should not be taken higher than that for C60/75 concrete without special investigation as higher strength concretes are more prone to brittle bond failure

2-1-1/clause 8.10.2.3(4)

The total anchorage length for anchoring a tendon with stress σ_{pd} is obtained from *2-1-1/clause 8.10.2.3(4)*:

$$l_{bpd} = l_{pt2} + \alpha_2 \phi (\sigma_{pd} - \sigma_{pm\infty})/f_{bpd} \qquad \text{2-1-1/(8.21)}$$

where $\sigma_{pm\infty}$ is the prestress after all losses and the other terms are defined above.
Determination of this anchorage length is illustrated in Fig. 8.10-4.

8.10.2.4. Transverse stresses from bursting and spalling (additional sub-section)

No guidance is given in EN 1992 section 8.10 on the calculation of transverse stresses at anchorage zones of pre-tensioned members, although they do still occur and must be considered in design. The behaviour is similar to that for post-tensioned members, described in 8.10.3 below, but the lengths of the primary prisms for bursting are longer due to the transfer of prestress over an anchorage length rather than through a mechanical anchorage device. The length of the overall equilibrium zone is similarly increased and typically a strut-and-tie diagram similar to that in Fig. 8.10-7(d) will be relevant. Consideration of these effects typically leads to the link reinforcement being increased at beam ends. *Model Code 90*[6] gives greater detail for the analysis of pre-tensioned beams, including determination of the lengths of prisms and calculation methods for the reinforcement necessary to control bursting and spalling.

8.10.3. Anchorage zones of post-tensioned members

An anchorage zone is the area where the concentrated prestress force spreads out over the full member cross-section. In design, attention has to be given to:

- the highly stressed compression concrete in the immediate vicinity of the anchorages;
- bursting stresses generated in the localized area of the anchorage;
- transverse tensile forces arising from any further spread of load outside this localized area.

2-1-1/clause 8.10.3(2)

2-1-1/clause 8.10.3(2) requires the design prestress force used in the local verifications to be in accordance with 2-1-1/clause 2.4.2.2(3), which contains a recommended partial load

CHAPTER 8. DETAILING OF REINFORCEMENT AND PRESTRESSING STEEL

factor of 1.2. **2-1-1/clause 8.10.3(3)** requires that the bearing stress behind anchor plates should be checked in accordance with the relevant ETA. In addition, 2-2/Annex J.104.2(102) requires the minimum spacings and edge distances recommended for the anchorages according to the relevant ETA to be observed. The determination of the tensile forces from the dispersal of the prestress can be done entirely with strut-and-tie models and if the reinforcement design stress is limited to 250 MPa, **2-2/clause 8.10.3(104)** permits no check on serviceability crack widths to be made. It is not clear whether this stress limitation should be met with the ULS partial load factor required by 2-1-1/clause 8.10.3(2) applied, but obviously it is conservative to include it.

2-2/clause 8.10.3(106) requires 'particular consideration' to be given to the design of anchorage zones with two or more anchored tendons and then invokes some additional rules in 2-2/Annex J. The need for these supplementary rules is not clear (the strut-and-tie rules suffice) and they are not required for the designing of anchorage zones in buildings.

2-1-1/clause 8.10.3(3)

2-2/clause 8.10.3(104)

2-2/clause 8.10.3(106)

8.10.3.1. Bursting reinforcement (additional sub-section)

Reinforcement is required in a zone local to the anchorage (called a primary prism) to prevent local bursting. This may be determined from local strut-and-tie models of a type indicated in Fig. 6.5-1. A method is also given in Informative Annex J of EN 1992 which is described below. It has many shortcomings, including both potential double-counting of reinforcement from bursting and overall equilibrium and lack of guidance on where to place spalling reinforcement. It only applies, however, to zones where there are two or more anchorages – **2-2/Annex J.104.2(101)**. This is intended to overcome the problems of double-counting reinforcement if applied to single anchorages, as discussed in section 8.10.3.3 below. It is still not always successful in this respect and, as the check does not depend on the shape of the member, the reinforcement requirements cannot always be justified from strut-and-tie analysis or finite element modelling.

2-2/Annex J.104.2(101)

As a consequence of the above, and as Annex J is informative only, it is recommended here that bursting reinforcement be determined following the method of CIRIA Guide 1.[23] The latter is consistent with the strut-and-tie approach of clause 6.5 as required by 2-2/clause 8.10.3(104). It may then be checked that Annex J does not require more reinforcement within the localized bursting area it defines. If necessary, the reinforcement can be locally increased to accommodate the Annex J requirements.

Method of EC2-2 Annex J

In 2-2/Annex J, the cross-sectional dimensions of the 'primary regularization prism' are determined from the spread of load needed to reduce the stress to a reasonable pressure for uniaxial compression according to **2-2/Annex J.104.2(102)**:

2-2/Annex J.104.2(102)

$$\frac{P_{\max}}{c \times c'} = 0.6 f_{ck}(t) \qquad \text{2-2/(J.101)}$$

where $f_{ck}(t)$ is the compressive strength of the concrete at the time of stressing and $P_{\max}$ is the maximum force applied to the tendon. (The hole for the duct and anchorage means that the calculated stress based on the rectangle in 2-2/Expression (J.101) is not the true stress.) The rectangular cross-section of the primary regularization prism should be approximately the same as that of the bearing plate (or the enclosing rectangle if the bearing plate is not itself rectangular) and must be centred on the bearing plate. The similarity of shape is defined by the need to satisfy equation (D8.10-1):

$$c/a \leq 1.25\sqrt{\frac{c \times c'}{a \times a'}} \text{ and } c'/a' \leq 1.25\sqrt{\frac{c \times c'}{a \times a'}} \qquad \text{(D8.10-1)}$$

where a, a', c and c' are the dimensions of the anchor plate and primary prism cross-section as shown in Fig. 8.10-5, which also shows the length of the primary prism. Where there are several anchorages, the cross-sectional dimensions of the prism must be chosen so as not to overlap at the stressing face but they may overlap away from the stressing face where adjacent tendons, and hence prisms, are not parallel.

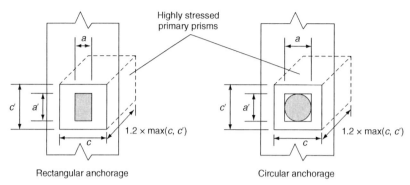

Fig. 8.10-5. Dimensioning of primary regularization prisms to 2-2/Annex J

2-2/Annex
J.104.2(103)

To prevent bursting and spalling local to the anchorage, a minimum reinforcement needs to be provided which *2-2/Annex J.104.2(103)* states must not be less than:

$$A_s = 0.15 \frac{P_{max}}{\sigma_s} \gamma_{p,unfav} \qquad \text{2-2/(J.102)}$$

P_{max} is the greatest load at the time of stressing and $\gamma_{p,unfav}$ is the load factor which is 1.2 for local design. The reinforcement should be distributed evenly over the length of the primary regularization prism. The significance of 'not less than' this amount of reinforcement is that this represents a minimum area of reinforcement in the primary regularization prism. If another strut-and-tie analysis, which better allows for anchorage zone geometry (as in 'Method of CIRIA Guide 1' below), indicates more reinforcement is necessary in the bursting zone, this should be provided.

Although the reinforcement from 2-2/Expression (J.102) is described as covering the effects of spalling at the end face, there is a lack of guidance on the location for placement of this reinforcement. 2-2/Annex J.104.2(103) also goes on to require a minimum surface reinforcement at the loaded face of $0.03 P_{max}/f_{yd} \gamma_{p,unfav}$. This is covered in section 8.10.3.2 below.

For tendons in external post-tensioned construction, it would be more appropriate to use the characteristic breaking load of the tendon to size the reinforcement, rather than $\gamma_{p,unfav} P_{max}$, unless a detailed non-linear analysis is performed to predict the increase in cable force with global deflection. This is because it is possible for short tendons located within highly stressed areas to reach stresses in excess of yield and the anchor block itself must resist this force.

Method of CIRIA Guide 1
In CIRIA Guide 1, the cross-sectional dimensions of the primary prism are determined from the geometry of the member in each transverse direction. The prism is symmetrical about the anchorage and the width in each direction is taken as the lesser of twice the distance to a free edge or the distance to an adjacent anchorage centreline, as illustrated in Fig. 8.10-6. The length of the prism for each direction equals the width.

Bursting reinforcement in each prism is essentially designed with strut-and-tie models for the plan and elevation directions separately. The guidance for full or partial discontinuities discussed in section 6.5 of this guide (Figs 6.5-1 and 6.5-2) could be applied to this situation (with spread width 'b' chosen so that stresses from adjacent tendons do not overlap or lie outside the cross-section). CIRIA Guide 1 does this directly by way of a table, relating the reinforcement tension T, to the prestressing force P_{max}. This is reproduced here as Table 8.10-3. The reinforcement is distributed in a region from $0.2y_1$ to $2y_1$ from the stressing face in the '1' direction and from $0.2y_2$ to $2y_2$ from the stressing face in the '2' direction. Tests have shown that the most effective reinforcement is closed links or spirals around the prisms, rather than mat reinforcement placed through the prism. Spirals and circular links should have a diameter at least 50 mm greater than the side of the anchorage plate.

CHAPTER 8. DETAILING OF REINFORCEMENT AND PRESTRESSING STEEL

Table 8.10-3. Tensile forces related to primary prism geometry

y_{pi}/y_i	≤0.3	0.4	0.5	0.6	≥0.7
$T/(\gamma_{p,unfav} P_{max})$	0.23	0.20	0.17	0.14	0.11

8.10.3.2. Spalling near the loaded face (additional sub-section)
Additionally, reinforcement to control spalling should be placed adjacent to the loaded face. According to CIRIA Guide 1, this reinforcement has to carry at least $0.04P_{max}$ in each direction, with a greater amount in some geometrical situations. This is the equivalent requirement to that in 2-2/Annex J.104.2(103) above, which requires a surface reinforcement area of $0.03(P_{max}/f_{yd})\gamma_{p,unfav}$ in each direction. Only one of these very similar provisions need be met.

8.10.3.3. Overall equilibrium (additional sub-section)
Where the load has not spread out uniformly across the section by the end of the primary prism, strut-and-tie analysis should be used to derive the reinforcement needed to control the remaining spread. 2-2/Annex J does not cover this further spread. Typical situations and reinforcement zones are as shown in Fig. 8.10-7, which also shows the primary prism locations according to CIRIA Guide 1. (The beams have been rotated so that their axes are vertical.) The figure does not show the primary regularization prisms according to 2-2/Annex J. These figures illustrate the spread of load that is still taking place outside the primary prisms.

In cases (a), (c), (d) and (e), the size of primary prisms according to Annex J and CIRIA Guide 1 would be similar. In case (e), the primary prism is shown less than the full width of the section on the assumption that the prism width for bursting design would be controlled by the thickness of the section. In case (b), the primary prism to CIRIA Guide 1 incorporates the entire zone as shown and the strut-and-tie models for bursting and 'overall' equilibrium are one and the same. The primary regularization prism to Annex J would be much smaller.

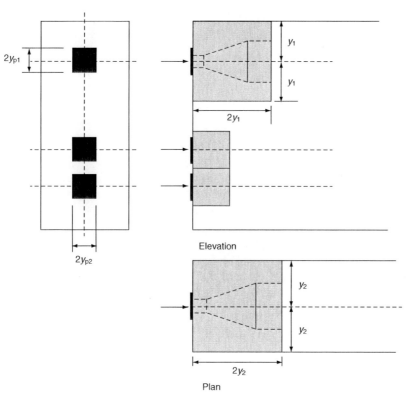

Fig. 8.10-6. Definition of primary prisms in CIRIA Guide 1

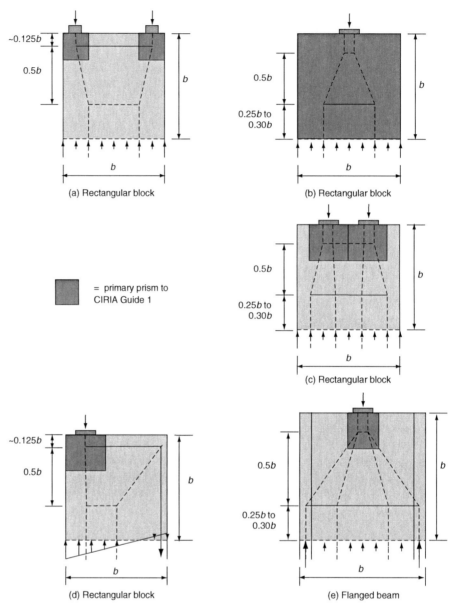

Fig. 8.10-7. Typical strut-and-tie models for anchorage zones

This highlights the problem with Annex J, which could lead to steel being provided for bursting, and then more steel provided for overall equilibrium outside the primary prism. This would double-count reinforcement requirements if the two were superimposed. This is the reason that EC2-2 Annex J only applies to 2 or more tendons anchored together, which generally avoids this problem. This situation is not entirely satisfactory as there is then a danger that case (e) could be designed considering only the overall strut-and-tie model shown and not the localized bursting forces in the vicinity of the anchorage that CIRIA Guide 1 would identify. Where Annex J does not apply, bursting should still always be checked using a local strut-and-tie model (e.g. Fig. 8.10-7(b)) or from the requirements of CIRIA Guide 1.

The internal lever arm of $0.5b$ in each model in Fig. 8.10-7 is consistent with that discussed in section 6.5 for discontinuity zones. The position of the ties is not prescribed but must follow the elastic flow of force as required by general strut-and-tie analysis principles. The tie positions in (a) and (d) follow from the recommendations in CIRIA Guide 1. This required the reinforcement to be provided within a length of $0.25b$ from the loaded face.

CHAPTER 8. DETAILING OF REINFORCEMENT AND PRESTRESSING STEEL

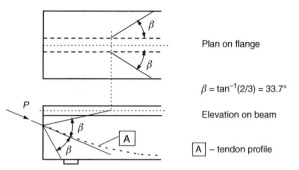

Fig. 8.10-8. Dispersion of prestress

(It is acceptable to place reinforcement nearer to the loaded face than this.) In (b) and (c), the tie position $0.25b$ from the end of the block was implied in CIRIA Guide 1, while other references (such as *Model Code 90*[6]) would lead to a dimension of $0.30b$. As an alternative to strut-and-tie analysis, beam theory can be used to determine reinforcement, but this would need to be based on the same assumptions of lever arm and reinforcement zone placement as above.

Where the beam is flanged, the strut-and-tie design for the web should take into account the concentrated reactions acting at the flange positions, as well as the distributed load in the web as indicated in Fig. 8.10-7(e). The flanges themselves should also be designed for the transverse spread out across their width by strut-and-tie analysis. Figure 8.10-7(b) can be used for this purpose, assuming the equilibrium force in the flange is applied as a concentrated load over a width equal to that of the web. The distance from the stressing face of this 'point' of application of this force can be determined from the idealized spread of force according to Fig. 8.10-8, which comes from the recommendations of *2-1-1/clause 8.10.3(5)*. This distance approximately corresponds to the position of the tie in Fig. 8.10-7(b) at which the load has essentially spread out across the section.

2-1-1/clause 8.10.3(5)

In addition to the above rules, a sliding wedge mechanism should be checked for any anchorages close to the edge of the section as required by *2-2/Annex J.104.1(104)* and discussed in the commentary to 2-2/Annex J.

2-2/Annex J.104.1(104)

Worked example 8.10-1. Reinforcement needed in post-tensioned anchor zone

Two tendons per web, each comprising 19 no. 15.7 mm diameter strands ($A_p = 150$ mm² per strand) with a characteristic tensile strength, f_{pk}, of 1860 MPa, are anchored on a box girder as shown in Fig. 8.10-9 with a square anchor plate of side 310 mm. For this specification of strand, the 0.1% proof stress (from Table 3 of EN 10138) is given as $f_{p0.1k} = 1600$ MPa. It is assumed here that the tendons initially remain horizontal and are to be stressed to $0.85 f_{p0.1k}$ when the concrete cylinder strength reaches 30 MPa. The centroid of the tendons is at the centroid of the cross-section for simplicity in this example.

(a) Bursting reinforcement – primary prisms

Annex J of EN 1992-2 does apply here as there is more than one tendon anchored on the web. The requirements of Annex J are compared to those from a strut-and-tie model and from CIRIA Guide 1 and the most onerous requirement taken. Annex J of EN 1992-2 is deemed to provide only a minimum reinforcement.

(1) Horizontal direction (through thickness of web)
EC2-2 Annex J
From 2-2/Expression (J.101), the cross-sectional area of the primary regularization prism is:

$$cc' = P_{max}/0.6 f_{ck}(t)$$

where P_{max} is the maximum force applied to the tendon where $P_{max} = 19 \times 150 \times 0.85 \times 1600 \times 10^{-3} = 3876\,kN$.

Thus: $cc' = 3876 \times 10^3/(0.6 \times 30) = 215\,333\,mm^2$

For a primary regularization prism with square cross-section, the side is therefore 464 mm. The length of the prism is therefore $= 1.2 \times 464 = 556$, say 560 mm. The minimum allowable area of bursting and spalling reinforcement in this zone is therefore given by 2-2/Expression (J.102):

$$A_s = 0.15 \frac{P_{max}}{f_{yd}} \gamma_{p,unfav} = 0.15 \times \frac{3876 \times 10^3}{500/1.15} \times 1.2 = 1605\,mm^2$$

Strictly, a check of crack widths would then be needed. To avoid such a check, the reinforcement stress should be limited to 250 MPa in accordance with 2-2/clause 8.10.3(104) whereupon an area of steel of 2791 mm^2 would be required. (It is here, and elsewhere in this example, conservatively assumed that the stress limit is verified using the factored prestressing force.) This reinforcement must be provided over the length of the prism, i.e. 560 mm and is needed both horizontally and vertically, so would best be provided in the form of spirals or closed links. **2×5 legs of 20 mm diameter bars within this length would suffice in providing the required 4.98 mm^2/mm**. If bars were provided at 100 mm centres (as is typical) over this length, a slight excess of steel would result, i.e. probably 2×6 legs.

Although this reinforcement is described as covering the effects of spalling at the end face, there is a lack of guidance on the location for the reinforcement. This is covered in item (c) below.

The above reinforcement to Annex J must be regarded as a minimum for this zone. The bursting reinforcement calculation of Annex J takes no account of the relative geometry of anchor plate and receiving section so a further check of bursting must be made to consider this. Either CIRIA Guide 1 or a strut-and-tie model in accordance with 2-1-1/clause 6.5 could be used. Both are considered below. Where they give more onerous reinforcement requirements than Annex J, either throughout the whole primary prism or over part of it, additional reinforcement should be added to meet these requirements.

Method of CIRIA Guide 1
The ratio of end plate side to primary prism side according to CIRIA Guide 1 and Fig. 8.10-6 is:

$2y_{p2}/2y_2 = 310/600 = 0.52$

From Table 8.10-3, the tensile force is $T = 0.17\gamma_{unfav}P_{max} = 0.17 \times 1.2 \times 3876 = 791\,kN$. This is equivalent to a steel area of 3164 mm^2 working at 250 MPa and to be provided over a length of $2y_2 = 600$ mm (5.27 mm^2/mm), which is a similar requirement (by coincidence) to that required by Annex J. **2×5 legs of 20 mm diameter bars within this length would suffice again**. In reality, 2×6 legs at 100 mm centres would probably be provided. Steel would best be provided in the form of spirals or closed links.

Strut-and-tie model of section 6.7 – Fig. 8.10-7(b)
An alternative is the use of a local strut-and-tie model which is essentially the only guidance in EC2-1-1. Using the above strut-and-tie idealization for the transverse direction leads to a transverse tensile force:

$$T = \frac{600 - 310}{600} \times 1/4 \times 3876 \times 10^3 \times 1.2 = 562\,kN$$

which equates to a reinforcement area (working at 250 MPa) of 2248 mm^2. This is less total reinforcement than calculated by the other methods, but needs to be provided in a shorter zone of $0.6b = 0.6 \times 600 = 360$ mm starting 240 mm from the end face, i.e.

6.24 mm²/mm. It would be more normal practice to continue this reinforcement over the entire 600 mm length (and this would be necessary to meet the Annex J requirements), whereupon the total steel required would then be $600/360 \times 2248 = 3747 \text{ mm}^2$ which would be equivalent to **2 × 6 legs of 20 mm diameter bars**.

Of the two latter possibilities, both require slightly more reinforcement than Annex J. It is recommended here that the reinforcement from the CIRIA Guide method should be adopted as this has been shown to produce adequate results in practice. The above calculations, however, show that similar reinforcement is provided by the three methods.

(2) Vertical direction (direction of beam depth)
EC2-2 Annex J
The reinforcement is as calculated above for the horizontal direction.

Method of CIRIA Guide 1
As the anchorages are spaced at 600 mm centres, the primary prism for each anchor is also 600 mm in width so the reinforcement provision would be the same as for the horizontal direction.

Strut-and-tie model of section 6.7 – Fig. 8.10-7(b)
For the reason above, the reinforcement provision would be the same as the horizontal direction.

Once again, as the two latter methods give a greater reinforcement requirement than does Annex J, it is recommended that the reinforcement from the CIRIA Guide method should be adopted as this has been shown to produce adequate results in practice.

(b) Overall equilibrium outside primary prism

(1) Horizontal direction (through thickness of web)
The stresses do not continue to spread beyond the primary prisms in the horizontal directions, so no further reinforcement is required. One problem with the EC2-2 Annex J method is that its primary prism is not actually quite the full width of the web, so there could be the misapprehension of an apparent further spread of stress and thus double-counting of reinforcement. This is why the bursting calculation using a strut-and-tie model or CIRIA Guide 1 method is also required as above.

(2) Vertical direction (over beam depth)
The stress is not uniform across the cross-section at the end of this primary prism, so a strut-and-tie model must be used to determine the transverse reinforcement. Reinforcement in the web can be determined from the strut-and-tie model in Fig. 8.10-7(e). (This model would need to be amended if the tendon was angled within the length of the equilibrium zone.) The tendons are anchored at the neutral axis so the stress remote from the anchorage is uniform across the section (a convenient simplification for this example). Considering half the box cross-section:

Uniform stress $= 2 \times 3.876 \times 10^3 \times 1.2/4.14 \times 10^6 = 2.25 \text{ MPa}$

Force in half width of bottom flange $= 3000 \times 300 \times 2.25 = 2022 \text{ kN}$

Force in web below centreline of lower anchorage $= 1380 \times 600 \times 2.25 = 1863 \text{ kN}$

Ignoring the width of the anchor plate and taking the truss internal lever arm as $0.5 \times 3500 = 1750 \text{ mm}$ gives a tensile tie force at the level of the bottom tendon of $(2022 \times 1.530 + 1863 \times 1.380/2)/1.750 = 2502 \text{ kN}$.

This force must be carried by reinforcement. The allowable stress is here taken as 250 MPa so that no check of crack widths is required and hence 10 008 mm² of reinforcement is needed. This reinforcement should be provided as U-bars and distributed over a maximum length of $0.6b$ from the strut-and-tie idealization (only $0.5b$ in CIRIA Guide 1) centred on the tension tie, as shown in Fig. 8.10-7(e). This equates to 32 no. 20 mm legs in a length $= 0.6 \times 3500 = 2100$ mm. (2 legs at 150 mm centres provides 30 legs in the 2100 mm length, which might be accepted given the conservative use of a 250 MPa stress limit and factored ultimate prestressing force. A check at serviceability could be made to justify this.) In practice, these legs would usually be carried back to the stressing face over the whole equilibrium zone length of 3500 mm. Reinforcement would be needed for shear in this zone in any case.

(c) Spalling from end face
There are two possibilities for design against spalling:
(1) 2-2/Annex J.104.2(103) requires:

$$0.03 \frac{P_{max}}{f_{yd}} \gamma_{p,unfav} = 0.03 \times \frac{3876 \times 10^3}{500/1.15} \times 1.2 = 320 \, \text{mm}^2$$

or

$$0.03 \times \frac{3876 \times 10^3}{250} \times 1.2 = 558 \, \text{mm}^2$$

for reinforcement working at a stress of 250 MPa to avoid a check of crack widths.
(2) CIRIA Guide 1 requires reinforcement to carry $0.04 P_{max}$ in each direction. In this case, this equates to:

$$0.04 \times \frac{3876 \times 10^3}{250} = 620 \, \text{mm}^2$$

Both options lead to 2 no. 20 mm legs in each direction, working at a design stress of 250 MPa.

(d) Spread into flanges
A similar check of transverse tension in the flange is required as discussed in the main text above, based on the strut-and-tie model of Fig. 8.10-7(b). This is not included here. The total reinforcement for all these effects is shown in Fig. 8.10-9. Note that reinforcement for shear and transverse bending in this vicinity would also need to be designed.

Note that the sliding wedge mechanism of Annex J will not be critical as the anchorages are remote from the section edge. If a construction joint was used between web and flange then the rules of 2-2/clause 6.2.5 for shear across joints would also need to be checked.

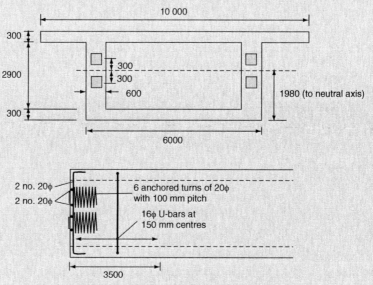

Fig. 8.10-9. Geometry and reinforcement for Worked example 8.10-1

8.10.4. Anchorages and couplers for prestressing tendons

Couplers often temporarily act as anchorages during construction, before the tendons from the next stage are coupled on, and must therefore be designed as such as discussed in section 8.10.3 above.

EN 1992 considers it to be good detailing practice to keep couplers away from intermediate supports (*2-1-1/clause 8.10.4(4)*) and to avoid coupling more than 50% of the tendons at any one section (*2-2/clause 8.10.4(105)*). The latter is because finite element modelling and physical model testing have shown that the stress state around couplers is very complex and uniform compressive stress is not achieved. The compressive stress from prestress in the vicinity of a coupled tendon can be significantly reduced from that obtained with a continuous tendon. Where all tendons are coupled at one location, this can lead to a very large reduction in the compressive stress expected. (The reduction was shown to be up to 70% by Oh and Chae in reference 24.) This can lead to cracking in webs and flanges under load, typically over a length of the order of the section depth each side of the coupler. Prestress from any tendons which are continuous helps to reduce the chances of cracking and additional longitudinal reinforcement can be used to control cracking at the serviceability limit state.

2-1-1/clause 8.10.4(4)
2-2/clause 8.10.4(105)

The 'good practice' of staggering couplers has often not been observed in the UK for practical reasons. To avoid anchoring tendons at one section, it is usually necessary to lap tendons across the construction joint using anchorages on ribs on the outside of webs or flanges. However, sometimes this will not be practical and all the tendons will be coupled at the same location in adjacent webs. Typically this will occur in cellular structures with small depth where it would be impractical to provide anchorages inside the voids (and impossible to access them). An alternative would be to provide top-pocket anchorages in the upper surface of the top flange but these are highly undesirable from durability considerations and should be avoided, particularly where de-icing salts are used or the environment is aggressive in some other way. This is noted in *2-2/clause 8.10.4(107)*.

2-2/clause 8.10.4(107)

If more than 50% of the tendons must be coupled at one section then 2-2/clause 8.10.4(105) requires either the minimum reinforcement requirements of 2-1-1/clause 7.3.2 to be provided to limit cracking or the section at the couplers should everywhere have a compressive stress of 3 MPa under the characteristic combination of actions. It will not generally be difficult to comply with this stress check since construction joints are usually positioned in areas of low moment, typically around span quarter points. 2-2/clause 8.10.4(105) recommends that no more than two-thirds of the tendons at one section should ever be coupled, but this may not always be practical as discussed above. Recommendations are given in 2-2/Table 8.101N on the minimum longitudinal distance between couplers required such that the couplers are not considered to occur at one location. These recommendations can be modified in the National Annex.

When deck slabs are transversely post-tensioned, it is necessary to check that the stress from the post-tensioning is uniform at the section where it is required. This is noted by *2-2/clause 8.10.4(106)*. The longitudinal spacing of anchorages can be determined from the angle of prestress dispersion in Fig. 8.10-7, such that adjacent zones just overlap at the section being considered.

2-2/clause 8.10.4(106)

Where a tendon is anchored within the body of a member (a typical detail is shown in Fig. 8.10-10), cracking may occur behind the anchorage due to restraint from the concrete behind the anchor resisting the deformations caused by stressing. It is usual to anchor some part of the prestressing force back into the concrete behind the anchor with reinforcement to control this cracking, unless there is sufficient residual compressive stress in this area from continuous tendons. *2-2/clause 8.10.4(108)* suggests there should be a residual compression of at least 3 MPa under the frequent combination, before such reinforcement can be omitted. No guidance on reinforcement provision is made in EN 1992-2. CIRIA Guide 1[23] recommends that 50% of the prestressing force should be anchored back into the concrete behind the anchor regardless of prestress, while Schlaich and Scheef[25] recommend that 25% of the prestressing force is anchored back, but this can be reduced with increasing compressive stress in the region of the anchorage produced by other continuous prestressing tendons.

2-2/clause 8.10.4(108)

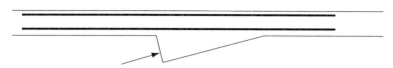

Fig. 8.10-10. Longitudinal reinforcement for anchorages in the body of a member

8.10.5. Deviators

The term 'deviator' refers to a discrete structural element which allows an external post-tensioning cable to be deviated from an otherwise straight line between adjacent anchorages. The protective duct surrounding the tendon strands is normally continuous through a deviator without direct connection to it, so as to facilitate replacement of a tendon. The structural element providing deviation may be concrete or structural steel and normally supports a deviator pipe through which the ducted tendon passes.

2-1-1/clause 8.10.5(1)P requires deviators to fulfil a number of requirements. Their basic purpose is to carry all forces applied to them and transmit them to the rest of the structure. It will often be necessary to use strut-and-tie models to ensure this. The effects of longitudinal forces from friction and asymmetry in the incoming and outgoing bend angle must also be carried by the deviator into the structure. The principal force is usually the transverse force due to the intended change in direction of the cable. However, in designing for this force, the effects of tolerance should also be considered. This includes unintentional deviation from tolerances in setting out the deviator pipes and also in setting out the whole bridge. These tolerances should be declared in the Project Specification. The design force should also include deviations from the effects of deflection and precamber during construction, together with the anticipated accuracy of the precamber.

EC2 does not specify the tendon design force to use in these calculations of deviation force, but it is suggested here that the breaking load of the cable should be used unless detailed second-order analysis considering material and geometric non-linearities has been used. This is because, although it is both conservative and common to assume in the global design that the tendon strain does not increase at the ultimate limit state, it is possible, particularly for short cables, for significant strain and hence stress increase to occur at the ultimate limit state as the structure deflects. At the serviceability limit state, the design load should be that of the cable before long-term losses.

The radius of deviation should also not lead to damage to the tendon, its protective system or the deviator pipe itself. This is the basis of *2-1-1/clause 8.10.5(2)P*. Tight radii give rise to large transverse pressures which can damage the strands, ducts and deviator pipes, while the longitudinal movement during stressing of a tendon on a tight radius can cause the strands to cut through the protective duct or deviator pipe (if plastic). To avoid such problems, *2-1-1/clause 8.10.5(3)P* requires appropriate minimum bend radii to be obtained from the prestressing supplier. Table 8.10-4 provides some preliminary guidance taken from BD 58/94.[26] The groupings in the table are based on similar characteristic strength for the two different tendon systems given. For systems with larger strands than given in Table 8.10-4, the radii should be determined on the basis of the group into which the system fits in terms of similar characteristic strength.

2-1-1/clause 8.10.5 does not specifically cover deflectors in pre-tensioned beams, but the same problems of damage to strand could be encountered. The recommendations of BD 58/94 could again be followed here. This would mean that for single tendons, the

Table 8.10-4. Guidance on minimum deviation radii for tendons

Tendon (strand no. and size)	Minimum radius (m)
19–13 mm and 12–15 mm	2.5
31–13 mm and 19–15 mm	3.0
53–13 mm and 37–15 mm	5.0

CHAPTER 8. DETAILING OF REINFORCEMENT AND PRESTRESSING STEEL

bend radius should not be less than five tendon diameters for wire or 10 diameters for strand. The total deviation angle should not exceed 15°.

A common problem with concrete deviators is spalling at the entrance and exit due to a misaligned tendon. The causes of such 'unexpected' angles of approach are as discussed above. In order to prevent this, some method of providing angular tolerance should be provided in the deviator pipe. Typically this can be done by providing a bell mouth, by overbending the pipe in the plane of principal curvature so that the tangent point for the tendon's contact with the deviator pipe is some way inside the pipe, or by using a compressible material around the pipe entrance. The designer will have to decide which measure might be appropriate in a given situation as each method has its shortcomings.

Care should be taken in the positioning of deviators so as to avoid a loss of eccentricity from second-order effects as the bridge deflects (or from inaccurate precamber). This is particularly important for spans with quarter point deviators only as the tendon will remain straight between quarter points, while the mid-span position deflects a greater amount with a consequent loss of eccentricity at mid-span. In such situations, the loss of eccentricity with deflection can offset any beneficial increase in tendon force caused by the deflection.

2-1-1/clause 8.10.5(4) appears to allow design tendon deviations of 0.01 radians to be accommodated without providing a deviator. However, given the potential for significant unintended deviations as discussed above, it is recommended here that a physical deviator detail is always provided where a deviation is to be made.

2-1-1/clause 8.10.5(4)

CHAPTER 9

Detailing of members and particular rules

This chapter deals with the detailing of members and particular rules as covered in section 9 of EN 1992-2 in the following clauses:

• General	*Clause 9.1*
• Beams	*Clause 9.2*
• Solid slabs	*Clause 9.3*
• Flat slabs	*Clause 9.4*
• Columns	*Clause 9.5*
• Walls	*Clause 9.6*
• Deep beams	*Clause 9.7*
• Foundations	*Clause 9.8*
• Regions with discontinuity in geometry or action	*Clause 9.9*

9.1. General

In addition to the general detailing rules discussed in section 8, 2-2/clause 9 gives additional rules for specific members including beams, slabs, columns, walls and foundations. **2-1-1/clause 9.1(1)** notes that the validity of general rules elsewhere for safety, serviceability and durability are dependent on the rules for detailing in this section being followed. **2-1-1/clause 9.1(2)** is a reminder that detailing should also be consistent with the design model adopted. For example, if strut-and-tie analysis is used to determine reinforcement requirements, it is essential that the position, orientation and anchorage of the detailed reinforcement is consistent with the assumptions in the analysis. The subject of minimum reinforcement (see section 9.2.1.1 below for beams) is introduced by way of **2-2/clause 9.1(103)**. Minimum reinforcement ensures that when the moment resistance of the un-cracked concrete section is exceeded, the reinforcement is able to provide a moment resistance at least as large as that of the gross concrete so that a sudden (brittle) failure is not initiated on cracking. It also ensures that wide cracks do not form when the concrete's tensile strength is overcome.

2-1-1/clause 9.1(1)
2-1-1/clause 9.1(2)

2-2/clause 9.1(103)

9.2. Beams

9.2.1. Longitudinal reinforcement

9.2.1.1. Minimum and maximum areas of reinforcement

All beams should be detailed with a minimum quantity of bonded reinforcement to avoid brittle failure or wide crack formation upon concrete cracking. The value of $A_{s,min}$ for

2-1-1/clause 9.2.1.1(1)

beams may be defined in the National Annex. **2-1-1/clause 9.2.1.1(1)** recommends the following:

$$A_{s,min} = 0.26 \frac{f_{ctm}}{f_{yk}} b_t d \geq 0.0013 b_t d \qquad \text{2-1-1/(9.1N)}$$

where:

- f_{ctm} is the mean tensile strength of the concrete
- f_{yk} is the characteristic yield strength of the reinforcement
- b_t is the mean width of the tension zone (excluding the compression flange for non-rectangular sections)
- d is the effective depth to the tension reinforcement

This requirement is derived from the development of an expression to ensure the reinforcement does not yield as soon as cracking in the concrete occurs. Considering a rectangular beam, the cracking moment is given by:

$$M_{cr} = f_{ctm} b h^2 / 6$$

where b and h are the breadth and overall depth of the concrete beam respectively. If the beam is reinforced with an area of steel A_s, of yield strength f_{yk} at an effective depth d, the ultimate moment of resistance is given by:

$$M_u = f_{yk} A_s z$$

where z is the lever arm. For the cracked strength of the section to exceed the mean un-cracked strength at first cracking, $M_u \geq M_{cr}$. This requires $f_{yk} A_s z \geq f_{ctm} b h^2 / 6$ which rearranges to:

$$A_s \geq \frac{1}{6} \frac{f_{ctm}}{f_{yk}} \frac{b h^2}{z}$$

Introducing the effective depth, d, gives:

$$A_s \geq \frac{1}{6} \frac{f_{ctm}}{f_{yk}} \left(\frac{h}{d}\right)^2 \frac{d}{z} bd$$

A typical value of h/d might be 1.1. A conservative value of d/z would be 1.25, corresponding to a heavily reinforced section with $z = 0.80d$, which is unlikely to arise with minimum reinforcement in a rectangular section, but could arise with other cross-section shapes, such as T beams. These values lead to:

$$A_s \geq 0.25 \frac{f_{ctm}}{f_{yk}} bd$$

2-1-1/Expression (9.1N) above is of the same form as this expression for a rectangular beam, but the 0.25 factor has been replaced by 0.26, which makes some allowance for other section geometries. The relatively small increase for other geometries reflects the conservative value of d/z already assumed above. Elements containing less than the minimum area of reinforcement, $A_{s,min}$, should be considered as un-reinforced and designed in accordance with 2-2/clause 12 as required by **2-1-1/clause 9.2.1.1(2)**.

2-1-1/clause 9.2.1.1(2)

In addition to the minimum reinforcement requirements detailed above, EC2-2 requires that all elements in bridge design should contain a minimum quantity of reinforcement to control cracking as discussed in section 7.3. For prestressed concrete sections, additional rules to prevent brittle failure are given in clause 6.1 and discussed in section 6.1.5 of this guide. Similarly, **2-1-1/clause 9.2.1.1(4)** requires the designer to check that the ultimate bending resistance of members with permanently un-bonded or external prestressing tendons exceeds the flexural cracking moment by a recommended factor of 1.15.

2-1-1/clause 9.2.1.1(4)

In order to ease the placing and compacting of concrete and to prevent cracking from excessive internal restraint to concrete shrinkage, reinforcement in beams should be limited to a maximum value, $A_{s,max}$. The value of $A_{s,max}$ for beams, outside lap locations, can be specified in the National Annex and *2-1-1/clause 9.2.1.1(3)* recommends the value of $0.04A_c$, where A_c is the gross cross-sectional concrete area of the element considered.

2-1-1/clause 9.2.1.1(3)

9.2.1.2. Other detailing arrangements

2-1-1/clause 9.2.1.2(1) requires that for monolithic construction, even where simple supports have been assumed in the design, support sections of beams should be reinforced to provide a minimum hogging moment resistance. This should be a proportion, β_1, of the maximum sagging moment in the span. The value of β_1 may be provided in the National Annex and EC2 recommends taking a value of 0.15. This relatively small minimum design moment could require considerable redistribution of moment to occur and it is likely that the National Annex will specify a larger value. For bridges, however, where plastic analysis is generally not permitted, it will not usually be acceptable to treat a monolithic connection as a simple support. The results of an elastic analysis, with or without any limited redistribution allowed in clause 5.5, are then likely to determine minimum hogging moment requirements. Crack width checks would additionally have to be performed using the results of the unmodified elastic analysis.

2-1-1/clause 9.2.1.2(1)

For flanged beams, *2-1-1/clause 9.2.1.2(2)* requires the total area of tension reinforcement required to be spread over the effective width of the flange, although part of the reinforcement may be concentrated over the web width. This is presumably to limit cracking across the width of the flange at SLS (including from early thermal cracking), even though such a distribution might not be necessary at ULS.

2-1-1/clause 9.2.1.2(2)

Where the design of a section has included the contribution of any longitudinal compression reinforcement in the resistance calculation, *2-1-1/clause 9.2.1.2(3)* requires the reinforcement to be effectively held in place by transverse reinforcement. This is intended to prevent buckling of the bars out through the cover. The spacing of this transverse reinforcement should be no greater than 15 times the diameter of the compression bar. The minimum size of the transverse reinforcement is not given, but it is recommended here that the requirements of 2-1-1/clause 9.5.3 for columns be followed. 2-1-1/clause 9.2.1.2(3) also does not define specific requirements for how the transverse reinforcement should enclose the compression reinforcement. The requirements for columns in 2-1-1/clause 9.5.3, as discussed in section 9.5.3 of this guide, could also be applied to compression flanges of beams.

2-1-1/clause 9.2.1.2(3)

9.2.1.3. Curtailment of longitudinal tension reinforcement

EC2 allows longitudinal bars to be curtailed beyond the point which they are no longer required for design, provided sufficient reinforcement remains to adequately resist the envelope of tensile forces acting at all sections (including satisfying minimum requirements). Bars should extend at least an anchorage length beyond this point. The contribution to the section resistance of bars within their anchorage length may be taken into account assuming a linear variation of force (as indicated in 2-1-1/Fig. 9.2-2 and allowed by *2-1-1/clause 9.2.1.3(3)*) or may conservatively be ignored.

2-1-1/clause 9.2.1.3(3)

In determining the point beyond which the reinforcement is no longer required, appropriate allowance must be given to the requirements to provide additional tensile force for shear design. This leads to the 'shift method' described in 2-1-1/clause 9.2.1.3 and discussed in section 6.2.3.1 of this guide. It should be noted that a further shift is required for position within wide flanges as discussed in section 6.2.4.

2-1-1/clause 9.2.1.3(4) also specifies an increased anchorage length requirement for bent-up bars which contribute to the shear resistance of a section. This ensures the bar can reach its design strength at the point it is required as shear reinforcement.

2-1-1/clause 9.2.1.3(4)

9.2.1.4. Anchorage of bottom reinforcement at an end support

For end supports, where little or no end fixity is assumed in the design, the designer should provide at least a proportion of the reinforcement provided in the span. This proportion, β_2, is recommended by EC2 to be taken as 0.25. This longitudinal reinforcement should not be less than that required for the shear design together with any axial force and must be adequately anchored beyond the face of the support.

9.2.1.5. Anchorage of bottom reinforcement at intermediate supports
For intermediate supports, EC2 allows the minimum bottom reinforcement calculated in accordance with 2-1-1/clause 9.2.1.4 to be curtailed after extending at least 10 bar diameters (for straight bars) into the support measured from the support face. For bars with bends or hooks, EC2 requires the bar to be anchored a length at least equal to the mandrel diameter beyond the support face, to ensure the bend of the bar does not begin until well within the support. Where sagging moments can develop, the bottom reinforcement should obviously be continuous through intermediate supports, using lapped bars if necessary. Considerations of early thermal cracking may also lead to the need for continuity of the longitudinal reinforcement.

9.2.2. Shear reinforcement

2-2/clause 9.2.2(101) requires shear reinforcement to form an angle, α, of between 45° and 90° to the longitudinal axis of the structural member. In most bridge applications, shear reinforcement is provided in the form of links enclosing the longitudinal tensile reinforcement and anchored in accordance with 2-1-1/clause 8.5. The inside of link bends should be provided with a longitudinal bar of at least the same diameter as the link as discussed in section 8.5. *2-1-1/clause 9.2.2(3)* allows links to be formed by lapping legs near the surface of webs (it is often convenient to form links from two U-bars, for example) only if the links are not required to resist torsion. It has, however, been common practice in the UK to form outer links in bridges in this way and they have performed adequately as torsion links.

2-2/clause 9.2.2(101) allows shear reinforcement to comprise a combination of links and bent-up bars, but a certain minimum proportion of the required shear reinforcement must be provided in the form of links. This minimum proportion, β_3, is recommended by *2-1-1/clause 9.2.2(4)* to be taken as 0.50 but it may be varied in the National Annex. The restriction is largely due to the lack of test data for combinations of different types of shear reinforcement. If strut-and-tie design was used in accordance with 2-1-1/clause 6.5, this restriction would not apply, providing all struts, nodes and concrete stresses at bends of bars were checked accordingly. The use of the other types of shear assemblies allowed in 2-1-1/clause 9.2.2(2) for buildings, which do not enclose the longitudinal reinforcement, is not recommended for bridges according to 2-2/clause 9.2.2(101), although the National Annex may permit their use.

2-1-1/clause 9.2.2(5) defines a minimum shear reinforcement ratio for beams (reproduced below) to ensure that the failure load exceeds the shear cracking load:

$$\rho_w = A_{sw}/(sb_w \sin \alpha) \geq \rho_{w,min} \qquad \text{2-1-1/(9.4)}$$

where:

- A_{sw} is the area of shear reinforcement within the length s
- s is the spacing of the shear reinforcement measured along the longitudinal axis of the member
- b_w is the breadth of the web of the member
- α is the angle between the shear reinforcement and the longitudinal axis

The value of the minimum shear reinforcement ratio, $\rho_{w,min}$, is recommended by EC2 to be at least that given by the following expression:

$$\rho_{w,min} = (0.08\sqrt{f_{ck}})/f_{yk} \qquad \text{2-1-1/(9.5N)}$$

where f_{ck} and f_{yk} are the characteristic cylinder strength of the concrete and characteristic yield strength of the shear reinforcement respectively (both in MPa).

Clauses 9.2.2(6), (7) and (8) of EN 1992-1-1 give additional recommendations for the detailing of shear reinforcement. These rules cover maximum longitudinal spacing

CHAPTER 9. DETAILING OF MEMBERS AND PARTICULAR RULES

between shear assemblies ($s_{l,max}$), maximum longitudinal spacing of bent-up bars ($s_{b,max}$) and maximum transverse spacing of the legs in a series of shear links ($s_{t,max}$). Values for each of these limits can be found in the National Annex and EC2 recommends the following:

$$s_{l,max} = 0.75d(1 + \cot \alpha) \qquad \text{2-1-1/(9.6N)}$$

$$s_{b,max} = 0.6d(1 + \cot \alpha) \qquad \text{2-1-1/(9.7N)}$$

$$s_{t,max} = 0.75d \leq 600\,\text{mm} \qquad \text{2-1-1/(9.8N)}$$

The recommended values for $s_{l,max}$ and $s_{b,max}$ have been chosen from a conservative consideration of test data to ensure that a shear failure plane cannot be formed between two adjacent sets of shear reinforcement. The expression for $s_{l,max}$ is intended to apply to links.

9.2.3. Torsion reinforcement

Torsion links should comply with the requirements of 2-1-1/Fig. 9.6, which require bars to be either fully lapped or anchored by means of hooks, and be fixed perpendicular to the member axis. **2-1-1/clause 9.2.3(2)** states that the provisions for minimum shear reinforcement in beams discussed above are generally sufficient to cover the minimum requirements for torsion links. However, **2-1-1/clause 9.2.3(3)** specifies further requirements for longitudinal spacing of torsion links, specifically that the spacing should not exceed $u/8$ (where u is the outer circumference of the cross-section) or the minimum dimension of the beam cross-section. This limit is to ensure that torsion cracks do not develop between planes of torsion reinforcement and to prevent premature spalling of the corners of the section under the action of the spiralling compression struts.

2-1-1/clause 9.2.3(4) requires that longitudinal reinforcement for torsion be arranged so that there is at least one bar in each corner of the links. The rest of the longitudinal reinforcement should be spaced evenly around the remaining perimeter, enclosed by the links, at a maximum spacing of 350 mm. The need for approximately uniform spacing of longitudinal bars is discussed in section 6.3.2 of this guide.

2-1-1/clause 9.2.3(2)
2-1-1/clause 9.2.3(3)

2-1-1/clause 9.2.3(4)

9.2.4. Surface reinforcement

Surface reinforcement is discussed in section J1 of this guide.

9.2.5. Indirect supports

Where beams frame into each other, such as might occur at the intersection between cross girders and main beams, the reaction between the two elements needs to be carried by reinforcement or prestressing steel. The need for 'suspension' reinforcement can be seen from a truss model and is illustrated in Fig. 9.2-1 for the case of a single-cell box girder diaphragm supported on a central bearing.

EN 1992-1-1 gives guidance on the area at the web–diaphragm intersection where this reinforcement should be provided, and this is also shown in Fig. 9.2-1 for the box girder diaphragm case. In general, suspension reinforcement will add to reinforcement for other effects and a truss model will often clarify this. This is stated in **2-1-1/clause 9.2.5(1)**. However, in the particular web–diaphragm region of Fig. 9.2-1, it can be seen that the suspension reinforcement need not be added to the normal web shear reinforcement within this intersection region as the respective ties do not coincide. In other words, the reinforcement for ordinary shear design starts outside this intersection zone. The density of links in the web should not, however, be reduced near to the diaphragm. **2-1-1/clause 9.2.5(2)** gives some guidance on placement of suspension reinforcement. Work by Leonhardt *et al.* (reference 27) suggests that 70% of the suspension reinforcement should always be provided within the box girder web, within the intersection region of Fig. 9.2-1, as experiments showed that reinforcement in this area was the most effective. The reinforcement should extend over the whole depth of the beam and be adequately anchored.

2-1-1/clause 9.2.5(1)

2-1-1/clause 9.2.5(2)

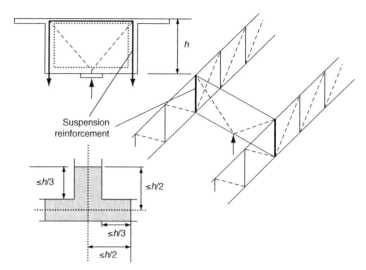

Fig. 9.2-1. Suspension reinforcement in a box girder and intersection region (plan) where it should be provided

Worked example 9.2-1: Box girder web–diaphragm region

A multi-span continuous box girder has the section shown in Fig. 9.2-2 and is supported on a single bearing. The bearing reaction, $R = 22\,\text{MN}$, is equally distributed between the webs and between adjacent spans. The suspension reinforcement that must be provided in the web–diaphragm interface is designed. It is assumed that the reinforcement has a yield strength of 500 MPa and the diaphragm is 1800 mm wide with a height of 2800 mm.

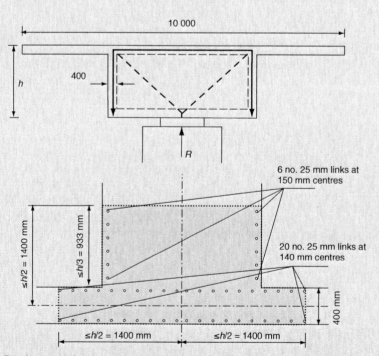

Fig. 9.2-2. Diaphragm and suspension reinforcement for Worked example 9.2-1

The force from each web in each span is:

$R/4 = 22.0/4 = 5.5\,\text{MN}$

CHAPTER 9. DETAILING OF MEMBERS AND PARTICULAR RULES

> Considering one-half of the diaphragm, the web–diaphragm interface is loaded from one web from each span and the force in this region is:
>
> $F = 2 \times 5.5 = 11.0 \, \text{MN}$
>
> This force must be carried by suspension reinforcement in the form of links. These links can be distributed in a region in accordance with 2-1-1/Fig. 9.7, as illustrated in Fig. 9.2-2 for this example. The total suspension reinforcement required in each half of the box is therefore given by:
>
> $A_s = F/f_{yd} = 11.0 \times 10^6/(500/1.15) = 25\,300 \, \text{mm}^2$
>
> which can be provided by 52 legs of 25 mm diameter bars.

9.3. Solid slabs

This section of the code covers specific detailing rules for the design of one-way and two-way spanning solid slabs, defined as elements where the breadth, b, and effective length, l_{eff}, are not less than 5 times the overall depth, h. This section covers general rules for flexural reinforcement in slabs as well as edge, corner and shear reinforcement. The subsequent section in the code on flat slabs (see section 9.4) gives additional recommendations for the detailing of slabs supported by columns and punching shear reinforcement.

9.3.1. Flexural reinforcement

The requirements for maximum and minimum areas of main flexural reinforcement in slabs are the same as those for longitudinal reinforcement in beams discussed in section 9.2.1 above. For two-way spanning slabs, both reinforcement directions should be considered as main reinforcement directions in determining minimum steel. *2-1-1/clause 9.3.1.1(2)* requires that the area of secondary reinforcement in one-way spanning slabs is at least 20% of the area provided in the principal direction, to ensure adequate transverse distribution but transverse reinforcement is not required in the top of slabs near supports where there is no chance of developing transverse bending moments. A suitable minimum quantity of transverse slab reinforcement will always, however, be needed to resist early thermal and shrinkage cracking.

2-1-1/clause 9.3.1.1(2)

Maximum spacing of flexural reinforcement in slabs is covered by *2-1-1/clause 9.3.1.1(3)*. The rules covering curtailment of reinforcement in beams also apply to slabs as noted in *2-1-1/clause 9.3.1.1(4)*, but with the 'shift' of bending moment taken as the effective depth, d, as for beams without shear reinforcement. In the rare circumstances where slabs do contain shear reinforcement, it would be reasonable to use the shift for beams with shear reinforcement in accordance with 2-1-1/clause 9.2.1.3(2), which is only $d/2$ for vertical links and a 45° truss.

2-1-1/clause 9.3.1.1(3)
2-1-1/clause 9.3.1.1(4)

Minimum reinforcement at supports is covered by 2-1-1/clause 9.3.1.2. This requires that at least half the calculated span reinforcement continues up to the face of a support and is then fully anchored. Where partial fixity occurs along the edge of a slab but is not considered in the analysis, *2-1-1/clause 9.3.1.2(2)* requires that top reinforcement is provided which is sufficient to resist at least 25% of the maximum moment in the adjacent span. These are more onerous requirements than the corresponding ones for beams and, unlike for beams, are not nationally determined parameters.

2-1-1/clause 9.3.1.2(2)

2-1-1/clause 9.3.1.4(1) requires that free edges of slabs should normally be detailed with suitable closing reinforcement comprising transverse U-bars enclosing the longitudinal bars. For two-way spanning slabs without edge stiffening, one set of edge U-bars will inevitably have to be placed inside the perpendicular set which will not comply with 2-1-1/Fig. 9.8. This problem will not often arise in practice in bridge decks, however, as deck edges usually terminate with parapet edge beams and expansion joint details at the ends of the deck often require additional longitudinal U-bars to be detailed on the outside, extending into joint nibs.

2-1-1/clause 9.3.1.4(1)

9.3.2. Shear reinforcement

2-1-1/clause 9.3.2(1)

2-1-1/clause 9.3.2(1) requires slabs with shear reinforcement to be at least 200 mm deep in order for the links to contribute to shear resistance. The general detailing rules for shear reinforcement are as for beams in 2-1-1/clause 9.2.2 with a few exceptions:

- The rules for minimum links only apply where designed links are necessary.
- If the applied shear force, V_{Ed}, is no greater than one-third of the design value of the maximum shear force, $V_{Rd,max}$, all of the shear reinforcement may be provided by either bent-up bars or links.
- The maximum longitudinal spacing of bent-up bars can be increased to $s_{max} = d$.
- The maximum transverse spacing of shear reinforcement can be increased to $1.5d$.

9.4. Flat slabs

The rules in clause 9.4 supplement those in 9.3 and cover reinforcement detailing for flexure at column supports and for punching shear in general. Only punching shear is considered here.

The punching shear reinforcement rules given in this part of the code are intended primarily for concentrated loads from columns, but also apply to moving loads on bridge decks, as discussed in section 6.4 of this guide. Since bridge deck slabs are usually detailed to avoid shear reinforcement, these problems will seldom arise. Detailing of punching shear reinforcement in bridge design is most common when considering pad foundations and pile caps.

2-1-1/clause 9.4.3(1)

Where reinforcement is required to resist punching shear, *2-1-1/clause 9.4.3(1)* requires that it should be placed between the loaded area or column and a distance kd within the control perimeter at which shear reinforcement is no longer required. The value of k is recommended to be 1.5 in clause 6.4.5(4) of EN 1992-1-1. The reinforcement should be provided in at least two perimeters, with spacing between the link perimeters not exceeding $0.75d$. Bent-up bars can be distributed on a single perimeter. Additional rules for bent-up bars are also given in clauses 9.4.3(3) and (4).

2-1-1/clause 9.4.3(2)

Where punching shear reinforcement is required, *2-1-1/clause 9.4.3(2)* defines a minimum value based on the same recommendation for minimum areas of links in beams for flexural shear. The difficulty in applying this rule to punching shear links is in determining an appropriate reinforcement ratio. The following equation is given:

$$A_{sw,min}(1.5\sin\alpha + \cos\alpha)/(s_r s_t) \geq (0.08\sqrt{f_{ck}})/f_{yk} \qquad \text{2-1-1/(9.11)}$$

where:

- $A_{sw,min}$ is the minimum area of one link leg (or equivalent)
- A is the angle between the shear reinforcement and the main reinforcement
- s_r is the spacing of shear links in the radial direction
- s_t is the spacing of shear links in the tangential direction
- f_{ck}, f_{yk} are the characteristic cylinder strength of the concrete and characteristic yield strength of the shear reinforcement respectively (in MPa)

Where punching shear reinforcement is provided in orthogonal grids and not on radial perimeters, the above minimum reinforcement requirements may be rather conservative as not all the bars within the outer perimeter will be considered. This is the same as the ULS strength problem discussed in section 6.4.5 of the guide.

9.5. Columns

9.5.1. General

The detailing requirements given in this section of EC2 are applicable to elements classed as columns where the larger cross-sectional dimension, b, is not greater than 4 times the smaller dimension, h. For solid columns, where $b > 4h$, the column is classed as a wall. For hollow columns, where $b > 4h$, no guidance is given.

9.5.2. Longitudinal reinforcement

A minimum percentage of longitudinal reinforcement is required in columns to cater for unintended eccentricities and to control creep deformations. Under long-term application of service loading conditions, load is transferred from the concrete to the reinforcement because the concrete creeps and shrinks. If the area of reinforcement in a column is small, the reinforcement may yield under long-term service loads. The recommended minimum area requirements in *2-1-1/clause 9.5.2(2)* therefore accounts for the design axial compressive force, as well as the column gross cross-sectional area such that:

$$A_{s,\min} = 0.10 N_{Ed}/f_{yd} \geq 0.002 A_c \qquad \text{2-1-1/(9.12N)}$$

2-1-1/clause 9.5.2(2)

where:

N_{Ed} is the maximum design axial compression force
f_{yd} is the design yield strength of the reinforcement
A_c is the gross cross-sectional concrete area

The National Annex may also specify minimum bar sizes to be used for longitudinal reinforcement in columns (recommended by *2-1-1/clause 9.5.2(1)* to be 8 mm) and maximum areas of reinforcement. The maximum areas have been chosen partly from practical considerations of placing and compacting the concrete and partly to prevent cracking from excessive internal restraint to concrete shrinkage caused by the reinforcement. *2-1-1/clause 9.5.2(3)* recommends a maximum reinforcement content of $0.04 A_c$ outside lap locations, increasing to $0.08 A_c$ at laps.

2-1-1/clause 9.5.2(1)

2-1-1/clause 9.5.2(3)

For columns with a polygonal cross-section, *2-1-1/clause 9.5.2(4)* requires at least one longitudinal bar to be placed at each corner with a minimum of four bars used in circular columns. It is recommended, however, that, as in current UK practice (and as recommended in *Model Code 90*[6]), a minimum of six bars is used for circular columns in order to ensure the stability of the reinforcement cage prior to casting.

2-1-1/clause 9.5.2(4)

9.5.3. Transverse reinforcement

'Transverse reinforcement' generally refers to links, loops or spirals enclosing the longitudinal reinforcement. The purpose of transverse reinforcement in columns is to provide adequate shear resistance and to secure longitudinal compression bars against buckling out through the concrete cover. Although not explicitly stated in EN 1992, it is recommended here that minimum links are provided for columns using the same requirements as those for beams. To ensure the stability of the reinforcement cages in columns prior to casting, *2-2/clause 9.5.3(101)* recommends the minimum diameter of the transverse reinforcement to be the greater of 6 mm or one-quarter the size of the maximum diameter of the longitudinal bars.

2-2/clause 9.5.3(101)

The maximum spacing of transverse reinforcement in columns, $s_{cl,\max}$, may be given in the National Annex, but *2-1-1/clause 9.5.3(3)* recommends taking a value of the least of the following:

2-1-1/clause 9.5.3(3)

- 20 times the minimum diameter of the longitudinal bars;
- the lesser dimension of the column;
- 400 mm.

2-1-1/clause 9.5.3(4) further recommends that these maximum spacing dimensions are reduced by multiplying them by a factor of 0.6 in column sections close to a beam or slab, or near lapped joints in the longitudinal reinforcement where a minimum of three bars evenly placed along the lap length is required.

2-1-1/clause 9.5.3(4)

2-1-1/clause 9.5.3(6) requires that every longitudinal bar or bundle of bars placed in a corner should be held by transverse reinforcement. In addition, no bar within a compression zone should be further than 150 mm from a 'restrained' bar. There is, however, no definition provided of what constitutes a restrained bar. It could be interpreted as requiring all compression bars in an outer layer to be within 150 mm of a bar held in place by a link, with links passing around every alternate bar. For box sections with wide flanges, this would

2-1-1/clause 9.5.3(6)

require additional link reinforcement in the flanges in addition to web links over the depth of the section. This interpretation was the one used in BS 5400 Part 4[9] for bars contributing to the section resistance. It is not practical to provide this detail in all situations – links in flanges is an obvious example. Where this detailing cannot be achieved, it is recommended here that transverse bars should still be provided on the outside of the longitudinal reinforcement (which 2-1-1/clause 9.6.3 for walls describes as 'horizontal' reinforcement rather than transverse reinforcement), but the longitudinal compression bars in an outer layer should not then be included in the resistance calculation. This detailing problem does not arise in circular columns with perimeter links as these will be sufficient to restrain the compression bars.

It should be noted that the inclusion of reinforcement in compression zones is implicit in other expressions elsewhere in the code (such as the nominal stiffness method for analysis of slender columns in 2-1-1/clause 5.8.7.2 and the interaction formula for biaxial bending of columns in 2-1-1/clause 5.8.9). For stiffness calculation (as in 2-1-1/clause 5.8.7.2), it would be reasonable to use the code formulae without enclosing every other compression bar by a link. For strength calculation (as in 2-1-1/clause 5.8.9), it is important that compression bars are properly held by links as above.

9.6. Walls

2-1-1/clause 9.6.1(1) defines a wall as having a length to thickness ratio of at least 4. Where a wall is subjected to predominantly out-of-plane bending, clause 9.3 for slabs applies. The amount of reinforcement in a wall and appropriate detailing for it may be determined from a strut-and-tie model.

Leaf piers may fall into the 'wall' category. 2-1-1/clause 9.6.3 and 2-1-1/clause 9.6.4 deal with 'horizontal' and 'transverse' reinforcement requirements respectively. Horizontal reinforcement lies parallel to the long face of the wall while transverse reinforcement passes through the width of the wall in the form of links. It is recommended here that the requirements for columns in clause 9.5 should also be met, which can be more onerous, e.g. minimum vertical reinforcement.

Considerations of early thermal cracking may give rise to greater reinforcement requirements. The UK National Annex is likely to give guidance here.

2-1-1/clause 9.6.1(1)

9.7. Deep beams

A deep beam is formally defined in EN 1992 as a member whose span is less than 3 times its overall section depth. In bridge design, this will most frequently apply to transverse diaphragms in box girders or between bridge beams. Strut-and-tie modelling, as discussed in section 6.5, will be the normal method of design and all the rules on anchorage of reinforcement at nodes and limiting concrete pressure will apply.

The main reinforcement determined for a deep beam from a strut-and-tie analysis may not require any surface reinforcement. A recommended minimum reinforcement ratio of 0.1% (but not less than 150 mm^2/m) for each face and each orthogonal direction is therefore required to be placed near each face in accordance with *2-1-1/clause 9.7(1)*. This amount can, however, be varied by the National Annex. From *2-2/clause 9.7(102)*, bar centres should not exceed the lesser of 300 mm or the web thickness (which may also be varied by the National Annex). This reinforcement is intended to control cracking from effects not directly modelled in the strut-and-tie analysis, such as transverse tension from bulging of compression struts as discussed in section 6.5, and surface strains resulting from tensile ties within the concrete section. (The nominal reinforcement is not intended to be sufficient to fully restrain the tensile forces perpendicular to a bulging compression strut and thus increase its allowable compressive stress. If this is required, reinforcement should be explicitly designed for this purpose in accordance with 2-1-1/clause 6.5.)

For deep beams, it is likely that reinforcement to control early thermal cracking will exceed the above minimum requirements. The UK National Annex is likely to give guidance here.

2-1-1/clause 9.7(1)
2-2/clause 9.7(102)

9.8. Foundations

2-1-1/clause 9.8 gives additional detailing guidance in section 9.8 for the following types of foundations: pile caps, footings, tie beams and bored piles. Pile caps and footings are discussed below.

Pile caps

2-1-1/clause 9.8.1 gives additional requirements for the design of reinforced concrete pile caps, including:

- Reinforcement in a pile cap should be determined by using either strut-and-tie or flexural methods.
- The distance from the edge of the pile to the edge of the pile cap should be sufficient to enable the tie forces in the pile cap to be properly anchored.
- The main tensile reinforcement to resist the action effects should be concentrated above the tops of the piles. This is a harsh requirement. Where flexural design is appropriate, it has been common practice in the UK to adopt a uniform distribution of reinforcement across the pile cap unless the pile spacings exceeded three pile diameters. Even where strut-and-tie analysis is used, it may be possible to consider reinforcement outside the pile width, provided transverse reinforcement is able to distribute the forces as shown in Fig. 9.8-1.
- If the concentrated reinforcement placed above the pile tops is at least equal to the minimum reinforcement requirements of the full section, *2-2/clause 9.8.1(103)* allows evenly distributed bars along the bottom surface of a member to be omitted. This is not recommended here for the design of bridge pile caps in order to control early thermal cracking in such areas. Note the provision in 2-1-1/clause 9.8.1(3) for buildings, which allows the side faces and tops of pile caps to be un-reinforced where there is no risk of tension developing, is omitted in EN 1992-2 for similar reasons.
- Welded bars are allowed to provide anchorage to the tensile reinforcement.
- The compression caused by the support reaction from the pile may be assumed to spread at 45° from the edge of the pile (2-1-1/Fig. 9.11) and may be taken into account when calculating anchorage lengths.
- The tops of the piles should extend a minimum of 50 mm into the pile cap.

2-2/clause 9.8.1(103)

Footings

2-1-1/clause 9.8.2 gives additional guidance on the design of column and wall footings including:

- The main tensile reinforcement should be provided with bars of a minimum diameter, d_{min}, with recommended value of 8 mm.
- The main reinforcement of circular footings may be orthogonal and concentrated in the middle half of the footing (±10%). If adopted, the remainder of the footing should be considered as plain concrete.
- The analysis should include checks for any tensile stresses resulting on the upper surface of the footing, and these should be adequately reinforced.

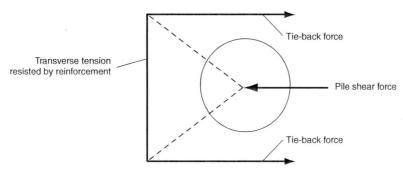

Fig. 9.8-1. Spread of load from a pile to adjacent ties

2-1-1/clause 9.8.2.2 describes a strut-and-tie method (Fig. 9.8-2) for calculating the forces along the length of the tension reinforcement in order to determine anchorage requirements. This model is necessary to account for the effects of 'inclined cracks', as noted in **2-1-1/clause 9.8.2.2(1)**. Anchorage is particularly important at the edges of the footing in determining whether or not the main tension reinforcement requires bends, hooks or laps onto side face reinforcement.

2-1-1/clause 9.8.2.2(1)

The tensile force to be anchored at any distance x from the edge of the base is given in **2-1-1/clause 9.8.2.2(2)** as follows:

2-1-1/clause 9.8.2.2(2)

$$F_s = Rz_e/z_i \qquad \text{2-1-1/(9.13)}$$

where:

- F_s is the force to be anchored at a distance x from the edge of the footing
- R is the resultant force from the ground pressure within distance x
- N_{Ed} is the vertical force corresponding to the total ground pressure between sections A and B (illustrated in Fig. 9.8-2)
- z_e is the external lever arm (the distance between R and N_{Ed})
- z_i is the internal lever arm (the distance between the reinforcement and the horizontal force, F_c)
- F_c is the compressive force corresponding to the maximum tensile force, $F_{s,max}$

The lever arms, z_e and z_i, can be readily determined from considerations of the compression zones from N_{Ed} and F_c respectively, but **2-1-1/clause 9.8.2.2(3)** gives simplifications where z_e may be calculated assuming $e = 0.15b$ and z_i is taken as $0.9d$. In practice, the value of z_i will already be known from the ULS bending analysis and this simplification is unnecessary. The value of N_{Ed}, however, depends on the distance between A and B, and, since B is not initially known, the process of finding a compatible value of e is iterative.

2-1-1/clause 9.8.2.2(3)

Where the available anchorage length, denoted l_b in Fig. 9.8-2, is not sufficient to anchor the force F_s at the distance x, **2-1-1/clause 9.8.2.2(4)** allows additional anchorage to be provided by bending the bars or providing suitable end anchorage devices. Theoretically, the anchorage of the bar should be checked at all values of x. For straight bars, the minimum value of x would be the most critical in determining anchorage requirements. This is because the diagonal compression strut is flatter for small x, and a greater proportion of the length x is concrete cover. If x is taken less than the cover then the reinforcement clearly cannot operate at all. **2-1-1/clause 9.8.2.2(5)** therefore recommends using a minimum value of $h/2$ as a practical simplification. For other types of anchorage, such as bends or mechanical devices, higher values of x may be more critical, since doubling the distance x, for example, will not double the available anchorage resistance. The calculation procedure is illustrated in Worked example 9.8-1.

2-1-1/clause 9.8.2.2(4)

2-1-1/clause 9.8.2.2(5)

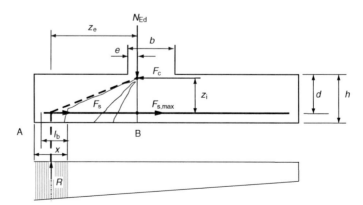

Fig. 9.8-2. Model for calculating required anchorage forces in tensile reinforcement

Worked example 9.8-1: Reinforced concrete pad footing

Anchorage of the main bars for the reinforced concrete pad footing in Worked example 6.4-1 is checked. 25 mm bars at 200 mm centres are provided in both directions in the bottom layers. The bending analysis for the inner layer gives $d = 687.5$ mm and $z_i = 659.8$ mm for an ultimate moment, $M_{z,Rd} = 704.1$ kNm/m, at the deepest section. The concrete grade is C35/45 and the cover to the end of the reinforcement is 50 mm. The pad base pressure is conservatively taken at its previously calculated maximum value of 300 kN/m² throughout. This pressure would be generated by a uniform axial force of 4800 kN in the absence of column moment.

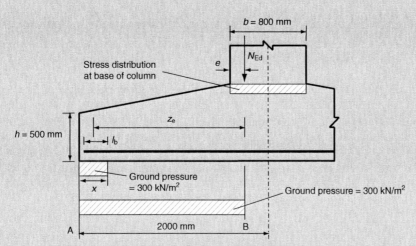

Fig. 9.8-3. Ground pressure distribution beneath pad foundation

Figure 9.8-3 illustrates the ground pressure distribution beneath half of the pad foundation. First, the position of B needs to be determined. Taking the column load to be uniformly distributed across its width, it is initially assumed that half of the column total force acts at $e = 0.25b = 200$ mm. Assuming this to be the location of N_{Ed} (equal to $4800/2 = 2400$ kN), the distance from A to B is then $2000 - (400 - 200) = 1800$ mm.

Therefore, the load from the soil pressure between A and B is:

$300 \times 4000 \times 1800 \times 10^{-6} = 2160$ kN

This is $2160/4800 = 0.45$ of the total applied load, therefore the position of N_{Ed} requires adjusting to $e = 0.45b/2 = 0.225b$.

This iteration converges with $e = 0.2223b$ giving $e = 177.8$ mm, distance from A to B as 1777.8 mm and $N_{Ed} = 300 \times 4000 \times 1777.8 \times 10^{-6} = 2133.4$ kN (44.45% of total).

For straight bars, 2-1-1/clause 9.8.2.2(5) allows the check of anchorage to be performed at $x = h/2$. Considering the minimum height of h, $x = 500/2 = 250$ mm:

$$z_e = \frac{4000}{2} - \frac{800}{2} - \frac{250}{2} + 177.8 = 1652.8 \text{ mm}$$

and $R = 300 \times 4000 \times 250 \times 10^{-6} = 300$ kN:

From 2-1-1/Expression (9.13): $F_s = 300 \times \dfrac{1652.8}{659.8} = 752$ kN

This force is to be provided over a width of 4000 mm by $4000/200 = 20$ bars; therefore force per bar is $752/20 = 37.58$ kN.

For a 25 mm bar, its maximum force at ULS is:

$$= \pi \times 12.5^2 \times \frac{500}{1.15} \times 10^{-3} = 213.4 \text{ kN}$$

which, for good bond conditions and C35/45 concrete, requires a basic anchorage length of 806 mm – see Table 8.4-1 in section 8.4 of this guide.

Therefore, at $x = 250$ mm, the required anchorage length, $l_b = 37.58/213.4 \times 806 = 142$ mm.

The available anchorage length $= 250 - 50 = 200$ mm $>$ required l_b of 142 mm; therefore the bars are adequate as straights. Note that, in practice, the main bars should be lapped with side face reinforcement to provide minimum reinforcement against early thermal cracking. The above calculation illustrates that the main reinforcement need not be extended up the side face; the smaller side face reinforcement can be lapped with the main bottom layers of reinforcement by bending them down into the bottom face.

9.9. Regions with discontinuity in geometry or action

The definition of a 'D-region', together with typical examples of such regions, is given in section 6.5.1 of this guide. D-regions have to be designed using strut-and-tie models (see section 6.5), unless specific rules are given elsewhere in EC2. (Such exceptions include beams with short shear span which are covered in section 6.2.)

CHAPTER 10

Additional rules for precast concrete elements and structures

This chapter deals with the design of precast concrete elements and structures as covered in section 10 of EN 1992-2 in the following clauses:

- General   Clause 10.1
- Basis of design, fundamental requirements   Clause 10.2
- Materials   Clause 10.3
- Structural analysis   Clause 10.5
- Particular rules for design and detailing   Clause 10.9

Comment on EN 15050: *Precast Concrete Bridge Elements* and its applicability to design are made in section 1.1.1 of this guide.

10.1. General
The design rules in section 10 of EC2-2 apply to bridges made partly or entirely of precast concrete, and are supplementary to the design recommendations discussed in the previous chapters and detailed in the corresponding sections of EC2-2. EC2-2 addresses the additional rules for the use of precast concrete elements in relation to each of the main general clauses in sections 1 to 9. The headings in section 10 are numbered 10, followed by the number of the corresponding main section. For example, rules supplementary to 'Section 3 – Materials' are given under clause '10.3 Materials'. The sub-sections in this guide follow the same format.

2-1-1/clause 10.1.1 provides definitions of terms relating to precast concrete design. 'Transient situation' is perhaps most relevant to bridge design. Examples of transient situations include de-moulding, transportation, storage, erection and assembly. Transient design situations during transportation and storage are particularly important in the design of pretensioned beams, which can experience moment reversal in a transient situation but not in the permanent situation.

2-1-1/clause 10.1.1

10.2. Basis of design, fundamental requirements
In the design and detailing of precast concrete elements and structures, *2-1-1/clause 10.2(1)P* requires the designer to specifically consider connections and joints between elements, temporary and permanent bearings and transient situations as above. *2-1-1/clause 10.2(2)* requires dynamic effects in transient situations to be taken into account where relevant.

2-1-1/clause 10.2(1)P
2-1-1/clause 10.2(2)

Dynamic actions are particularly likely to arise during erection when lifting and landing beams, but the clause is not intended to apply to accidental situations such as a dropped beam. No specific guidance is given in EC2-2 on dynamic load calculation, other than to permit representation of dynamic effects by magnification of static effects by an appropriate factor. Factors of 0.8 or 1.2 would be reasonable in such calculations, depending on whether the static effects were favourable or unfavourable for the effects being checked. These factors were recommended in *Model Code 90*.[6]

An analogous rule for amplification of static effects is given in clause 3.2.6 of EN 1993-1-11 for the dynamic effects of the case of a sudden cable failure on a cable-supported structure, but this represents an extreme case.

It should be noted that in situ construction may also have transient situations which need to be checked and these are covered by 2-2/clause 113. Some of the recommendations in 2-2/clause 113 also apply to precast concrete.

10.3. Materials
10.3.1. Concrete
The rules in this section are mainly concerned with the effects of heat curing on the rate of gain of compressive strength and the creep and shrinkage properties of concrete. Where heat curing is applied, the 'maturity function' of 2-1-1/Annex B expression (B.10) is used to produce a fictitious older age of loading which leads to reduced creep when used in conjunction with the other formulae in 2-1-1/Annex B. The time at which the creep effect is calculated should also be similarly adjusted according to expression (B.10). Expression (B.10) can also be used in conjunction with expression (3.2) to calculate an accelerated gain of compressive strength.

10.3.2. Prestressing steel

2-1-1/clause 10.3.2

2-1-1/clause 10.3.2 requires the accelerated relaxation of prestressing steel to be considered where heat curing of the concrete is undertaken. An expression is given to calculate an equivalent time which may then be used in the standard relaxation equations of 2-1-1/clause 3.3.2(7).

10.4. Not used in EN 1992-2

10.5. Structural analysis
10.5.1. General

2-2/clause 10.5.1(1)P

The structural analysis of precast bridges usually involves staged construction, which is covered by 2-2/clause 113. *2-2/clause 10.5.1(1)P* gives further requirements for analysis of precast members. In particular, the designer must specifically consider:

- development of composite action (e.g. superstructures comprising pretensioned beams with in situ deck slabs);
- behaviour of connections (e.g. in situ stitch joints between precast deck panels);
- tolerances on geometry and position that may affect load distribution (e.g. box beam landed on twin bearings at each end is susceptible to possible uneven sharing of bearing reactions due to the torsional stiffness of the box).

2-2/clause 10.5.1(2)

2-2/clause 10.5.1(2) allows beneficial horizontal restraint caused by friction to be used in design, providing the element is not in a seismic zone and the possibility of significant impact loading is eliminated – both of these could temporarily eliminate the compressive reactions that led to the frictional restraint. This is particularly applicable to bearings but applies equally to precast and in situ construction. Further restrictions are that friction must not provide the sole means of attaining structural stability and that it should not be relied

CHAPTER 10. ADDITIONAL RULES FOR PRECAST CONCRETE ELEMENTS AND STRUCTURES

upon if the element geometry and bearing arrangements can lead to irreversible sliding translations. In the absence of a mechanical fixed bearing, the latter could occur between superstructure and substructure under cycles of temperature expansion and contraction, causing alternate sticking and sliding.

10.5.2. Losses of prestress
2-1-1/clause 10.5.2(1) gives a method for calculating losses in tendons from temperature differences during the heat curing of precast concrete elements.

2-1-1/clause 10.5.2(1)

10.6, 7, 8. Not used in EN 1992-2

10.9. Particular rules for design and detailing
10.9.1. Restraining moments in slabs
The rules in this section are of little relevance to bridges and are not discussed here.

10.9.2. Wall to floor connections
The rules in this section are of little relevance to bridges and are not discussed here.

10.9.3. Floor systems
The rules in this section are of little relevance to bridges and are not discussed here.

10.9.4. Connections and supports for precast elements
General
The detailing of connections for precast concrete elements is a critical aspect of their design. Materials used for connections must be stable and durable for the lifetime of the structure and must possess adequate strength. The principles in *2-1-1/clauses 10.9.4.1* and *10.9.4.2* seek to ensure this. EC2-1-1 gives specific rules for the detailing of connections transmitting compressive forces, shear forces, bending moments and tensile forces as discussed in the following sections. It also covers the detailing of half joints and the anchorage of reinforcement at supports.

2-1-1/clauses 10.9.4.1 and 10.9.4.2

Connections transmitting compressive forces
In the design of connections transmitting compressive forces, *2-1-1/clause 10.9.4.3(1)* allows shear forces to be ignored if they are less than 10% of the compressive force. Where greater shear forces exist, the vector resultant force could be used in the check of bearing pressure, as suggested in section 6.7 of this guide.

2-1-1/clause 10.9.4.3(1)

2-1-1/clause 10.9.4.3(2) reminds the designer that appropriate measures should be taken to prevent relative movements that might disrupt bedding materials between elements during setting. This applies also to stitch joints between precast elements. The prevention of relative movement may require restrictions to adjacent construction activities or provision of temporary clamping systems.

2-1-1/clause 10.9.4.3(2)

Bearing areas should be reinforced to resist transverse tensile stresses in adjacent elements, as discussed in sections 6.5 and 6.7 of this guide. The rules in clause 6.7 generally dictate the maximum achievable bearing pressure. *2-1-1/clause 10.9.4.3(3)*, however, restricts bearing pressures to $0.3f_{cd}$ for dry connections without bedding mortar. Additionally, concrete faces with rubber bearings are susceptible to splitting caused by transverse expansion of the rubber and *2-1-1/clause 10.9.4.3(5)* gives a method of calculating the required surface reinforcement.

2-1-1/clause 10.9.4.3(3)

2-1-1/clause 10.9.4.3(5)

Connections transmitting shear forces
Interface shear at construction joints between two concrete elements (for example, a precast beam and in situ deck slab) should be checked in accordance with clause 6.2.5.

291

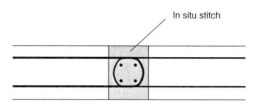

Fig. 10.9-1. Connection of precast units by overlapping reinforcement loops

Connections transmitting bending moments or tensile forces

For connections transmitting bending moments and tensile forces, *2-1-1/clause 10.9.4.5(1)* requires reinforcement to be continuous across the connection and be adequately anchored into the adjacent concrete elements. *2-1-1/clause 10.9.4.5(2)* suggests methods of achieving continuity. The methods listed, together with some comments on their use, are as follows:

- lapping of bars – requires a large in situ plug, so is often not a suitable option for precast deck panels landed on beams, for example where limited connection width is available;
- grouting reinforcement bars into holes – requires accurate setting out and placement;
- overlapping reinforcement loops (see Fig. 10.9-1) – useful for minimizing the size of the in situ joint, but physical testing may be needed to demonstrate adequate serviceability performance;
- welding of reinforcement bars or steel plates – useful for minimizing the size of the in situ joint, but bars will require fatigue checks in accordance with 2-1-1/clause 6.8.4 and the fatigue verifications are more onerous for welded reinforcement than un-welded reinforcement;
- prestressing – as in precast segmental box girder construction. Although the use of prestressing in this manner can eliminate the need for in situ concrete plugs, the joint may require glue to seal the interface;
- couplers – threaded types of coupler are usually impossible to use to join bars in infill bays while mechanically bolted couplers are usually of larger diameter, requiring greater concrete cover.

Half joints

2-1-1/clause 10.9.4.6(1) provides two alternative strut-and-tie models for the design of half joints which may be used either separately or combined. It is common to use prestressing tendons to provide the ties for the half-joint nib, particularly for those ties angled across the corner from nib to main body of the member. This helps to limit crack sizes in an area which is hard to inspect and ensures the ties are adequately anchored. The lack of provision for inspection and maintenance and the difficulties of excluding contaminants, such as de-icing salts, dictate that half joints should not generally be used for bridge applications. There have been many examples in the UK of half-joint details which have been adversely affected by corrosion.

Anchorage of reinforcement at supports

2-1-1/clause 10.9.4.7(1) reminds designers that reinforcement in supported and supporting members should be detailed to provide adequate anchorage, allowing for construction and setting-out tolerances.

10.9.5. Bearings

2-1-1/clause 10.9.5 covers the detailing of bearing areas. *2-1-1/clause 10.9.5.2(2)* gives recommendations for sizing of bearings based on the following allowable bearing pressures:

- $0.4 f_{cd}$ for dry connections, i.e. those without bedding mortar;
- the design strength of the bedding material for all other cases but limited to a maximum of $0.85 f_{cd}$.

CHAPTER 10. ADDITIONAL RULES FOR PRECAST CONCRETE ELEMENTS AND STRUCTURES

These pressures are only intended to be used to determine minimum bearing dimensions. Reinforcement in the bearing areas must still be determined in accordance with clause 2-1-1/6.5. 2-1-1/Table 10.2 gives absolute minimum bearing lengths. These are clearly intended for building structures and bridge bearing dimensions will always be significantly greater, dictated by limitation of bearing stress as above.

10.9.6. Pocket foundations

2-1-1/clause 10.9.6 covers pocket foundations which are capable of transferring vertical actions, bending moments and horizontal shear forces from columns to foundations. For bridges, precast pockets are sometimes used for pile to pile cap connections and pile cap to column connections, with in situ concrete used to form a plug between elements.

For connections where shear keys are provided, **2-1-1/clause 10.9.6.2** allows the connection to be designed as monolithic, but a check on the shear connection is still required. In particular, only if adequate interface shear resistance is provided under bending and axial force (see 2-1-1/clause 6.2.5) can punching shear be checked assuming a monolithic column/pocket interface. Lap lengths need to be increased due to the distance between lapping bars in adjacent elements, as shown in Fig. 10.7(a) of EC2-1-1.

2-1-1/clause 10.9.6.2

Where there are no shear keys and the interface between precast elements is smooth, **2-1-1/clause 10.9.6.3(1)** allows force and moment transfer to be achieved by compressive reactions on the sides of the pocket (through the in situ plug concrete) and corresponding frictional forces, as shown in Fig. 10.7(b) of EC2-1-1. This behaviour is very much that of a dowel in a socket, rather than of a monolithic connection. **2-1-1/clause 10.9.6.3(3)** therefore requires the reinforcement in column and pocket to be detailed accordingly for the forces acting on them individually. In particular, care is needed with checking shear in the element in the pocket as high shear forces can be generated in producing the fixity moment.

2-1-1/clause 10.9.6.3(1)

2-1-1/clause 10.9.6.3(3)

CHAPTER 11

Lightweight aggregate concrete structures

This chapter deals with the design of lightweight aggregate concrete structures as covered in section 11 of EN 1992-2 in the following clauses:

- General   Clause 11.1
- Basis of design   Clause 11.2
- Materials   Clause 11.3
- Durability and cover to reinforcement   Clause 11.4
- Structural analysis   Clause 11.5
- Ultimate limit states   Clause 11.6
- Serviceability limit states   Clause 11.7
- Detailing of reinforcement – general   Clause 11.8
- Detailing of members and particular rules   Clause 11.9

No comments are made on the 'Additional rules for precast concrete elements and structures' in clause 11.10 as it makes no modifications to the rules of 2-1-1/clause 10.

11.1. General

The design recommendations discussed in the previous chapters and detailed in the corresponding sections of EC2-2 have been developed for concrete made from normal-weight aggregates. As naturally occurring aggregates become less abundant and increasingly expensive, manufactured aggregates are increasingly used and most manufactured aggregates are lightweight. The use of lightweight aggregate concretes (LWAC) also has obvious advantages where it is desirable to reduce dead loads, such as in long spans that are dead load dominated.

Lightweight aggregate concrete has been used throughout the world although less so in the UK, particularly in bridge construction. There is extensive test data verifying the properties of lightweight aggregate concrete and the implications its use has on the design verifications of concrete structures. Section 11 addresses these implications on the use of the main general sections for normal-weight aggregate concrete. All the clauses given in sections 1 to 10 and 12 of EC2 are generally applicable to lightweight aggregate concrete unless they are substituted by special clauses given in section 11. The headings in section 11 are numbered 11, followed by the number of the corresponding main section that it modifies, e.g. section 3 of EC2-2 is 'Materials' so 11.3 is similarly called 'Materials' and makes specific material requirements for lightweight aggregate concrete.

2-1-1/clause 11.1.1(3) clarifies that section 11 does not apply to air-entrained concrete or lightweight concrete with an open structure. *2-1-1/clause 11.1.1(4)* defines lightweight

2-1-1/clause 11.1.1(3)
2-1-1/clause 11.1.1(4)

aggregate concrete as concrete having a closed structure with an oven-dry density of no more than 2200 kg/m³, made with a proportion of artificial or natural lightweight aggregates with a particle density of less than 2000 kg/m³.

11.2. Basis of design

2-2/clause 2 is valid for lightweight aggregate concrete without modifications. This includes the material factors for concrete.

11.3. Materials

11.3.1. Concrete

The strength classes of lightweight aggregate concrete are designated by adopting the symbol LC in place of C for normal-weight aggregate concrete, thus a lightweight aggregate concrete with characteristic cylinder strength of 40 MPa and cube strength of 44 MPa is designated LC40/44. It should be noted that the ratio of cylinder strength to cube strength is generally higher for LWAC than for normal-weight concrete. EC2 adopts an additional subscript (*l*) to designate mechanical properties for lightweight aggregate concrete. In general, where strength values originating from 2-1-1/Table 3.1 have been used in expressions elsewhere in the code, those values should be replaced by the corresponding values for lightweight aggregate concrete given in 2-1-1/Table 11.3.1 (reproduced here as Table 11.3-1 for convenience).

Lightweight aggregate is classified in EN 206-1 according to its density, as illustrated in 2-1-1/Table 11.1. Densities are given for both plain and reinforced concrete, the latter assuming a reinforcement content of 100 kg/m³. **2-1-1/clause 11.3.1(1)** permits the quoted reinforced concrete densities to be used for weight calculation. In heavily reinforced bridges with low-density concrete, it may be more appropriate to calculate the reinforced concrete weight more accurately according to **2-1-1/clause 11.3.1(2)**.

The tensile strength of concrete is affected by the moisture content, since drying reduces tensile strength and low-density concretes undergo greater moisture loss. **2-1-1/clause 11.3.1(3)** introduces a coefficient, η_1, to take account of the reduction in tensile strength with density:

$$\eta_1 = 0.40 + 0.60\rho/2200 \qquad \text{2-1-1/(11.1)}$$

where ρ is the upper limit of the oven-dry density in kg/m³ from 2-1-1/Table 11.1. For a given cylinder strength, the tensile strength for a lightweight concrete should be obtained by multiplying the tensile strength given in 2-1-1/Table 3.1 by η_1, as indicated in Table 11.3-1.

11.3.2. Elastic deformation

The elastic modulus is strongly influenced by the relative oven-dry density of the aggregate used. **2-1-1/clause 11.3.2(1)** therefore introduces a coefficient, η_E, to take account of the reduced modulus of elasticity for lightweight aggregate concrete. For a given cylinder strength, the values for E_{cm} given in 2-1-1/Table 3.1 should be multiplied by η_E for lightweight aggregate concrete where:

$$\eta_E = (\rho/2200)^2 \qquad \text{2-1-1/(11.2)}$$

Tests give considerable scatter for E_{cm}, so, where more accurate determination of concrete stiffness is needed, 2-1-1/clause 11.3.2(1) requires tests to be carried out on the specific mix proposed to determine the modulus of elasticity in accordance with ISO 6784.[28]

The coefficient of thermal expansion for lightweight aggregate concretes varies widely depending on the type of aggregate, but is typically less than that of normal-weight concrete. **2-1-1/clause 11.3.2(2)** permits a value of 8×10^{-6}/K to be used in design where thermal movements are 'not of greatest importance'. This value should be of general applicability. Exceptions might include bridges where movement ranges are predicted to be at the limit of

CHAPTER 11. LIGHTWEIGHT AGGREGATE CONCRETE STRUCTURES

Table 11.3-1. Stress and deformation characteristics for lightweight concrete according to 2-1-1/Table 11.3.1

	Strength classes for lightweight aggregate concrete													Formulae/notes
f_{lck} (MPa)	12	16	20	25	30	35	40	45	50	55	60	70	80	
$f_{lck,cube}$ (MPa)	13	18	22	28	33	38	44	50	55	60	66	77	88	
f_{lcm} (MPa)	17	22	28	33	38	43	48	53	58	63	68	78	88	For $f_{lck} \geq 20$ MPa; $f_{lcm} = f_{lck} + 8$ (MPa)
f_{lctm} (MPa)	←—————— $f_{lctm} = f_{ctm} \times \eta_1$ ——————→													$\eta_1 = 0.40 + 0.60\rho/2200$
$f_{lctk,0.05}$ (MPa)	←—————— $f_{lctk,0.05} = f_{ctk,0.05} \times \eta_1$ ——————→													5% fractile
$f_{lctk,0.95}$ (MPa)	←—————— $f_{lctk,0.95} = f_{ctk,0.95} \times \eta_1$ ——————→													95% fractile
E_{lcm} (GPa)	←—————— $E_{lcm} = E_{cm} \times \eta_E$ ——————→													$\eta_E = (\rho/2200)^2$
ε_{lc1} (‰)	←— $\varepsilon_{lc1} = k f_{lcm}/(E_{lci}\eta_E)$ — $k = 1.1$ for sanded lightweight aggregate concrete; $k = 1.0$ for all other lightweight aggregate concrete —→													
ε_{lcu1} (‰)	←—————— $\varepsilon_{lcu1} = \varepsilon_{lc1}$ ——————→													
ε_{lc2} (‰)	←—————— 2.0 ——————→								2.2	2.3	2.4	2.5		
ε_{lcu2} (‰)	←—————— $3.5\eta_1$ ——————→								$3.1\eta_1$	$2.9\eta_1$	$2.7\eta_1$	$2.6\eta_1$		$|\varepsilon_{lcu2}| \geq |\varepsilon_{lc2}|$
n	←—————— 2.0 ——————→								1.75	1.6	1.45	1.4		
ε_{lc3} (‰)	←—————— 1.75 ——————→								1.8	1.9	2.0	2.2		
ε_{lcu3} (‰)	←—————— $3.5\eta_1$ ——————→								$3.1\eta_1$	$2.9\eta_1$	$2.7\eta_1$	$2.6\eta_1$		$|\varepsilon_{lcu3}| \geq |\varepsilon_{lc3}|$

an expansion joint's capacity or for integral bridges where temperature movement is restrained. In such cases, a range of coefficients of thermal expansion could be considered in the design. 2-1-1/clause 11.3.2(2) also allows the differences between the coefficients of thermal expansion of reinforcing steel and lightweight aggregate concrete to be ignored in design.

The reduced elastic deformation properties for lightweight aggregate concrete have the following implications on bridge design:

- Stresses arising from restrained thermal or shrinkage movements are generally less than for normal-weight aggregate concrete.
- Elastic losses in prestressed concrete members can be significantly greater than for normal-weight aggregate concrete members, although this loss is often a relatively small fraction of the total loss.
- Deflections of members and hence also second-order effects will be greater than for normal-weight aggregate concrete members. This can be significant for slender columns.

11.3.3. Creep and shrinkage

Test data on the creep characteristics of lightweight aggregate concrete shows considerable scatter with some tests suggesting that LWAC exhibits more creep than normal-weight aggregate concretes and others suggesting the opposite. ***2-1-1/clause 11.3.3(1)*** recommends that the creep coefficient, ϕ, is taken as the value for normal-weight aggregate concrete multiplied by a factor of $(\rho/2200)^2$. Since the elastic modulus is also reduced by the same factor in 2-1-1/clause 11.3.2(1), the calculated creep strains produced for a given stress are, therefore, the same for lightweight concrete and normal-weight concrete (as creep strain = elastic strain × creep coefficient). The creep strains so derived have to be multiplied by a further factor, η_2, but this is 1.0 for concrete grades of LC20/22 and above (i.e. all structural bridge concretes).

2-1-1/clause 11.3.3(1)

Shrinkage strains for lightweight aggregate concretes also vary greatly. ***2-1-1/clause 11.3.3(2)*** allows the final drying shrinkage values for lightweight aggregate concrete to be obtained by multiplying the values for normal density concrete by a factor, η_3, which is 1.2 for concrete grades of LC20/22 and above (i.e. all structural bridge concretes). ***2-1-1/clause 11.3.3(3)*** allows the autogenous shrinkage strain to be taken equal to that for normal density concrete, but it notes that this can be an over-estimate for concretes with water-saturated aggregates.

2-1-1/clause 11.3.3(2)

2-1-1/clause 11.3.3(3)

11.3.4. Stress strain relations for non-linear structural analysis
2-1-1/clause 11.3.4(1) requires the appropriate strain limits for LWAC from 2-1-1/Table 3.1 to be used with 2-1-1/Fig. 3.2. LWAC values should also be used for f_{lcm} and E_{lcm}.

2-1-1/clause 11.3.4(1)

11.3.5. Design compressive and tensile strengths
The values of the design compressive and tensile strengths of lightweight aggregate concrete are defined in 2-1-1/clauses 11.3.5(1) and 11.3.5(2) as:

$$f_{lcd} = \alpha_{lcc} f_{lck}/\gamma_c \qquad \text{2-1-1/(11.3.15)}$$

and

$$f_{lctd} = \alpha_{lct} f_{lctk}/\gamma_c \qquad \text{2-1-1/(11.3.16)}$$

respectively, where γ_c is the partial safety for concrete and α_{lcc} and α_{lct} are coefficients to account for the long-term effects on the compressive strength. α_{lcc} and α_{lct} are equivalent to α_{cc} and α_{ct}, i.e. in 2-1-1/clause 3.1.6. The values of α_{lcc} and α_{lct} may be given in the National Annex to EC2-1-1 – there is no equivalent National Annex provision in EC2-2. Both are recommended to be taken as 0.85. It is not recommended in this guide to take α_{lcc} as 1.0 for shear, as is appropriate for normal-weight concrete, until more test evidence is available to support this for LWAC.

11.3.6. Stress strain relations for the design of sections
2-1-1/clause 11.3.6(1) requires the appropriate strain limits for LWAC to be used with 2-1-1/Figs 3.3 and 3.4. This should also apply to the rectangular stress block in 2-1-1/Fig. 3.5, but this has been omitted. LWAC values should also be used for f_{lcd}.

2-1-1/clause 11.3.6(1)

11.3.7. Confined concrete
An increased compressive strength is allowed for lightweight aggregate concrete elements under triaxial stress, but a given confining pressure gives less strength increase for LWAC than for normal density concrete. This, together with the reduction of tensile strength for LWAC, leads to the need for modification to the partially loaded area rules, as given in 2-1-1/clause 11.6.5.

11.4. Durability and cover to reinforcement
The same environmental exposure classes used for normal-weight aggregate concrete are appropriate for use with lightweight aggregate concrete. However, 5 mm needs to be added to the minimum cover obtained from 2-1-1/Table 4.2, which relates to the cover necessary to provide adequate bond stress for anchorage and laps.

11.5. Structural analysis
The structural analysis of lightweight aggregate concrete structures is affected mainly by the reduced elastic deformation properties discussed above. In addition, *2-1-1/clause 11.5.1* requires that the rotation capacities, $\theta_{pl,d}$, given in 2-1-1/Fig. 5.6N are multiplied by a factor $(\varepsilon_{lcu2}/\varepsilon_{cu2})$. Since rotation capacity depends on the limiting strain for reinforcement as well as that of the concrete (see section 5.6.3.1 of this guide), multiplying the rotation capacity by this ratio is conservative as the limiting reinforcement strain is unchanged.

2-1-1/clause 11.5.1

11.6. Ultimate limit states
This section makes direct modifications to most of the ULS resistance rules with the exception of bending.

Bending

The rules for bending are only indirectly modified for the use of lightweight aggregate concrete; the limiting concrete strains and the stress–strain relationship for concrete is adjusted by 2-1-1/clause 11.3.6. The stress block comparisons illustrated in Table 3.1-4 of section 3.1.7 of this guide are not therefore applicable to lightweight aggregate concrete, although the formulae presented there for average stress in the compression zone, f_{av}, and depth of centroid of compressive force, β, can still be used for LWAC. For the parabolic-rectangular and bi-linear stress blocks, the reduction in failure strain, ε_{lcu2}, for LWAC leads to a reduction in f_{av} (and a relatively smaller reduction in β) compared to the values for normal-weight concrete. These values are unaffected for the rectangular block, making it even more relatively economic for bending calculation than for normal-weight concrete. This was probably not intended and it is therefore safer to use one of the other two more realistic stress blocks, although they will only produce a significant difference to the rectangular block for heavily over-reinforced members.

Shear and torsion (2-1-1/clauses 11.6.1 to 11.6.3)

Significant research has been undertaken in assessing the shear behaviour of lightweight aggregate concrete. The tests indicate that, where shear cracks develop in lightweight aggregate concrete members, they often pass through the aggregate rather than around it, resulting in significantly smoother shear surfaces. This results in less shear force being transmitted by aggregate interlock and thus the shear strength of LWAC is reduced compared to normal-weight concrete. The expressions defined in section 6.2 of EN 1992-2 are therefore modified for use with lightweight aggregate reinforcement as discussed below.

In **2-1-1/clause 11.6.1**, the design value of shear resistance of a member without shear reinforcement is replaced by:

$$V_{lRd,c} = (C_{lRd,c}\eta_1 k(100\rho_1 f_{lck})^{1/3} + k_1\sigma_{cp})b_w d \geq (v_{l,min} + k_1\sigma_{cp})b_w d \qquad \text{2-1-1/(11.6.2)}$$

2-1-1/clause 11.6.1

where:

$C_{lRd,c}$ is a nationally determined parameter. EN 1992 recommends taking a value of $0.15/\gamma_c$

η_1 is defined in 2-1-1/clause 11.3.1(3)

$v_{l,min}$ is a nationally determined parameter. EN 1992 recommends taking a value of $v_{l,min} = 0.030k^{3/2}f_{lck}^{1/2}$

k_1 is a nationally determined parameter with recommended value = 0.15

All other parameters are as for normal-weight concrete. Note that the reduced shear strength of lightweight aggregate concrete discussed above is reflected in lower recommended values of $C_{lRd,c}$ and $v_{l,min}$ and the use of the reduction factor η_1. It should be noted that 2-1-1/Expression (11.6.2) does not include the parameter η_1 with $v_{l,min}$, whereas it is present in 2-1-1/Expression (11.6.47) for punching shear. This omission was not intended.

For members with shear reinforcement, **2-1-1/clause 11.6.2(1)** simply states that the reduction factor for the crushing resistance of concrete struts for LWAC is ν_1, a nationally determined parameter with recommended value $\nu_1 = 0.5\eta_1(1 - f_{lck}/250)$. (Note that the density-dependent reduction factor η_1 is again used here.) This compares with $\nu_1 = 0.6(1 - f_{ck}/250)$ from 2-1-1/clause 6.2.3(3) for normal-weight concrete. This LWAC value of ν_1 is then used with expression (6.9) of EC2-1-1. No reduction is necessary for the shear resistance in expression (6.8) of EC2-1-1, as it contains only a reinforcement contribution. For members without shear reinforcement, **2-1-1/clause 11.6.1(2)** explicitly gives the shear crushing resistance for LWAC as $0.5\eta_1 b_w d\nu_1 f_{lcd}$ compared to the normal-weight concrete value of $0.5b_w d\nu f_{cd}$ from 2-1-1/clause 6.2.2. Since η_1 appears both explicitly in the expression for shear strength itself and also within ν_1, the strength is reduced by η_1^2, which was not intended.

2-1-1/clause 11.6.2(1)

2-1-1/clause 11.6.1(2)

2-1-1/clause 11.6.3.1(1) also applies the reduction factor $\nu_1 = 0.5\eta_1(1 - f_{lck}/250)$ to the crushing resistance of concrete struts in torsion calculations, by replacing ν with ν_1 in 2-1-1/Expression (6.30).

2-1-1/clause 11.6.3.1(1)

Punching shear (2-1-1/clause 11.6.4)
Similar modifications to those for the flexural shear resistances are made to the punching shear resistances. The punching shear design stress of a slab is from **2-1-1/clause 11.6.4.1**:

2-1-1/clause 11.6.4.1

$$v_{lRd,c} = C_{lRd,c}\eta_1 k(100\rho_1 f_{lck})^{1/3} + k_2\sigma_{cp} \geq (\eta_1 v_{l,min} + k_2\sigma_{cp}) \qquad \text{2-1-1/(11.6.47)}$$

and k_2 is recommended to be 0.08. The values of $C_{lRd,c}$, η_1, f_{lck} and $v_{l,min}$ are as given above for flexural shear. All other parameters are as for normal-weight concrete. Unlike for flexural shear, EC2 considers a proportion of the concrete resistance component in the punching shear design of elements with punching shear reinforcement. Therefore the above reductions in punching shear strength are also made in **2-1-1/clause 11.6.4.2(1)** for the concrete terms in the design resistance of slabs with shear reinforcement.

2-1-1/clause 11.6.4.2(1)
2-1-1/clause 11.6.4.2(2)

2-1-1/clause 11.6.4.2(2) makes a reduction to the maximum punching shear stress in the same way as for flexural shear, such that:

$$v_{Ed} = \frac{V_{Ed}}{u_0 d} \leq v_{lRd,max} = 0.5\nu_1 f_{lcd} \qquad \text{2-1-1/(11.6.53)}$$

Partially loaded areas (2-1-1/clause 11.6.5)
The mechanism for the development of enhanced bearing pressures in partially loaded areas is discussed in section 6.7 of this guide. The maximum bearing pressure depends on both the concrete's tensile strength (to generate confinement to the loaded area) and the concrete's compressive resistance when confining pressure is present. Both of these are lower for a given concrete grade for LWAC than for normal-weight concrete. **2-1-1/clause 11.6.5(1)** therefore modifies the expression in 2-1-1/clause 6.7 as follows:

2-1-1/clause 11.6.5(1)

$$F_{Rdu} = A_{c0} f_{lcd} [A_{c1}/A_{c0}]^{\rho/4400} \leq 3.0 f_{lcd} A_{c0} \left(\frac{\rho}{2200}\right) \qquad \text{2-1-1/(11.6.63)}$$

Strut-and-tie models
No modifications are made in 2-1-1/clause 11 to the resistances of concrete struts in 2-1-1/clause 6.5.2. However, for consistency with the LWAC shear modification in 2-1-1/clause 11.6.3.1(1), it is recommended here that the design strength given in 2-1-1/clause 6.5.2(2) for struts with transverse tension be reduced from $0.6\nu' f_{cd}$ to $0.5\eta_1 \nu' f_{lcd}$, where $\nu' = (1 - f_{lck}/250)$. A similar inclusion of η_1 might be considered appropriate for the resistance of node types (b) and (c) in 2-1-1/clause 6.5.4 due to the presence of transverse tension.

Worked example 11.1: Reinforced lightweight aggregate concrete deck slab
The bending resistance and shear resistance of the 250 mm-thick deck slab considered in Worked examples 6.1-1 and 6.2-1 is found. Cover to the 20ϕ bars at 150 mm centres is 40 mm. Concrete class is LC35/38 with lightweight aggregate of class 1.6.

From 2-1-1/Table 11.1, the upper-bound oven-dry density for class 1.6 lightweight aggregate, $\rho = 1600 \text{ kg/m}^3$; thus from 2-1-1/Expression (11.1):

$$\eta_1 = 0.40 + 0.60\rho/2200 = 0.40 + 0.60 \times 1600/2200 = 0.836$$

First, the bending resistance is considered using the parabolic-rectangular stress–strain block (which is less economic than the rectangular block):

From 2-1-1/Table 11.3.1, for LC35/38 concrete, $f_{lck} = 35$ MPa.

From 2-1-1/Expression (11.3.15): $f_{lcd} = \alpha_{lcc} f_{lck}/\gamma_c$ and using $\alpha_{lcc} = 0.85$ and $\gamma_c = 1.5$ as recommended gives $f_{lcd} = 0.85 \times 35/1.5 = 19.833$ MPa.

From 2-1-1/Table 11.3.1, $\varepsilon_{lcu2} = 3.5\eta_1 = 3.5 \times 0.836 = 2.927\text{‰}$, $\varepsilon_{lc2} = 2.0\text{‰}$ and $n = 2$. Using these values in equations (D3.1-4) and (D3.1-5), it can be shown that the average

CHAPTER 11. LIGHTWEIGHT AGGREGATE CONCRETE STRUCTURES

concrete stress is $f_{av} = 15.316\,\text{MPa}$ centred at a ratio of depth to neutral axis from the compression fibre of $\beta = 0.403$ (compared to $f_{av} = 16.056\,\text{MPa}$ and $\beta = 0.416$ for equivalent normal-weight aggregate concrete):

$$d = 250 - 40 - 10 = 200\,\text{mm}, b = 1000\,\text{mm} \text{ and } A_s = \pi \times 10^2 \times \frac{1000}{150} = 2094.4\,\text{mm}^2$$

Therefore, from equation (D6.1-4):

$$\rho = \frac{A_s}{bd} = \frac{2094.4}{1000 \times 195} = 0.0105$$

From equation (D6.1-3):

$$\frac{x}{d} = \frac{f_{yk}}{f_{av}\gamma_s}\rho = \frac{500}{15.316 \times 1.15} \times 0.0105 = 0.297$$

Check against limit from equation (D6.1-8) to ensure reinforcement yields:

$$\frac{1}{\left(\frac{f_{yk}}{\gamma_s E_s \varepsilon_{lcu2}} + 1\right)} = \frac{1}{\left(\frac{500}{1.15 \times 200 \times 10^3 \times 0.002927} + 1\right)} = 0.574 \geq \frac{x}{d}$$

therefore the steel yields.

Thus $x = 0.297 \times 200 = 59.5\,\text{mm}$ and, from equation (D6.1-5), $z = 200 - 0.403 \times 59.5 = 176\,\text{mm}$. From equation (D6.1-1):

$$M = \frac{f_{yk}}{\gamma_s}A_s z = \frac{500}{1.15} \times 2094.4 \times 176 \times 10^{-6} = \mathbf{160.2\,kNm}$$

This is a 0.2% reduction in bending resistance compared to the resistance of 160.6 kNm/m, which is obtained for an equivalent normal-weight aggregate concrete.

Second, the shear resistance is considered:

$$\text{Again, } \rho_l = \frac{A_s}{bd} = \frac{2094.4}{1000 \times 200} = 0.0105 < 0.02 \text{ limit as required}$$

Taking $\gamma_c = 1.5$ and $C_{lRd,c} = 0.15/\gamma_c$ as recommended gives $C_{lRd,c} = 0.15/1.5 = 0.10$.

From equation (D6.2-1): $k = 1 + \sqrt{200/200} = 2.0$ (the limit).

The shear resistance is given by 2-1-1/Expression (11.6.2):

$$V_{lRd,c} = C_{lRd,c}\eta_1 k(100\rho_l f_{lck})^{1/3}b_w d \text{ in the absence of axial load, thus}$$

$$V_{lRd,c} = 0.10 \times 0.836 \times 2.0 \times (100 \times 0.0105 \times 35)^{1/3} \times 1000 \times 200 \times 10^{-3}$$

$$= \mathbf{111.2\,kN/m}$$

It is checked that this exceeds the minimum shear strength for lightweight aggregate concrete considering:

$$v_{l,min} = 0.030k^{3/2}f_{lck}^{1/2} = 0.030 \times 2.0^{3/2} \times 35^{1/2} = 0.502\,\text{MPa}$$

so the minimum value of resistance according to 2-1-1/Expression (11.6.2) is:

$$v_{l,min}b_w d = 0.502 \times 1000 \times 200 \times 10^{-3} = 100\,\text{kN/m} < 111.2\,\text{kN/m}$$

(but see the comment in the main text regarding the absence of the parameter η_1).

This shear resistance of 111.2 kN/m represents a 30% reduction compared to the resistance of 159.9 kN/m from Worked example 6.2-1 for an equivalent normal-weight aggregate concrete.

11.7. Serviceability limit states

2-1-1/clause 11.7(1) modifies the span-to-depth ratios in 2-1-1/clause 7.4.2(2) that are deemed to satisfy deflection criteria. This is necessary because of the reduction in the modulus of elasticity for LWAC. 2-1-1/clause 7.4.2(2) is, however, intended for use in building design.

2-1-1/clause 11.7(1)

11.8. Detailing of reinforcement – general

2-1-1/clause 11.8.1(1) effectively requires the minimum mandrel sizes in 2-1-1/Table 8.1N to be increased by 50%. This is necessary because the reduced tensile strength of LWAC leads to splitting of the concrete inside bends of bars at a lower bearing pressure than for normal-weight aggregate concrete.

2-1-1/clause 11.8.1(1)

2-1-1/clause 11.8.2(1) requires the ultimate bond stress to be calculated using f_{lctd} in place of f_{ctd}, as bond stress is strongly dependent on concrete tensile resistance. There appears to be no need for this clause as 2-1-1/clause 11.1.1(1) states that, unless clause 11 provides otherwise, all normal-weight concrete expressions apply to LWAC when the relevant strength parameters are taken from 2-1-1/Table 11.3.1. A similar modification should therefore apply to the anchorage and transmission lengths of pre-tensioned tendons in 2-1-1/clause 8.10.2, but this is not specifically identified in 2-1-1/clause 11.8.2(1).

2-1-1/clause 11.8.2(1)

11.9. Detailing of members and particular rules

2-2/clause 11.9(101) requires that reinforcement bars should not normally exceed 32 mm diameter and bundles should not contain more than 2 bars (with a maximum equivalent diameter of 45 mm). These restrictions reflect the lack of test data for the use of larger diameter bars in lightweight aggregate concrete.

2-2/clause 11.9(101)

CHAPTER 12

Plain and lightly reinforced concrete structures

This chapter deals with the design of plain and lightly reinforced concrete structures (where the reinforcement provided is less than the minimum required for reinforced concrete) as covered in section 12 of EN 1992-2 in the following clauses:

- General   Clause 12.1
- Materials   Clause 12.3
- Structural analysis: ultimate limit states   Clause 12.5
- Ultimate limit states   Clause 12.6
- Serviceability limit states   Clause 12.7
- Detailing of members and particular rules   Clause 12.9

The clauses given in sections 1 to 11 of EN 1992-2 are generally applicable unless they are substituted or modified by specific clauses in section 12. The headings in section 12 of EN 1992-1-1 are numbered 12, followed by the number of the corresponding main section after the decimal point. This format is not followed in this section of the guide.

The use of plain concrete is not common in bridge design and there are no bridge-specific provisions in EN 1992-2, so the comments on this section are limited. **2-1-1/clause 12.1(2)** states that the provisions of section 12 do not apply to members resisting 'effects such as those from rotating machines and traffic loads'. The restriction in applicability for traffic loading clearly applies to members which are directly trafficked, but it is less clear how far the restriction extends to supporting members. The rules could certainly be applied to elements whose load effects are not influenced directly by traffic actions, such as wing walls. They could probably also be applied to other foundation elements where dynamic effects from traffic are small.

2-1-1/clause 12.1(2)

Materials and structural analysis

Plain concrete, in the absence of any confining reinforcement, is less ductile than reinforced concrete. **2-1-1/clause 12.3.1(1)** therefore requires the values of α_{cc} and α_{ct} for plain concrete members (denoted $\alpha_{cc,pl}$ and $\alpha_{ct,pl}$) to be less than those taken for reinforced concrete. The limited ductility of plain concrete also means that many of the assumptions for structural analysis in section 5 of EN 1992-2 may no longer be valid. Linear elastic analysis with redistribution cannot generally be used because the real behaviour of plain concrete approximates to that of linear elastic un-cracked sections until cracking occurs, whereupon failure may occur before moment can be redistributed. Plastic analysis is similarly not generally allowable. **2-1-1/clause 12.5(1)** states that these methods should not be used 'unless their application can be justified'. Justification would involve a check of deformation capacity.

2-1-1/clause 12.3.1(1)

2-1-1/clause 12.5(1)

Cracking need not, however, trigger ultimate failure for members with axial compressive force. For a wall, for example, decompression will occur when the load moves outside the middle third. However, a plastic stress block can develop in the concrete to permit greater eccentricity of load and this is the basis of the expression for axial resistance in 2-1-1/clause 12.6.1.

2-1-1/clause 12.5(2)

2-1-1/clause 12.5(2) allows structural analysis to be based on linear elastic theory (the simplest) or non-linear theory. For non-linear analysis (such as fracture mechanics) the deformation capacity must be checked to ensure ductility is sufficient to achieve the analysis assumptions.

Bending and axial force

2-1-1/clause 12.6.1(3)

2-1-1/clause 12.6.1(3) provides an expression for axial resistance for eccentrically loaded plain concrete walls and columns that are not slender. The tensile strength of the concrete is neglected and the rectangular stress block for concrete in compression is used as shown in Fig. 12-1. The ultimate axial resistance, N_{Rd}, is determined from the stress block:

$$N_{Rd} = \eta f_{cd} \times b \times 2(h_w/2 - e) \tag{D12-1}$$

which rearranges to give the expression in 2-1-1/clause 12.6.1(3):

$$N_{Rd} = \eta f_{cd} b h_w (1 - 2e/h_w) \qquad \text{2-1-1/(12.2)}$$

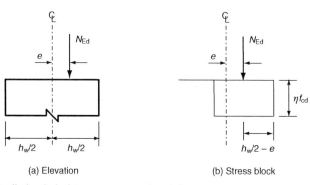

Fig. 12-1. Eccentrically loaded plain concrete wall at failure

2-1-1/clause 12.6.2(1)P

2-1-1/clause 12.6.2(1)P requires the eccentricity of the applied force to be limited to avoid large cracks forming, unless other measures have been taken to avoid local tensile failure of the cross-section. No guidance is given in EN 1992-2 on either the limit of eccentricity or suitable control measures. Cracking from bursting stresses adjacent to the load can be controlled by limiting the bearing pressure in accordance with the recommendations for bursting in section 6.7 of this guide; plain concrete can be considered by way of Expression (D6.7-4). Control of cracking is, however, generally a serviceability issue, so a suitable approach might be to limit all tensile stresses in serviceability calculations to $f_{ct,eff}$ in accordance with 2-1-1/clause 7.1(2). *2-1-1/clause 12.7(2)* supports this by suggesting control of cracking through the 'limitation of concrete tensile stresses to acceptable values'.

2-1-1/clause 12.7(2)

Shear and torsion

2-1-1/clause 12.6.3(1)
2-1-1/clause 12.6.3(2)

2-1-1/clause 12.6.3(1) allows the tensile strength of concrete to be considered in the ultimate limit state for shear, provided brittle failure can be avoided and adequate resistance ensured. The calculation method given in *2-1-1/clause 12.6.3(2)* is based on limiting principal stresses to the design tensile strength of the concrete, f_{ctd}, in a similar manner to the shear design of un-cracked sections for prestressed concrete members, as discussed in section 6.2.2.2 of this guide. In the calculation, the shear stress is taken as:

$$\tau_{cp} = k V_{Ed}/A_{cc} \qquad \text{2-1-1/(12.4)}$$

CHAPTER 12. PLAIN AND LIGHTLY REINFORCED CONCRETE STRUCTURES

The value of k may be given in the National Annex; the recommended value is 1.5. It accounts for the non-uniformity of the shear stress distribution through the section. A value of 1.5 is appropriate for an elastic distribution of shear for a rectangular cross-section and is conservative for other cross-section shapes.

2-1-1/clause 12.6.4(1) states that cracked sections of plain members should not normally be designed to resist torsional moments, 'unless it can be justified otherwise'. For un-cracked members, the torsional resistance could be calculated by again limiting the principal tensile stress to the design tensile strength.

2-1-1/clause 12.6.4(1)

Buckling of slender sections

The slenderness of a wall or column is given by $\lambda = l_0/i$, where i is the minimum radius of gyration and l_0 is the effective length of the member. The effective length is defined in EN 1992-1-1 as $l_0 = \beta l_w$, where l_w is the clear height of the member and β is a coefficient to account for the end restraint conditions. *2-1-1/clause 12.6.5.1(1)* and 2-1-1/Table 12.1 provide values of β for members with different edge restraint conditions. 2-1-1/clause 12.6.5.1(5) generally restricts the slenderness of walls in plain concrete to $l_0/h_w \leq 25$.

2-1-1/clause 12.6.5.1(1)

Having determined the slenderness, *2-1-1/clause 12.6.5.2(1)* gives the following simplified expression for the resistance to axial compression:

2-1-1/clause 12.6.5.2(1)

$$N_{Rd} = b h_w f_{cd} \Phi \qquad \text{2-1-1/(12.10)}$$

where Φ is a factor taking into first- and second-order eccentricities and the normal effects of creep. Φ is defined in EC2 as:

$$\Phi = (1.14(1 - 2e_{tot}/h_w)) - 0.02 l_0/h_w \leq (1 - 2e_{tot}/h_w) \qquad \text{2-1-1/(12.11)}$$

where e_{tot} is the sum of the first-order eccentricity of the load and additional eccentricity from geometrical imperfections. The upper limit on Φ in 2-1-1/Expression (12.11) equates to that implicit in 2-1-1/Expression (12.2) for stocky members.

CHAPTER 13

Design for the execution stages

This chapter discusses design for the execution stages as covered in section 113 of EN 1992-2 in the following clauses:

- General Clause 113.1
- Actions during execution Clause 113.2
- Verification criteria Clause 113.3

13.1. General

Section 113 gives design rules covering structures during construction (termed 'execution' in the Eurocodes). For bridges built in more than one stage, it is necessary to allow for the build-up of forces and stresses from the construction sequence. *2-2/clause 113.1(101)* identifies four circumstances where the construction sequence should be considered in design. These are essentially where:

- elements experience temporary forces, e.g. bearing friction on piers during launching of a deck or out of balance forces on piers during balanced cantilever construction;
- redistribution of stresses throughout the structure or across local cross-sections is possible due to creep, shrinkage or steel relaxation;
- the construction sequence affects the build-up of stresses and the geometry of the final structure;
- the construction sequence affects the temporary stability of the structure.

It is often the case that the designs of some bridge elements are governed by action combinations during construction rather than on the completed structure, such as the piers during balanced cantilever construction for example. *2-2/clause 113.1(102)* therefore requires ultimate and serviceability limit state verifications to be carried out at each construction stage.

Creep can have a significant effect in modifying the stress state for serviceability checks for bridges built by staged construction. Creep tends to cause action effects built up from staged construction to move towards the action effects that would have been produced had the structure been constructed monolithically. (Creep redistribution is discussed in Annex KK of this guide.) This is the basis of *2-2/clause 113.1(103)* which requires the designer to consider the effects of creep in global analysis as well as in the local section design – particularly significant in the design of beam-and-slab-type construction for example.

2-2/clause 113.1(104) is a reminder that where the erection procedure has a significant influence on the stability of a structure during construction (such as in balanced cantilever construction for example) or on the built-up forces in design, the construction sequence assumed in design should be detailed on the drawings.

Clause 113.1 is not exhaustive. The designer should always consider the construction sequence and the possible interaction with the temporary works. Flexible falsework, which

2-2/clause 113.1(101)

2-2/clause 113.1(102)

2-2/clause 113.1(103)

2-2/clause 113.1(104)

allows stresses to develop in the permanent works as they are constructed, is one particular example of a situation to consider.

13.2. Actions during execution

The actions to take into account in the design of structures during construction are covered in EN 1991-1-6. These actions include the following:

- self-weight;
- soil movement and earth pressures;
- prestressing actions;
- pre-deformations (such as loosening of cables or supports);
- temperature effects;
- shrinkage and hydration effects;
- wind actions;
- snow loads;
- actions due to water and ice;
- accidental actions (from failures of auxiliary construction equipment for example);
- seismic loads;

as well as the additional temporary construction loads from:

- personnel;
- storage of movable items;
- movable heavy equipment in position or during movement (such as travelling formwork, gantries or launching noses);
- other equipment free to move (such as cranes);
- variable loads from parts of the structure (such as wet concrete);
- impacts from plant.

EN 1991-1-6 is intended for use by contractors as well as designers, and the magnitudes of many of the construction-related plant loads may need to be agreed with proposed contractors during the design phase. The magnitudes of the characteristic values of some of these construction loads are recommended in EN 1991-1-6 and can be modified in its accompanying National Annex.

2-2/clause 113.2(102)

Clause 113.2 gives additional requirements. ***2-2/clause 113.2(102)*** recommends taking a minimum horizontal or uplift wind pressure of 200 N/m^2 on one of the cantilevers in balanced cantilever construction. This pressure should only be considered as an absolute minimum value. It can be significantly exceeded if a structure is susceptible to excitation by wind turbulence (which becomes increasingly likely with reducing natural frequency) or to vortex shedding. Dynamic analysis can be performed to establish the wind loading or more conservative static pressures used.

2-2/clause 113.2(103)

2-2/clause 113.2(103) requires the design to allow for an accidental fall of formwork in in situ balanced cantilever construction. This would be an accidental action. Loss of the formwork traveller (together with any concrete being placed at the time) would unbalance the cantilever and put additional moment into the pier or supporting temporary moment restraint. The clause does not specifically require consideration of the loss of the whole traveller but if this is not considered, the risk of its loss would have to be addressed by other means, such as the provision of fail-safe systems in the design of the traveller itself.

2-2/clause 113.2(104)

2-2/clause 113.2(104) requires similar considerations for the drop of a precast unit.

2-2/clause 113.2(105)

2-2/clause 113.2(105) is a reminder to consider the deformations imposed by launching a bridge deck.

For balanced cantilever construction, another significant consideration is the out of balance moments from deck self-weight on the piers. This comprises out of balance moments from the sequence of casting and also from unintentional variations in self-weight (due to dimensional tolerances and concrete density variations). For the latter, it is

CHAPTER 13. DESIGN FOR THE EXECUTION STAGES

Table 13.2-1. Recommended return periods for the assessment of characteristic values of climatic actions

Duration	Return period
≤3 days	2 years
≤3 months	5 years
≤1 year	10 years
>1 year	50 years

common to adjust the nominal self-weight on either side of a pier (say +3% on one side and −2% on the other side). Out of balance moments due to construction loads and the sequence of traveller movements also need to be considered. Any restrictions assumed in design need to be agreed with the constructor and clearly indicated on construction drawings.

The combination of actions is covered by the general rules given in EN 1990 and the structure should be designed for appropriate persistent, transient, accidental and seismic design situations.

The nominal duration of any transient design situations should be taken as being equal to or greater than the anticipated duration of the appropriate construction stage in accordance with EN 1991-1-6. The design situations can take into account the reduced likelihood of the occurrence of any variable action (such as wind or temperature effects) by considering reduced return periods for the actions. Recommended return periods for the assessment of characteristic values of such climatic actions are given in Table 3.1 of EN 1991-1-6 and are summarized here in Table 13.2-1 for convenience. Use of the reduced return periods for transient design situations is generally covered in the Annexes to the relevant loading Eurocode (e.g. EN 1991-4 for wind and EN 1991-5 for temperature).

13.3. Verification criteria
13.3.1. Ultimate limit state
The ultimate limit state verifications required by the code for design during construction are the same as those given in section 6 for completed structures.

13.3.2. Serviceability limit states
The serviceability limit state verifications required by the code for design during construction are generally the same as those given in EN 1992-2 section 7 for completed structures, but some exceptions are given in 2-2/clauses 113.3.2(102) to (104).

In general, the criteria associated with the serviceability limit states during construction should take into account the requirements for the completed structure and should not be detrimental to the permanent works. Construction operations which can cause excessive cracking or early deflections which may adversely affect the durability, fitness for use or aesthetic appearance of the completed structure should be avoided. Conversely, operations which will not affect the durability or appearance of the final bridge need not be assessed, such as temporary deflections during construction. This is the basis of *2-2/clause 113.3.2(102)*.

For temporary conditions, *2-2/clause 113.3.2(103)* and *2-2/clause 113.3.2(104)* relax the allowable tensile stress limits and criteria for crack width verification for certain concrete members. Such relaxations are made on the basis that small tensile stresses are unlikely to cause cracking but, where they do, the cracks will close again upon removal of the temporary actions.

2-2/clause 113.3.2(102)
2-2/clause 113.3.2(103)
2-2/clause 113.3.2(104)

ANNEX A

Modification of partial factors for materials (informative)

A1. General
The partial factors for materials defined in clause 2.4 correspond to the permitted geometrical tolerances of Class 1 to ENV 13670-1 and a normal level of workmanship and inspection (equivalent to Inspection Class 2 to ENV 13670-1). If these tolerances are tightened up in the project specification, these partial factors may be reduced under certain circumstances in accordance with EC2-1-1 Annex A.

A2. In situ concrete structures
For in situ reinforced or prestressed concrete structures, 2-1-1/clause A.2 allows the material partial factor for reinforcement to be reduced by doing any of the following:

- specifying reduced setting-out tolerances to 2-1-1/Table A.1 (clause A.2.1);
- taking dimensional tolerances explicitly into account in the design calculations, e.g. in effective depth calculation (clause A.2.2);
- using measured dimensions from the final structure (clause A.2.2).

The material partial factor for concrete may be reduced by doing any of the following:

- specifying reduced setting-out tolerances to 2-1-1/Table A.1 and limiting the coefficient of variation of the concrete strength (clause A.2.1);
- taking dimensional tolerances into account in the design calculations, with or without limiting the coefficient of variation of the concrete strength (clause A.2.2);
- using measured dimensions from the final structure, with or without limiting the coefficient of variation of the concrete strength (clause A.2.2);
- using measured concrete strengths in the final structure (clause A.2.3). This may be considered in conjunction with the above.

The National Annex may give the reduced values of the material factors under the above conditions. Any of the above methods which rely on measuring dimensions and strengths in the final structure (rather than by tightening up tolerances in the project specification) will only be relevant in verifying the adequacy of an element which has perhaps not been constructed as intended; they obviously cannot be relied upon at the design stage.

A3. Precast elements and products

The above provisions may also be applied to precast elements, provided suitable quality control and assurance measures are in place. Specific recommendations for the factory production control required to enable reduced material partial factors are given in the appropriate product standards, but general recommendations may be found in EN 13369.

ANNEX B

Creep and shrinkage strain (informative)

Annex B in EC2-2 incorporates all the rules in EC2-1-1 and adds some additional sections. It covers the following 4 areas in the treatment of creep and shrinkage:

(1) B.1 and B.2 from EC2-1-1 provide mathematical formulae behind the creep and shrinkage figures and tables in 2-1-1/clause 3.1.4, together with information on the rate of creep with time;
(2) B.103 provides alternative formulae for high-strength concrete with Class R cements, distinguishing between concretes with and without silica fume;
(3) B.104 provides a means of determining creep and shrinkage parameters from tests;
(4) B.105 recommends values of additional partial factors on calculated long-term creep and shrinkage strains to allow for uncertainties in the formulae arising from lack of available long-term testing data.

B1. Creep
2-1-1/clause B.1 provides the formulae behind the figures and tables in 2-1-1/clause 3.1.4 and provides a formula for the rate of development of creep strain which is not otherwise covered in the main body of EC2. The use of the formulae is illustrated in Worked example 3.1-1 in section 3.1.4.1 of this guide, where some of the formulae are reproduced. This part of Annex B also gives a method for accounting for elevated temperatures in the creep calculation.

2-1-1/clause B.1

B2. Shrinkage
2-1-1/clause B.2 gives a formula for nominal drying shrinkage strain which forms the basis of the simplified table in 2-1-1/clause 3.1.4. It is self-explanatory and is not discussed further here.

2-1-1/clause B.2

B3. Creep and shrinkage in high-strength concrete
For high-strength concrete with grade greater than or equal to C55/67, *2-2/clause B.103* gives alternative rules for creep and shrinkage calculation, which were considered by their drafters to be more accurate than those in EC2-1-1. However, the rules can produce spurious results for relative humidities in excess of around 70%, which is likely to be due to the fact that they were produced with lower relative humidities in mind. As a result, the UK National Annex does not allow the use of 2-2/Annex B, preferring to stick with the more established rules in

2-2/clause B.103

2-1-1/Annex B. Concretes with and without silica fume are treated separately. There can be significant benefit using these formulae with silica fume concretes as they can give significantly smaller creep strains. Concretes without silica fume can, however, give greater creep strains.

A silica fume concrete is one with a mass of silica fume of at least 5% of that of the cementitious content. Creep strains are considered to come from two mechanisms. 'Basic' creep depends only on the concrete strength at the time of loading and the 28-day compressive strength and does not depend on the movement of water out of the section. 'Drying creep' depends on concrete strength, relative humidity and effective section thickness and the mechanism is governed by the squeezing of water out of the concrete. Shrinkage strains are similarly split into 'autogeneous' shrinkage, which occurs during hardening of the concrete and does not depend on movement of water out of the section, and 'drying' shrinkage. The formulae for calculating these strains are self-explanatory and are not discussed further here.

B4. Determination of creep and shrinkage parameters by experiment

2-2/clause B.104

In general, the scatter of test results from the formulae predictions of Annex B can easily be ±30%. For greater accuracy, *2-2/clause B.104* therefore provides a method of determining creep parameters experimentally, for use in the formulae provided in clauses B.1, B.2 and B.103. This can be used where greater precision is required in the creep and shrinkage predictions for the particular concrete mix to be used.

Guidance is not given on when one might need to resort to testing. The formulae in B.1 to B.103 are not valid for early loading where the mean strength at time of loading is less than 60% of the mean strength at 28 days (2-2/clause B.100(103) refers). In this situation, 2-2/clause B.104 could clearly be applied. Most prestressed structures are reasonably sensitive to assumed creep and shrinkage strains, due to the loss of prestress which occurs as a result. Externally post-tensioned bridges with un-bonded tendons are arguably more sensitive to predicted long-term strains than other prestressed bridges, as their ultimate bending resistance can be reduced by increasing loss of prestress force. Testing may therefore be of benefit in such cases.

Performing testing will still not reduce the uncertainty associated with extrapolating the results of tests, typically carried out over a matter of months, to a long-term strain appropriate to the bridge's design life. This is dealt with in 2-2/clause B.105.

B5. Long-term values of creep and shrinkage

2-2/clause B.105(102)

2-2/clause B.105 addresses the uncertainty in the formulae used in B.1, B.2 and B.103 for determining long-term creep and shrinkage strains. The uncertainty arises because the tests upon which the formulae are based have generally only been performed over relatively short durations of up to a few years. To allow for uncertainty in extrapolating to long-term values, *2-2/clause B.105(102)* suggests that an additional safety factor should be applied to the strains derived 'when safety would be increased by an overestimation of delayed strains, and when it is relevant in the project'.

The circumstances when this factor should be applied are not clear. The use of the word 'safety' in 2-2/clause B.105(102) implies considerations of ultimate limit states. For bridges with bonded tendons, flexural resistance is only marginally influenced by pre-strain in the tendons and hence by long-term creep. The shear resistance may be influenced to a slightly greater degree by increased creep strains where shear is partially carried by inclined tendons. The flexural resistance of externally post-tensioned bridges with un-bonded tendons is, however, potentially more significantly affected by long-term creep as discussed in section B4 above. Since the Annex is informative and commercial pressures will often prevent the use of a systematically conservative approach, the decision on when to include this additional

ANNEX B. CREEP AND SHRINKAGE STRAIN

partial factor will probably need to be made on a project by project basis after consultation with the Client.

2-2/clause B.105(103) suggests that suitable values for the safety factor are given in 2-2/Table B.101. These vary from 1.0 for calculation of strains at an age of 1 year (which tests are deemed to have adequately covered) to 1.25 at 300 years (where there is greater uncertainty due to lack of test evidence).

2-2/clause B.105(103)

ANNEX C

Reinforcement properties (normative)

2-1-1/clause C.1(1)

Reinforcement specified to EN 10080 has properties which are compatible with the design assumptions in EN 1992. For other reinforcement, and for some supplementary requirements, 2-1-1/clause 3.2.1(3)P requires properties to be checked in accordance with 2-1-1/clause 3.2.2 to 3.2.6 and 2-1-1/Annex C. Annex C therefore gives requirements for the mechanical and geometrical properties of reinforcing steel suitable for use with EN 1992; it is the only normative annex in EN 1992. Some of these properties (which *2-1-1/clause C.1(1)* states are valid for temperatures between −40°C and 100°C) are summarized in 2-1-1/Table C.1 and are consistent with those in EN 10080. The reproduction of information in EN 1992-1-1 mainly affects specification and generally does not need to be considered in design itself. The provisions of Annex C are not therefore discussed in detail here. Some information, however such as minimum characteristic tensile strength and strain at maximum force, can be used in calculations.

In 2-1-1/Table C.1, the Classes A, B and C refer to the ductility characteristics of the reinforcement. Classes A and B will already be familiar to UK designers and are identical to those specified in BS 4449: 1997. Class C specification defines reinforcement with greater ductility, which can permit greater rotation capacity to be achieved – see 2-1-1/Table 5.6N. Rotation capacity is important for the justification of plastic global analysis methods and for the neglect of imposed deformations – section 5.6 of this guide refers. The former is restricted for bridge design and the latter will normally be found to be possible with Class B reinforcement. The use of Class C reinforcement will therefore not typically bring much economic benefit in bridge design, other than a small potential increase in bending resistance for under-reinforced sections. 2-2/clause 3.2.4(101)P recommends that only reinforcement of ductility Class B or C should be used for bridge design; the choice is a matter for the National Annex.

Additional requirements for fatigue performance and relative rib area of bars can be defined in the National Annex and 2-1-1/Table C.2 gives recommended values. The fatigue criteria should be compatible with the stress ranges given in 2-1-1/Table 6.3N, which is also the subject of national choice in the National Annex. The bond requirements for the minimum relative rib area of bars are given to ensure the validity of the minimum bond strengths given in section 8. Definitions of the surface geometry of reinforcing bars, together with the requirements for measuring and calculating rib geometry and projected rib areas, are given in EN 10080.

2-1-1/clause C.3(1)P

2-1-1/clause C.3(1)P requires bendability of high-yield reinforcement to be verified by bend and re-bend tests in accordance with EN 10080. This requires bar specimens to be bent through approximately 90° over a mandrel of defined diameter, aged and then bent back by at least 20°. After the tests the specimens must show no signs of fracture or cracking.

ANNEX D

Detailed calculation method for prestressing steel relaxation losses (informative)

D1. General

The formulae given in 2-1-1/clause 3.3.2 for calculating relaxation losses of prestressing steel are intended to be used to determine long-term relaxation losses. As discussed in section 3.3.2 of this guide, however, the formulae tend to produce conservative values for the long-term losses. Since the relaxation loss is itself affected by the variation in stress in the tendons over time, a better approximation of the total relaxation loss can be obtained by considering the reduction of stress over time due to other time-dependent effects, such as creep and shrinkage. This reduction is considered approximately by way of the 0.8 factor in 2-1-1/Expression (5.46), but greater benefit can be obtained using 2-1-1/Annex D. Relaxation losses derived from Annex D should not be used in conjunction with the 0.8 factor when using 2-1-1/Expression (5.46). It should also be noted that if the stress in the tendons increases with time due to other effects, the relaxation loss can be increased from that predicted by 2-1-1/clause 3.3.2.

2-1-1/clause D.1 provides an 'equivalent time' method for determining prestressing steel relaxation losses where the stress in the tendon varies with time due to effects other than just the steel relaxation. It is an iterative method, as illustrated in Worked example D-1. It is based around 2-1-1/Expressions (3.28) to (3.30) and introduces the following notation (illustrated in Fig. D-1) to enable a stage by stage approach to be adopted.

2-1-1/clause D.1

t_i the actual time at the start of the stage considered
$\sigma_{p,i}^-$ the tensile stress in the tendon just before t_i
$\sigma_{p,i}^+$ the tensile stress in the tendon just after t_i (to allow for an instantaneous reduction in prestress)
$\Delta\sigma_{pr,i}$ the absolute value of relaxation loss in the stage considered
$\sum_1^{i-1} \sigma_{pr,j}$ the sum of all the relaxation losses of the preceeding stages up to time t_i

The equivalent time, t_e, is calculated as the time taken to achieve this total relaxation loss $\sum_1^{i-1} \sigma_{pr,j}$ at time t_i, using the appropriate expression 2-1-1/Expressions (3.28) to (3.30), with a modified 'initial' stress equal to $\sigma_{p,i}^+ + \sum_1^{i-1} \Delta\sigma_{pr,j}$. This modified 'initial' stress reflects the fact that the tendon stress may also reduce (or increase) with time due to effects other than relaxation, thus reducing the relaxation itself. Using this stress:

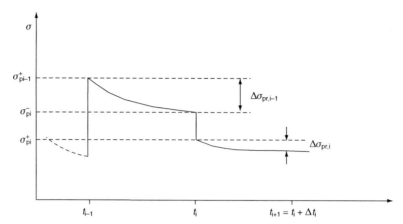

Fig. D-1. Equivalent time method

$$\mu = \left(\sigma_{p,i}^+ + \sum_{1}^{i-1} \Delta\sigma_{pr,j}\right) \Big/ f_{pk} \tag{D.1}$$

Thus for a Class 2 prestressing steel, for example, 2-1-1/Expression (3.29) becomes:

$$\sum_{1}^{i-1} \Delta\sigma_{pr,j} = 0.66\rho_{1000}\, e^{9.1\mu} \left(\frac{t_e}{1000}\right)^{0.75(1-\mu)} \left(\sigma_{p,i}^+ + \sum_{1}^{i-1} \Delta\sigma_{pr,j}\right) \times 10^{-5} \qquad \text{2-1-1/(D.1)}$$

which can be rearranged to find the equivalent time. The relaxation loss for the stage considered is then obtained by adding this equivalent time to the interval of time considered for the current stage (Δt_i), thus:

$$\Delta\sigma_{pr,i} = 0.66\rho_{1000}\, e^{9.1\mu} \left(\frac{t_e + \Delta t_i}{1000}\right)^{0.75(1-\mu)} \left(\sigma_{p,i}^+ + \sum_{1}^{i-1} \Delta\sigma_{pr,j}\right) \times 10^{-5}$$

$$- \sum_{1}^{i-1} \Delta\sigma_{pr,j} \qquad \text{2-1-1/(D.2)}$$

Worked example D-1 shows that including the loss of stress from a representative creep loss history results in a reduction of long-term relaxation losses from that derived using 2-1-1/Expressions (3.28) to (3.30). The long-term losses so obtained for Class 2 prestressing strands become close to the ρ_{1000} hour value, which was the value taken for long-term losses in calculation in previous UK practice to BS 5400 Part 4[9]. If the equivalent time method is used to determine a total relaxation loss, the 0.8 factor applied to the $\Delta\sigma_{pr}$ term in 2-1-1/Expression (5.46) should be omitted as discussed above.

Worked example D-1: Relaxation loss of low-relaxation prestressing tendon using equivalent time method

A prestressing tendon has the following properties:

- 19 no. 15.7 mm diameter low-relaxation strands (Class 2) with $\rho_{1000} = 2.5\%$.
- $A_p = 19 \times 150 = 2850\,\text{mm}^2$ per tendon.
- Characteristic tensile strength of strand, $f_{pk} = 1860\,\text{MPa}$.

It is installed with all the other tendons in the bridge in a single stage. The long-term relaxation loss is found assuming the tendon is tensioned to an initial stress of 1339.2 MPa (which includes a 10% loss from friction, wedge slip and elastic deformation of the concrete immediately after the stressing process). An illustrative creep and shrinkage loss profile is assumed, where 50% of the total creep and shrinkage occurs in the first six months. It is assumed here that creep and shrinkage would lead to a 15% total loss of

ANNEX D. DETAILED CALCULATION METHOD FOR PRESTRESSING STEEL RELAXATION LOSSES

prestressing force in the absence of other losses. (This should obviously be calculated on a bridge-specific basis.) Clearly, the shrinkage loss does not relate to the concrete stress in the same way as does the creep loss, but the two losses are lumped together here for simplicity. The important consideration is the total loss from creep and shrinkage during the different time intervals.

Time (t)		Δt_i to next time (hours)	Percentage of total creep and shrinkage completed
(hours)	(days)		
$t_1 = 0$	0	$\Delta t_1 = 24$	0
$t_2 = 24$	1	$\Delta t_2 = 144$	5
$t_3 = 168$	7 (1 week)	$\Delta t_3 = 504$	10
$t_4 = 672$	28 (1 month)	$\Delta t_4 = 1344$	20
$t_5 = 2016$	84 (3 months)	$\Delta t_5 = 2016$	35
$t_6 = 4032$	168 (6 months)	$\Delta t_6 = 4728$	50
$t_7 = 8760$	365 (1 year)	$\Delta t_7 = 491\,240$	65
$t_8 = 500\,000$	(approx. 57 years)		100

Iterative calculation for the long-term relaxation loss is now carried out up to 500 000 hours in accordance with 2-1-1/clause 3.2.3(8). Stage i covers the time from t_i to t_{i+1}.

Stage 1: Consider the first time interval from $t_1 = 0$ hours to $t_2 = 24$ hours, so $\Delta t_1 = 24$ hours.

Stress in the tendon at the start of the stage = 1339.2 MPa. The creep and shrinkage loss in this interval = 5% of a total 15% loss = 0.75%, equivalent to 10 MPa for a tendon stress of 1339.2 MPa. Conservatively, this loss is not included until the end of the current stage, after relaxation losses have been accounted for using higher stress levels. The increment of creep loss is therefore treated as a step change in stress of the form in 2-1-1/Fig. D-1.

Thus $\sigma_{p,i}^+ = \sigma_{p,1}^+ = 1339.2$ MPa and $\sum_1^{i-1} \Delta\sigma_{pr,j} = 0.0$ MPa (since there are no previous stages at $t = 0$).

From equation (D.D1), $\mu = (\sigma_{p,i}^+ + \sum_1^{i-1} \Delta\sigma_{pr,j})/f_{pk} = (1339.2 + 0)/1860 = 0.72$.

2-1-1/Expression (D.1) can be rearranged as follows to give the equivalent time for the current stress level to be reached under relaxation losses alone:

$$t_e = 1000 \times \left[\frac{\sum_1^{i-1} \Delta\sigma_{pr,j} \times 10^5}{\sigma_{p,i}^+ + \sum_1^{i-1} \Delta\sigma_{pr,j}} \times \frac{1}{0.66\rho_{1000}\, e^{9.1\mu}} \right]^{1/(0.75(1-\mu))}$$

thus

$$t_e = 1000 \times \left[\frac{0 \times 10^5}{1339.2 + 0} \times \frac{1}{0.66 \times 2.5 \times e^{9.1 \times 0.72}} \right]^{1/(0.75(1-0.72))} = 0.0\,\text{hours}$$

as expected, as there are no losses from creep at $t_1 = 0$.

The relaxation losses for the current stage can now be calculated from 2-1-1/Expression (D.2) using the equivalent time calculated above:

$$\Delta\sigma_{pr,i} = \Delta\sigma_{pr,1} = 0.66\rho_{1000}\, e^{9.1\mu} \left(\frac{t_e + \Delta t}{1000} \right)^{0.75(1-\mu)} \left(\sigma_{p,i}^+ + \sum_1^{i-1} \Delta\sigma_{pr,j} \right) \times 10^{-5}$$

$$- \sum_1^{i-1} \Delta\sigma_{pr,j}$$

thus:

$$\Delta\sigma_{\text{pr},1} = 0.66 \times 2.5 \times e^{9.1 \times 0.72} \left(\frac{0+24}{1000}\right)^{0.75(1-0.72)} \times (1339.2 + 0) \times 10^{-5} - 0$$

$$= 7.1 \text{ MPa}$$

The tensile stress in the tendon at the end of the stage, including creep, shrinkage and relaxation losses, is therefore $1339.2 - 7.1 - 10.0 = 1322.1$ MPa.

Stage 2: Consider the second time interval from $t_2 = 24$ hours to $t_3 = 168$ hours so $\Delta t_2 = 144$ hours.

Stress in the tendon at the start of the stage = 1322.1 MPa.

The creep and shrinkage loss in this interval = 5% of a total 15% loss = 0.75%, equivalent to 9.9 MPa for a tendon stress of 1322.1 MPa. (This creep and shrinkage loss has been based on the reduced tendon stress allowing for previous losses, rather than on the initial tendon stress, as a smaller creep and shrinkage loss is conservative for calculation of total relaxation loss.) Again, this loss is not included until the end of the current stage as this is conservative.

Thus:

$$\sigma_{\text{p},i}^+ = \sigma_{\text{p},2}^+ = 1322.1 \text{ MPa} \quad \text{and} \quad \sum_1^{i-1} \Delta\sigma_{\text{pr},j} = \sum_1^1 \Delta\sigma_{\text{pr},j} = 7.1 \text{ MPa}$$

(sum of previous losses due to relaxation only).

From 2-1-1/Expression (D.D1):

$$\mu = \left(\sigma_{\text{p},i}^+ + \sum_1^{i-1} \Delta\sigma_{\text{pr},j}\right) \bigg/ f_{\text{pk}} = (1322.1 + 7.1)/1860 = 0.715$$

and the equivalent time for the start of the current stage is:

$$t_e = 1000 \times \left[\frac{7.1 \times 10^5}{1322.1 + 7.1} \times \frac{1}{0.66 \times 2.5 \times e^{9.1 \times 0.715}}\right]^{1/(0.75(1-0.715))} = 33.6 \text{ hours}$$

(compared to the actual 24 hours passed).

The relaxation losses for the current stage using the new equivalent time is thus:

$$\Delta\sigma_{\text{pr},2} = 0.66 \times 2.5 \times e^{9.1 \times 0.715} \left(\frac{33.6 + 144}{1000}\right)^{0.75(1-0.715)} \times (1322.1 + 7.1) \times 10^{-5} - 7.1$$

$$= 3.0 \text{ MPa}$$

and the tensile stress in the tendon at the end of the stage (including creep, shrinkage and relaxation losses) is therefore $1322.1 - 3.0 - 9.9 = 1309.1$ MPa.

The relaxation from the remaining time intervals can be evaluated similarly, to obtain the following results:

Stage	Time interval (hours)	μ	t_e (hours)	Relaxation $\Delta\sigma_{\text{pr},i}$ (MPa)	Creep and shrinkage $\Delta\sigma_{\text{pc,s},i}$ (MPa)
1	0 to 24	0.72	0	7.1	10.0
2	24 to 168	0.715	33.6	3.0	9.9
3	168 to 672	0.709	236.9	2.9	19.6
4	672 to 2016	0.699	1224.1	2.4	28.9
5	2016 to 4032	0.683	4892.2	1.3	28.2
6	4032 to 8760	0.668	12 055.2	1.4	27.6
7	8760 to 500 000	0.653	27 242.7	20.8	62.8

ANNEX D. DETAILED CALCULATION METHOD FOR PRESTRESSING STEEL RELAXATION LOSSES

Thus, the total sum of relaxation losses is

$$\sum_1^i \Delta\sigma_{pr,j} = \sum_1^7 \Delta\sigma_{pr,j} = 38.9\,\text{MPa}$$

This represents a loss of $38.9/1339.2 = \mathbf{2.9\%}$ (compared to 4.3% loss calculated using 2-1-1/clause 3.3.2 in Worked example 3.3-1).

ANNEX E

Indicative strength classes for durability (informative)

E1. General

As discussed in section 4 of this guide, the choice of adequately durable concrete for corrosion protection of reinforcement and for resistance to attack of the concrete itself depends on the composition of the concrete. This consideration can result in a higher compressive strength of concrete being required than that necessary for structural design. 2-1-1/Annex E defines 'indicative' strength classes for concrete dependent on environmental class. They are effectively the minimum acceptable concrete classes and provide the benchmark for assessing cover requirements (discussed in section 4). The recommended values of indicative strength classes may be varied in the National Annex. The EC2 recommended values are summarized here in Table E-1 for convenience.

2-1-1/clause E.1(2) reminds the designer that where the chosen concrete strength is higher than that required for the structural design, it is necessary to check minimum reinforcement using the design value of the tensile concrete strength (f_{ctm}) associated with the higher strength concrete. This is because increased tension stiffening elevates section cracking moment and hence reinforcement requirements. Other situations may also require

2-1-1/clause E.1(2)

Table E-1. Indicative concrete strength classes

	Exposure classes (from 2-1-1/Table 4.1)									
	Carbonation-induced corrosion				Chloride-induced corrosion			Chloride-induced corrosion from sea water		
Corrosion	XC1	XC2	XC3	XC4	XD1	XD2	XD3	XS1	XS2	XS3
Indicative strength class	C20/25	C25/30	C30/37		C30/37		C35/45	C30/37	C35/45	

	Exposure classes (from 2-1-1/Table 4.1)						
	No risk	Freeze/thaw action			Chemical attack		
Damage to concrete	X0	XF1	XF2	XF3	XA1	XA2	XA3
Indicative strength class	C12/15	C30/37	C25/30	C30/37	C30/37		C35/45

consideration of the stronger concrete. For example, if non-linear analysis has been used to determine the ultimate bending resistance of a beam with unbonded tendons, tension stiffening away from the critical section can have an adverse effect on tendon strain increase and hence strength – section 5.10.8 of this guide refers.

ANNEX F

Tension reinforcement expressions for in-plane stress conditions (informative)

The provisions of Annex F are discussed in section 6.9 of this guide. This includes amending the sign convention used in the Annex F formulae to match that used in 2-2/clause 6.109 and Annexes LL and MM. Section 6.9 also includes some proposed design equations for skew reinforcement.

ANNEX G

Soil–structure interaction

G1. General

2-1-1/clause 5.1.1 and *2-1-1/clause G.1.1(1)* both require that soil–structure interaction be taken into account where the interaction has a significant influence on the action effects in the structure. 2-1-1/Annex G gives informative guidance on soil–structure interaction for shallow foundations and piles. The basic statement in *2-1-1/clause G.1.1(2)* requires soil and structure displacements and reactions to be compatible and the remainder of Annex G adds little to this. It could be added that serviceability limit state requirements should be met for both structure and soil, with realistic stiffnesses employed in analysis. At the ultimate limit state, allowable soil pressures should not be exceeded and all members should be sufficiently strong and possess sufficient rotation capacity to justify the distribution of forces assumed.

2-1-1/clause G.1.1(1)

2-1-1/clause G.1.1(2)

Given the lack of normative rules, the consideration of soil–structure interaction for bridge foundations need not be different to previous UK practice and other general points apply:

- Where soil–structure interaction is included in the analysis, the model support conditions must be chosen to realistically model the actual soil stiffness. A sensitivity analysis may need to be carried out where results are very sensitive to assumed soil stiffness.
- If the soil stiffness is non-linear under applied load, the analysis will require a degree of iteration to obtain the 'mean' stiffness for the loading conditions (if the non-linearity of the soil cannot be modelled directly).

For piled foundations, a further relevant consideration is that simple spring elements will often not suffice for modelling pile groups, since a moment applied to a pile cap produces both a rotation and a displacement. (Similarly, a shear force applied produces both a displacement and a rotation.) If the piles are not themselves modelled together with springs representing the soil, suitable modelling techniques include use of a flexibility (or stiffness) matrix as a boundary condition at the pile cap level or the use of an equivalent cantilever beneath the pile cap to emulate the correct displacement behaviour. Modelling pile group behaviour correctly is particularly important in the design of integral bridges as the forces and moment attracted can be sensitive to changes in foundation stiffness and modelling assumptions.

Annex H. Not used in EN 1992-2

ANNEX I

Analysis of flat slabs (informative)

Annex I of EN 1992-2 covers the analysis of flat slabs and is based on the requirements of Annex I in EN 1992-1-1 with some of its provisions deleted, including those concerning shear walls.

2-2/clause I.1.1(2) requires 'proven methods' of analysis to be used in the design of flat slabs. These include lower-bound methods, based on elastic grillage or shell finite element models, or upper-bound methods, such as yield line analysis. The latter is an example of plastic analysis which is generally restricted for bridge design, as discussed in section 5.6.1 of this guide. Non-linear analysis (using grillage or shell finite element models) would also be appropriate. A further alternative is the simplified 'equivalent frame analysis' outlined in *2-2/clause I.1.2*, based on sub-division of the slab into column strips and middle strips for the purpose of flexural design. This is a technique more suited to regular repeating column layouts in building structures and will rarely be used in bridge design.

Regardless of the method employed for the flexural design, the punching shear provisions of EN 1992-2 clause 6.4 apply in addition to checks of flexural shear and bending moment.

2-2/clause I.1.1(2)

2-2/clause I.1.2

ANNEX J

Detailing rules for particular situations (informative)

J1. Surface reinforcement

2-1-1/clause J.1 gives informative rules on the use of surface reinforcement, as called up by 2-1-1/clause 8.8 and 2-1-1/clause 9.2.4. The provision of surface reinforcement serves two purposes:

(1) Reduces crack widths through use of small bar diameter and bar spacing.
   Where surface reinforcement is used to control cracking in beams with large diameter bars (as defined in 2-1-1/clause 8.8), the supplementary provisions of 2-1-1/clause 8.8(8) apply for minimum surface reinforcement areas. Surface reinforcement does not *have* to be used to control cracks, however, as explicit crack width calculations can still be used in accordance with 2-1-1/clause 8.8(2).
(2) Prevents concrete spalling when large diameter bars are used.

2-1-1/clause J.1(1) recommends the use of surface reinforcement to resist concrete spalling where the main reinforcement is made up of large diameter bars, greater than 32 mm diameter. The definition of 'large' diameter is a nationally determined parameter in 2-1-1/clause 8.8, but the recommended value is again 32 mm. The UK National Annex is likely to choose 40 mm as the definition of 'large' diameter, with the intention of avoiding the need for surface reinforcement where 40 mm bars are used. Surface reinforcement should be placed outside the main links.

Where provided, *2-1-1/clause J.1(2)* requires the area of surface reinforcement to not be less than $A_{s,surf}$, a nationally determined parameter with a recommended value of $0.01 A_{ct,ext}$, where $A_{ct,ext}$ is defined as the area of tensile concrete external to the links. The surface reinforcement should be placed in two directions – parallel and orthogonal to the main reinforcement direction. If surface reinforcement is provided, it may be included in the bending and shear resistance calculations in addition to the main reinforcement, providing it is suitably detailed to the standard requirements for arrangement and anchorage (see sections 8 and 9 of this guide).

2-1-1/clause J.1(3) recommends that surface reinforcement is also used where covers to main reinforcement are greater than 70 mm. However, minimum cover in accordance with 2-2/clause 4.4.1 must still be provided to the surface reinforcement.

UK bridge designers have rarely specified surface reinforcement in the past, even when 40 mm bars have been used, for example in piles (where its provision would be difficult) and retaining walls. The fact that Annex J is only informative gives some room for manoeuvre in this respect and it is likely that the UK National Annex will not require

2-1-1/clause J.1(1)

2-1-1/clause J.1(2)

2-1-1/clause J.1(3)

J2. Frame corners

2-1-1/clause J.2

2-1-1/clause J.2 gives examples of concrete frame corners, provides possible strut-and-tie models for analysing their behaviour and recommends suitable reinforcement arrangements that are consistent with the analysis. The concrete strength, $\sigma_{Rd,max}$, should be determined in accordance with 2-1-1/clause 6.5.2.

In bridge design, frame corners are most commonly encountered in box girder web–flange junctions and certain substructure elements. Completely different strut-and-tie models are required for corners with closing moments and opening moments. It should be noted that in bridge design, the moments in typical frame corners may be reversible and therefore the element design and reinforcement arrangements must be able to accommodate both sets of strut-and-tie models. 2-1-1/Fig. J.2 shows typical models for closing moments while Figs J.3 and J.4 cover moderate opening and large opening moments respectively. (The figures are not reproduced here.)

It will be noted that Fig. J.2 indicates some link reinforcement in the main members, which does not result from the corner model, while Figs J.3 and J.4 do not. This is not intended to imply that such reinforcement is unnecessary for opening moment cases. A further observation is that the diagonal bars across the re-entrant corners in Fig. J.4 are very effective at controlling cracks and therefore consideration should always be given to their provision in corners with opening moments, even though equilibrium and adequate ULS performance can be achieved without them as in Fig. J.3. This is the subject of *2-1-1/clause J.2.3(2)*. Particular care should be taken when detailing corner regions using overlapping U-bars (as in Fig. J.3(b) of EC2-1-1) to ensure satisfactory serviceability limit state performance. If insufficient overlap is provided and diagonal bars are not present, wide cracks may open up under service loading as the hinge detail formed by the reinforcement rotates.

2-1-1/clause J.2.3(2)

A further consideration not mentioned in Annex J is that bend radii in bent bars should be detailed to limit the bearing stresses in bends in accordance with 2-1-1/clause 8.3.

J3. Corbels

2-1-1/clause J.3 gives recommendations for the design of concrete corbels based on the strut-and-tie model shown in Fig. J-1 (Fig. J.5 in EN 1992-1-1). Two cases are covered:

(a) $a_c \leq 0.5 h_c$

2-1-1/clause J.3(2)

In addition to the main reinforcement provided at the top of the corbel (with a total area of $A_{s,main}$), *2-1-1/clause J.3(2)* requires the designer to provide closed horizontal or inclined links distributed within the depth of the corbel. They should be centred on the tie labelled F_{wd}. The total area of this reinforcement is recommended to be taken as a minimum of $k_1 A_{s,main}$, where k_1 is a nationally determined parameter defined in the National Annex with a recommended value of 0.25 in EC2-2. A value of 0.5 would be more in line with previous UK practice for corbel design. Regardless of the amount of this secondary tie steel, the force coming into the top and bottom nodes is the same. It therefore appears that the purpose of the secondary tie is to reinforce what is effectively a bulging compression strut between top and bottom nodes, thus increasing its resistance to match that of the nodes – sections 6.5.2 and 6.7 of this guide refer. In this respect, the steel would be more effective if placed perpendicular to the compression diagonal.

The check of the compression strut can effectively be made by limiting the shear stress such that $F_{Ed} \leq 0.5 b_w d \nu f_{cd}$ in accordance with 2-1-1/clause (6.5).

(b) $a_c > 0.5 h_c$

In addition to the main reinforcement provided at the top of the corbel (with a total area of $A_{s,main}$), vertical links are required where the shear force exceeds the concrete shear

ANNEX J. DETAILING RULES FOR PARTICULAR SITUATIONS

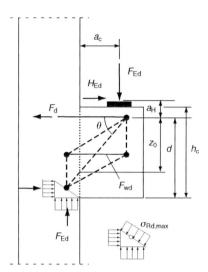

Fig. J-1. Corbel strut-and-tie model

resistance according to 2-1-1/clause 6.2.2. **2-1-1/clause J.3(3)** expresses this latter condition as $F_{Ed} > V_{Rd,c}$, but it would be reasonable to interpret this as $\beta F_{Ed} > V_{Rd,c}$ in accordance with 2-1-1/clause 6.2.2(6), where β is the reduction factor to allow for shear enhancement. Where link reinforcement is required, its provision clearly no longer relates to the strut-and-tie model of Fig. J-1. 2-1-1/clause J.3(3) requires the link force provided to be a minimum of 50% of the applied vertical force. This should be viewed as an absolute minimum requirement, with the link area determined more generally in accordance with 2-1-1/clause 6.2.3(8). The check of the compression strut can again effectively be made by checking that the shear is limited such that $F_{Ed} \leq 0.5 b_w d \nu f_{cd}$ in accordance with 2-1-1/Expression (6.5).

2-1-1/clause J.3(3)

For all sizes of corbel, the anchorage of the main reinforcement projecting into the supporting member should be checked for adequacy in accordance with the rules in sections 8 and 9 of EN 1992-2.

J4. Partially loaded areas
J4.1. Bearing zones of bridges
The design for allowable bearing pressure and reinforcement design is covered in sections 6.5 and 6.7 of this guide. **2-2/clause J.104.1** gives some additional requirements as follows:

2-2/clause J.104.1

- the minimum distance between the edge of a loaded area and the edge of the section should not be less than 50 mm or less than 1/6 of the corresponding dimension of the loaded area;

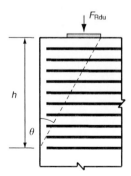

Fig. J-2. Sliding wedge mechanism under concentrated load

- for concrete grades in excess of C55/67, f_{cd} needs to be replaced by $[0.46 f_{ck}^{2/3}/(1 + 0.1 f_{ck})] f_{cd}$. This formula doesn't actually reduce the value of f_{cd} until the concrete grade exceeds C60/75;
- an additional sliding wedge mechanism, illustrated in Fig. J-2, also needs to be checked. The 'failure' plane is defined by $\theta = 30°$ and an amount of reinforcement given by $A_r f_{yd} \geq F_{Rdu}/2$ must be uniformly distributed over the height h. The provided reinforcement should be suitably detailed and anchored in accordance with 2-2/clause 9, which will normally necessitate the use of closed links.

J4.2. Anchorage zones of post-tensioned members

2-2/clause J.104.2

2-2/clause J.104.2 gives rules for anchor zones of post-tensioned members which are supplementary to those given in 2-1-1/clause 8.10.3. These provisions are discussed in section 8.10.3 of this guide.

ANNEX K

Structural effects of time dependent behaviour (informative)

Annex KK of EN 1992-2 is mainly concerned with the redistribution of internal actions and stresses that occur in bridges built in stages. This includes, for example, box girder bridges built span by span with striking of formwork at each stage and precast composite members in un-propped construction where a deck slab is cast after erection of the main precast beams. This section of the guide does not follow the clause headings of EN 1992 Annex KK and is structured as follows:

General considerations	*Section K1*
General method	*Section K2*
Simplified methods	*Section K3*
Application to pre-tensioned composite members	*Section K4*

K1. General considerations

Creep will tend to cause action effects built up from staged construction to redistribute towards the action effects that would have been produced had the structure been constructed monolithically all at the same time.

The effect of creep redistribution is illustrated in Fig. K-1 for the dead load moments in a three-span bridge. The bridge is built span-by-span in the stages shown and the dead load moment creeps from its built-up distribution towards that for monolithic construction. In prestressed members, the prestress secondary moments are similarly redistributed. Similar creep redistribution also occurs in simply supported pre-tensioned beams which are subsequently made continuous.

For prestressed members (whether post-tensioned or pre-tensioned) this redistribution of moments and stresses is particularly important for the serviceability limit state design, as it can lead to unacceptable cracking and serviceability stresses if not considered properly – *2-2/clause KK.2(101)* refers. Consideration of the redistribution is often less important at the ultimate limit state and can be ignored where there is sufficient rotation capacity available to shed the restraint moments, unless any of the bridge members are susceptible to significant second-order effects – the Note to 2-2/clause KK.2(101) refers.

2-2/clause KK.2(101)

Providing the concrete stress under quasi-permanent loads does not exceed $0.45f_{ck}(t)$, linear creep behaviour may be assumed where the creep strain varies linearly with the creep – *2-2/clause KK.2(102)*. Where the concrete stress exceeds $0.45f_{ck}(t)$, non-linear creep has to be considered, whereupon the creep strain varies exponentially with stress.

2-2/clause KK.2(102)

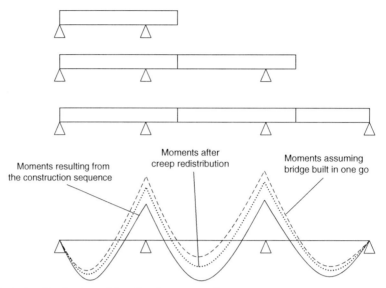

Fig. K-1. Typical redistribution of dead load moments due to creep

Non-linear creep may need to be considered for pre-tensioned beams which are stressed to high loads at a young age. This is discussed in section 3.1.4 of this guide.

Annex KK of EC2-2 does not cover the effects of differential shrinkage. This is covered in K4.2 of this guide.

Five methods of considering the effects of creep redistribution are presented in EN 1992-2:

(a) general step-by-step analysis – 2-2/clause KK.3;
(b) differential version of the above – 2-2/clause KK.4;
(c) application of the theorems of visco-elasticity – 2-2/clause KK.5;
(d) coefficient of ageing method – 2-2/clause KK.6;
(e) simplified method based on coefficient of ageing method – 2-2/clause KK.7.

Generally, only method (e) will need to be explicitly considered by designers as this will usually predict long-term effects adequately and is the only method that lends itself to simple hand calculation. Where it is important to predict losses and deflections at intermediate stages of construction, as would be the case in balanced cantilever construction, it will usually be necessary to use proprietary software which is likely to use variations on method (a). Only methods (a) and (e) are therefore considered below and their use in the context of both composite and non-composite beams is discussed.

K2. General method

The general method involves a step by step calculation of the strain according to 2-2/Expression (KK.101), which will typically be performed by iterative computer analysis. Such a method will also include losses due to the primary effects of creep and shrinkage, as well as due to creep redistribution effects:

$$\varepsilon_c(t) = \frac{\sigma_0}{E_c(t_0)} + \phi(t,t_0)\frac{\sigma_0}{E_c(28)} + \sum_{i=1}^{n}\left(\frac{1}{E_c(t_i)} + \frac{\phi(t,t_i)}{E_c(28)}\right)\Delta\sigma(t_i) + \varepsilon_{cs}(t,t_s) \qquad \text{2-2/(KK.101)}$$

The application of this formula to unrestrained concrete (ignoring shrinkage strain) under a series of additive externally applied axial forces is illustrated in Fig. K-2. In general for real members, the concrete stress will not change in discrete steps as shown in Fig. K-2, as the stress itself will constantly vary with the changing strain due to the presence of external restraint or internal restraint (from reinforcement or prestressing). A computer analysis is therefore required to split the analysis into a series of time steps such that the effects of

ANNEX K. STRUCTURAL EFFECTS OF TIME DEPENDENT BEHAVIOUR

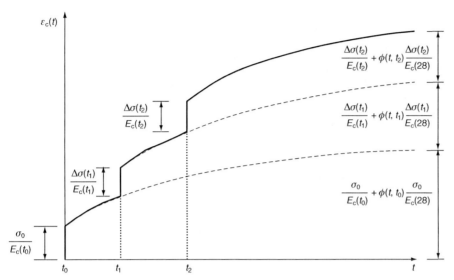

Fig. K-2. Creep strain according to 2-2/(KK.101) for unrestrained concrete with several load increments

the constantly varying stress can be transformed into small discrete steps similar to those in Fig. K-2. A differential version of this procedure is provided in 2-2/clause KK.4.

The effects of prestressing steel relaxation also need to be considered. Computer methods will normally also include the effects of relaxation loss in a way similar to that discussed in Annex D.

K3. Simplified methods

A simple method for calculating the long-term creep redistribution effects is given in *2-2/clause KK.7* based on the ageing coefficient method. The basic procedure is to calculate the distribution of moments (or other internal actions) built up by fully modelling the construction sequence and to then recalculate the distribution of moments assuming that the structure was built in one go with all dead loads, superimposed dead load and prestressing applied to the final structure. The actual long-term moments accounting for creep redistribution can then be found by an interpolation between these two sets of moments according to 2-2/Expression (KK.119), without the need for calculation by a time-step method. 2-2/Expression (KK.119) also applies to stresses:

2-2/clause KK.7

$$S_\infty = S_0 + (S_c - S_0)\frac{\phi(\infty, t_0) - \phi(t_c, t_0)}{1 + \chi\phi(\infty, t_c)} \qquad \text{2-2/(KK.119)}$$

The redistribution of actions due to creep is therefore:

$$\Delta S = (S_c - S_0)\frac{\phi(\infty, t_0) - \phi(t_c, t_0)}{1 + \chi\phi(\infty, t_c)} \qquad \text{(DK-1)}$$

where:

S_0 are the internal actions obtained from the construction sequence build-up

S_c are the internal actions obtained assuming that the whole structure is built in one go and then all the permanent load applied to it. The phrase 'constructed on centering' is used in EN 1992-2 to convey this idea of the structure being constructed in its entirety on continuous supports, with the supports or 'centering' being released only after the entire structure is completed. Strictly, the word 'centering' is normally used to describe temporary supports to arches

$\phi(\infty, t_0)$ is the final creep coefficient for a concrete age t_0 at time of first loading

$\phi(t_c, t_0)$ is the creep coefficient for a concrete age t_0 at time of first loading up to an age t_c, where t_c is the concrete age at which the structural system is changed, such as closure of a stitch between adjacent spans. It is therefore the amount of creep factor 'used up' before the structural system is changed. Where the structural system is changed in this way a number of times, a representative average age should be used based on the average age at which each stage is connected to the next. For example, if each stage takes 30 days to construct (including stripping of falsework) and is then immediately connected to the previous span, then $t_c = t_0 = 30$ days would be a reasonable approximation as the previous stage concrete would be older than 30 days, but some of the current stage concrete would be younger

$\phi(\infty, t_c)$ is the final creep coefficient for a concrete age t_c at time of first loading

The creep coefficients can be calculated as discussed in section 3.1.4 of this guide.

The ageing coefficient χ can be thought of as representing the reduction in creep, and therefore increase in stiffness, for restraint of deformations occurring after the time t_c. It can be taken as 0.8, which is a good representative value for most construction, but in reality it varies with age at time of loading and creep factor.

The redistribution in equation (DK-1) can be thought of as follows. If the initial moments in a structure are $M_{0,i}$, resulting from load applied at time t_0, then the free creep curvature from these moments occurring after a time t_c will be $M_{0,i}/EI(\phi(\infty, t_0) - \phi(t_c, t_0))$. If the structure were to then be fully restrained everywhere (a purely theoretical rather than practical situation) at time t_c, restraint moments would be developed. The effective Young's modulus for this load case would be $E/(1 + \chi\phi(\infty, t_c))$ and hence the fully restrained moments developed would be $-M_{0,i}(\phi(\infty, t_0) - \phi(t_c, t_0))/(1 + \chi\phi(\infty, t_c))$. These restraint moments represent the redistribution moments for this particular change of structural system. Equation (DK-1) predicts the same result. In this case, $S_0 = M_{0,i}$. S_c is obtained assuming the structure to be in its final condition (i.e. fully restrained everywhere in this case) prior to initial loading. This leads to $S_0 = M_{c,i} = 0$ and thus equation (DK-1) gives redistribution moments of $-M_{0,i}(\phi(\infty, t_0) - \phi(t_c, t_0))/(1 + \chi\phi(\infty, t_c))$.

For non-prestressed concrete bridges, there is no problem with the interpretation of 2-2/Expression (KK.119) and the 'S' terms contain only the dead load and superimposed dead load. The internal actions are then redistributed as in Fig. K-1. For prestressed bridges, however, the magnitude of the prestressing force itself changes during the life of the bridge so the value to use in the interpolation needs careful consideration. The secondary effects of prestress are themselves altered by creep (due to the loss of prestressing force) even without changing the structural system.

There are several interpretations possible for the prestressing force to use in the interpolation. This is a result of the fact that the method is not exact. Four interpretations (others are possible) include:

(a) The prestress force after all short-term and time-dependent long-term losses is considered in deriving S_c, but the prestress force including only whatever losses that have occurred up to 'closing' the structure is considered in deriving S_0. In each case, redistribution effects are ignored in calculating S_c and S_0 (other than the small change in secondary effects of prestress caused by the loss of prestressing force). This appears to be the literal interpretation of the definitions of S_c and S_0 given in KK.7/Expression (101). This is not the intended interpretation, however, as S_∞ then does not include all the loss of prestressing force and the effects on reducing the primary prestress moments. This is because the final stress state is effectively then obtained from an interpolation between values of S_c which include all losses of prestress, and values of S_0 which include only the immediate losses and a small fraction of the long-term time-dependent losses. This approach is therefore inappropriate as it underestimates losses and overestimates the actual final prestress force.

(b) The prestress force, including all short-term and time-dependent long-term losses, is used to determine both S_c and S_0. The long-term prestress losses in each case can be determined from the concrete stresses after application of the initial prestressing force, including the immediate losses. In each case, redistribution effects are ignored (other than the small change in prestress secondary moment resulting from the loss of prestressing force). S_∞ from 2-2/Expression (KK.119) then includes all the long-term loss of prestress as well and can be taken to represent the final internal actions including all the long-term losses.

(c) The prestress force including only short-term losses (i.e. not considering time-dependent long-term losses) is used to determine both S_c and S_0. The term $(S_c - S_0)\dfrac{\phi(\infty, t_0) - \phi(t_c, t_0)}{1 + \chi\phi(\infty, t_c)}$ from equation (DK-1) is then calculated, which represents the redistribution effects only. The long-term loss of prestress is then subsequently calculated from the built-up internal actions S_0 and these built-up internal actions modified accordingly to allow for long-term losses of prestress. This leads to a set of long-term internal actions derived from following the construction sequence but without considering redistribution, say $S_{0,\infty}$. The term $(S_c - S_0)\dfrac{\phi(\infty, t_0) - \phi(t_c, t_0)}{1 + \chi\phi(\infty, t_c)}$ derived above is added to $S_{0,\infty}$ to give the final set of internal actions allowing for both creep redistribution and all long-term losses of prestress, i.e. $S_\infty = S_{0,\infty} + (S_c - S_0)\dfrac{\phi(\infty, t_0) - \phi(t_c, t_0)}{1 + \chi\phi(\infty, t_c)}$.

(d) As (c) but the prestress force used to calculate $(S_c - S_0)\dfrac{\phi(\infty, t_0) - \phi(t_c, t_0)}{1 + \chi\phi(\infty, t_c)}$ is that immediately at the time of the change of structural system, e.g. connection of adjacent spans. $S_{0,\infty}$ is derived considering the build-up of stresses and all long-term losses of prestress.

These approaches are all approximate but method (d) often gives results closest to the results of a time-step analysis. This approach is summarized below under the heading 'Application to post-tensioned construction'. Only method (a) above is completely inappropriate to use.

Given the inherent uncertainty in creep calculations, this interpolation approach is normally of satisfactory accuracy for predicting long-term redistribution effects. Where there are many stages of construction with concretes of different ages before a structure is made statically indeterminate (such as occurs in balanced cantilever construction), or where there are many changes to the structural system (such as occurs in span-by-span construction), one difficulty is deciding a representative single value for the creep coefficients $\phi(\infty, t_0) - \phi(t_c, t_0)$ (which is the creep remaining after the structural system has been modified) and $\phi(\infty, t_c)$. It will usually be adequate to select an average residual creep value for the structure. In most cases for post-tensioned structures, the value of $\dfrac{\phi(\infty, t_0) - \phi(t_c, t_0)}{1 + \chi\phi(\infty, t_c)}$ will typically be between 0.65 and 0.8.

A simpler version of 2-2/Expression (KK.119) often used (and which was effectively used in BS 5400 Part 4[9]) is:

$$S_\infty = S_0 + (S_c - S_0)(1 - e^{-\phi}) \tag{DK-2}$$

Equation (DK-2) has only the one creep factor which can be taken as $\phi = \phi(\infty, t_0) - \phi(t_c, t_0)$, but it does not contain the ageing coefficient. It is therefore less accurate than 2-2/Expression (KK.119). A value for ϕ of between 1.5 and 2.0 will be representative in most cases for post-tensioned bridges.

With the above expression, the redistribution of actions due to creep is therefore:

$$\Delta S = (S_c - S_0)(1 - e^{-\phi}) \tag{DK-3}$$

Equations (DK-2) and (DK-3) tend to over-predict the amount of redistribution that will occur, particularly in structures where the concrete is quite old at the time of modifying the structural system (such as closing a structure with a stitch).

Application to post-tensioned construction
For post-tensioned beams, a suggested approximate procedure, following EC2-2 formulae, is as follows:

(1) Build up the internal actions for the bridge from the construction sequence considering only immediate losses – see 2-2/clause 5.10.5.
(2) Calculate long-term losses in prestressing force using 2-2/clause 5.10.6, based on the concrete stresses obtained from above.
(3) Determine the change in primary and secondary prestress moments from this loss of prestress.
(4) Modify the internal actions in (1) by the effects of the losses in (3) to give long-term effects excluding creep redistribution.
(5) Modify the internal actions in (4) by adding the effects of redistribution according to equation (DK-1) with S_0 and S_c, based on the loss at the time of modifying the structural system.

It is always possible to use a computer time-step analysis as an alternative.

Worked example K-1: Calculation of creep factors for three-span bridge
The three-span post-tensioned bridge in Fig. K-3 is constructed in situ in two stages, each taking 30 days to complete. The creep coefficients for use in the creep redistribution calculation are calculated and it is determined by how much the internal action effects would creep from the as-built conditions towards those which would be achieved if the structure was built and loaded in one stage. The relative humidity is 70%, the concrete grade is C40/50, the cement is Ordinary Portland and the average effective section thickness is 500 mm.

For each stage, the age at loading (i.e. prestressing and striking falsework) will be less than 30 days on average and similarly the average age of the second-stage concrete will be less than 30 days on stitching to the first. However, by the time the first stage is connected to the second, the age of the first stage concrete will be greater than 30 days and will therefore be stiffer for the effects of creep redistribution. Even this simple example illustrates the difficulty in producing a single set of creep coefficients. The saving grace is that the results will not be very sensitive to the assumptions made for the concrete ages. It will be assumed, therefore, that $t_c = t_0 = 30$ days.

The creep coefficients can most easily be calculated from 2-1-1/Annex B as the relative humidity of 70% is not given in Fig. 3.1 and interpolation would be required:

The creep coefficient $\phi(\infty, t_0)$ is calculated from 2-1-1/Expression (B.1):

$$\phi(t, t_0) = \phi_0 \times \beta_c(t, t_0) \qquad \text{2-1-1/(B.1)}$$

$\beta_c(t, t_0)$ is a factor that describes the amount of creep that occurs with time and at time $t = \infty$ it equals 1.0. The total creep is therefore given by ϕ_0 from 2-1-1/Expression (B.2):

$$\phi_0 = \phi_{RH} \times \beta(f_{cm}) \times \beta(t_0) \qquad \text{2-1-1/(B.2)}$$

ϕ_{RH} is a factor to allow for the effect of relative humidity on the notional creep coefficient. Two expressions are given depending on the size of f_{cm}.

From Table 3.1, $f_{cm} = f_{ck} + 8 = 40 + 8 = 48$ MPa

For $f_{cm} > 35$ MPa, ϕ_{RH} is given by 2-1-1/Expression (B.3b):

$$\phi_{RH} = \left[1 + \frac{1 - RH/100}{0.1 \times \sqrt[3]{h_0}} \times \alpha_1\right] \times \alpha_2 \qquad \text{2-1-1/(B.3b)}$$

RH is the relative humidity of the ambient environment in % = 70% here.

$\alpha_{1/2}$ are coefficients to consider the influence of the concrete strength from 2-1-1/Expression (B.8c):

$$\alpha_1 = \left[\frac{35}{f_{cm}}\right]^{0.7} = \left[\frac{35}{48}\right]^{0.7} = 0.80$$

$$\alpha_2 = \left[\frac{35}{f_{cm}}\right]^{0.2} = \left[\frac{35}{48}\right]^{0.2} = 0.94 \qquad \text{2-1-1/(B.8c)}$$

From 2-1-1/Expression (B.3b):

$$\phi_{RH} = \left[1 + \frac{1 - 70/100}{0.1 \times \sqrt[3]{500}} \times 0.80\right] \times 0.94 = 1.22$$

$\beta(f_{cm})$ is a factor to allow for the effect of concrete strength on the notional creep coefficient from 2-1-1/Expression (B.4):

$$\beta(f_{cm}) = \frac{16.8}{\sqrt{f_{cm}}} = \frac{16.8}{\sqrt{48}} = 2.42$$

The effect of concrete age at loading on the notional creep coefficient is given by the factor $\beta(t_0)$ according to 2-1-1/Expression (B.5). For loading at 30 days this gives:

$$\beta(t_0) = \frac{1}{(0.1 + t_0^{0.20})} = \frac{1}{(0.1 + 30^{0.20})} = 0.48$$

The final creep coefficient from 2-1-1/Expression (B.2) is then:

$$\phi(\infty, t_0) = \phi_0 = 1.22 \times 2.42 \times 0.48 = \mathbf{1.43}$$

(Note that the creep coefficient for individual parts of the deck will often be significantly greater than this as they will be younger than 30 days old at first loading.)

The creep coefficient $\phi(t_c, t_0)$ is equal to zero as it was assumed that $t_c = t_0$. Similarly $\phi(\infty, t_c) = \phi(\infty, t_0) = 1.43$.

The redistribution factor according to equation (DK-1) is:

$$\frac{\phi(\infty, t_0) - \phi(t_c, t_0)}{1 + \chi\phi(\infty, t_c)} = \frac{1.43}{1 + 0.8 \times 1.43} = 0.67$$

The moments in the structure would therefore creep 67% of the way from the as-built conditions to the monolithic conditions, assuming the structure to have been built and loaded in one stage. (The similar redistribution factor according to equation (DK-3) is $(1 - e^{-\phi}) = (1 - e^{-1.423}) = 0.76$, which would be a slight overestimate in reality.) The main text above describes how the redistribution moments would be determined from here.

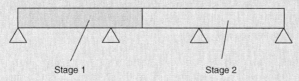

Fig. K-3. Worked example K-1

K4. Application to pre-tensioned composite members
K4.1. Differential creep in composite beams
When a deck slab is cast on a pre-tensioned beam, the change in cross-section after the beam has been prestressed will generate restraint stresses, as the loaded pre-tensioned beam tries to deflect further with creep and this is resisted by the deck slab. The stresses will redistribute from the built-up values towards those obtained if the dead load and prestress were applied to the final composite section. Even if the bridge is not built in stages span-wise, redistribution will take place internal to each cross-section.

This redistribution is illustrated in Fig. K-4 for a simply supported beam. It illustrates that the internal actions and stresses should be calculated for the as-built case and monolithic case and then interpolation between them performed using either equation (DK-1) or equation (DK-3) to determine the redistribution effects. As discussed above, equation (DK-3) may overestimate the magnitude of redistribution. The prestress forces to use should be those immediately after casting the deck slab, so an assumption has to be made with respect to this timing and thus how much creep has therefore already occurred. Determination of the creep factors to use in equation (DK-1) or equation (DK-3) also requires estimation of this timing. Worked example K-2 illustrates the calculation.

If the simply supported pre-tensioned beam is subsequently made continuous, which is typically achieved by connecting adjacent simply supported spans together, the creep deformation is further restrained by the modified structural system and the support reactions are modified. This leads to the development of moments at the supports due to the redistribution of dead load moments and prestress. These continuity moments develop in the same way as discussed for post-tensioned beams due to the changed support conditions. Usually, sagging moments develop at the supports and additional reinforcement is required across the joint to prevent excessive cracking. This problem can be treated by interpolation between the as-built case (1) and monolithic case (2) (with all loads applied to the composite section) as in Fig. K-4, but case (2) will also contain the secondary effects of prestress due to the continuity and the dead load moments will also be modified by continuity. Either equation (DK-1) or equation (DK-3) can again be used. This is shown in Fig. K-5.

Alternatively, the internal locked-in stresses can first be determined exactly as in Fig. K-4 without considering the bridge to be made continuous. Restraint of the creep from the

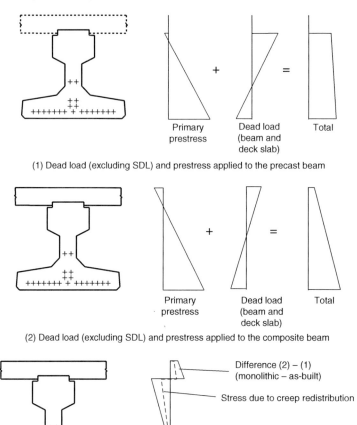

Fig. K-4. Redistribution of stresses in simply supported composite beam due to creep

ANNEX K. STRUCTURAL EFFECTS OF TIME DEPENDENT BEHAVIOUR

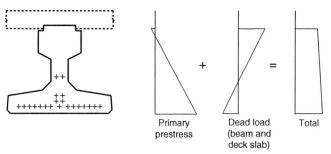

(1) Dead load (excluding SDL) and prestress applied to the precast beam

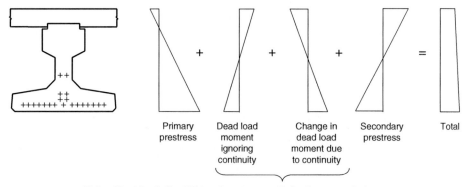

(2) Dead load (excluding SDL) and prestress applied to the composite beam

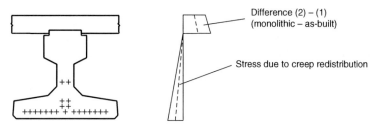

(3) Stress from creep redistribution from (1) towards (2)

Fig. K-5. Redistribution of stresses due to creep in simply supported composite beam subsequently made continuous

prestress and dead load leads to the development of support moments. These restraint moments are linear between supports and the crept value at the support can be found using equation (DK-1) or (DK-3). In this case, $S_0 = 0$ at the supports and S_c should consider the secondary effects of prestress and the dead load support hogging moment on the continuous beam only, as other effects are included in the initial step.

In either case, the stress from these redistribution effects should be added to the built-up stresses for the beam including all losses.

Worked example K-2: Creep redistribution in simply supported composite deck

A simply supported span comprises pre-tensioned beams as shown in Fig. K-6 with in situ deck slab. The deck slab is cast when the precast beams are 120 days old, which leads to a residual creep factor after the deck is cast of $\phi = \phi(\infty, t_0) - \phi(t_c, t_0) = 0.90$. The creep factor for loading at 120 days is $\phi = \phi(\infty, t_c) = 1.5$. The prestress force at midspan after losses at the time of casting the top slab is 3274 kN. The dead load moment from beam self-weight and deck slab wet concrete weight is 494 kNm (sagging) at midspan. The section properties are as follows:

Precast beam:

$A = 349 \times 10^3 \, \text{mm}^2$

$Z_{\text{bot}} = 74.31 \times 10^6 \, \text{mm}^3$

$Z_{\text{top}} = 46.96 \times 10^6 \, \text{mm}^3$

Eccentricity of strands from neutral axis $= 142 \, \text{mm}$.

Composite beam:

$A = 500 \times 10^3 \, \text{mm}^2$

$Z_{\text{bot}} = 114.9 \times 10^6 \, \text{mm}^3$

$Z_{\text{top}} = 119.6 \times 10^6 \, \text{mm}^3$

Eccentricity of strands from neutral axis $= 306 \, \text{mm}$.
The stresses from creep redistribution are calculated for the midspan section.

First determine the as-built stresses at midspan just after casting the slab

The moment due to prestress $= 3274 \times 0.142 = 464.9 \, \text{kNm}$

Bottom fibre stress from prestress $= \dfrac{3274}{349} + \dfrac{464.9}{74.31} = 15.64 \, \text{MPa}$

Top fibre stress of precast unit from prestress $= \dfrac{3274}{349} - \dfrac{464.9}{46.96} = -0.52 \, \text{MPa}$

Bottom fibre stress from dead load $= -\dfrac{494}{74.31} = -6.65 \, \text{MPa}$

Top fibre stress of precast unit from dead load $= \dfrac{494}{46.96} = 10.5 \, \text{MPa}$

Bottom fibre total stress from dead load and prestress $= 8.99 \, \text{MPa}$

Top fibre stress of precast unit from dead load and prestress $= 9.98 \, \text{MPa}$

Next determine the midspan stresses assuming the prestress, beam self-weight and slab weight were applied monolithically to the whole beam

The moment due to prestress $= 3274 \times 0.306 = 1001.8 \, \text{kNm}$

Bottom fibre stress from prestress $= \dfrac{3274}{500} + \dfrac{1001.8}{114.9} = 15.27 \, \text{MPa}$

Top fibre stress from prestress $= \dfrac{3274}{500} - \dfrac{1001.8}{119.6} = -1.83 \, \text{MPa}$

Bottom fibre stress from dead load $= -\dfrac{494}{114.9} = -4.30 \, \text{MPa}$

Top fibre stress from dead load $= \dfrac{494}{119.6} = 4.13 \, \text{MPa}$

Bottom fibre total stress from dead load and prestress $= 10.97 \, \text{MPa}$

Top fibre stress from dead load and prestress $= 2.30 \, \text{MPa}$

Next interpolate between the as-built and monolithic stresses

The difference in stresses between the two cases is shown in Fig. K-6. Using equation (DK-1) gives the following stresses at midspan due to creep redistribution:

$\Delta \sigma_{\text{top,slab}} = (2.3)\left(\dfrac{0.9}{1 + 0.8 \times 1.5}\right) = 0.94 \, \text{MPa}$

$$\Delta\sigma_{\text{bot,slab}} = (3.51)\left(\frac{0.9}{1+0.8\times1.5}\right) = 1.44\,\text{MPa}$$

$$\Delta\sigma_{\text{top,precast}} = (-6.47)\left(\frac{0.9}{1+0.8\times1.5}\right) = -2.65\,\text{MPa}$$

$$\Delta\sigma_{\text{bot,precast}} = (1.98)\left(\frac{0.9}{1+0.8\times1.5}\right) = 0.81\,\text{MPa}$$

Where these stresses are beneficial, they could be safely ignored. Differential shrinkage stresses would also need to be calculated.

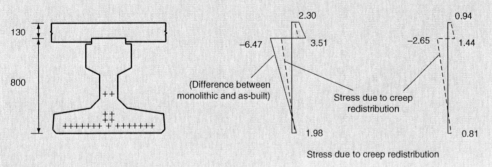

Fig. K-6. Composite beam for Worked example K-2

K4.2. Differential shrinkage

EN 1992-2 does not explicitly cover the calculation of the effect of differential shrinkage, but it must be allowed for at the serviceability limit state in composite beams where the residual shrinkage strain of deck slab and precast beam differ after the beam is made composite. Usually the precast beam will have undergone a significant part of its total shrinkage strain by the time the deck slab is cast, so the deck slab will shrink by a relatively greater amount. This relative shrinkage will compress the top of the precast beam (causing axial force and sagging moment in it) while generating tension in the deck slab itself.

The differential shrinkage strain can be estimated from 2-2/clause 3.1.4 if the approximate age of the beam at casting is known so that:

$$\varepsilon_{\text{diff}} = \varepsilon_{\text{sh,slab}}(\infty) - (\varepsilon_{\text{sh,beam}}(\infty) - \varepsilon_{\text{sh,beam}}(t_1)) \qquad \text{(DK-4)}$$

where $\varepsilon_{\text{sh,beam}}(\infty)$ is the total shrinkage strain of the precast beam, $\varepsilon_{\text{sh,slab}}(\infty)$ is the total shrinkage strain of the slab and $\varepsilon_{\text{sh,beam}}(t_1)$ is the shrinkage strain of the precast beam before casting the slab.

If the beam was fully restrained and the shrinkage occurred instantaneously, the restraint axial force in the slab would be $F_{\text{sh}} = \varepsilon_{\text{diff}} E A_{\text{slab}}$. However, as the shrinkage strain occurs slowly, this shrinkage force is modified by creep so that the fully restrained force is actually more realistically estimated from:

$$F_{\text{sh}} = \varepsilon_{\text{diff}} E A_{\text{slab}} \frac{(1 - e^{-\phi})}{\phi} \qquad \text{(DK-5)}$$

Once again, it is not simple to estimate a single creep ratio to apply in this case covering both slab and precast beam. Given the uncertainties involved in the calculation as a whole, a value of 2.0 will generally suffice. This restrained force can be separated into a restraint axial force and moment acting on the whole cross-section, together with a locked-in self equilibriating stress as shown in Fig. K-7. The axial stress and bending stress can be determined from the axial force in equation (DK-5) acting on the composite section.

If the bridge is statically determinate, then the restraint moment component can be released without generating any secondary moments. Similarly, if the deck has no restraint

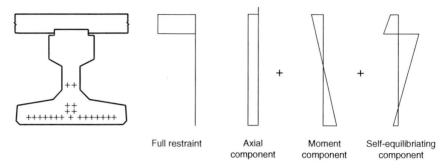

Fig. K-7. Components of differential shrinkage restrained stress

to contraction, then the axial component can similarly be released without generating any restraining tension. If the bridge is statically indeterminate, however, the release of the restraint moment component will generate secondary moments. The determination of secondary moments is illustrated in Fig. K-8 for a simple two-span beam of constant cross-section. In Fig. K-8, the restraint moment is hogging, so to release this moment a sagging moment of equal magnitude must be applied to the beam in the analysis model. A calculation of differential shrinkage effects is given in Worked example 5.10-3.

K4.3. Summary procedure for pre-tensioned composite beams

For pre-tensioned composite beams, a suggested approximate procedure using formulae in EC2-2 is as follows:

(1) Build up the internal actions for the beams from the construction sequence considering only immediate losses – see 2-2/clause 5.10.4.
(2) Calculate long-term losses in prestressing force using 2-2/clause 5.10.6 with the above built-up stresses.
(3) Determine the change in prestress moments from this loss of prestress. That part of the loss occurring prior to casting the slab should be used to reduce the prestressing force on the precast beam alone. The part of the loss occurring after the slab has been cast can be estimated by applying the remaining loss of force as a series of tensile forces to the composite section along the line of the prestress centroid. (If the strands are straight then a single force equal and opposite at each of the beam will suffice.) It is quite common, however, to apply all the loss to the precast beam as it will generally make little difference – see Worked example 5.10-3 in section 5.10 of this guide.
(4) Modify the internal actions in (1) by the effects of the losses in (3).

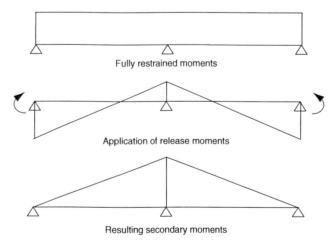

Fig. K-8. Secondary moments from differential shrinkage

(5) Modify the internal actions in (4) by adding the effects of redistribution according to equation (DK-1) with S_0 and S_c based on the prestressing forces at the time of constructing the slab, as described in section K4.1 of this guide.
(6) Add the effects of differential shrinkage.

ANNEX L

Concrete shell elements (informative)

EN 1992-2 Annex LL, in conjunction with 2-2/Annex F and 2-2/clause 6.109 (invoked by paragraph (112) of EN 1992-2 Annex LL), gives a method of designing concrete elements subject to in-plane axial and shear forces, together with out-of-plane moments and shear forces. It is most likely that this annex will be used when such a stress field has been determined from a finite element model, although the equations are presented in terms of stress resultants (force or moment per metre) rather than stresses. The annex can also be used where simpler analysis has been performed, as illustrated in the discussions on Annex MM in this guide, covering the design of box girder webs in bending and shear.

A sandwich model is employed to convert out-of-plane moments and twisting moments into stress resultants acting in a single plane in each of the outer layers of the sandwich. The outer layers also carry the direct and shear stresses acting in the plane of the element. The core section of the sandwich is used only to carry shear forces which are transverse to the element's thickness. Paragraphs (109) and (110) require the principal transverse shear force in the core to be verified separately by means of the beam rules for shear in 2-2/clause 6.2. Expression (LL.123) in paragraph (110) recommends that the reinforcement ratio for this check be based on the strength of the reinforcement layers resolved in the direction of the principal shear, whereas previous practice (e.g. reference 6) has based the reinforcement ratio on the stiffness in this direction. As the latter is more established, it is recommended here that Expression (LL.123) is modified in line with this assumption such that:

$$\rho_1 = \rho_x \cos^4 \phi_0 + \rho_y \sin^4 \phi_0 \tag{DL-1}$$

Having determined the stresses in each outer layer, the membrane rules in 2-2/clause 6.109 and 2-2/Annex F can be used to design orthogonal reinforcement and check concrete stress fields. Skew reinforcement is not covered in EC2, but section 6.9 of this guide provides some guidance in such cases. It should be noted that the use of the sandwich model and membrane rules in design do not make any allowance for plastic redistribution across a cross-section and so can be very conservative. Consequently, it is always better to use the member resistance rules in the main body of EC2, where applicable, as such redistribution is implicit within those rules.

The first part of Annex LL deals with checking whether or not the element will crack under the loading considered. If the element is predicted to be un-cracked, the only check required is that the principal compressive stress is below $\alpha_{cc} f_{ck}/\gamma_C$ or a higher limit based on $f_{ck,c}$ where there is triaxial compression. The formula presented in paragraph (107) for checking cracking is a general triaxial expression (based on principal stresses σ_1, σ_2 and σ_3). It is only one of several existing possible crack prediction formulae. More background on this subject

ANNEX L. CONCRETE SHELL ELEMENTS

can be found in reference 29. It is inappropriate that mean values of concrete strengths should be used in this verification, as implicit in Expression (LL.101), as the calculations relate to ultimate limit state strength; design values of concrete strength would be more appropriate. It is therefore recommended here that f_{cm} and f_{ctm} should be replaced by f_{cd} and f_{ctd} respectively in Expressions (LL.101) and (LL.112). A simpler alternative for biaxial stress states only would be to use the cracking verification of Annex QQ, but replacing the characteristic tensile strength with the design tensile strength in Expression (QQ.101).

The sandwich model and equations are fairly self-explanatory and are much simplified when the reinforcement in each direction in a layer is assumed to have the same cover. This assumption may not be appropriate in very thin elements. The use of the equations is not discussed further here, but the special case of shear and transverse bending in beam webs, with some further simplifications, is investigated in detail in Annex MM of this guide. Use of the sandwich model is complicated slightly where reinforcement is not centred on its respective sandwich layer. It may sometimes be necessary to choose layer thicknesses such that this occurs where the in-plane compressive stress field is high. Annex LL gives a method to account for this eccentricity in its paragraph (115) and this is again discussed in Annex M of this guide.

A further use of Annex LL would be for the design of slabs subjected principally to transverse loading. In previous UK practice, such cases would have been designed using the Wood–Armer equations[19,20] or the more general capacity field equations.[21] The combined use of Annex F, 2-2/clause 6.109 and Annex LL to design slab reinforcement does not necessarily lead to conflict with these approaches. The reinforcement produced is usually the same, other than minor differences due to assumptions for lever arms. However, 2-2/clause 6.109 sometimes limits the use of solutions from the Wood–Armer equations or the more general capacity field equations through its limitation of $|\theta - \theta_{el}| = 15°$. It also requires a check of the plastic compression field, which references 19, 20 and 21 do not require. Despite neglect of these requirements, the Wood–Armer equations[19,20] have, however, successfully been used in the past.

ANNEX M

Shear and transverse bending (informative)

M1. Sandwich model

Although the maximum allowable shear stress, determined by crushing of the web concrete within the diagonal compression struts, is generally significantly higher in EN 1992 than previously used in the UK, there will be occasions when this higher limit cannot be mobilized. In webs of box girders, transverse bending moments can lead to significant reductions in the maximum permissible coexistent shear force because the compressive stress fields from shear and from transverse bending have to be combined. The stresses from the two fields are not, however, simply additive because they act at different angles. In the UK, it has been common practice to design reinforcement in webs for the combined action of transverse bending and shear, but not to check the concrete itself for the combined effect. The lower limit in shear used in the UK made this a reasonable approximation, but it is potentially unsafe if a less conservative (and more realistic) crushing strength is used.

2-2/clause 6.2.106 formally requires consideration of the above shear–moment interaction, but if the web shear force is less than 20% of $V_{Rd,max}$ or the transverse moment is less than 10% of the maximum transverse moment resistance then the interaction does not need to be considered. These criteria are unlikely to be satisfied for box girders, but the allowance for coexisting moment will often be sufficient to negate the need for a check of webs in typical beam and slab bridges. Where the interaction has to be considered, Annex MM can be used.

2-2/Annex MM uses the rules for membrane elements in 2-2/clause 6.109 and a sandwich model (based on Annex LL) to idealize the web as two separate outer layers subject to in-plane forces only. Such calculations are potentially lengthy because the longitudinal direct stress varies over the height of the beam and the vertical direct stresses vary through the thickness, making the angle of the elastic principal compressive stress vary everywhere. Also, the membrane rules of 2-2/clause 6.109 apply to the design of plates with a general stress field obtained from finite element (FE) analysis. The check of transverse bending and shear in webs is usually going to be done without reference to an FE analysis. Consequently, some simplifications are made in **2-2/MM(101)** to facilitate the web design as follows:

- The shear per unit height may be considered as having constant value along Δy in Fig. M-1: $v_{Ed} = V_{Ed}/\Delta y$.

    The intent is to permit the shear stress to be taken as constant on all sides of the element. The use of v_{Ed} as a shear flow is unfortunate as it is used as a shear stress in the rest of EC2. There is also no guidance given on the length Δy used for averaging

ANNEX M. SHEAR AND TRANSVERSE BENDING

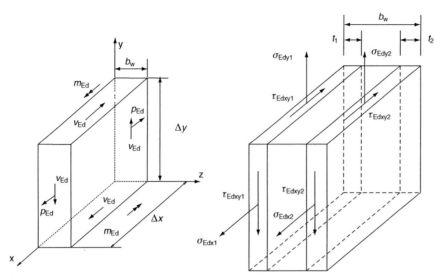

Fig. M-1. Conversion of stress resultants into layer stresses

the shear stress. When the design for shear is based on overall member behaviour, the shear flow could sensibly be taken as $v_{Ed} = V_{Ed}/z$, where z is the lever arm according to clause 6.2 and V_{Ed} is the shear in one web. This makes the shear stress compatible with that assumed in the main shear section.

- The transverse bending moment per unit length should be considered as having constant value along Δy: $m_{Ed} = M_{Ed}/\Delta x$

This can be a conservative simplification, effectively making the transverse moment uniform over the whole height of the web, based on the greatest moment.

- Longitudinal axial force (e.g. from prestressing) can be given a constant value in Δy: $p_{Ed} = P_{Ed}/\Delta y$

Again, no guidance is given on limits for the length Δy. A reasonable interpretation is to consider only the uniform axial component of the prestress, i.e. the stress at the centroid of the section providing the centroid lies in the web. This is consistent with the approach used in 2-2/clause 6.2.3 for the determination of σ_{cp}.

- The transverse shear within the web, due to the variation of the corresponding bending moment, can be neglected in Δy.

The transverse shear corresponding to variations in the transverse moment are, in any case, usually fairly small.

One final simplification, which is not explicitly mentioned in EC2-2, but is reasonable as it is compatible with the member shear design rules, is to ignore the effects of the main beam moment on longitudinal forces in the web.

With the above assumptions, it is possible to use a sandwich model as shown in Fig. M-1 and the membrane rules of 2-2/clause 6.109 and Annex F to design the reinforcement. The designer is free to decide on the thicknesses of the layers. Expressions (MM.101) to (MM.106) are given in **2-2/MM(102)** to determine stresses in the layers and they are reproduced as equations (DM1-1) to (DM1-6) below. The notation and sign convention for σ_{Edy} have been amended in both the equations below and in Fig. M-1 to give compatibility with clause 6.109 and Annex LL in EC2-2. If Annex MM is used as presented in EC2-2, it is important to note that its sign convention is not consistent with other parts of EC2-2:

2-2/MM(102)

$$\tau_{Edxy1} = v_{Ed} \frac{b_w - t_2}{(2b_w - t_1 - t_2)t_1} \quad \text{(DM-1)}$$

347

$$\tau_{Edxy2} = v_{Ed} \frac{b_w - t_1}{(2b_w - t_1 - t_2)t_2} \qquad \text{(DM-2)}$$

$$\sigma_{Edy1} = \frac{-m_{Ed}}{(b_w - (t_1 + t_2)/2)t_1} \qquad \text{(DM-3)}$$

$$\sigma_{Edy2} = \frac{m_{Ed}}{(b_w - (t_1 + t_2)/2)t_2} \qquad \text{(DM-4)}$$

$$\sigma_{Edx1} = p_{Ed} \frac{b_w - t_2}{(2b_w - t_1 - t_2)t_1} \qquad \text{(DM-5)}$$

$$\sigma_{Edx2} = p_{Ed} \frac{b_w - t_1}{(2b_w - t_1 - t_2)t_2} \qquad \text{(DM-6)}$$

One remaining problem is the determination of θ_{el} (defined in 2-2/clause 6.109) which varies through the thickness of the element in the un-cracked elastic state. The choice of location to determine the un-cracked flexural stress is therefore critical. It is recommended here, somewhat arbitrarily, that θ_{el} be calculated from the stresses determined *after* the element has been split into the layers of the sandwich. This appears to be what is required in Annex MM from the order of the paragraphs, but it does not then strictly relate to the initial principal compression angle in the un-cracked elastic section.

The simplest application of the rules can be done where the layer thicknesses are based on twice the cover so that the reinforcement is at the centre of the layer. This will usually then lead to an 'un-used' gap between the layers, which reduces the maximum shear resistance. In bridges, the webs will often be highly stressed in shear, so use of the full width of web thickness will often be necessary. Consequently, it will often be necessary to make the layer thicknesses total either the full width of the web or a significant fraction of it. This, however, leads to the added problem that the derived layer forces act eccentrically to the actual reinforcement forces. In this case, the procedure is to perform the concrete verification ignoring this eccentricity and then to make a correction in the calculation of the reinforcement forces in accordance with **2-2/MM(103)**. Formulae for this correction are given in Annex LL as equations (LL.149) and (LL.150). They are derived here as equations (DM-7) and (DM-8) with notation again changed to suit Fig. M-1.

The stress resultants in each layer are shown in Fig. M-2. n_{Ed} represents the force from the sandwich model acting at the centre of the layer and n_{Ed}^* represents the force to be used in the reinforcement design so as to maintain equilibrium.

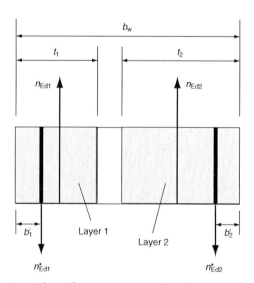

Fig. M-2. Internal equilibrium with reinforcement eccentric to layer centre

ANNEX M. SHEAR AND TRANSVERSE BENDING

Taking moments about the reinforcement in layer 2 gives:

$$(b_w - b'_1 - b'_2)n^*_{Ed1} = n_{Ed1}\left(b_w - \frac{t_1}{2} - b'_2\right) + n_{Ed2}\left(\frac{t_2}{2} - b'_2\right)$$

Thus:

$$n^*_{Ed1} = \frac{n_{Ed1}\left(b_w - \frac{t_1}{2} - b'_2\right) + n_{Ed2}\left(\frac{t_2}{2} - b'_2\right)}{(b_w - b'_1 - b'_2)} \quad \text{(DM-7)}$$

From force equilibrium:

$$n^*_{Ed2} = n_{Ed1} + n_{Ed2} - n^*_{Ed1} \quad \text{(DM-8)}$$

The above equations also allow different covers to the two faces which (LL.149) and (LL.150) do not. Using the reinforcement design formulae and axes convention in section 6.9 of this guide, the reinforcement requirements n^*_{sy} for the y direction in terms of stress resultant are from equations (DM-7) and (DM-8):

$$n^*_{sy1} = \frac{n_{sy1}\left(b_w - \frac{t_1}{2} - b'_2\right) + n_{sy2}\left(\frac{t_2}{2} - b'_2\right)}{(b_w - b'_1 - b'_2)} \quad \text{(DM-9)}$$

with $n_{sy1} = |n_{Edxy1}|\tan\theta_1 + n_{Edy1}$ and $n_{sy2} = |n_{Edxy2}|\tan\theta_2 + n_{Edy2}$

$$n^*_{sy2} = n_{sy1} + n_{sy2} - n^*_{sy1} \quad \text{(DM-10)}$$

Similar equations can be produced for the x direction but noting that $n_{sx1} = |n_{Edxy1}|\cot\theta_1 + n_{Edx1}$ and $n_{sx2} = |n_{Edxy2}|\cot\theta_2 + n_{Edx2}$.

The use of these equations and the sandwich model is illustrated in Worked example M-1. It illustrates the process of making the correction where reinforcement is not centred on the layers. It shows little difference in the reinforcement produced from models with layers greater than twice the cover and equal to twice the cover. This will generally be the case where layers are equal, but it is especially important to make the correction where the layers are different sizes to avoid violation of equilibrium between overall web stress resultants and internal actions in concrete and reinforcement.

The application of the membrane rules to cases of shear and transverse bending will often lead to tensile forces in the longitudinal direction. For shear acting alone, the longitudinal force produced is $V_{Ed}\cot\theta$ and is the same force as predicted in the shear model of section 6.2; there it is shared between tension and compression chords, increasing the tension by $0.5V_{Ed}\cot\theta$ and reducing the compression by $0.5V_{Ed}\cot\theta$. In applying the membrane rules to webs in shear and transverse bending, it is reasonable to distribute the reinforcement (or forces) between chords in the same way, rather than providing continuous longitudinal reinforcement up the webs. This is the subject of *2-2/MM(104)*.

2-2/MM(104)

M2. Alternative approach based on longitudinal shear rules

2-2/MM(105) allows an alternative simpler method for combining the effects from transverse bending and shear in the absence of axial force, based on the rules for longitudinal shear in 2-2/clause 6.2.4. The depth of the compression zone required for transverse bending, h_f, can first be determined and then this width discounted for the purposes of calculating the maximum permissible shear force, $V_{Rd,max}$, as shown in Fig. M-3. For concrete verification, this is conservative as the compressive stress fields for bending and shear do not act at the same angle and therefore add vectorially rather than algebraically.

2-2/MM(105)

Figure M-3 implies that the centroid of the shear resistance in this simple model is displaced from the centre of the web and therefore the reinforcement for shear would be slightly eccentric to maintain this same centroid of loading. There is no requirement to consider this effect in the longitudinal shear rules in 2-2/clause 6.2.4. In the membrane

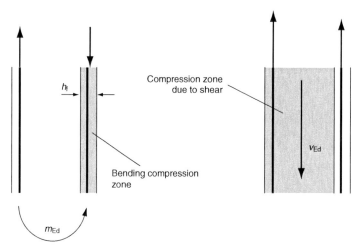

Fig. M-3. Simplified combination of bending and shear in webs

rules, this effect is minimized by having a different compression angle on each face of the web, with a flatter truss on the side in flexural tension. The resulting shear steel requirement can then be added to that for bending on the tension face. The longitudinal shear rules in 2-2/clause 6.2.4 do not require full combination of bending and shear forces in the reinforcement design. Something less than full addition will also be achieved by application of the method in Annex MM, but it will not give as much reduction as allowed in 2-2/clause 6.2.4. Worked example M-1 illustrates this. For web design, 2-2/Annex MM(105) therefore requires full combination of the reinforcement for shear and bending.

The drafters of EC2-2 have specifically intended the rules for flanges with longitudinal shear and transverse bending to *not* be used for the design of webs with axial force. This is because the longitudinal shear rules ignore the presence of any axial compression that is present. The membrane rules of 2-2/clause 6.109 give a reduction in shear crushing resistance as soon as any axial load is applied. By contrast, the web shear rules of 2-1-1/clause 6.2.3(3) allow an *enhancement* of concrete crushing strength in shear for average web compressive stress up to 60% of the design cylinder strength by way of the recommended value of α_{cw}. It is only beyond this value of compression that the shear resistance is actually reduced. In normal prestressed beams the axial force will not be this high, in which case it would generally appear reasonable to ignore the axial load and to use the modified longitudinal shear rules above for checking web crushing in combined shear and transverse bending. This would, however, depart from the recommendations of Annex MM.

It is worth noting that various clauses in EC2 can be applied to shear in an element and none are fully consistent:

(1) The web design rules for shear in 2-1-1/clause 6.2 have a reduced crushing strength under very high axial compression by way of the factor α_{cw}. The resistance is, however, allowed to be enhanced in the presence of low and moderate axial stress.
(2) A similar reduction is made to the allowable torsional shear stress in flanges under very high flange compression by way of α_{cw}.
(3) Flanges in longitudinal shear, however, are not reduced in strength by high compression as there is no α_{cw} term in the relevant formula in 2-2/clause 6.2.4.
(4) The membrane rules of 2-2/clause 6.109 implicitly reduce maximum shear strength as soon as any axial load is present. Stricter limits are also imposed on the direction of the compression struts.

These differences are, however, inevitable as each rule for a particular situation (such as a web in shear, a plate element in shear, a flange in longitudinal shear) have only been verified in that particular specific application. Generally, the member resistance rules make allowance for plastic redistribution across the cross-section on the basis of test results. The more generalized membrane rules do not.

ANNEX M. SHEAR AND TRANSVERSE BENDING

Worked example M-1: Shear and transverse bending in a box girder web

The reinforced concrete box girder in Fig. M-4 has concrete with 40 MPa cylinder strength, and link reinforcement with characteristic yield strength of 500 MPa. The shear in each web is 3021 kN and there is a coexisting transverse bending moment of 122 kNm/m. The lever arm, z, for main beam bending $= 0.9 \times 1900 = 1710$ mm. The web vertical reinforcement needed is determined and the adequacy of the concrete compressive stress field is checked. The results are compared with the application of the rules for longitudinal shear in 2-2/clause 6.2.4, as modified by 2-2/Annex MM(105):

$$v_{Ed} = \frac{V_{Ed}}{z} = \frac{3021 \times 10^3}{1710} = 1767 \, \text{Nmm}^{-1}$$

$$m_{Ed} = 122 \times 10^3 \, \text{Nmm/mm}$$

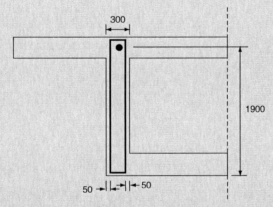

Fig. M-4. Box girder for Worked example M-1

(a) Initially try a sandwich with two 100 mm thick layers centred on the reinforcement. Using the equations in the main text above:

$$\tau_{Edxy1} = v_{Ed} \left[\frac{b_w - t_2}{(2b_w - t_1 - t_2)t_2} \right]$$

$$= 1767 \times \left[\frac{300 - 100}{(2 \times 300 - 100 - 100) \times 100} \right] = 8.84 \, \text{MPa}$$

$$\tau_{Edxy2} = 8.84 \, \text{MPa}$$

It is already clear that the concrete compressive stress will be unacceptable from the size of the shear stress, but the calculation is continued.

Layer 1 will be taken as the layer where the transverse moment induces compression:

$$\sigma_{Edy1} = \left[\frac{-m_{Ed}}{(b_w - (t_1 + t_2)/2)t_1} \right] = \left[\frac{-122 \times 10^3}{(300 - (100 + 100)/2) \times 100} \right] = -6.1 \, \text{MPa}$$

$$\sigma_{Edy2} = 6.1 \, \text{MPa}$$

From the Mohr's circles in Fig. M-5, the elastic angles of the principal compression from the x axis are determined as follows:

$$\phi_1 = \tan^{-1} \frac{8.84}{6.1/2} = 70.96°$$

$$\theta_{el1} = 90 - 70.96/2 = 54.52°$$

$$\phi_2 = 70.96°$$

$$\theta_{el2} = 70.96/2 = 35.48°$$

Equations (D6.9-1) to (D6.9-3) in section 6.9 of this guide are now used to check each layer. The angles of compressive stress fields are taken equal to the elastic angles.

Layer 1:

$$\sigma_{cd1} = -|\tau_{Edxy1}|(\tan\theta_1 + \cot\theta_1) = -8.84(\tan 54.52 + \cot 54.52) = -18.7\,\text{MPa}$$

Assuming the reinforcement is to be designed to be fully stressed, then the allowable stress is from 2-2/Expression (6.111):

$$\sigma_{cd\,max} = \nu f_{cd} = 0.6(1 - 40/250) \times \frac{40}{1.5} = 13.44\,\text{MPa} < 18.7\,\text{MPa}$$

so the concrete stress is not acceptable:

$$\rho_{y1}\sigma_{sy1} = |\tau_{Edxy1}|\tan\theta_1 + \sigma_{Edy1} = 8.84 \times \tan 54.52 - 6.1 = 6.30\,\text{MPa}$$

Taking the reinforcement at its design strength of 435 MPa gives a steel area

$$= \frac{6.30 \times 100}{435} = \mathbf{1.45\,mm^2/mm}$$

Layer 2:

$$\sigma_{cd2} = -|\tau_{Edxy2}|(\tan\theta_2 + \cot\theta_2) = -8.84(\tan 35.48 + \cot 35.48) = -18.7\,\text{MPa}$$

The compressive stress is greater than 13.44 MPa *so the concrete stress is not acceptable*:

$$\rho_{y2}\sigma_{sy2} = |\tau_{Edxy2}|\tan\theta_2 + \sigma_{Edy2} = 8.84 \times \tan 35.48 + 6.1 = 12.40\,\text{MPa}$$

Taking the reinforcement at its design strength of 435 MPa gives a steel area

$$= \frac{12.40 \times 100}{435} = \mathbf{2.85\,mm^2/mm}$$

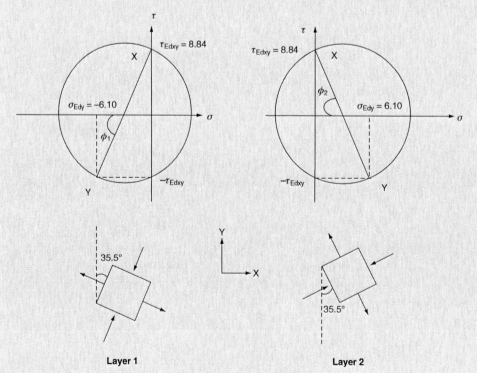

Fig. M-5. Mohr's circle for case (a) with layers = 100 mm thick

(b) Now try a sandwich with two 150 mm thick layers which are now no longer centred on the reinforcement. Using the equations in the main text above:

$$\tau_{Edxy1} = v_{Ed}\left[\frac{b_w - t_2}{(2b_w - t_1 - t_2)t_2}\right] = 1767\left[\frac{300 - 150}{(2 \times 300 - 150 - 150) \times 150}\right]$$

$$= 5.89 \, \text{MPa}$$

$\tau_{Edxy2} = 5.89 \, \text{MPa}$

Layer 1 will be taken as the layer where the moment induces compression:

$$\sigma_{Edy1} = \left[\frac{-m_{Ed}}{(b_w - (t_1 + t_2)/2)t_1}\right] = \left[\frac{-122 \times 10^3}{(300 - (150 + 150)/2) \times 150}\right] = -5.42 \, \text{MPa}$$

$\sigma_{Edy2} = 5.42 \, \text{MPa}$

From the Mohr's circles in Fig. M-6, the elastic angles of the principal compression from the x axis are calculated as follows:

$$\phi_1 = \tan^{-1}\frac{5.89}{5.42/2} = 65.28°$$

$\theta_{el1} = 90 - 65.28/2 = 57.36°$

$\phi_2 = 65.28°$

$\theta_{el2} = 65.28/2 = 32.64°$

The angles of compressive stress fields are again taken equal to the elastic angles.

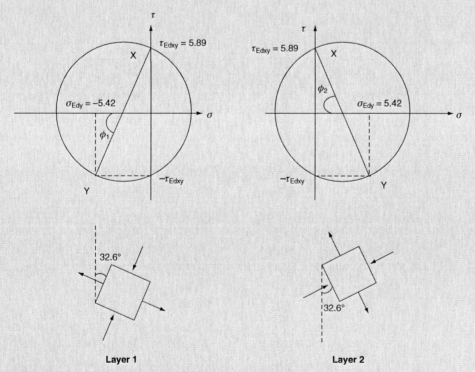

Fig. M-6. Mohr's circle for case (b) with layers = 150 mm thick

Layer 1:

$$\sigma_{cd1} = -|\tau_{Edxy1}|(\tan\theta_1 + \cot\theta_1) = -5.89(\tan 57.36 + \cot 57.36) = -12.97 \, \text{MPa}$$

Assuming the reinforcement is to be designed to be fully stressed, then the allowable stress is:

$$\sigma_{\text{cd max}} = \nu f_{\text{cd}} = 0.6(1 - 40/250) \times \frac{40}{1.5} = 13.44\,\text{MPa} > 12.97\,\text{MPa}$$

so the concrete stress is just acceptable.

The reinforcement design cannot now be done directly with the Annex F equations (as modified in section 6.9 of this guide) as the reinforcement is not co-planar with the centre of the layer. Annex LL Expression (LL.149), reproduced as equation (DM-7), needs to be used to determine the stress resultants for use in the reinforcement design.

Stress resultants are:

$n_{\text{Edxy1}} = \tau_{\text{Edxy1}} \times t_1 = 5.89 \times 150 = 883.5\,\text{Nmm}^{-1}$

$n_{\text{Edxy2}} = 883.5\,\text{Nmm}^{-1}$

$n_{\text{Edy1}} = \sigma_{\text{Edy1}} \times t_1 = -5.42 \times 150 = -813.3\,\text{Nmm}^{-1}$

$n_{\text{Edy2}} = 813.3\,\text{Nmm}^{-1}$

Ignoring any eccentricity initially, the equations for reinforcement design in the y direction from section 6.9 of this guide, modified to be in terms of stress resultants, are:

Layer 1: $n_{\text{sy1}} = |n_{\text{Edxy1}}|\tan\theta_1 + n_{\text{Edy1}} = 883.5 \times \tan 57.36° - 813.3 = 566.1\,\text{Nmm}^{-1}$

Layer 2: $n_{\text{sy2}} = |n_{\text{Edxy2}}|\tan\theta_2 + n_{\text{Edy2}} = 883.5 \times \tan 32.64° + 813.3 = 1379.2\,\text{Nmm}^{-1}$

Substituting these into equation (DM-9):

$$n^*_{\text{sy1}} = \frac{n_{\text{sy1}}\left(b_{\text{w}} - \frac{t_1}{2} - b'_2\right) + n_{\text{sy2}}\left(\frac{t_2}{2} - b'_2\right)}{(b_{\text{w}} - b'_1 - b'_2)}$$

$$= \frac{566.1\left(300 - \frac{150}{2} - 50\right) + 1379.2\left(\frac{150}{2} - 50\right)}{300 - 50 - 50} = 667.7\,\text{Nmm}^{-1}$$

Taking the reinforcement at its design strength of 435 MPa gives a steel area

$$= \frac{667.7}{435} = 1.53\,\text{mm}^2/\text{mm}$$

Layer 2:

$$\sigma_{\text{cd2}} = -|\tau_{\text{Edxy2}}|(\tan\theta_2 + \cot\theta_2) = -5.89(\tan 32.64 + \cot 32.64) = -12.97\,\text{MPa}$$

The compressive stress is less than 13.44 MPa *so the concrete stress is just acceptable.*

The reinforcement design again cannot now be done directly with the Annex F equations, as the reinforcement is not co-planar with the centre of the layer. Annex LL equation (LL.150) needs to be used to determine the stress resultants for use in the reinforcement design.

From equation (DM-10):

$$n^*_{\text{sy2}} = n_{\text{sy1}} + n_{\text{sy2}} - n^*_{\text{sy1}} = 566.1 + 1379.2 - 667.7 = 1277.6\,\text{Nmm}^{-1}$$

Taking the reinforcement at its design strength of 435 MPa gives a steel area

$$= \frac{1277.6}{435} = 2.94\,\text{mm}^2/\text{mm}$$

The reinforcement is essentially the same as from the first sandwich model (the differences arising because of the different angle used for the compressive stress field), but the concrete stress field is now acceptable. A check in accordance with

Annex LL(115) should now also be made of the core for the shear stress resulting from the force transferred between the layers. The shear could be determined from Fig. M-1 and the core checked using Annex LL(109) and (110). This check is, however, not made in normal shear design where the compressive diagonal force in each layer will typically also not be co-planar with the reinforcement.

(c) **Check the web using the longitudinal shear rules**

Assuming the lever arm for transverse bending to be $0.95d$, the depth of the flexural compression zone is:

$$h_f = \frac{1}{0.8} \times \frac{122 \times 10^6 / 225}{1000 \times 0.85 \times 40/1.5} = 29 \text{ mm}$$

The shear stress on the reduced width is therefore $3021 \times 10^3 / 1710 \times ((300 - 29)) = 6.52 \text{ MPa}$ compared to a maximum allowable with a 45° truss of $13.44/2 = 6.72 \text{ MPa}$. The compressive resistance is 97.0% mobilized. Note the compressive resistance according to (b) above was 96.5% mobilized, so the longitudinal shear rules are slightly conservative for crushing in this instance.

The reinforcement needed for bending on the tension face

$$= \frac{122 \times 10^3}{0.95 \times 250 \times 435} = 1.18 \text{ mm}^2/\text{mm}$$

The reinforcement needed for shear per face assuming a 45° truss from 2-2/clause 6.2.3

$$= \frac{3021 \times 10^3}{0.90 \times 1900 \times 435} \times 0.5 = 2.03 \text{ mm}^2/\text{mm}$$

If the reinforcement is added on the tension face, the requirement is **3.21 mm²/mm** which is slightly greater than from the membrane rules. The reinforcement requirement for the compression face is **2.03 mm²/mm**. This is slightly more conservative than the membrane rules. (Note that the possibility of only adding half of the reinforcement needed for shear allowed in 2-2/clause 6.2.4 is not allowed for web design here on the basis that it gives less reinforcement than the membrane elements method.)

ANNEX N

Damage equivalent stresses for fatigue verification (informative)

N1. General
Annex NN of EN 1992-2 is used in conjunction with 2-2/clause 6.8.5 and gives a simplified procedure to calculate the damage equivalent stresses for fatigue verification of reinforcement and prestressing steel in concrete road and railway bridge decks. It is based on fatigue load models given in EN 1991-2. Although the annex is informative, there is no other simple alternative for the fatigue verification of reinforcement and prestressing steel, as discussed in section 6.8.4 of this guide. Worked example 6.8-1 illustrates the damage equivalent stress calculation for a road bridge. Annex NN also provides a damage equivalent stress method for the verification of concrete in railway bridges, but not for road bridges.

N2. Road bridges
The damage equivalent stress method for road bridges is based on fatigue Load Model 3, defined in EN 1991-2 clause 4.6.4. This model is illustrated in Fig. N-1. The weight of each axle is equal to 120 kN. Where required by EN 1991-2 and its National Annex, two vehicles in the same lane should be considered.

2-2/clause NN.2.1(101)

For calculation of the damage equivalent stress ranges for steel verification, *2-2/clause NN.2.1(101)* requires the axle loads of the fatigue model to be multiplied by the following factors:

- 1.75 for verification at intermediate support locations;
- 1.40 for verification at all other locations.

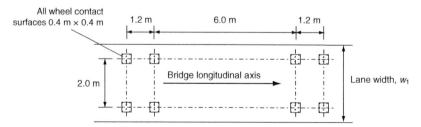

Fig. N-1. Fatigue Load Model 3

ANNEX N. DAMAGE EQUIVALENT STRESSES FOR FATIGUE VERIFICATION

The modified vehicle is then moved across the bridge along each notional lane and the stress range determined for each cycle of stress fluctuation.

2-2/clause NN.2.1(102) gives the following expression for calculating the damage equivalent stress range for steel verification:

$$\Delta\sigma_{s,equ} = \Delta\sigma_{s,Ec}\lambda_s \qquad \text{2-2/(NN.101)}$$

where:

$\Delta\sigma_{s,Ec}$ is the stress range caused by fatigue Load Model 3, modified as above and applied in accordance with clause 4.6.4 of EN 1991-2, assuming the load combination given in 2-1-1/clause 6.8.3

λ_s is a factor to calculate the damage equivalent stress range from the stress range caused by the modified fatigue load model

The factor, λ_s, includes the influences of span, annual traffic volume, service life, multiple lanes, traffic type and surface roughness, and is calculated from:

$$\lambda_s = \phi_{fat}\lambda_{s,1}\lambda_{s,2}\lambda_{s,3}\lambda_{s,4} \qquad \text{2-2/(NN.102)}$$

Each of these factors is discussed in turn below.

ϕ_{fat} is a damage equivalent impact factor influenced by road surface roughness. The factor is defined in Annex B of EN 1991-2, together with recommendations for selecting the appropriate value, as follows:

- $\phi_{fat} = 1.2$ for surfaces of good roughness (recommended for new and maintained roadway layers such as asphalt);
- $\phi_{fat} = 1.4$ for surfaces of medium roughness (recommended for old and unmaintained roadway layers).

In addition to the above factors, where the section under consideration is within a distance of 6.0 m from an expansion joint, a further impact factor should be applied to the whole loading, as defined in Annex B of EN 1991-2. This clause refers to EN 1991-2 Fig. 4.7 (reproduced as Fig. N-2).

The $\lambda_{s,1}$ factor is obtained from 2-2/Fig. NN.1 (for the intermediate support area) or 2-2/Fig. NN.2 (for the span and local elements) as appropriate. For main longitudinal reinforcement in continuous beams, it would be reasonable to use 2-2/Fig. NN.1 for a length of 15% of the span each side of an intermediate support and to use 2-2/Fig. NN.2 elsewhere. This is the approach in EN 1993-2. The design of shear reinforcement is based on 2-2/Fig. NN.2.

$\lambda_{s,1}$ accounts for the critical length of the influence line or surface and the shape of the S–N curve, so its value depends on the type of element under consideration. In ENV 1992-2,[30] the same graphs were provided but the horizontal axis related to span length. The latter is often appropriate, but not always. For example, for continuous beams in hogging bending at intermediate supports, there are two positive lobes to the influence line from each adjacent span, each causing a cycle of stress variation. The length of the span is therefore appropriate. The same is true for shear. For reaction, however the greatest positive lobe length of the

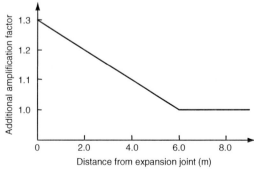

Fig. N-2. Additional amplification factor for proximity to expansion joints

influence line, defining a cycle of stress variation, covers two spans, so the total length of the two spans is more appropriate in this case. EN 1993-2 uses the above considerations to determine the base length to use in calculation and its clause 9.5.2(2) can be used for guidance in determining the appropriate length here.

The $\lambda_{s,2}$ correction factor takes account of the annual traffic volume and traffic type (i.e. weight). It is calculated from the following expression:

$$\lambda_{s,2} = \bar{Q} \times \sqrt[k_2]{N_{\text{obs}}/2.0} \qquad \text{2-2/(NN.103)}$$

where:

$\bar{Q}$ is a factor taken from 2-2/Table NN.1. It accounts for the damage done by the actual mix of traffic weights compared to that done by the fatigue load model. $\bar{Q}$ need not, however, be determined from traffic spectra (as the equivalent parameter in EN 1993-2 needs to be) as it may be determined from 2-2/Table NN.1 for a given 'traffic type', as defined in Note 3 of EN 1991-2 clause 4.6.5(1). The definitions given there are not easy to interpret:

- 'long distance' means hundreds of kilometres;
- 'medium distance' means 50 to 100 km;
- 'local traffic' means distances less than 50 km.

'Long distance' will typically apply to motorways and trunk roads. The use of the other cases will need to be agreed with the Client

k_2 is the slope of the appropriate S–N curve for the element under consideration

N_{obs} is the number of lorries per year (in millions) in accordance with Table 4.5 of EN 1991-2, reproduced here as Table N-1 (these values may be modified by the National Annex)

The $\lambda_{s,3}$ factor takes account of the required service life:

$$\lambda_{s,3} = \sqrt[k_2]{N_{\text{Years}}/100} \qquad \text{2-2/(NN.104)}$$

where N_{Years} is the design life of the bridge (in years).

The $\lambda_{s,4}$ correction factor takes account of the influence of traffic from adjacent lanes. Since most bridge decks are able to distribute load transversely, elements will also pick up fatigue loading from vehicles passing in lanes remote from those directly above the element. $\lambda_{s,4}$ is calculated from the following:

$$\lambda_{s,4} = \sqrt[k_2]{\frac{\sum N_{\text{obs},i}}{N_{\text{obs},1}}} \qquad \text{2-2/(NN.105)}$$

where:

$N_{\text{obs},i}$ is the number of lorries expected on lane i per year
$N_{\text{obs},1}$ is the number of lorries expected on the slow lane per year

This expression is very approximate and does not contain any direct measure of the relative influence of traffic in each lane, unlike the equivalent parameter in EN 1993-2.

Table N-1. Indicative number of heavy vehicles expected per year per slow lane

Traffic categories	N_{obs} per year per slow lane
(1) Roads and motorways with two or more lanes per direction with high flow rates of lorries	2.0×10^6
(2) Roads and motorways with medium flow rates of lorries	0.5×10^6
(3) Main roads with low flow rates of lorries	0.125×10^6
(4) Local roads with low flow rates of lorries	0.05×10^6

N3. Railway bridges

The damage equivalent stress method for fatigue verification of railway bridges is split into two sections; one for reinforcing and prestressing steel and one for concrete elements. The key aspects of both are discussed below.

N3.1. Reinforcing and prestressing steel

The damage equivalent stress method for reinforcement and prestressing steel in railway bridges follows a similar format to that for road bridges, but is based on using the normal rail traffic models (Load Model 71 and Load Model SW/0) defined in EN 1991-2, but excluding the α factor defined therein. The characteristic (static) values for Load Model 71 consist of a load train of four 250 kN axles and uniformly distributed loads of 80 kN/m. The geometry of this load model is illustrated in Fig. N-3. In addition, for continuous span bridges (including single spans with integral abutments), the effects from Load Model SW/0 should also be considered if worse than Load Model 71. The characteristic (static) values for Load Model SW/0 consist of two uniformly distributed loads of 133 kN/m with the geometry illustrated in Fig. N-4. Reference should be made to EN 1991-2 and its relevant National Annex for full details of the load models including transverse eccentricity of vertical loads.

For structures with multiple tracks, *2-2/clause NN.3.1(101)* requires the relevant load model to be applied to a maximum of two tracks to determine the steel stress range. The following expression for calculating the damage equivalent stress range for steel verification is given:

2-2/clause NN.3.1(101)

$$\Delta\sigma_{s,equ} = \lambda_s \times \phi \times \Delta\sigma_{s,71} \qquad \text{2-2/(NN.106)}$$

where:

- $\Delta\sigma_{s,71}$ is the stress range caused by Load Model 71 or SW/0 as above. It should be calculated in the load combination given in 2-1-1/clause 6.8.3
- ϕ is a dynamic factor which enhances the static load effects obtained from the above load models. This factor is defined in EN 1991-2 clause 6.4.5 (as Φ) and should be taken as either Φ_2 for carefully maintained track or Φ_3 for track with standard maintenance. Using the formulae defined in EN 1991-2, Φ_2 lies within the range of 1.0 to 1.67 and Φ_3 lies within the range of 1.0 to 2.0
- λ_s is a correction factor to calculate the damage equivalent stress range from the stress range caused by above load models

The correction factor, λ_s, includes the influences of span, annual traffic volume, service life and multiple tracks and is calculated from:

$$\lambda_s = \lambda_{s,1}\lambda_{s,2}\lambda_{s,3}\lambda_{s,4} \qquad \text{2-2/(NN.107)}$$

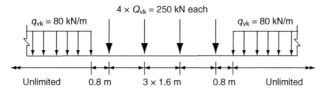

Fig. N-3. Load Model 71 and characteristic (static) values for vertical loads

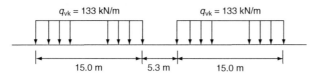

Fig. N-4. Load Model SW/0 and characteristic (static) values for vertical loads

The correction factor $\lambda_{s,1}$ takes into account the influence of the length of the influence line or surface and traffic mix. The values of $\lambda_{s,1}$ for standard or heavy traffic mixes (defined in EN 1991-2) are given in 2-2/Table NN.2 and 2-2/(NN.108). The values given for mixed traffic correspond to the combination of trains given in Annex F of EN 1991-2. No values for a light traffic mix are provided. In such circumstances, either $\lambda_{s,1}$ values can be based on the standard traffic mix or calculations on the basis of the actual traffic spectra can be undertaken.

The $\lambda_{s,2}$ correction factor is used to include the influence of the annual traffic volume:

$$\lambda_{s,2} = \sqrt[k_2]{\frac{Vol}{25 \times 10^6}} \qquad \text{2-2/(NN.109)}$$

where:

Vol is the volume of traffic (tonnes per year per track)
k_2 is the slope of the appropriate S–N curve for the element under consideration

The $\lambda_{s,3}$ correction factor is used to include the influence of the service life and is identical to that for road bridges above.

2-2/clause NN.3.1(106)

The $\lambda_{s,4}$ correction factor is used to account for the influence of loading from more than one track. The expression given in *2-2/clause NN.3.1(106)* includes the effects of stress ranges from load on tracks other than from that which produces the greatest stress range. It is therefore more rational than the corresponding expression for road bridges, which considers only the traffic flow in adjacent lanes.

N3.2. Concrete subjected to compression

The damage equivalent stress method for concrete under compression in railway bridges is similar to that for reinforcement and prestressing steel above; the damage equivalent stresses are based on the stress ranges obtained from analysis using Load Model 71.

2-2/clause NN.3.2(101)

2-2/clause NN.3.2(101) states that adequate fatigue resistance may be assumed for concrete elements in compression if the following expression is satisfied:

$$14 \times \frac{1 - E_{cd,max,equ}}{\sqrt{1 - R_{equ}}} \geq 6 \qquad \text{2-2/(NN.112)}$$

where:

$$R_{equ} = \frac{E_{cd,min,equ}}{E_{cd,max,equ}}$$

$$E_{cd,min,equ} = \gamma_{sd} \frac{\sigma_{cd,min,equ}}{f_{cd,fat}}$$

$$E_{cd,max,equ} = \gamma_{sd} \frac{\sigma_{cd,max,equ}}{f_{cd,fat}}$$

where:

γ_{sd} is the partial factor for modelling uncertainty, defined in clause 6.3.2 of EN 1990. Values are likely to be in the range of 1.0 to 1.15 and are set in the National Annex to EN 1990 (γ_{sd} is not defined in EC2-2 other than in 2-2/clause 5.7)
$f_{cd,fat}$ is the design fatigue compressive strength of concrete from 2-2/clause 6.8.7

$\sigma_{cd,max,equ}$ and $\sigma_{cd,min,equ}$ are the upper and lower stresses of the damage equivalent stress spectrum with 10^6 number of cycles. These upper and lower stresses should be calculated from the following expressions:

$$\sigma_{cd,max,equ} = \sigma_{c,perm} + \lambda_c(\sigma_{c,max,71} - \sigma_{c,perm})$$

$$\sigma_{cd,min,equ} = \sigma_{c,perm} + \lambda_c(\sigma_{c,perm} - \sigma_{c,min,71})$$

2-2/(NN.113)

where:

$\sigma_{c,perm}$ is the compressive concrete stress caused by the characteristic combination of permanent actions (excluding the effects of the fatigue load models)

$\sigma_{c,max,71}$ is the maximum compressive concrete stress caused by the characteristic combination of actions including Load Model 71, and the appropriate dynamic factor, ϕ, from EN 1991-2 (as discussed above)

$\sigma_{c,min,71}$ is the minimum compressive concrete stress caused by the characteristic combination of actions including Load Model 71, and the appropriate dynamic factor, ϕ

λ_c is a correction factor to calculate the upper and lower stresses of the damage equivalent stress spectrum from the stresses caused by the fatigue load model

The λ_c correction factor covers the same effects that λ_s covers for steel verifications, but additionally includes the influence of permanent compressive stress by way of the factor $\lambda_{c,0}$. The factors $\lambda_{c,1}$, $\lambda_{c,2,3}$ and $\lambda_{c,4}$ are equivalent to $\lambda_{s,1}$ to $\lambda_{s,4}$ for steel verifications.

ANNEX O

Typical bridge discontinuity regions (informative)

Annex OO of EN 1992-2 provides guidance on the design of diaphragms for box girders with:

- twin bearings;
- single bearings; or
- a monolithic connection to a pier.

Guidance is also given on the design of diaphragms in decks with Double Tee cross-sections.

The inclusion of this material was made at a very late stage in the drafting of EN 1992-2. It differs in style and content to the rest of the document in that it provides no Principles or Application Rules, but rather gives guidance on suitable strut-and-tie idealizations that can be used in design. As such, the material requires no further commentary here other than to note that other idealizations are possible and may sometimes be dictated by constraints on the positioning of reinforcement and/or prestressing steel. The layouts provided can, however, be taken to be examples of good practice.

ANNEX P

Safety format for non-linear analysis (informative)

The provisions of Annex PP of EN 1992-2 are discussed in section 5.7 of this guide.

ANNEX Q

Control of shear cracks within webs (informative)

2-2/clause 7.3.1(110) states that 'in some cases it may be necessary to check and control shear cracking in webs' and reference is made to 2-2/Annex QQ. Some suggestions for when and why such a check might be appropriate are made in section 7.3.1 of this guide.

2-2/Annex QQ gives a procedure for checking shear cracking, although it is prefaced by the statement that 'at present, the prediction of shear cracking in webs is accompanied by large model uncertainty'. It also states that the check is particularly necessary for prestressed members. This is probably more because of the perceived need to keep a tighter control on crack widths where prestressing tendons are involved, rather than due to any inherent vulnerability of prestressed sections to web shear cracking (although thinner webs are often used in prestressed sections, which will be more likely to crack). Worked example Q-1 shows the benefits of prestressing in reducing web crack widths due to shear.

In the procedure, the larger principal tensile stress in the web (σ_1) is compared to the cracking strength of the concrete, allowing for a biaxial stress state, given as:

$$f_{ctb} = \left(1 - 0.8 \frac{\sigma_3}{f_{ck}}\right) f_{ctk,0.05} \qquad \text{2-2/(QQ.101)}$$

where σ_3 is the larger compressive principal stress (compression taken as positive) but not greater than $0.6 f_{ck}$.

Where the greatest principal tensile stress $\sigma_1 < f_{ctb}$ (in this check, tension is positive), the web is deemed to be un-cracked and no check of crack width is required. Minimum longitudinal reinforcement should, however, be provided in accordance with 2-2/clause 7.3.2.

Where $\sigma_1 \geq f_{ctb}$, EC2 states that cracking in the web should be controlled either by the method of 2-2/clause 7.3.3 or the direct calculation method of 2-2/clause 7.3.4. In both cases, it is necessary to take account of the deviation angle between principal tensile stress and reinforcement directions. Since the reinforcement directions do not, in general, align with the direction of principal tensile stress and the reinforcement is likely to be different in the two orthogonal directions, there is difficulty in deciding what unique bar diameter and spacing to use in the simple method of 2-2/clause 7.3.3. If the method of 2-2/clause 7.3.4 is used, the effect of deviation angle between principal tensile stress and reinforcement directions on crack spacing, $s_{r,max}$, can be calculated using 2-1-1/Expression (7.15):

$$s_{r,max} = \left(\frac{\cos\theta}{s_{r,max,y}} + \frac{\sin\theta}{s_{r,max,z}}\right)^{-1} \qquad \text{2-1-1/(7.15)}$$

The effect of the deviation angle between principal tensile stress and reinforcement directions on calculation of reinforcement stress is not given in EN 1992. One approach is to calculate the stress in the reinforcement, σ_s, by dividing the stress in the concrete in the

ANNEX Q. CONTROL OF SHEAR CRACKS WITHIN WEBS

direction of the principal tensile stress by an effective reinforcement ratio, ρ_l, where $\rho_l = \sum_{i=1}^{N} \rho_i \cos^4 \alpha_i$ and ρ_i is the reinforcement ratio in layer i at an angle α_i to the direction of principal tensile stress. This reinforcement ratio is based on equivalent stiffness in the principal stress direction, which was the approach used in BS 5400 Part 4.[9] σ_s is then used in 2-1-1/Expression (7.9) to calculate the strain for use in the crack calculation formula of 2-1-1/Expression (7.8). In applying 2-1-1/Expression (7.9), $\rho_{p,eff}$ could be taken as ρ_l but, in view of the uncertainty of the effects of tension stiffening where the reinforcement direction is skew to the cracks, it is advisable to ignore the tension stiffening term and take $\varepsilon_{sm} - \varepsilon_{cm} = \sigma_s/E_s$.

The direction of the principal tensile stress in the un-cracked condition will vary over the height of the web due to the variation of both shear stress and flexural stress. In determining whether or not the section is cracked, all such points should be checked. However, a similar treatment after cracking would also require crack checks at all locations throughout the depth of the beam. To avoid this situation, and since the subject of the calculation is 'cracking due to shear', one simplification for doubly flanged beams might be to ignore flexural stresses when checking web crack widths and to take the shear stress as uniform over a depth z, where z is an appropriate lever arm, such as the ULS flexural lever arm. This latter assumption is not conservative for shear stress, but is a reasonable approximation given the uncertainties of the method. The axial force component of any prestressing should also be considered in the crack width calculation.

The neglect of the flexural stresses above can partly be justified for doubly flanged beams by the fact that, on cracking, flexural tensile stresses will shed to the flanges where the main flexural reinforcement is provided. Despite the shedding of stress, longitudinal strain will still be produced in the web in order for the flange reinforcement to take up the load. While strictly this strain will add to that calculated from the shear stresses, it may be reasonable to ignore this as the flexural crack widths must still be checked separately. Care with this approach should, however, be exercised with T beams when the stem is in tension. Since, in this case, there is no discrete tension flange containing the main flexural reinforcement, the flexural forces in the web cannot be shed on cracking. In such cases, it is more conservative to consider the full stress field in checking web cracking.

> **Worked example Q-1: Check of shear cracking in a box girder web**
> Figure Q-1 shows a box girder and its web reinforcement. The shear force per web in the appropriate SLS combination is 1800 kN. The cover to the shear links is 50 mm. Cracking is checked with and without prestress.
>
> **Without prestress**
> The average shear stress in the web, based on a lever arm taken as the distance between the flanges, is $\tau_{yzEd} = 1800 \times 10^3/(2000 \times 300) = 3.0$ MPa.
>
> Ignoring the flexural stress in the web, the principal compressive stress is $\sigma_3 = 3.0$ MPa and the principal tensile stress is $\sigma_1 = 3.0$ MPa at an angle θ of 45° to the horizontal, measured anticlockwise from the y axis. The directionally dependent concrete tensile strength is given by 2-2/Expression (QQ.101):
>
> $$f_{ctb} = \left(1 - 0.8\frac{\sigma_3}{f_{ck}}\right) f_{ctk,0.05} = \left(1 - 0.8 \times \frac{3}{40}\right) 2.5 = 2.4 \text{ MPa}$$
>
> which is less than the principal tensile stress, so crack widths must be checked. In verifying whether or not the concrete is cracked in this way using 2-2/Expression (QQ.101), strictly flexural stresses should be included as well. However, as the section is found to be cracked here under the shear stresses alone, there is no need to repeat the check considering the flexural stresses.
>
> The reinforcement ratios are:
>
> $$\rho_y = \frac{2 \times \pi \times 12^2/4}{150 \times 300} = 0.0050$$

$$\rho_z = \frac{2 \times \pi \times 25^2/4}{150 \times 300} = 0.0218$$

The effective reinforcement ratio in the direction of principal stress is:

$$\rho_1 = \sum_{i=1}^{N} \rho_i \cos^4 \alpha_i = 0.0050 \times \cos^4 45° + 0.0218 \times \cos^4 45° = 0.0067$$

Ignoring flexural stresses (as discussed in the main text), the stress in the reinforcement is therefore:

$$\sigma_s = \frac{\sigma_1}{\rho_1} = \frac{3.0}{0.0067} = 448 \, \text{MPa}$$

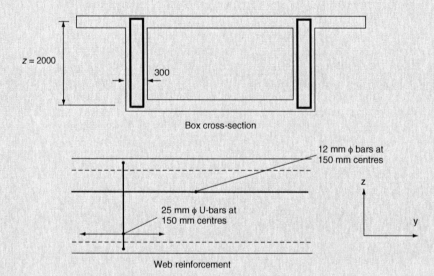

Fig. Q-1. Web reinforcement for Worked example Q-1

From 2-1-1/Expression (7.11):

$$s_{r,max,y} = k_3 c + k_1 k_2 k_4 \phi / \rho_{p,eff,y} = 3.4 \times 75 + 0.8 \times 1.0 \times 0.425 \times 12/0.0050 = 1071 \, \text{mm}$$

$$s_{r,max,z} = k_3 c + k_1 k_2 k_4 \phi / \rho_{p,eff,z} = 3.4 \times 50 + 0.8 \times 1.0 \times 0.425 \times 25/0.0218 = 560 \, \text{mm}$$

From 2-1-1/Expression (7.15):

$$s_{r,max} = \left(\frac{\cos\theta}{s_{r,max,y}} + \frac{\sin\theta}{s_{r,max,z}}\right)^{-1} = \left(\frac{\cos 45°}{1071} + \frac{\sin 45°}{560}\right)^{-1} = 520 \, \text{mm}$$

From 2-1-1/Expression (7.9), ignoring tension stiffening:

$$\varepsilon_{sm} - \varepsilon_{cm} = \frac{\sigma_s}{E_s} = \frac{448}{210 \times 10^3} = 2.133 \times 10^{-3}$$

From 2-1-1/Expression (7.8) the crack width is:

$$w_k = s_{r,max}(\varepsilon_{sm} - \varepsilon_{cm}) = 520 \times 2.133 \times 10^{-3} = 1.1 \, \text{mm!}$$

With prestress

Prestressing is now added such that the average compressive stress in the web from the prestress is 6 MPa. The calculations are repeated:

The average shear stress in the web remains as $\tau_{yzEd} = 3.0 \, \text{MPa}$

Ignoring the flexural stresses in the web, but including the uniform axial component of the prestress, the principal compressive stress is now $\sigma_3 = 7.24 \, \text{MPa}$ and the principal

ANNEX Q. CONTROL OF SHEAR CRACKS WITHIN WEBS

tensile stress reduces to $\sigma_1 = 1.243$ MPa at an angle θ of 67.5° to the horizontal, measured anticlockwise from the y axis. The directionally dependent concrete tensile strength is given by:

$$f_{\text{ctb}} = \left(1 - 0.8\frac{\sigma_3}{f_{\text{ck}}}\right)f_{\text{ctk},0.05} = \left(1 - 0.8 \times \frac{7.24}{40}\right)2.5 = 2.1 \text{ MPa}$$

With the prestress added therefore, the cracking strength exceeds the principal tensile strength when the flexural stresses are ignored. However, for the check of whether the concrete cracks, flexural stresses should be included. For the purpose of this example, it is assumed that when flexural stresses are included, the direct stress at the top of the web reduces to 0 MPa and then the principal tensile stress increases to the value above without prestress i.e. 3 MPa; the concrete is then found to be cracked and crack widths then need to be checked.

The reinforcement ratios in the y and z directions are as above. The effective reinforcement ratio in the direction of the principal tensile stress is:

$$\rho_1 = \sum_{i=1}^{N} \rho_i \cos^4 \alpha_i = 0.0050 \times \cos^4 67.5° + 0.0218 \times \cos^4 22.5° = 0.0160$$

Ignoring flexural stresses (as discussed in the main text), the stress in the reinforcement is therefore:

$$\sigma_s = \frac{\sigma_1}{\rho_1} = \frac{1.243}{0.0160} = 77 \text{ MPa}$$

The maximum crack spacings in the y and z directions are as above.

From 2-1-1/Expression (7.15):

$$s_{r,\max} = \left(\frac{\cos\theta}{s_{r,\max,y}} + \frac{\sin\theta}{s_{r,\max,z}}\right)^{-1} = \left(\frac{\cos 67.5°}{1071} + \frac{\sin 67.5°}{560}\right)^{-1} = 498 \text{ mm}$$

From 2-1-1/Expression (7.9), ignoring tension stiffening:

$$\varepsilon_{\text{sm}} - \varepsilon_{\text{cm}} = \frac{\sigma_s}{E_s} = \frac{77}{210 \times 10^3} = 3.667 \times 10^{-4}$$

From 2-1-1/Expression (7.8) the crack width is:

$$w_k = s_{r,\max}(\varepsilon_{\text{sm}} - \varepsilon_{\text{cm}}) = 498 \times 3.667 \times 10^{-4} = 0.18 \text{ mm}$$

The crack width is therefore reduced when prestress is applied due to both the reduction in principal tensile stress and the rotation of the principal tensile stress towards the direction of the vertical link reinforcement.

References

1. Gulvanessian, H., Calgaro, J.-A. and Holický, M. (2002) *Designers' Guide to EN 1990 – Eurocode: Basis of Structural Design*. Thomas Telford, London.
2. European Commission (2002) *Guidance Paper L (Concerning the Construction Products Directive – 89/106/EEC). Application and Use of Eurocodes*. EC, Brussels.
3. International Organization for Standardization (1997) *Basis of Design for Structures – Notation – General symbols*. ISO, Geneva, ISO 3898.
4. Regan, P. E., Kennedy-Reid, I. L., Pullen, A. D. and Smith, D. A. (2005) *The influence of aggregate type on the shear resistance of reinforced concrete, structural engineer*, Vol. 83, No. 23/24, pp. 27–32.
5. Concrete Society (1996) *Durable Bonded Post-tensioned Concrete Bridges*. Technical Report No. 47.
6. CEB-FIP (1993) *Model Code 90*. Thomas Telford, London.
7. British Standards Institution (2000) *Steel, Concrete and Composite Bridges Part 3: Code of Practice for the Design of Steel Bridges*. London, BSI 5400.
8. Schlaich, J., Schafer, K. and Jennewein, M. (1987) Toward a consistent design of structural concrete, *PCI journal* Vol. 32, No. 3, pp. 74–150.
9. British Standards Institution (1990) *Steel, Concrete and Composite Bridges Part 4: Code of Practice for the Design of Concrete Bridges*. London, BSI 5400.
10. Cranston, W. B. (1972) *Analysis and Design of Reinforced Concrete Columns*. Cement and Concrete Association, London, Research Report 20.
11. Leung, Y. W., Cheung, C. B. and Regan, P. E. (1976) *Shear Strength of Various Shapes of Concrete Beams Without Shear Reinforcement*. Polytechnic of Central London.
12. Regan, P. E. (1971) *Shear in Reinforced Concrete – an Experimental Study*. CIRIA, London, Technical Note 45.
13. Hendy, C. R. and Johnson, R. P. (2006) *Designers' Guide to EN 1994-2 – Eurocode 4, Design of Composite Steel and Concrete Structures. Part 2, General Rules and Rules for Bridges*. Thomas Telford, London.
14. Comité Européen de Normalisation (1992) *Design of Concrete Structures. Part 1-1, General Rules and Rules for Buildings*. CEN, Brussels, ENV 1992-1-1.
15. Leonhardt, F. and Walther, R. (1964) *The Stuttgart Shear Tests 1961*, C&CA Translation No. 111. Cement and Concrete Association, London.
16. Asin, M. (2000) *The Behaviour of Reinforced Concrete Deep Beams*. PhD Thesis, Delft University of Technology, the Netherlands.
17. Chalioris, C. E. (2003) *Cracking and Ultimate Torque Capacity of Reinforced Concrete Beams*, Role of concrete bridges in sustainable development. Thomas Telford, London.
18. Young, W. C. (1989) *Roark's Formulas for Stress and Strain*, 6th edition. McGraw-Hill International Editions, New York.
19. Wood, R. H. (1968) *The reinforcement of slabs in accordance with a pre-determined field of moment, concrete*, 2, February, pp. 69–76.

20. Armer, G. S. T. (1968) Correspondence. *Concrete*, 2, Aug., pp. 319–320.
21. Denton, S. R. and Burgoyne, C. J. (1996) The assessment of reinforced concrete slabs, *The Structural Engineer*, Vol. 74, No. 9, pp. 147–152.
22. American Association of State Highway and Transportation Officials (2004) *AASHTO LRFD Bridge Design Specifications*. 3rd edition.
23. CIRIA Guide 1 (1976) *A Guide to the Design of Post-tensioned Prestressed Concrete Members*, CIRIA, London.
24. Oh, B. H. and Chae, S. T. (2003) *Stress Distribution Around Tendon Coupling Joints in Prestressed Concrete Girders, Role of Concrete Bridges in Sustainable Development.* Thomas Telford, London.
25. Schlaich, J. and Scheef, H. (1982) *Concrete Box Girder Bridges*. Structural engineering Documents 1st edtion, IABSE, Switzerland.
26. Highways Agency (1994) *The Design of Concrete Highway Bridge and Structures with External and Unbonded Prestressing*. London, BD 58/94.
27. Leonhardt, F., Koch, R. and Rostásy, F. S. (1971) Suspension reinforcement for indirect load application to prestressed concrete beams: test report and recommendations, *Beton und Stahlbetonbau*, Vol. 66, Issue 10, pp. 233–241.
28. International Organization for Standardization (1982) *Concrete – Determination of Static Modulus of Elasticity in Compression*. ISO, Geneva, ISO 6784.
29. *Concrete Under Multiaxial States of Stress – Constitutive Equations for Practical Design*, *CEB Bulletin*, 156, Lausanne.
30. Comité Européen de Normalisation (1996) *Design of Concrete Structures. Part 2, Concrete Bridges*. CEN, Brussels, ENV 1992-2.
31. Muttoni, A., Burdet, O. L. and Hars, E. (2006) Effect of Duct Type on Shear Strength of Thin Webs, *ACI Structural Journal*, September–October 2006.

Index

Page numbers in *italics* denote figures

accelerated relaxation, 290
actions, 9–10, 288, 308–309
actions combinations, 209, 226
additional longitudinal reinforcement, 143–145
aggregate concrete structures, 295–302
allowable maximum strain, *107*
alternative methods, 349–350
amplification factors, *357*
analysis, 95–97, 326, 363
   see also structural analysis
anchorage, 29, 245, 247–251, *247–248*, *251*, *272*
   bottom reinforcement, 277-8
   force calculation model, *286*
   prestressing, 87, *89–90*, 271
   pretensioned tendons, 259–262, *262*
   support reinforcement, 292
   tension, 261–262, *286*
   ultimate limit state, 201, 261–262
   zones, 262–270, *262*, *265–267*, *270*, 330
angular imperfections, *43*
application rules, 5
arches, 43
assumptions, 4–5, 25, 28–29
autogenous shrinkage, 18
axial forces, *78*, 304
axial loads, 62–80, 123, 125–126

'balanced' sections, *78*
bars, 246–247, 250–258, *254*
bases, 179–185, *184–185*
beams, 275–281
   bearings, 47
   composite, 337–339, *338–339*
   creep, 337–339, *338–339*
   doubly reinforced rectangular, 112–114, *113*
   effective spans, 47, *48*
   end stress transfer, 99–100
   idealization for space frame analysis, 172
   K values, *251*
   lateral instability, 80–81
   midspan stress transfer, *100*
   monolithic with supports, 47
   plastic analysis, 52, *53*
   precast pretensioned, 160–164
   rectangular, notation, *227*
   shear, *136*, 137, 278–279
   slender, 80–81
   time-dependent losses, *91*
   with uniform loads, 51
   see also slabs
bearings, 47, *204*, 205–207, 292–293, 329–330
bending, 105–131, 299, 304, *350*
biaxial bending, 80, 125–126
biaxial compression, 218
biaxial shear, 218
bilinear diagrams, 21
bonds, 96, 99–103, *100*, *102*, 245, 250
bottom reinforcement, 277–278
box girders, 280–281, *280*, 362
   discontinuity regions, 362
   effective flange widths, 46–47, *47*
   losses, *87*, 94–95
   post-tensioned, 129–131, 139–140, 148–151
   shear cracks, 365–367, *366*
   tendon profiles, *151*
   three span, losses, *87*
   time-dependent losses, 94–95
   torsion, 172–173
   without tendon drape, 149–151
   worked examples, 87–90, 94–95, 172–173, 280–281, *280*
braced columns, *74*
braced members, *63*
bridge piers *see* piers
brittle failure, 126–131
buckling, 43, *63*, *65*, 305
bundled bars, 258
bursting, 203–205, *204*, 262–263

cables, 126–127
cantilevering piers, 67, *70*, 75–76, 79–80
carbonation corrosion, 34

CCC nodes, 197
CCT nodes, 198
centreline distances, 260
chemical attack corrosion, 35
chloride corrosion, 34
CIRIA Guide 1 method, 264, *265*
classes, 24, 34–35, 37, 322–323
classification, 37
climatic actions, 309
close to slab edges/corners, *176*
close to supports, 135–136, 145
coefficients of friction, 86
columns, *185*, 282–284
   see also piers
combinations, 157–158, 169–170, *350*
composite beams, *161*, 337–341, *338–39*, *341*
composite construction, 93, 160–165
composite deck slabs, 162–164
composite members, 342–343
compressive strengths, 11–12, 20–21, 298, 360–361
compressive stresses, *115*
concrete, 11–23
   cast at different times, 158–160
   confined, 298
   cover verification methods, 36–38
   creep, 14–17, *333*
   damage, 322
   deformation characteristics, 13
   elastic deformation, 14
   fatigue verification, 213–215
   flexural tensile strengths, 22–23
   heat curing effects, 290
   indicative strength classes, 322
   lightweight aggregate structures, 295–302
   shear, 213–215
   shell elements, 344–345
   shrinkage, 14, 18–19
   strengths, 11–14, 20–21, 296, 360–361
   stress characteristics, 13
   stress limitation, 82–83
   stress/strain relationships, 19, *19–20*, 21–22
   worked examples, 16–17
contraflexure points, 147
control, cracks, 230–242, *237*, *240*, 364–367
control perimeters, 176–177, *176*, *184*
corbels, 328–329, *329*
corner piles, *185*, *189*, *191*
corrosion, 32–35, 126–127, 322
couplers, 29, 252–257, 271
cover, 31–38, 261, 298
cracks
   cables, 126–127
   checks, 236–237, 241–242
   control, 230–242, 364–367
   early thermal, 243
   flexure, 140, 148–149
   prediction, 240–241
   sections, *228*
   strains, *237*, *240*
   widths, 237–242, *237*, *240*
   worked examples, 236–237, 241–242
creep, 69–70, 313–315, *332*, 333–341, *333*, *338–339*, *341*
   concrete, 14–17, 297, 313–314, *333*
   time-dependent losses, 92–94
   worked examples, 336–337, 339–341, *341*
cross sections *see* sections
crushing, 169–170, 201–203, *202*
CTT nodes, 198–199
curvature forces, *259*
curvature methods, 76–80

damage, 210–212, 322, 356–361
dead load moments, *332*
deck slabs, 228–230, 300–301
   cover example, 38
   crack control, 236–237, 241–242
   shear, 134
   ultimate limit states, 110–111
   voided reinforced concrete, 146–147
deep beams, 284
definitions, 5, *155*, *167*, *265*
deflection, *70*, 243
deformation, 13, 297
design, 20–21, 298
   anchorage lengths, 249–251
   assumptions, 25, 28–29
   basis, 7–10, 289–290, 296
   execution stages, 307–309
   particular rules, 291
   shear reinforcement, 133–153, *140–141*, *145*, *153*
   torsion, 167–171
detailed calculation method, 317–321
detailing rules, 245–273, 275–288, 291–293, 302, 327–330
deviation allowances, 38
deviators, 272–273
diaphragms, 199–201, *200–201*, 280–281, *280*, 362
differential creep, 337–339, *338–339*
differential shrinkage, *102*, 341–342, *342*
discontinuity, 195–196, *196*, 288, 362
dispersion, *267*
doubly reinforced concrete beams/slabs, 112–114, *113*
D-regions, *57*, 288
   see also strut-and-tie model
drying shrinkage, 18
ductility, 24, 27–28
ducts, 147–148, *148*, 258–261, *259*
durability, 31–38, 298, 322–323

early thermal cracking, 243
EC2-2 Annex J method, 263–264
eccentricity, 42, *304*, *348*
edge sliding reinforcement, 206–207
effective lengths, 65–67, *65*
effective thickness, *169*

INDEX

effective widths, 44–46, *47*, 147–148
elastic deformation, 14, 85–86, 296–297
end moments, *74*
end supports, 277–278
environmental conditions, 32–35
equivalent time method, 318–321, *318*
examples *see* worked examples
execution stages, 307–309
expansion joints, *357*
exposure, 32–35, 37
external non-bonded tendons, 29
external tendons, 29, *98*
external vs. internal post-tensioning, 95–96

failure, *54*, *123*, 126–127, *248*, *304*
fatigue, 25, 28, 98–99, 208–215, *209*, 356–361
   worked examples, 211–212, 214–215
finite element models, 344
flanged reinforced beams, 115–117, *115*, *117*
flanges, 44–47, *47*, *107*, *156*, 157–158
flat slabs, 104, 282, 326
flexural reinforcement, 281
flexural shear, *187*, *189*
flexural tensile strengths, 22–23
floor systems, 291
footings, 285–286
   see also pad footings
force calculation model, *286*
foundations, 10, 285–288, *285*, *287*, 293
frames, 52, 328
free body diagrams, *144*
freeze/thaw corrosion, 34–35
friction, 86–87, *89*

general method, 332–333
geometric data, 8, 44–48
geometric imperfections, 40–44, *41*–*44*, 74
geometry, discontinuity regions, 288

half joints, 292
heat curing, 290
heavy vehicles, 358
high-strength concrete, 313–314
hollow piers, *204*

'I' beams, *154*
idealization, *170*, 172
imperfections *see* geometric imperfections
imposed curvature, *55*
indicative strength classes, 322–323
indirect supports, 279–280
in-plane buckling, 43, *43*
in-plane stress conditions, 324
in situ concrete structures, 311
in situ deck slabs, 160–164
interaction diagrams, *122*
intermediate supports, 278
internal vs. external post-tensioning, 95–96
isolated members, *41*, 64–67, *65*
ISO standard 3898: 1997 symbols, 5

joints, *153*, *357*

K values, anchorage, 249, *251*

laps, 201, 252–257, *252*, *254*, *256*
large diameter bars, 257–258
lateral instability, 80–81
layer centres, *348*
layer stresses, *347*
layouts, tendons, 258–259
leaf piers, 205–207, *207*
lightly reinforced structures, 303–305
lightweight aggregate concrete structures, 295–302
   analysis, 298
   cover, 298
   deformation, 297
   design, 296, 298
   elastic deformation, 296–297
   materials, 296–298
   non-linear analysis, 298
   shear/torsion, 299
   shrinkage, 297
   stress, 297–298
   strut-and-tie models, 300
   ultimate limit states, 298–300
   worked examples, 300–301
limited redistribution, 49–51
limiting slenderness checks, 68–69
limit states, 7, 58–60, 62, 82–83, 98–103
   see also ultimate limit states
linear elastic analysis, 48–51, *51*
links, 251
loads
   close to slab edges/corners, *176*
   close to supports, 145
   dispersal, *207*
   distribution, 176–177, *176*, *202*
   factors, 125
   models, *359*
   spread, *285*
longitudinal reinforcement, 143–145, 275–278, 283
   anchorage, 247–251, *247*–*248*, *251*, *272*
longitudinal shear, 154–158, 349–350, 355
longitudinal tension reinforcement, 277
losses
   anchorage, 87
   elastic deformation, 85–86
   post-tensioning, 85–90
   prestress, 84–90, *89*, 291, 317–321, *318*
   pretensioning, 84–85
   steel relaxation, 317–321, *318*
   worked examples, 87–90
low-relaxation prestressing tendons, 318–321

mandrels, 246–247
materials, 8, 10–29, 290, 296–298, 303–304
   partial factor modification, 311–312
maximum moments, 42–43

maximum punching shear stress, 185
M beams, 162–165, 173–175, *175*
members, 140–153, *140–141, 145*
   detailing, 275–288, 302
   not requiring design shear reinforcement, 133–140
   in tension, 198–199
membrane elements, 215–223, *216, 222*
membrane rules, *216*, 218–222, *222*
minimum cover, 36–38
minimum reinforcement, 127–128
minimum shear reinforcement, 137
modification, partial factors, 311–312
Mohr's circles, *222, 352–353*
moments, 42–43, *70, 74, 332, 342*
monitoring facility provision, 128

nodes, 196–201, *197–198*
nominal curvature methods, 76–80
nominal stiffness, 71–76
non-linear analysis, 19, *19*, 58–62, *61*, 70–71, 298, 363
normative references, 4
notation, rectangular beams, *227*

open sections, 170–171
out-of-plane buckling, 43
overlapping, *292*

pad footings, 180–182, *180*, 287–288, *287*
parabolic-rectangular diagrams, 21
parameters, creep/shrinkage, 314
Part 2 scope, 4
partial factor method, 9–10
partial factor modification, 311–312
partially loaded areas, 201–207, *202, 204, 207*, 300, 329–330
partial sheared truss models, *141*
particular rules, 275–288, 291, 302
particular situations, 327–330
piers, *43, 63*, 68–69, *68, 207*
   worked examples, 16–17, *17*
   see also cantilevering piers; columns
pile caps, 185–193, *186–193*, 285
piles *see* corner piles
pin-ended struts, *42*
placement of reinforcement, *158*
plain reinforced concrete structures, 303–305
plastic analysis, 52–58, *53, 55*
plastic hinges, *55*
pocket foundations, 293
post-tensioning, 85–90
   box girders, 129–131, 139–140, 148–153
   cables, 272–273
   construction, 336
   ducts, 258–259
   members, 262–270, *265, 270*, 330
precast concrete, *153*, 160–165, 289–293, 312
prestress dispersion, *267*
prestressed beams, 118–131, *119*

prestressed members, 137, 140, 147–153, *148*
prestressed steel, 25–29, *28*
prestressed structures, 81–103
   bonded tendons, 99–103, *100, 102–103*
   creep, 92–94
   fatigue limit states, 98–99
   limit states, 97–99
   losses, 84–95, *89*, 291
   post-tensioning, 95–96
   prestress considerations, 95–97
   prestressing forces, 82–84
   primary/secondary effects, 96–97
   serviceability limit states, 98–103
   unbonded prestress, 96
   worked examples, 99–103, *100, 102–103*
prestressing devices, 29
prestressing forces, 82–90
prestressing steel, 209–210, 245–273, 290, 317–321, *318*, 359–360
prestressing systems, 32–35, 38
prestress transfer, 260–261
pretensioning, 84–85
   beams, 131, 160–164
   composite members, 337–339, 342–343
   concrete M beams, 162–165, 173–175, *175*
   strands, *259*
   tendons, 258–262, *262*
primary prestress effects, 96–97, *96*
primary regularization prism, 263–264, *264–265*
principles, 5, 7, 40
properties of materials, 8, 23, 26–27, 316
provision of monitoring facilities, 128
proximity to expansion joints, *357*
punching, 175–193
   control perimeters, 176–177, *176, 184*
   corner piles, *189, 191*
   load distribution, 176–177, *176*
   pile caps, 185–193
   shear, 177–179, 185, *192–193*, 300
   shear resistance, 179–185, *184*
   worked examples, 180–182, *180*, 186–193, *186–193*

railway bridges, 359–361
rectangular beam notation, *227*
rectangular stress blocks, *117*
redistribution, *332, 338–339, 339–341*
regions with discontinuity, 288, 362
reinforced concrete, 105–117, *108*, 121–126, *122–123*, 133–147, 303–305
   see also lightly reinforced . . .
reinforcement, 222–223, *222*, 245–273, *247–248, 251, 270, 272, 277*, 300–2
   corrosion potentials, 32–35
   cover, 31–38, 298
   crack control, 232–234
   durability, 31–38
   eccentric, *348*
   edge sliding, 206–207

INDEX

fatigue verification, *209*
foundations, 287–288, *287*
in-plane stress conditions, 324
loop overlapping, *292*
minimum, 37, 127–128, 137
normative properties, 316
pad footings, 287–288, *287*
piers, *68*, *207*
properties, 316
punching, *184*
shear, 133–153, 251
surfaces, 327–328
suspension, *280*
tension, 277, 324
torsion, 167–169, *167*
transverse, 254–255, *254*
webs, *366*
worked examples, 255–257, *256*, 300–301
reinforcing steel, 23–25, *24–25*, 209–210, *209*, 359–360
relaxation, 27, 92, 94, 317–321, *318*
resultant conversions, stress, *347*
return periods, 309
ribbed wires, 257
road bridges, 211–212, 214–215, 356–358
rotation capacity, 53–56, *54–55*
rules *see* detailing rules

safety format, *61–62*, 363
St Venant torsion, 166
sandwich model, 346–349, *347–348*, 353–355
scalar combinations, 60–61, *61*
scope, Eurocode 2, 3–4
sea water corrosion, 34
secondary moments, *342*
secondary prestress effects, 96–97, *96*
second-order analysis, 62–80
    biaxial bending, 80
    braced columns, *74*
    creep, 69–70
    definitions, 62–63
    methods, 70–71, 76–80
    nominal stiffness, 71–76
    simplified criteria, 64–69
    soil-structure interaction, 64
    stiffness, 63–64, 71–76
    unbraced columns, 73
sections, 21–22, *78*, 154–158, *169*, 170, *170*, 298
    see also prestressed sections
segmental construction, 153, *153*, 171
serviceability limit states, 225–243
    actions combinations, 226
    crack control, 230–243
    deck slabs, 228–230, 302
    execution stages, 309
    prestressed structures, 98–103
    reinforced concrete deck slabs, 228–230
    stress limitation, 226–230
    worked examples, 228–230
shear, 131–165

alternative methods, 349–350
beam enhancement comparisons, *136*
bending, 160, 346–355, *347–348*, *350*
between web/flanges, 154–158
composite construction, 160–165
concrete cast at different times, 158–160
contraflexure points, 147
cracking, 364–367, *366*
deck slabs, 134
flexure failures, 137
flow calculation definitions, *155*
lightweight structures, 299
members not requiring reinforcement, 133–140
members requiring reinforcement, 140–153
post-tensioned box girders with tendon drape, 151–153
precast concrete, 160–165
reinforced concrete, 133–140, 304–305
reinforcement, *145*, *190*, *192–193*, 251, 278–279, 282
tension, 137–139
torsion, 169–170
total transverse reinforcement, *157*
T-sections, 154–158
verification procedure rules, 132–133
wall analysis, 104
worked examples
    bending, 351–355
    box girders, 139–140, 148–153
    deck slabs, 134, 146–147
    flexure, 139–140
    tendon drape, 151–153
    without tendon drape, 149–151
see also longitudinal shear; punching
shear resistance, 179–185, *184*
shear stress, *161*, 185
shell elements, 344–345
shift method, 143–145
short spans, *145*
shrinkage, 313–315
    concrete, 14, 18–19
    differential, *102*, 341–342, *342*
    lightweight structures, 297
    time-dependent losses, 92, 94
sign convention, *216*
simplified criteria, 64–69
simplified methods, 333–336
simplified rectangular diagrams, 21
singly reinforced concrete, 107–112, *108*
sinusoidal imperfections, *43*
skew reinforcement, 222–223, *222*
slabs
    edges/corners, *176*
    K values, *251*
    plastic analysis, 52
    punching, *176*, 179–185, *184*
    shear resistance, 179–185, *184*
    shear stress distribution, *161*
    torsion, 172–175, *175*

slabs (*continued*)
    see also beams; flat slabs; solid slabs
slenderness, 64–67, *65*, 68–69, 80–81, 305
    see also lightweight...
sliding wedge mechanism, *329*
SLS see serviceability limit states
soil structure interactions, 64, 325
solid sections, *169–170*, 281–282
space frame analysis, 172
spacing, 246, *259*
spalling, 201–203, *202*, 262, 265
spans, beams, 47, *48*
square sections, 170, *170*
staged construction, 93
static values see characteristic values
steel, 25–9, *28*, 228, 359–360
    see also prestressing steel
stiffness, 63–64, 71–76
strain
    compatibility, 106–107, 118, 122–123
    cracks, *237*, 240–241, *240*
    creep, 313–315, *333*
    distributions, *117*
    external tendons, *98*
strength
    classes, 296, 322–323
    concrete, 11–14
    curve, *209*
    prestressed steel, 27
    reinforcing steel, 23–24
    see also individual strengths
stress
    anchorage zones, *262*
    block idealization comparisons, 22
    concentrated loads, *202*
    concrete characteristics, 13
    damage equivalent, 356–361
    differential shrinkage, *102*
    distances, uniformity, *261*
    fatigue verification, 356–361
    flanged beams, *117*
    lightweight structures, 297–298
    limitation, 226–230
    Mohr's circles, *222*
    non-linear structural analysis, 298
    prestressed steel, *28*
    redistribution, creep, *338–339*
    reinforced concrete deck slabs, 228–230
    reinforcement fatigue verification, *209*
    resultant conversions, *347*
    section design, 298
    serviceability limit states, 226–230
    strain relations, 298
    temperature differentials, 230
    transfer in beams, 99–100, *100*
    transverse, 262
stress–strain diagrams, 24–25, *28*
stress–strain profiles, *119*
stress–strain relationships, 19, *19–20*, 21–22
structural analysis, 39–104
    geometric imperfections, 40–44, *41–44*
    global, 39, 45–46
    lateral instability, 80–81
    lightweight structures, 298
    linear elastic analysis, 48–51
    local, 39–40
    non-linear analysis, 58–62, 298
    plastic analysis, 52–58
    post-tensioning, 95–96
    precast structures, 290–291
    prestressed members/structures, 81–103
    principles, 40
    reinforced structures, 303–304
    second-order effects analysis, 62–80
    structure idealization, 44–48
    time-dependent behaviour, 331–343, *332*
    worked examples, 67–69, 75–76, 79–80
structural classification, 37
structure idealization, 44–48
strut-and-tie models, 56–58, *57*, 195–196, *196*
    anchorage zones, *266*
    corbels, *329*
    ducts, *148*
    hollow piers, *204*
    lightweight structures, 300
    nodes, 196–201, *197–198*
    struts, 193–195
    ultimate limit states, 193–201
    worked examples, 199–201, *200–201*
struts
    tensile stress, 194–195
    transverse stress, 193–195
    see also pin-ended struts; strut-and-tie models
summary procedure, 342–343
supports, 135–136, 145, *187*, 291–292
surface reinforcement, 279, 327–328
suspension reinforcement, *280*
symbols, 5

temperature differentials, 230
tendons, 149–151, *151*
    external non-bonded, 29
    low-relaxation prestressing, 318–321
    post-tensioned, 86
    prestressing, 258–273, 291
    strain, *98*
    vertical profiles, *87*
tensile forces, 82–83, 261–262
tensile reinforcement, *286*, 324
tensile strength, 12–14, 20–21
testing, 10
thin walled sections, *170*
three-span bridges, *87*, 336–337
ties, 195–196, *196*
time-dependent behaviour, 90–95, *91*, 331–343, *332*
time-dependent losses, *91*, 92–95
torsion
    combined shear/torsion, 169–170
    definitions, *167*

INDEX

design procedure, 167–171, *170*
lightweight structures, 299
M beam bridges, 173–175, *175*
open sections, 170
punching, 175–193
reinforced structures, 167–169, *167*, 279, 304–305
St Venant torsion, 166
segmental construction, 171
slabs, 172–175, *175*
ultimate limit state, 166–175
warping torsion, 171–172
worked examples, 172–175
  pad footings, 180–182, *180*
  pile caps, 186–193, *186–193*
  slabs, 173–175, *175*
total transverse reinforcement, *157*
transfer, prestress, 260–261
transverse bending, 157–158, *157*, 160, 346–355, *347–348*
transverse forces, 42–43
transverse reinforcement, *158*, 254–255, *254*, 283–284
transverse stress, 193–195, 262
truss models, *140–141*, *144*, *154*
T-sections, 154–158
two-span beams, 51

ULS *see* ultimate limit states
ultimate bond strength, 248
ultimate limit states (ULS), 105–223
  anchorage, 201, 261–262
  beams, 105–117, *108*, 112–117, *113*
  bending, 105–131, *122–123*
  brittle failure, 126–131
  columns, 121–126, *122–123*
  deck slabs, 110–111
  doubly reinforced rectangular beams, 112–114, *113*
  execution stages, 309
  fatigue, 208–215
  flanged beams, 115–117
  laps, 201
  lightweight structures, 298–300
  membrane elements, 215–223
  methods, 58–60
  partially loaded areas, 201–207, *202*, *204*
  prestressed concrete beams, 118–121
  prestressed structures, 97–98
  shear, 131–165
  singly reinforced concrete beams/slabs, 107–112, *108*
  strain compatibility, 106–107
  strut-and-tie models, 193–201
  torsion, 166–175
  voided reinforced concrete slabs, 111–112
  worked examples, 110–111, 113–117, 129–131
unbonded prestress, 96
unbonded tendons, 121

unbraced columns, 73
unbraced members, *63*
un-cracked in flexure, 137–140
uniaxial bending, 123
uniaxial compression, 218–220
uniaxial shear, 218–220
uniaxial tension, 220–222
uniformity, stresses, *261*
uniform loads, 51
unreinforced concrete, *204*
unrestrained concrete, *333*

values
  drying shrinkage, 18
  material factors, 10
  vertical loads, *359*
variables, basic, 7–8
vector combinations, 61–62
vehicle expected numbers, 358
verification
  execution stage criteria, 309
  fatigue, 208–215, 356–361
  methods, 36–38
  partial factor method, 9–10
  procedure rules, 132–133
vertical profiles, *87*
voided reinforced concrete slabs, 111–112, 146–147

walls, 284, 291, *304*
warping torsion, 166, 171–172
webs
  bending/shear combinations, *350*
  box girders, 280–281, *280*
  control of shear cracks, 364–367, *366*
  diaphragm region, 280–281, *280*
  longitudinal shear rules, 355
  reinforcement, 365–367, *366*
  widths, 147–148
welding, 25, 251, 257
worked examples
  anchorage zones, 267–270, *270*
  bonded tendons, 99–103, *100*, *102–103*
  box girders, 87–90, 94–95, 139–140, 280–281, *280*, 365–366
  bridge piers, 16–17, *17*
  brittle failure, 129–131
  cantilevered piers, 67, 75–76, 79–80
  composite beams/slabs, 162–164, 339–341, *341*
  crack control, 236–237, 241–242
  creep, 339–341, *341*
  damage equivalent stress range, 211–212
  deck slabs, 38, 255–257, *256*
  diaphragm design, 199–201, *200–201*
  doubly reinforced concrete slabs, 113–114
  effective flange widths, 46–47, *47*
  equivalent time method, 318–321
  fatigue, 211–212, 214–215
  flanged beams, 115–117

377

worked examples (*continued*)
    foundations, 287–288, *287*
    lap lengths, 255–257, *256*
    leaf piers, 205–207, *207*
    lightweight structures, 300–301
    linear elastic analysis, 51, *51*
    losses, 87–90
    low-relaxation prestressing tendons, 318–321
    membrane elements, 218–222, *222*
    partially loaded areas, 205–207, *207*
    post-tensioned members, 139–140, 267–270, *270*
    prestressed beams, 118–121, 129–131
    prestressed steel, 27
    prestressed structures, 99–103, *100*, *102–103*
    pretensioned concrete M beams, 162–165
    punching, 180–182, *180*, 186–193, *186–193*
    reinforced concrete
        columns, 124–125
        slabs, 110–112, 228–230, 241–242
    relaxation losses, 27, 318–321
    sandwich model, 353–355
    serviceability limit states, 228–230
    shear
        cracking, 365–366
        post-tensioned box girders, 149–153
        reinforced concrete, 134–135
        transverse bending combinations, 351–355
    structural analysis, 67–69, 75–76, 79–80
    structure idealization, 46–47
    strut-and-tie models, 199–201, *200–201*
    three-span bridges, 336–337
    time-dependent losses, 94–95
    torsion, 172–175
    transverse bending, 351–355
    ultimate limit states, 110–112, 115–117
    voided reinforced concrete slabs, 111–112, 146–147
    web reinforcement, *366*

zones for shear reinforcement, *190*